Proceedings of the Scientific-Practical Conference
"Research and Development - 2016"

K.V. Anisimov · A.V. Dub
S.K. Kolpakov · A.V. Lisitsa
A.N. Petrov · V.P. Polukarov
O.S. Popel · V.A. Vinokurov
Editors

Proceedings of the Scientific-Practical Conference "Research and Development - 2016"

14–15 December 2016 Moscow, Russia

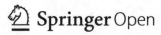 Springer Open

Editors

K.V. Anisimov
GmbH RusOnyx
Moscow
Russia

A.N. Petrov
Directorate of Scientific-Technical Programs
Moscow
Russia

A.V. Dub
JSC ≪Science and Innovations≫
Moscow
Russia

V.P. Polukarov
JSC ≪Military-engineering Corporation≫
Korolev, The Moscow Area
Russia

S.K. Kolpakov
JSC ≪Interdepartmental analytical Center≫
Moscow
Russia

O.S. Popel
Joint Institute for High Temperatures
Russian Academy of Sciences
Moscow
Russia

A.V. Lisitsa
Orekhovich Institute of Biomedical Chemistry
Moscow
Russia

V.A. Vinokurov
Gubkin Russian State University of Oil and Gas
Moscow
Russia

The Compiler of this Book and financial Support for Publishing is NTF-National Training Foundation
123022, Moscow, Russia, 1905 Goda st., 7, bld. 1
Phones: +7 (495) 274-03-90/91/92
Fax: +7 (495) 665-40-75
http://www.ntf.ru
E-mail: info@ntf.ru

ISBN 978-3-030-09681-6   ISBN 978-3-319-62870-7   (eBook)
https://doi.org/10.1007/978-3-319-62870-7

Composition of the Editorial Board

Foreword

Dear ladies and gentlemen!

The central goal of the government is to develop applied and experimental research aimed at producing advanced cross-sectoral technological advances in priority areas. The Ministry of Education and Science of the Russian Federation is making considerable efforts to support staff at research and educational institutions and personnel at manufacturing company's R&D Units at all levels of their careers.

One of the tools of this support is the Federal Targeted Programme "Research and Development in Priority Areas of Development of the Russian Scientific and Technological Complex for 2014–2020." This programme focusses on research that will lead to the development products in priority areas of the Russian economy and facilitate its competitiveness.

The research-to-practice conference "Research and development 2016" was held in Moscow on December 14 and 15. This conference, which was dedicated to applied and experimental research conducted in priority areas of development, discussed the results of the institutions' and scientists' research activities in high-tech industry and outlined innovative research trends in cross-disciplinary areas.

The articles published in this collection reflect the key findings that are of laboratory character at present, but will become a part of corporate life in the future. I firmly believe that they will be of interest for researchers as well as the research and business community as a source of valuable information on the development of applied science in Russia.

On behalf of the Ministry of Education and Science of the Russian Federation

Grigoriy V. Trubnikov
Deputy Minister of Education
and Science of the Russian Federation

Contents

Part I Computer Science

**Multimodal Control System of Active Lower Limb
Exoskeleton with Feedback** 3
S.A. Mineev

**Investigation and Development of Methods for Improving
Robustness of Automatic Speech Recognition Algorithms
in Complex Acoustic Environments** 11
M.L. Korenevsky, Yu.N. Matveev and A.V. Yakovlev

**Smart Endoscope—Firmware Complex for Real-Time
Analysis and Recognition of Endoscopic Videos** 21
K.U. Erendgenova, E.D. Fedorov, R.M. Kadushnikov, O.A. Kulagina,
V.V. Mizgulin, D.I. Starodubov and S.I. Studenok

**The Development of Constructive-Technological Decisions
on Creation of a Family of Microelectronic Elements on the «Silicon
on Insulator» (SOI) Structures to Provide the Ability to Create
Sensors of External Influences of a Various Functional Purpose** 31
M.I. Kakoulin, A.V. Leonov, A.A. Malykh,
V.N. Mordkovich, A.B. Odnolko and M.I. Pavlyuk

Thermopile IR Sensor Arrays 39
V.A. Fedirko, E.A. Fetisov, R.Z. Khafizov,
G.A. Rudakov and A.A. Sigarev

**Development Signal Processing Integrated Circuit
for Position Sensors with High Resolution** 49
G.V. Prokofiev, K.N. Bolshakov and V.G. Stakhin

**Brain-Controlled Biometric Signals Employed
to Operate External Technical Devices** 59
Vasily I. Mironov, Sergey A. Lobov, Innokentiy A. Kastalskiy,
Susanna Y. Gordleeva, Alexey S. Pimashkin, Nadezhda P. Krilova,
Kseniya V. Volkova, Alexey E. Ossadtchi and Victor B. Kazantsev

**Improving Talent Management with Automated Competence
Assessment: Research Summary** 73
N.S. Nikitinsky

**Educational Potential of Quantum Cryptography
and Its Experimental Modular Realization** 83
A.K. Fedorov, A.A. Kanapin, V.L. Kurochkin, Yu.V. Kurochkin,
A.V. Losev, A.V. Miller, I.O. Pashinskiy,
V.E. Rodimin and A.S. Sokolov

**Interactive Visualization of Non-formalized Data Extracted
from News Feeds: Approaches and Perspectives** 93
D.A. Kormalev, E.P. Kurshev, A.N. Vinogradov,
S.A. Belov and S.V. Paramonov

**Development of Pulsed Solid-State Generators of Millimeter
and Submillimeter Wavelengths Based on Multilayer
GaAs/AlGaAs Heterostructures** 101
V.A. Gergel, N.M. Gorshkova, R.A. Khabibullin, P.P. Maltsev,
V.S. Minkin, S.A. Nikitov, A.Yu. Pavlov, V.V. Pavlovskiy and
A.A. Trofimov

**Asymmetric Magnetoimpedance in Bimagnetic
Multilayered Film Structures** 107
A.S. Antonov and N.A. Buznikov

**The Model of the Cybernetic Network and Its Realization
on the Cluster of Universal and Graphic Processors** 117
A.E. Krasnov, A.A. Kalachev, E.N. Nadezhdin,
D.N. Nikolskii and D.S. Repin

**Autonomous Mobile Robotic System for Coastal Monitoring
and Forecasting Marine Natural Disasters** 129
V.V. Belyakov, P.O. Beresnev, D.V. Zeziulin, A.A. Kurkin,
O.E. Kurkina, V.D. Kuzin, V.S. Makarov, P.P. Pronin,
D.Yu. Tyugin and V.I. Filatov

**On Creation of Highly Efficient Micro-Hydraulic Power Plants
of Pontoon Modular Design in Conditions of Super-Low Flow
Parameters** 137
A.V. Volkov, A.A. Vikhlyantsev, A.A. Druzhinin,
A.G. Parygin and A.V. Ryzhenkov

**Development of Scientific and Technical Solutions to Create
Hybrid Power Source Based on Solid Oxide Fuel Cells
and Power Storage System for Responsible Consumers**.............. 149
A.I. Chivenkov, E.V. Kryukov, A.B. Loskutov and E.N. Sosnina

**Automated Control Unit of Power Flow in Intellectual
Electricity Distribution Network** 159
M.G. Astashev, D.I. Panfilov, P.A. Rashitov,
A.N. Rozhkov and D.A. Seregin

**The Partial Replacement of Diesel Fuel in Hot Water Boiler with
Syngas Obtained by Thermal Conversion of Wood Waste** 165
O.M. Larina, V.A. Lavrenov and V.M. Zaitchenko

**The Experimental Research on Independent Starting and
Autonomous Operation of HDTB Considered as a Basic Block
of AES Based on Supercritical Hydrothermal Destruction** 171
A.D. Vedenin, V.S. Grigoryev, Ya.P. Lobatchevskiy,
A.I. Nikolaev, G.S. Savelyev and A.V. Strelets

**Development of a Multifunctional All-Terrain Vehicle Equipped
with Intelligent Wheel-Drive System for Providing Increased
Level of Energy Efficiency and Improved Fuel Economy** 179
V.V. Belyakov, P.O. Beresnev, D.V. Zeziulin,
A.A. Kurkin, V.S. Makarov and V.I. Filatov

**Development and Implementation of an Integrated Approach
to Improving the Operating Cycle and Design
of an Energy-Efficient Forced Diesel Engine**....................... 189
K.V. Gavrilov, V.G. Kamaltdinov, N.A. Khozeniuk,
E.A. Lazarev and Y.V. Rozhdestvensky

**The Development of the New Type Universal Collective Survival
Craft with Unmanned Control Function for Evacuation of Personnel
in Emergency Situations of Natural and Technogenic Character
on the Arctic Shelf** 199
I.A. Vasilyev, R.A. Dorofeev, J.V. Korushova, A.A. Koshurina
and M.S. Krasheninnikov

**Development of Active Safety Software of Road Freight Transport,
Aimed at Improving Inter-City Road Safety, Based on Stereo Vision
Technologies and Road Scene Analysis**......................... 209
V.E. Prun, V.V. Postnikov, R.N. Sadekov and D.L. Sholomov

**Analysis of the Stress State in Steel Components Using
Portable X-Ray Diffraction**.................................. 219
S.A. Nikulin, S.L. Shitkin, A.B. Rozhnov, S.O. Rogachev
and T.A. Nechaykina

**The VLSI High-Level Synthesis for Building Onboard
Spacecraft Control Systems** 229
O.V. Nepomnyashchiy, I.V. Ryjenko, V.V. Shaydurov,
N.Y. Sirotinina and A.I. Postnikov

**A Concept of Robotic System with Force-Controlled Manipulators
for On-Orbit Servicing Spacecraft** 239
I. Dalyaev, V. Titov and I. Shardyko

**Development of Microlinear Piezo-Drives for
Spacecraft Actuators** 247
A.V. Azin, S.V. Rikkonen, S.V. Ponomarev and A.M. Khramtsov

**Design of Dynamic Scale Model of Long Endurance
Unmanned Aerial Vehicle** 255
V.S. Fedotov, A.V. Gomzin and I.I. Salavatov

Features of the Development of Regional Transport Models 263
P.V. Loginov, A.N. Zatsepin and V.A. Pavlov

Part II NanoScience and NanoTechnology

**The Influence of AlGaN Barrier-Layer Thickness on the GaN HEMT
Parameters for Space Applications** 273
A.G. Gudkov, V.D. Shashurin, V.N. Vyuginov, V.G. Tikhomirov,
S.I. Vidyakin, S.V. Agasieva, E.N. Gorlacheva and S.V. Chizhikov

**Application of Volume-Surface Hardening by High-Speed Water Flow
for Improving Static and Cyclic Strength of Large-Scale Castings
from Low-Carbon Steel** 281
S.A. Nikulin, A.B. Rozhnov, T.A. Nechaykina, V.I. Anikeenko,
V.Yu. Turilina and S.O. Rogachev

**Thermotropic Gel-Forming and Sol-Forming Systems
for Enhanced Oil Recovery and Technologies of Their Joint
Application with Thermal Methods for Oil Production** 287
L.K. Altunina and V.A. Kuvshinov

**The Mixture of Fatty Acids Conversion into Hydrocarbons
Over Original Pt-Sn/Al$_2$O$_3$ Catalyst** 297
A.E. Gekhman, A.V. Chistyakov, M.V. Tsodikov,
P.A. Zharova, S.S. Shapovalov and A.A. Pasynskii

**Beneficiation of Heat-Treated Crushed Brown Coal
for Energy Production and Utilities** 305
V.A. Moiseev, V.G. Andrienko, V.G. Piletskii,
V.A. Donchenko and A.I. Urvantsev

NiMo/USY-Alumina Catalysts with Different Zeolite Content
for Vacuum Gas Oil Hydrocracking Over Stacked Beds   319
P.P. Dik, V.P. Doronin, E.Yu. Gerasimov, M.O. Kazakov,
O.V. Klimov, G.I. Koryakina, K.A. Nadeina,
A.S. Noskov and T.P. Sorokina

Comparative Mechanical Tests of Samples Obtained
by the Domestic Experimental Unit Meltmaster^{3D}-550   329
A.V. Dub, V.V. Beregovsky, E.V. Tretyakov,
S.A. Schurenkova and A.V. Yudin

Development of Lithium-Ion Battery of the "Doped Lithium
Iron Phosphate–Doped Lithium Titanate" System for Power
Applications   341
A.A. Chekannikov, A.A. Kuz'mina, T.L. Kulova, S.A. Novikova,
A.M. Skundin, I.A. Stenina and A.B. Yaroslavtsev

Advanced Heat-Resistant TiAl (Nb,Cr,Zr)-Based Intermetallics
with the Stabilized β(Ti)-Phase   351
A.V. Kartavykh, M.V. Gorshenkov and A.V. Korotitskiy

Structural and Magnetic Properties of As-Cast Fe–Nd Alloys   363
V.P. Menushenkov, I.V. Shchetinin, M.V. Gorshenkov and
A.G. Savchenko

Laser Technology of Designing Nanocomposite Implants
of the Knee Ligaments   373
A.Yu. Gerasimenko, U.E. Kurilova, M.V. Mezentseva, S.A. Oshkukov,
V.M. Podgaetskii, I.A. Suetina, V.V. Zar and N.N. Zhurbina

Properties of Structural Steels with Nanoscale Substructure   385
T.V. Lomaeva, L.L. Lukin, L.N. Maslov,
O.I. Shavrin and A.N. Skvortsov

Near-Net Shapes Al_2O_3–SiC_w Ceramic Nanocomposites Produced
by Hybrid Spark Plasma Sintering   397
E. Kuznetsova, P. Peretyagin, A. Smirnov, W. Solis and R. Torrecillas

Development of Technical and Technological Solutions
in the Field of Multilayer Graphene for Creating Electrode
Nanomaterial Energy Storage Devices   405
N.R. Memetov, A.V. Schegolkov, G.V. Solomakho and A.G. Tkachev

Carbon Fiber-Reinforced Polyurethane Composites
with Modified Carbon–Polymer Interface   415
A.R. Karaeva, N.V. Kazennov, V.Z. Mordkovich,
S.A. Urvanov and E.A. Zhukova

**Development and Research of Multifrequency X-ray
Tube with a Field Nanocathode** 421
T.A. Gryazneva, G.D. Demin, M.A. Makhiboroda,
N.A. Djuzhev and V.E. Skvorcov

**Quasicrystalline Powders as the Fillers for Polymer-Based
Composites: Production, Introduction
to Polymer Matrix, Properties** 429
A.A. Stepashkin, D.I. Chukov, L.K. Olifirov, A.I. Salimon
and V.V. Tcherdyntsev

**Selection of Aluminum Matrix for Boron–Aluminum
Sheet Alloys** 439
N.A. Belov, K.Yu. Chervyakova and M.E. Samoshina

**Features of Carbide Precipitation During Tempering
of 15H2NMFA and 26HN3M2FA Steels** 449
S.V. Belikov, V.A. Dub, P.A. Kozlov, A.A. Popov, A.O. Rodin,
A.Yu. Churyumov and I.A. Shepkin

**Improvement of the Mechanical and Biomedical Properties of
Implants via the Production of Nanocomposite Based on
Nanostructured Titanium Matrix and Bioactive Nanocoating** 461
E.G. Zemtsova, A.Yu. Arbenin, R.Z. Valiev and V.M. Smirnov

**Nanopowders Synthesis of Oxygen-Free Titanium
Compounds—Nitride, Carbonitride, and Carbide
in a Plasma Reactor** 469
N.V. Alexeev, D.E. Kirpichev, A.V. Samokhin, M.A. Sinayskiy
and Yu.V. Tsvetkov

**The Technology and Setup for High-Throughput Synthesis
of Endohedral Metal Fullerenes** 481
D.I. Chervyakova, G.N. Churilov, A.I. Dudnik, G.A. Glushenko,
E.A. Kovaleva, A.A. Kuzubov, N.S. Nikolaev, I.V. Osipova and
N.G. Vnukova

**On Some Features of Nanostructural Modification
of Polymer-Inorganic Composite Materials for Light
Industry and for Building Industry** 491
M.V. Akulova, S.A. Koksharov, O.V. Meteleva and S.V. Fedosov

**High-Speed Laser Direct Deposition Technology: Theoretical
Aspects, Experimental Researches, Analysis of Structure,
and Properties of Metallic Products** 501
K.D. Babkin, V.V. Cheverikin, O.G. Klimova-Korsmik, M.O. Sklyar,
S.L. Stankevich, G.A. Turichin, A.Ya. Travyanov, E.A. Valdaytseva
and E.V. Zemlyakov

**Synthesis and Properties of Energetics Metal Borides
for Hybrid Solid-Propellant Rocket Engines** 511
S.S. Bondarchuk, A.E. Matveev, V.V. Promakhov, A.B. Vorozhtsov,
A.S. Zhukov, I.A. Zhukov and M.H. Ziatdinov

**Mechanical Treatment of ZrB_2–SiC Powders and Sintered
Ceramic Composites Properties** 521
S.P. Buyakova, A.G. Knyazeva, A.G. Burlachenko, Yu. Mirovoi and
S.N. Kulkov

Part III   Health and Ecology and Environment Sciences

**The Influence of DCs Loaded with Tumor Antigens on the Cytotoxic
Response of MNC Culture Patients with Oncology** 533
A.P. Cherkasov, J.N. Khantakova, S.A. Falaleeva, A.A. Khristin,
N.A. Kiryishina, V.V. Kozlov, E.V. Kulikova, V.V. Kurilin,
J.A. Lopatnikova, I.A. Obleukhova, S.V. Sennikov, J.A. Shevchenko,
S.V. Sidorov, A.V. Sokolov and A.E. Vitsin

**Establishment of a Technological Platform for Pre-Clinical Evaluation
of Biomedical Cellular Products in Russia** 543
P.I. Makarevich, Yu P. Rubtsov, D.V. Stambolsky, N.I. Kalinina,
Zh A. Akopyan, Y.V. Parfyonova and V.A. Tkachuk

**Combination of Functional Electrical Stimulation and Noninvasive
Spinal Cord Electrical Stimulation for Movement Rehabilitation
of the Children with Cerebral Palsy** 551
A.G. Baindurashvili, G.A. Ikoeva, Y.P. Gerasimenko,
T.R. Moshonkina, I.E. Nikityuk, I.A. Solopova,
I.A. Sukhotina, S.V. Vissarionov and D.S. Zhvansky

**Bifunctional Recombinant Protein Agent Based on Pseudomonas
Exotoxin A Fragment for Targeted Therapy of HER2-Positive
Tumors** 563
S.M. Deyev, O.M. Kutova, E.N. Lebedenko,
G.M. Proshkina, A.A. Schulga and E.A. Sokolova

**Development of Classification Rules for a Screening Diagnostics
of Lung Cancer Patients Based on the Spectral Analysis
of Metabolic Profiles in the Exhaled Air** 573
A.V. Borisov, Yu.V. Kistenev, D.A. Kuzmin,
V.V. Nikolaev, A.V. Shapovalov and D.A. Vrazhnov

**Antitumor Effect of Vaccinia Virus Double Recombinant
Strains Expressing Genes of Cytokine GM-CSF and Oncotoxic
Peptide Lactaptin** 581
G.V. Kochneva, O.A. Koval, E.V. Kuligina,
A.V. Tkacheva and V.A. Richter

**Genome-Wide Association Studies for Milk Production Traits in
Russian Population of Holstein and Black-and-White Cattle**.......... 591
A.A. Sermyagin, E.A. Gladyr, K.V. Plemyashov, A.A. Kudinov,
A.V. Dotsev, T.E. Deniskova and N.A. Zinovieva

**Overview of 17,856 Compound Screening for Translation Inhibition
and DNA Damage in Bacteria** 601
P.V. Sergiev, E.S. Komarova (Andreianova), I.A. Osterman, Ph.I. Pletnev,
A.Ya. Golovina, I.G. Laptev, S.A. Evfratov, E.I. Marusich, M.S. Veselov,
S.V. Leonov, Ya.A. Ivanenkov, A.A. Bogdanov and O.A. Dontsova

**Shape of the Voltage–Frequency Curve Depending on the Type
of the Object Detached from the QCM Surface** 609
F.N. Dultsev

Complex Technology of Oil Sludge Processing 617
A.V. Anisimov, V.I. Frolov, E.V. Ivanov,
E.A. Karakhanov, S.V. Lesin and V.A. Vinokurov

**Comprehensive Ground-Space Monitoring of Anthropogenic
Impact on Russian Black Sea Coastal Water Areas** 625
V.G. Bondur and V.V. Zamshin

**Determination of the Optimal Technological Conditions
of Processing of the Alkali Alumosilicate** 639
V.N. Brichkin, A.M. Gumenyuk, A.V. Panov and A.G. Suss

**New Highly Efficient Dry Separation Technologies
of Fine Materials**.. 649
V.A. Arsentyev, A.M. Gerasimov,
S.V. Dmitriev and A.O. Mezenin

**Hydrogenation Processing of Heavy Oil Wastes in the Presence of
Highly Efficient Ultrafine Catalysts** 659
A.E. Batov, Kh.M. Kadiev, M.Kh. Kadieva,
A.L. Maximov and N.V. Oknina

**Development of Unified Import-Substituting Energy-Saving
Technology for Purification of Roily Oils, Oil-Slimes,
and Chemical and Petrochemical Effluents** 669
V.V. Grigorov and G.V. Grigoriev

**Development of Remote and Contact Techniques for Monitoring
the Atmospheric Composition, Structure, and Dynamics** 679
B.D. Belan, Yu.S. Balin, V.A. Banakh, V.V. Belov,
V.S. Kozlov, A.V. Nevzorov, S.L. Odintsov,
M.V. Panchenko and O.A. Romanovskii

**Technology of Integrated Impact on the Low-Permeable
Reservoirs of Bazhenov Formation** 693
V.S. Verbitskiy, V.V. Grachev and A.D. Dmitrievskiy

**Development of the First Russian Anammox-Based Technology
for Nitrogen Removal from Wastewater** 699
A.M. Agarev, A.G. Dorofeev, A.Yu. Kallistova, M.V. Kevbrina,
M.N. Kozlov, Yu.A. Nikolaev and N.V. Pimenov

Pulse-Detonation Hydrojet 709
S.M. Frolov, K.A. Avdeev, V.S. Aksenov, F.S. Frolov,
I.A. Sadykov, I.O. Shamshin and R.R. Tukhvatullina

**Development of Technological Process of Matrix Conversion
of Natural and Associated Petroleum Gases into Syngas
with Low Content of Nitrogen** 721
V.S. Arutyunov, A.V. Nikitin, V.I. Savchenko, I.V. Sedov,
O.V. Shapovalova and V.M. Shmelev

Part I
Computer Science

Multimodal Control System of Active Lower Limb Exoskeleton with Feedback

S.A. Mineev

Abstract Current paper describes multimodal control system of active lower limb exoskeleton with feedback, which provides switching between manual and semi-automatic modes of exoskeleton motion control in the process of movement. Channel of proportional control of exoskeleton actuators and visual feedback allow exoskeleton pilot to overcome different kinds of obstacles on the move.

Keywords Lower limb robotic exoskeleton · Control architecture
Intention recognition · Volitional control · Proportional control
Sensory feedback · Biomechatronic

Introduction

Biofeedback is crucial for controlling human movement during walking [1]. Sensory function disorders resulting from injuries and diseases lead to a loss of movement ability, as well as motor function impairments. An issue of organizing feedback in a pilot-exoskeleton system in the process of developing robotic lower limb medical exoskeleton often is not considered. It is assumed that when exoskeleton pilot walks, retained sensory function is sufficient to organize movement, i.e. lost motor function is replaced by the exoskeleton function, but lost sensory function is not replaced.

A number of research projects devoted to the development of robotic lower limb exoskeleton interfaces attempted to organize feedback in a biomechanical pilot-exoskeleton system by means of vibrostimulation, acoustic signals, visual indicators [1], and electrostimulation [2]. However, despite the proven feasibility of real-time feedback on the pressure exerted on the foot by the surface [2], or on the current angles of joints, feedback systems did not find a practical application in the production of robotic lower limb exoskeletons. The main reason is inability to

S.A. Mineev (✉)
Lobachevsky State University of Nizhni Novgorod, Nizhny Novgorod, Russia
e-mail: sergm@nifti.unn.ru

© The Author(s) 2018
K.V. Anisimov et al. (eds.), *Proceedings of the Scientific-Practical Conference "Research and Development - 2016"*, https://doi.org/10.1007/978-3-319-62870-7_1

organize a direct channel for proportional control of exoskeleton drive for the pilots who lost both sensory and motor lower limb function (who are considered the main target audience for lower limb robotic exoskeletons). Without a direct proportional control, feedback channel does not provide any advantages. Direct proportional control of exoskeleton drive was achieved only in the exoskeleton HAL, which is based on multichannel EMG signals, registered on the lower limbs muscles of a pilot. Thus, such technology could not be used for pilots who lost both sensory and motor lower limb function. At the same time, command control systems have received wide application in mass-produced robotic lower limb exoskeletons [3–6]. By performing a certain action, for example by pressing a button on crutches, an operator informs the exoskeleton control system of his/her intention to take a step with the left foot. By performing another action, for example, by pressing another button, an operator can communicate his/her intention to take a step with the right foot. This approach allows the system to produce a movement of the pilot-exoskeleton system on a flat surface and does not require significant time spent on pilot training. Moreover, command control does not require significant concentration of a pilot, which conserves pilot's energy and has a positive effect on the duration of continuous walking. The main disadvantage of command control systems is the difficulty to adapt to changing environment (slopes, obstacles, doorstep, etc.). Lower limb movement in such conditions is executed according to a pre-pattern and, at best, can switch from one template to another during the walking mode.

Thus, we can state that a promising robotic lower limb medical exoskeleton must support at least two modes of operation:

- Command semi-automatic mode, which allows the movement of the pilot-exoskeleton system on a flat surface with constant parameters of walking mode, such as step length, feet elevation, etc.;
- Manual independent proportional control mode of each exoskeleton joint, providing biomechanical motion adaptation of the system to the complex environment, such as obstacles, doorsteps, slopes.

The latter should be implemented using feedback, allowing the pilot with impaired lower limb sensory function sense exoskeleton current status and orientation of its parts relative to each other.

Switching between modes in a multimode robotic exoskeleton should be executed in the process of movement, without stopping or any special procedures.

This work is devoted to developing and testing a multimode active control system of the lower limb exoskeleton with feedback.

Exoskeleton Structure

Active lower limb exoskeleton with four degrees of freedom developed in the Lobachevsky State University of Nizhni Novgorod was used as the basis for of the multimodal robotic exoskeleton. Active movement is executed in the sagittal plane with knee and hip joints. The exoskeleton is driven by brushless electric motors Maxon EC 45 flat 36 V (knee joints) and Maxon EC 90 flat 36 V (hip joints). Actuators are controlled by a group of 4 controllers Electriprivod BLSD-20 (Russia). Controllers are calibrated in torque units (Nm), which is exerted by the actuators of exoskeleton joints. Drive units of right and left limbs with spur gear-boxes provide a torque of 100 Nm in the knee joint and 150 Nm in the hip joint. The ankle is passive. A 36 V, 10 Ah lithium polymer battery provides 2 h of continuous motion of the pilot-exoskeleton system. Exoskeleton is attached to the pilot's torso via a segmented belt and to the limbs via U-shaped staples.

Exoskeleton Sensory Subsystem

Robotic lower limb exoskeleton sensory subsystem created in the framework of this study is based on the sensory subsystem described in [7]. Four force sensors were added to the original subsystem and placed on the soles of the feet (2 sensors on each foot). Tekscan FlexiForce A401 strain gauges were used as force sensors to provide an ability to measure the reaction force on the soles of the feet in the range of 0–1220 N at frequency of 100 Hz.

Given the improvements, exoskeleton sensory subsystem provides the following abilities:

- to measure angles of pilot's torso deviation from the vertical in sagittal and frontal planes with an accuracy of 1 degree or better;
- to measure flexion angles of knee and hip joints with an accuracy of 0.5° or better;
- to measure floor reaction force on the soles of the feet with an accuracy of 20 N or better.

Data acquisition from sensors and issuing control commands to exoskeleton actuators is executed on board the microprocessor board BeagleBoard-xM rev. C [8]. Sampling frequency of data acquisition is 100 Hz.

Human–Exoskeleton Interface with Feedback

In order to construct human–exoskeleton interface with feedback, we proposed a scheme that uses strain gauges placed on the phalanges of the exoskeleton pilot. Strain gauges provide establishment of the exoskeleton proportional drive control channel. It took eight strain gauges in total, four on each of the pilots' wrist, to control four actuators, each of which can rotate the shaft in two opposite directions.

In addition to strain gauges on the phalanges of the pilots' fingers, we placed buttons that provide an ability to promptly change the operational mode of exoskeleton control system. Location of sensors and buttons is illustrated in Fig. 1.

(a) **(b)**

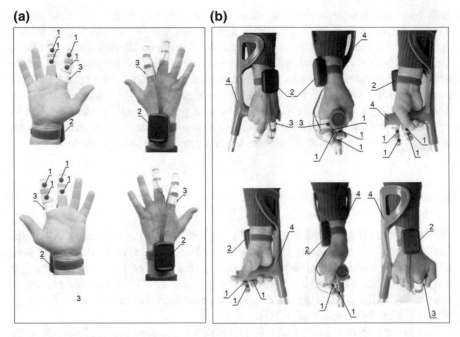

Fig. 1 Location of sensors and buttons on fingers of a pilot (**a**), method of capture of crutches and impact on strain gauges (**b**), where: *1* strain gauges, *2* preamplifier (4 channels), *3* buttons of switching of the modes, *4* crutches

To provide feedback, we decided to use visual stream from a display unit attached to the pilots' head and placed in his field of vision. The following information should be displayed on the unit block:

- Information on the current operational mode of exoskeletons' control system (a 7-segment single-character digital indicator is used);
- Information on the current flexion angles of the exoskeleton joints (two bar indicators are used);
- Information on the reaction force excreted on the exoskeleton feet (four LEDs, brightness of which depends on the absolute values of the forces excreted on the soles of the exoskeleton feet in heel and toe areas);
- Emergency information (red LED lights up in the event of threats to critical failure of exoskeleton).

A general view of a visual feedback display unit and its attachment method to the exoskeleton pilots' head are shown in Fig. 2.

Display unit is operated by a microcontroller, which executes the following tasks:

- Polls strain gauges of proportional control channel;
- Polls relay buttons of switching operational mode of exoskeleton;

(a)

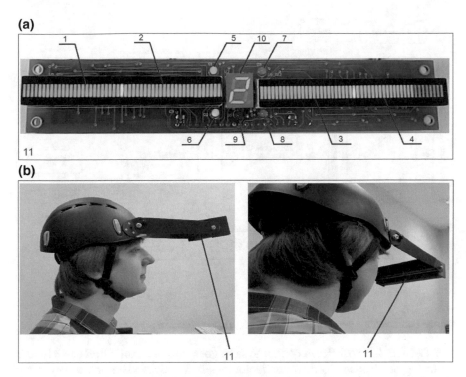

(b)

Fig. 2 A general view of a visual feedback display unit (**a**) and its attachment method to the exoskeleton pilots' head (**b**), where: *1* indicator of *left knee angle*, *2* indicator of *left hip angle*, *3* indicator of *right hip angle*, *4* indicator of *right knee angle*, *5* indicator of the force excreted on the sole of the exoskeleton *left foot* in toe area, *6* indicator of the force excreted on the sole of the exoskeleton *left foot* in heel area, *7* indicator of the force excreted on the sole of the exoskeleton *right foot* in toe area, *8* indicator of the force excreted on the sole of the exoskeleton *right foot* in heel area, *9* Emergency indicator, *10* single-character digital indicator, *11* visual feedback display unit

– Receives information from the control program functioning on BeagleBoard-xM microprocessor board on the status of sensors of flexion angle of the joints, on the forces acting on the feet, on the current operational mode of exoskeletons' control system and on the signs of emergency;
– Changes the state of display unit in accordance with received information;
– Sends status information of strain gauges and buttons to the control program.

Interaction of BeagleBoard-xM microprocessor board with the display unit is executed via RS485 interface by means of RS232-RS485 converter.

A scheme of developed human–machine interface is shown in Fig. 3.

Control system of exoskeleton supports six modes: STANDBY, BLOCKAGE, STAND UP, SIT DOWN, AUTO MOVEMENT and MANUAL MOVEMENT. Modes could be switched with non-locking buttons placed near the base of the index fingers of exoskeleton pilot. Button placed on the right hand is referred to as the SWR and button placed on the left hand is called SWL. Buttons are pressed with the thumbs. Let us review the operational modes:

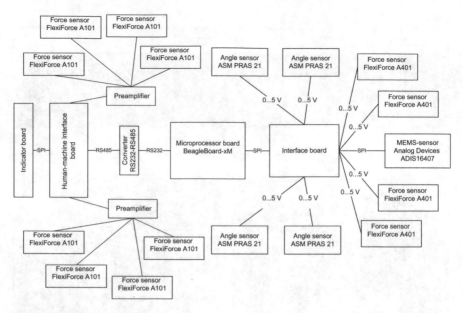

Fig. 3 Exoskeleton control system with the developed human–machine interface

STANDBY. Exoskeleton control system goes into this mode immediately after switching on. Voltage to actuators is not supplied, only sensors are surveyed and exoskeleton status is displayed on a display unit (symbol 'F' is displayed on a single-character indicator). Holding either SWR or SWL buttons will switch exoskeleton control system to STANDBY mode from any other mode.

BLOCKAGE. This mode is activated with a single press of either SWR or SWL buttons. In this mode, a small current flows through the motor winding, which blocks motor shafts and does not allow exoskeleton joints to flex or stretch. All the sensors are surveyed and exoskeleton status is displayed on a display unit (symbol 'L' is displayed on a single-character indicator).

STAND UP. This mode is activated by double pressing SWR button if pilot-exoskeleton system is currently in a sitting position. In this mode, control system executes a template movement to bring exoskeleton in a standing position. All the sensors are surveyed and exoskeleton status is displayed on a display unit (symbol 'U' is displayed on a single-character indicator). Upon completion of stand up command, BLOCKAGE mode is automatically activated.

SIT DOWN. This mode is activated by pressing the SWR button three times given that pilot-exoskeleton system is in a standing position. In this mode, control system executes a template movement to bring exoskeleton in a sitting position. All the sensors are surveyed and exoskeleton status is displayed on a display unit (symbol 'S' is displayed on a single-character indicator). Upon completion of this procedure, STANDBY mode is automatically activated.

AUTO MOVEMENT. This mode is activated by double pressing SWL button if pilot-exoskeleton system is currently in a standing position. In this mode, the control system monitors the deviation of the pilot torso from the vertical in frontal and sagittal planes, measures forces exerted on the exoskeleton feet and flexing angles of the joints. Exoskeleton status is displayed on the display unit (symbol 'A' is displayed on a single-character indicator). If sensors indicate that one of the legs is both a support leg and is in front of the other leg, a step forward is made with the other leg. This way, a pilot with impaired motor and sensory lower limb functions can walk helping himself with his hands relying on the crutches. In the AUTO MOVEMENT mode, movement can only be executed on a flat surface with a slope of 15 degrees or less.

MANUAL MOVEMENT. This mode is activated by pressing SWL button three times. In this mode exoskeleton control system polls all sensors and displays exoskeleton status on a display unit (symbol 'H' is displayed on a single-character indicator). Each strain gauge is associated with an exoskeleton joint and with the direction of motor shaft rotation. For example, a strain gauge mounted on the distal phalange of the index finger is associated with hip flexion, while a strain gauge mounted on the medial phalange of the same finger is associated with hip straightening. Movement of the knee joint is controlled by the strain gauges mounted to phalanges of the middle finger in a similar way to the index finger.

Pilot presses on the corresponding phalanx using handle crutches, while fingers not involved in the movement control hold the crutch. Pressing is performed with different strengths. Actuators develop small torque on the motor shafts if the force is weak and large torque up to maximal 100 Nm in the knee and 150 Nm in the hip joint if the force is strong. Torque changes according to a predetermined calibration characteristic, *strength of finger pressure/torque on the motor shaft of the joint.* A pilot controls torque on the motor shafts of the joints by focusing on the indicator readings and visually checking the surroundings. Performing the movements requires considerable concentration of the pilots' attention. This mode is used to perform voluntary movements for stepping over obstacles, climbing stairs, etc.

Described interface has been implemented and tested at National Research N.I. Lobachevsky State University of Nizhni Novgorod. The test results confirmed the possibility of creating a multimodal active control system of lower limbs exoskeleton with feedback, which allows prompt adapting of the exoskeleton movement to the environment in the process of movement.

On the basis of this research, we sent an application for invention titled "Methods for generating control signals and manual control of lower limb exoskeleton operation, as well as control interfaces for operation of this exoskeleton in manual and software control modes using this generation method", registered by the Federal Service for Intellectual Property of Russian Federation #2016144426 11.11.2016.

Conclusions

The challenge of developing a control interface for an active lower limb exoskeleton that is easy to use for pilots with significant impairments of sensory and motor lower limb functions is one of the main problems preventing active lower limb exoskeleton from widespread application in the medical field.

Solution to this problem lies in implementing proportional control and feedback mechanisms that would allow using the human brain plasticity for replacing lost sensory and motor lower limb functions.

Proposed interface has significant potential to improve consumer properties in terms of improving proportional control channel and feedback channel. In the near future we plan to replace bulky display unit with augmented reality glasses and implement feedback channel by means of cutaneous vibratory stimulation.

Acknowledgments Research are carried out (conducted) with the financial support of the state represented by the Ministry of Education and Science of the Russian Federation. Agreement (contract) no. 14.575.21.0031 30 Jun 2014. Unique project Identifier: RFMEFI57514X0031.

References

1. Tucker, M.R., Olivier, J., Pagel, A., Bleuler, H., Bouri, M., Lambercy1, O., del R Millán, J., Riener, R., Vallery, H., Gassert, R.: Control strategies for active lower extremity prosthetics and orthotics: a review. J. NeuroEng. Rehabil. **12**, 1 (2015)
2. Raj, A.K., Neuhaus, P.D., Moucheboeuf, A.M., Noorden, J.H., Lecoutre, D.V.: Mina: a sensorimotor robotic orthosis for mobility assistance. J. Robot. **2011**
3. Chen, G., Chan, C.K., Guo, Z., Yu, H.: A review on lower extremity assistive robotic exoskeleton in rehabilitation therapy. Crit. Rev. Biomed. Eng. 2013
4. Kawamoto, H., Sankai, Y.: Power assist method based on phase sequence driven by interaction between human and robot suit. In 13th IEEE International Workshop on Robot and Human Interactive Communication, pp. 491–496 (2004)
5. Medical exoskeleton for rehabilitation (Electronic resource). www.exoatlet.ru (site). URL http://www.exoatlet.ru/ (date of view 23.11.16) (2004)
6. Li, N., Yan, L., Qian, H., Wu, H., Wu, J., Men, S.: Review on lower extremity exoskeleton robot. Open Autom. Control Syst. J. **7**, 441–453 (2015)
7. Mineev, S.A., Novikov, V.A., Kuzmina, I.V., Shatalin, R.A., Grin, I.V.: Goniometric sensor interface for exoskeleton system control device. Biomed. Eng. **49**(6), 357–361 (2016)
8. BeagleBoard-xM Rev C System Reference Manual (Electronic resource). beagleboard.org (site). URL http://beagleboard.org/static/BBxMSRM_latest.pdf (date of view 23.11.16) (2010)

Investigation and Development of Methods for Improving Robustness of Automatic Speech Recognition Algorithms in Complex Acoustic Environments

M.L. Korenevsky, Yu. N. Matveev and A.V. Yakovlev

Abstract Aims and objectives of the study are described; state-of-the-art techniques in the study area are outlined. Several effective approaches proposed in the study and targeted at robustness improvement in complex acoustic environments are described. They are multichannel alignment algorithm, vector Taylor series-based features compensation with phase-term modeling, and environment adaptation method based on GMM-derived features. Experimental results analysis and comparison to state of the art are presented.

Keywords Speech recognition · Robustness · Distortion · Noise
Compensation · Adaptation · Beamforming · VTS · GMM-derived features

Introduction

The past decade has seen a rapid development of speech recognition technology, which has led to significant improvements in the recognition accuracy for all scenarios of its usage and the large introduction of recognition technologies into many spheres of human activity. The reasons for this development are primarily related to the widespread adoption of multilayer (deep) neural networks for acoustic modeling. In several tasks, this made it possible to closely approach to (or even exceed) the human level of recognition accuracy. However, under strong noises and acoustic distortions, especially nonstationary, automatic speech recognition (ASR) algorithms are still noticeably inferior to human abilities. Accuracy of the acoustic models which are almost error-free in recognition of clean speech,

M.L. Korenevsky (✉) · Yu.N. Matveev · A.V. Yakovlev
ITMO University, Saint Petersburg, Russia
e-mail: korenevsky@speechpro.com

Yu.N. Matveev
e-mail: matveev@speechpro.com

A.V. Yakovlev
e-mail: yakovlev@speechpro.com

© The Author(s) 2018   11
K.V. Anisimov et al. (eds.), *Proceedings of the Scientific-Practical Conference*
"Research and Development - 2016", https://doi.org/10.1007/978-3-319-62870-7_2

generally deteriorates when they are used in complex acoustic environments (strong noises, distant microphone, reverberation, etc.), i.e., even state-of-the-art ASR systems are mostly not sufficiently robust. This paper is devoted to research and development of new approaches to improve the robustness of ASR algorithms with the emphasis on using of neural network acoustic models.

Aims and Objectives

The aims of this study were to design new methods and software/engineering solutions for real-time automatic continuous speech recognition in complex acoustic environments. The designed methods should provide: noise suppression in the processed speech signal with a minimal distortion of its spectrum and as a consequence, its intelligibility improvement; voice activity detector (VAD) reliability improvement; accounting and compensation of the influence of noisy conditions on the recognition accuracy; speech recognition improvement in acoustic conditions different from those used for acoustic models training.

The relevance of research directions is confirmed by the fact that still there is no commercially successful speech recognition product, which would provide human-comparable recognition accuracy in the complex acoustic environment.

State of the Art in Study Area

Research in improving the robustness of ASR algorithms have a long history, first significant studies date back to 1970s, see for example [1]. At that time the main direction of increasing speech recognition accuracy was to preprocess speech signal itself to remove or at least suppress a noise component strongly in order to improve speech quality and intelligibility. This direction is still being actively developed, although conventional denoising techniques based on signal processing methods are gradually superseded with more sophisticated approaches which use non-negative matrix factorization (NMF) [2], filtering based on spectral masks generated by neural networks [3], missed data restoration techniques [4], and so on. The special place in this direction belongs to the processing of signals from several microphones (microphone array) [5]. Such approaches make it possible to take into account geometric features of relative positions of microphones and speaker and to "beamform" microphone array in such a way to amplify speech signal from a target direction and to suppress interference and noise from all other directions. Development of such processing methods is especially important for the applications like "smart home" when microphones are located in several parts of the room and both location and orientation of speaker's head are unknown in advance. In order to promote the development of multi-microphone approaches in robust speech recognition, several international competitions like CHiME (Computational

Hearing in Multisource Environments) Challenge [6–9] have been organized in recent years.

One more direction in improving ASR system robustness is using robust acoustic features, i.e., those whose distribution is distorted only slightly on changes of acoustic conditions (and which still keep good abilities to discriminate speech phones). In developing such features, researchers often refer to human auditory system which is able to recognize speech even in very adverse conditions. The examples of robust acoustic features based on auditory system processing are PNCC (Power-Normalized Cepstral Coefficients) [10] or gammatone filterbanks energies [11].

A similar problem of features variability reduction is also solved by various normalization methods like CMVN (Cepstral Mean and Variance Normalization) [12] or more general histogram equalization [13].

Besides, a large group of developed approaches are aimed at not in increasing features resistance to different distortions but instead try to explicitly remove the influence of these distortions—these are feature compensation methods. The most noticeable techniques among them are SPLICE (stereo-piecewise linear compensation for environment) [14], which uses statistics of joint speech and noisy features distribution to construct piecewise linear transform from noisy to clean speech features, and VTS (Vector Taylor Series) [15] which uses approximate linearization of nonlinear model of speech distortion by noise and channel to construct similar transform.

The idea of VTS is also applied in other groups of approaches for robustness improving, namely in adaptation of acoustic models to acoustic environment changes, i.e., adjusting models trained for clean speech recognition to the acoustic features distortions. The same linearized distortion model is used to modify features distribution parameterized as a GMMs (Gaussian Mixture Model). A number of other successful approaches to adapt GMM-based acoustic models were also developed such as MLLR, CMLLR (fMLLR) [16], MAP [17], PMC [18], and some their combinations.

However, in the past years, GMM-HMM acoustic models were almost everywhere superseded by acoustic models based on deep neural networks (DNNs), which provide much better accuracy in the vast majority of tasks. DNNs need completely different ways of adaptation, development of which was in a very initial stage when our study started. In order to keep the network architecture most of the methods developed till that moment modified weights of a trained neural net by fine-tuning them on adaptation data with a backpropagation algorithm. This is rather computationally demanding and needs to create a copy of initial network (which has millions of parameters) for each new acoustic conditions.

During this study, new approaches were developed within three of above-mentioned robustness improvement directions. The architecture and programming implementation of experimental software were also developed, where these approaches were integrated into single speech processing and recognition pipeline for complex acoustic environments. In the following sections, these approaches and results of their application are considered in more detail.

Multichannel Alignment (MCA)

This algorithm first described in [19] is an adaptive microphone array (MA) beamforming method using an output of well-known Delay-and-Sum beamforming method [5] as a "reference" signal. The algorithm computes adaptive transfer functions for each microphone channel signals by means of «aligning» their spectra relative to that of the reference signal. Signals passed through the transfer functions are then averaged to provide the resulting speech signal. This approach makes the width of the MA's directivity pattern main lobe narrower (i.e., improves spatial directivity) and significantly reduces the level of sidelobes (i.e., suppresses noises and interferences received on them). The scheme of MCA processing (for the case of 4 microphones) is depicted in Fig. 1. It is worth noting that the reference signal may be presumably obtained from any other beamforming algorithm as well, and better the reference more noticeable the effect of MCA should be.

STFT and IFT stand for Short-Time Fourier Transform and Inverse Fourier Transform respectively, $D_i(f)$ are channels' delays (steering) vectors.

MCA algorithm was successfully applied in our submission to the CHiME Challenge 2015 where it has demonstrated competitive results compared to several well-known beamforming algorithms [20]. The important characteristics of MCA include the implementation simplicity, low computational complexity and resistance to target direction errors.

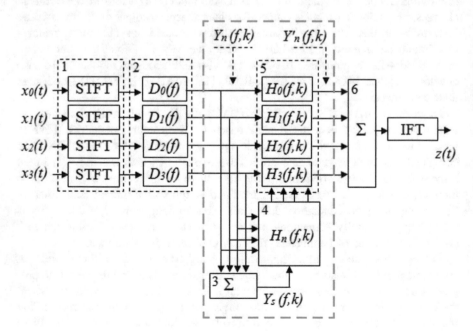

Fig. 1 Multichannel alignment algorithm

VTS Feature Compensation with a Phase-Term Modeling

VTS-based methods comprise an extremely important class of approaches to ASR robustness improvement: they are applied for features distortion compensation, adaptation and adaptive training acoustic models as well as for dealing with uncertainty remained after feature compensation during speech decoding. As it was already mentioned, VTS-based adaptation is applicable to only GMM-HMM acoustic models, therefore under widely used DNN-HMM framework its direct application is not possible. Thus, VTS-based acoustic features compensation (cleaning) becomes more important now.

The most widely used speech distortion model by noise and channel has the following form:

$y(t) = x(t) * h + n(t)$, where $x(t)$, $y(t)$, $n(t)$, and h denote clean speech, noisy speech, and noise signals as well as channel impulse response, respectively, and $*$ denotes convolution. When computing the most widespread MFCC (mel-frequency cepstral coefficients) features the signal is processed with several linear and non-linear transformations. This results in the following relation between the features of the above signals:

$$\mathbf{y} = \mathbf{x} + g(\mathbf{x}, \mathbf{h}, \mathbf{n}, \boldsymbol{\alpha}) = \mathbf{x} + \mathbf{h} + C \log\left(1 + e^{D(\mathbf{n}-\mathbf{x}-\mathbf{h})} + 2\boldsymbol{\alpha} \bullet e^{D(\mathbf{n}-\mathbf{x}-\mathbf{h})/2}\right),$$

where $\mathbf{x}, \mathbf{y}, \mathbf{n}, \mathbf{h}$ denote vectors of MFCCs for clean speech, noisy speech, noise and channel response, $\boldsymbol{\alpha}$ is a "phase" vector, C and D—are the matrices of direct and inverse discrete cosine transform (DCT) and, finally, $\bullet$ is an elementwise (Hadamard) product of vectors. This model involves nonlinear vector-function g and its presence makes estimation of the clean speech features from noisy speech features extremely difficult. The essence of VTS method is a linearization of this nonlinearity by means of Vector Taylor expansion up to the first-order terms around some set of points:

$$\mathbf{y} \approx \mathbf{x} + g(\mathbf{x}_0, \mathbf{h}_0, \mathbf{n}_0, \boldsymbol{\alpha}_0) + \nabla_{\mathbf{x}} g^T (\mathbf{x} - \mathbf{x}_0) + \nabla_{\mathbf{h}} g^T (\mathbf{h} - \mathbf{h}_0)$$
$$+ \nabla_{\mathbf{n}} g^T (\mathbf{n} - \mathbf{n}_0) + \nabla_{\boldsymbol{\alpha}} g^T (\boldsymbol{\alpha} - \boldsymbol{\alpha}_0).$$

This, of course, introduces some error into the model, but greatly facilitates the following inference.

The last nonlinearity term which contains phase vector $\boldsymbol{\alpha}$ was first taken into account in [21],[1] where it was demonstrated that this improves the model accuracy. However, in both just cited paper and subsequent ones, which use such distortion model phase vector was treated in some special ways: for example it was assumed to have equal components, which is not physically adequate, or its distribution

---

[1]This term was always discarded in previous papers as presumable being close to zero.

Table 1 Accuracy of the VTS in clean training scenario for different SNR values

SNR, dB	No VTS	VTS without phase	VTS with phase
Clean (>40)	99.08	99.01	99.04
20	94.32	98.22	98.47
15	84.65	96.67	97.49
10	64.07	92.57	94.52
5	36.46	82.47	86.79
0	16.18	56.71	64.58
−5	9.04	22.66	28.96

parameters were estimated from the trainset in advance and then considered as known. We proposed a new variant of VTS-based on the same model, where phase vector is treated as a multivariate Gaussian with the unknown parameters (as it is usually done for noise features vector $\mathbf{n}$), and these parameters are inferred based on maximum likelihood principle and using EM-algorithm. This approach is not limited to noises available in the training set and does not put tight constraints on the phase vector structure. We derived EM expressions to update $\mathbf{n}, \alpha$, and $\mathbf{h}$ distributions parameters and formula for estimating clean speech features [22].

Experiments for assessing effectiveness of proposed VTS variant were performed, inter alia, on the Aurora2 database [23], which contains utterances of sequences of English digits distorted with different noises and channels. Obtained results, part of which is shown in Table 1, clearly demonstrate that the proposed method significantly improves the recognition accuracy compared to both VTS without phase-term modeling and especially to unprocessed noisy speech recognition.

Adaptation Based on GMM-Derived Features

It was already mentioned that well-designed adaptation methods for GMM-HMM acoustic models appeared to be not applicable after the migration to neural network acoustic models. Possibility of using GMM models for feature compensation gave rise to the idea of using GMM for adaptation as well if the features for DNN are not raw MFCCs but their GMM-based likelihoods of simple GMM-HMM acoustic model.

This idea led to the proposed method of adaptation based on GMM-derived features, designed in details in the papers [24–26]. The scheme of method application in its original variant (for speaker adaptation[2]) is depicted on Fig. 2.

---

[2]Method can be easily applied to environment but not speaker adaptation.

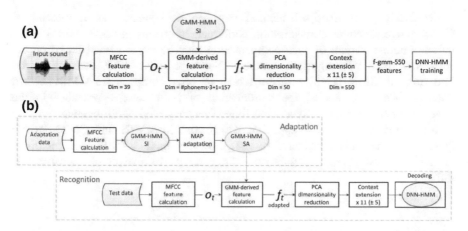

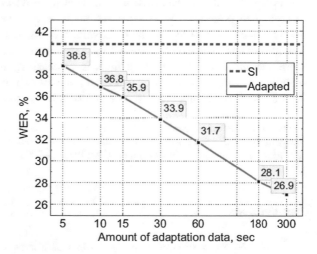

Fig. 2 Training and adaptation of DNN acoustic model based on GMM-derived features. **a** Training, **b** Adaptation and decoding

Fig. 3 Effect of adaptation depending on amount of adaptation data

As shown on the figure, speaker-independent (SI) GMM-HMM and DNN-HMM models are first trained. In this stage, the dimensionality of GMM likelihoods is reduced by PCA transform, and then features vector is extended by features from the neighboring frames. On the adaptation stage only GMM-HMM model is adapted (with MAP here) and then its (speaker adapted) outputs are fed into the same pipeline as for SI. As a result, the main DNN-HMM acoustic model remains unchanged. Number of experiments described in [24–26] show that although SI-DNN-HMM on GMM-derived features works worse than on MFCCs, its adaptation is extremely efficient. The illustration of the last statement (with application to speaker adaptation) is shown on Fig. 3.

Obviously, the described method may be also treated as a method of GMM-derived features compensation performed by means of GMM-HMM adaptation. Another variant of such compensation may be implemented based on the above-described VTS algorithm, where VTS-compensated MFCC features are fed into original GMM-HMM acoustic model to infer compensated GMM-derived features. We implemented the combination of both these approaches in the developed experimental software and found that they work well together.

Interestingly, the similar approach which combines VTS and GMM-derived features was recently considered in [27], however there VTS is used for direct GMM-HMM adaptation, therefore the direct comparison of these approaches is difficult.

Conclusions

Several different approaches which provide improvements of speech recognition accuracy in complex acoustic environments were developed in this study. They demonstrate competitive results on several well-known speech recognition benchmarks and have some advantages compared to many state-of-the-art analogues. The combination of the proposed methods is successfully implemented in the experimental software, which may be used as a basis for deployment of new ASD systems and devices, providing reliable speech recognition in adverse acoustic conditions.

Acknowledgments Research are carried out with the financial support of the state represented by the Ministry of Education and Science of the Russian Federation. Agreement no. 14.575.21.0033 27. June 2014. Unique project Identifier RFMEFI57514X0033.

References

1. Boll, S.F.: Suppression of acoustic noise in speech using spectral subtraction. IEEE Tran. Acoust. Speech Sig. Process. **27**(2), 113–120 (1979)
2. Wilson, K.W, Raj, B., Smaragdis, P., Divakaran, A.: Speech denoising using nonnegative matrix factorization with priors. Proceedings of IEEE International Conference on Acoustics, Speech and Signal Processing (ICASSP), Las Vegas, USA, 31 March–4 April 2008
3. Li, B., Sim, K.: An ideal hidden-activation mask for deep neural networks based noise-robust speech recognition. Proceedings of IEEE International Conference on Acoustics, Speech and Signal Processing (ICASSP), pp. 200–204, Florence, Italy, 4–9 May 2014
4. Raj, B., Seltzer, M., Stern, R.: Reconstruction of missing features for robust speech recognition. Speech Commun. **43**(4), 275–296 (2004)
5. Brandstein, M., Ward, D. (eds.): Microphone Arrays. Springer, Heidelberg (2001), 398 p
6. Barker, J., Vincent, E., Ma, N., Christensen, C., Green, P.: The PASCAL CHiME speech separation and recognition challenge. Comput. Speech Lang. **27**(3), 621–633 (2013)
7. Barker, J., Marxer, R., Vincent, E., Watanabe S.: The third 'chime' speech separation and recognition challenge: Dataset, task and baselines. Proceedings of IEEE Automatic Speech Recognition and Understanding Workshop (ASRU) (2015)

8. Proceedings of the 4th International Workshop on Speech Processing in Everyday Environments (CHiME 2016), San Francisco, 13 Sept 2016, 90 p
9. Vincent, E., Barker, J., Watanabe, S., Le Roux, J., Nesta, F., Matassoni, M.: The second CHiME speech separation and recognition challenge: datasets, tasks and baselines. Proceedings of IEEE International Conference on Acoustics, Speech, and Signal Processing (ICASSP), Vancouver, Canada, 26 May–31 May 2013 (2013)
10. Kim, C., Stern, R.M.: Power-normalized cepstral coefficients (PNCC) for robust speech recognition. IEEE Transactions on Audio, Speech and Language Processing (2012)
11. Shao, Y., Jin, Zh., Wang, D., Srinivasan, S.: An auditory-based feature for robust speech recognition. Proceedings of IEEE International Conference on Acoustics, Speech and Signal Processing (ICASSP), Taipei, Taiwan, 19 Apr–24 Apr 2009
12. Viikki, O., Laurila, K.: Cepstral domain segmental feature vector normalization for noise robust speech recognition. Speech Commun. 25(1) (1998)
13. de la Torre, Á., Peinado, A.M., Segura, J.C., Pérez-Córdoba, J.I.., Benítez, M.C., Rubio, A.J.: Histogram equalization of speech representation for robust speech recognition. IEEE Trans. Speech Audio Process. 13(3) (2005)
14. Droppo, J., Deng L., Acero A.: Evaluation of SPLICE on the Aurora 2 and 3 tasks. Proceedings of the International Conference on Speech and Language Processing (ICSLP), pp. 29–32 (2002)
15. Moreno, P., Raj, B., Stern, R.: A vector taylor series approach for environment-independent speech recognition. Proceedings of the IEEE International Conference on Acoustics, Speech and Signal Processing (ICASSP),vol. 2, pp. 733–736, Atlanta, Georgia, USA, 7–10 May 1996
16. Gales, M.J.F.: Maximum likelihood linear transformations for HMM-based speech recognition. Comput. Speech Lang. 12(2), 75–98 (1998)
17. Shinoda, K.: Speaker adaptation techniques for automatic speech recognition. Proceedings APSIPA ASC 2011 Xi'an (2011)
18. Gales, M.J.F., Young, S.J.: Robust continuous speech recognition using parallel model combination. IEEE Trans. Speech Audio Process. 4(5), 352–359 (1996)
19. Stolbov, M., Aleinik, S.: Speech enhancement with microphone array using frequency-domain alignment technique. Proceedings of 54-th International Conference on AES Audio Forensics, pp. 101–107, London (2014)
20. Prudnikov, A., Korenevsky, M., Aleinik, S.: Adaptive beamforming and adaptive training of dnn acoustic models for enhanced multichannel noisy speech recognition. Proceedings of 2015 IEEE Automatic Speech Recognition and Understanding Workshop (ASRU 2015), pp. 401–408 (2015)
21. Deng, L., Droppo, J., Acero, A.: Enhancement of log mel power spectra of speech using a phase-sensitive model of the acoustic environment and sequential estimation of the corrupting noise. IEEE Trans. Speech Audio Process. 12(2), 133–143 (2004)
22. Korenevsky, M., Romanenko, A.: Feature space VTS with phase term modeling. Speech Comput. Lect. Notes Comput. Sci. 9811, 312–320 (2016)
23. Hirsch, H., Pearce, D.: The aurora experimental framework for the performance evaluations of speech recognition systems under noisy conditions. Proceedings of ISCA ITRWASR2000 on Automatic Speech Recognition: Challenges for the Next Millennium (2000)
24. Tomashenko, N., Khokhlov, Y.: Speaker adaptation of context dependent deep neural networks based on MAP-adaptation and GMM-derived feature processing. Proceedings of Interspeech, pp. 2997–3001 (2014)
25. Tomashenko, N., Khoklov, Yu.: GMM-derived features for effective unsuper-vised adaptation of deep neural network acoustic models. Proceedings of Interspeech (2015)
26. Tomashenko, N., Khokhlov, Yu., Estève, Y.: On the use of gaussian mixture model framework to improve speaker adaptation of deep neural network acoustic models. Proceedings of Interspeech (2016)
27. Kundu, S., Sim, K.C., Gales, M.: Incorporating a generative front-end layer to deep neural network for noise robust automatic speech recognition. Inter-speech (2016)

Smart Endoscope—Firmware Complex for Real-Time Analysis and Recognition of Endoscopic Videos

K.U. Erendgenova, E.D. Fedorov, R.M. Kadushnikov,
O.A. Kulagina, V.V. Mizgulin, D.I. Starodubov and S.I. Studenok

Abstract The method for analyzing endoscopic video images, obtained with high-resolution endoscopes, and featuring gastric and colon mucosa microstructures is proposed. The method was implemented in the form of a highly productive "Smart Endoscope" firmware complex used for real-time endoscopic video analysis supported with neural network. Complex was tested, and the accuracy analysis of neoplasm recognition was performed.

Keywords Cancer diagnostics · Decision support · Real-time image analysis
Machine learning · Image recognition · Classification · High-resolution endoscopy
Narrow-band endoscopy

K.U. Erendgenova · E.D. Fedorov
Pirogov Russian National Research Medical University, Moscow, Russia
e-mail: kerendzhenova@gmail.com

E.D. Fedorov
e-mail: efedo@mail.ru

R.M. Kadushnikov · V.V. Mizgulin · D.I. Starodubov · S.I. Studenok (✉)
SIAMS Ltd, Yekaterinburg, Russia
e-mail: studenok@siams.com

R.M. Kadushnikov
e-mail: radi@siams.com

V.V. Mizgulin
e-mail: mizgulin@gmail.com

D.I. Starodubov
e-mail: starodubov@siams.com

O.A. Kulagina
Lomonosov Moscow State University, Moscow, Russia
e-mail: bu747@yandex.ru

K.V. Anisimov et al. (eds.), *Proceedings of the Scientific-Practical Conference*
"Research and Development - 2016", https://doi.org/10.1007/978-3-319-62870-7_3

Introduction

Endoscopic research is a minimally invasive medical procedure that allows examination of body cavities, including gastric tract. It is the most reliable method for revealing early stages of gastric mucosa malignant neoplasms, and pre-cancer diseases that increase chances for cancer development [2, P.88].

First flexible tract endoscopes appeared in early 1960s, and an ability to transform optical signal into electric impulses allowed displaying images, and storing them in analog, and later—digital form. In late 1980s—early 1990s, with the expanded availability of computers, and the development of programming languages and environments allowed development of information systems for search and classification of gastric surface epithelial neoplasms, referred to as CADs (computer-aided diagnosis systems). CADs produce additional diagnostic data or preliminary diagnosis (either posterior or real time), or series of marked images used by a specialist to select a case that best matches the observed one. Recent decade was marked with an increase of interest directed toward CADs [10, P.73], which can be related to accumulation of an extensive set of image and video files, transforming visual examination, analysis, interpretation, and classification of that data into an extremely labor-intensive process. Besides that, developments in the area of wireless capsule endoscopy made the task of creating automated image analysis and diagnostic methods extremely relevant.

Medic performing an endoscopic examination has to possess substantial experience and remain focused during the whole procedure. Endoscopic image recognition accuracy greatly depends upon the specialist qualification, and is usually considered to be between 80 and 100% [5, P.174; 1, P.A507]. Modern CADs come close to visual examination by this criterion; however, they require further improvement before implementation into clinical practice [6, P.15; 7, P.350; 4, P.7130]. It is necessary to refine the algorithms used to select, analyze, and classify characteristic image elements while medical community reaches consensus on the topic of visual criteria of endoscopic image assessment [3, P.471; 8, P.526; 11, P.17].

The purpose of research was to develop and implement methodology of making diagnostic solutions based upon high-resolution endoscopic images in a CAD system in order to improve quality of diagnosis for stomach and colon oncological diseases.

Research goals included the following:

- Formalization of visual criteria used to assess endoscopic image, and definition of computed microstructure parameters;
- Development of algorithms for analysis of gastric and colon mucosa microstructure;
- Assessment of gastric and colon neoplasm recognition algorithm accuracy;
- Development of a highly productive endoscopic complex for running real-time analysis of endoscopic videos.

Gastric and Colon Mucosa Microstructure Analysis and Recognition Algorithm

An algorithm used to analyze microstructure of stomach or colon, which consists of capillaries, grands, and their excretory ducts, includes the following stages:

- Selecting boundaries of interest areas—focused part of an image without background, flares, and other artifacts that complicate processing (Fig. 1a);
- Selecting pits and capillaries (Fig. 1b), skeletonizing (Fig. 1c), and thickness calculation (Fig. 1d).

Skeleton of the capillaries is built using Zhang–Suen algorithm [12, P.237]. Image of the pits is obtained by subtracting image of capillaries from the part of an initial image limited by an area interest. At each skeleton branch, secants that are

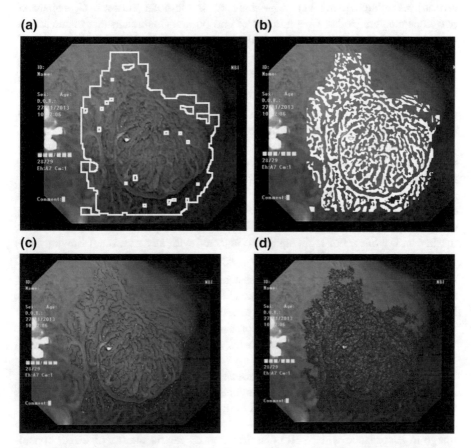

Fig. 1 Steps of processing gastric mucosa endoscopic image: **a** selecting interest area; **b** selecting mask of capillaries; **c** building skeleton of capillaries, and **d** the map of pit thickness (*red lines*), and thickness of capillaries (*blue lines*)

perpendicular to the branches were drawn with equal intervals. Intensity profiles are built along the secants, and used to determine thickness of pits and capillaries. Distortion degree for pit and capillary boundaries is characterized by a relative intensity drop that is calculated using formula (1)

$$\text{drop} = \frac{\min\left(1; \frac{\text{LeftMin}}{\text{Max}}\right) + \min\left(1; \frac{\text{RightMin}}{\text{Max}}\right)}{2},$$ (1)

where Max is the intensity of a central pixel; and LeftMin and RightMin are the closest local minimums located left and right of a central maximum.

In order to perform recognition the interest area is divided into squares with preset size. For each square six parameters are calculated: C_1—ratio of an average pit thickness to average capillary thickness; C_2—standard deviation of pit thickness; C_3—average value for relative intensity drop in secant profiles for the pits (distortion), calculated using (1); C_4—share of single-node clusters; C_5—share of two-node clusters (rods); C_6—number of end points. Coefficient that characterizes mucosa within a given square is calculated using the following formula:

$$K_a = \sum_{i=1\ldots6} B_i p_i,$$ (2)

where B_i is a coefficient equal to 0 if $C_i \geq C_{i2}$, and equal to 1 if $C_i < C_{i2}$; and p_i and C_{i2} are threshold values determined by an expert.

The range of K_a values is broken into four equal intervals. Depending upon the value of K_a coefficient, the boundaries of square area are color coded, and serve as local indicators of status for the parts of mucosa (Fig. 2a).

In order to characterize the overall state of a part of gastric or colon mucosa, presented on an image, as a whole, the value of C_1 for each square is additionally compared with expert-defined threshold value C_{10}. Squares with $C_1 \geq C_{10}$ are marked as *key ones*. For eight squares connected with key one, the following pairs of conditions are checked: (a) $C_2 \geq \tilde{N}_{20}$ AND $\tilde{N}_1 \geq \tilde{N}_{11}$, and (b) $C_3 \geq \tilde{N}_{30}$ AND $\tilde{N}_1 \geq \tilde{N}_{11}$. The squares, for each at least one pair of conditions (a or b) is true, together with key square form the risk zone R. Image can contain several risk zones. Indicative coloring of risk zones is performed based upon the results of comparing values of C_1, C_2, and C_3, averaged for a zone, with threshold values set by an expert. After examination, the boundaries of risk zone squares get colored with indicative colors (Fig. 2b).

Endoscopic images were also processed using GoogLeNet neural network. Training set of endoscopic images contained 200 neoplasm cases, and 300 cases where no neoplasms were present. All endoscopy images were loaded into a web atlas (http://endoscopy.siams.com). Histologic composition was specified for each image in accordance with WHO classification of gastric tumors of 2010, considering Vienna classification of gastrointestinal epithelial neoplasia of 2002 [9, P.52]. Medical practitioners applied web atlas tools to loaded histological images from the

(a) **(b)**

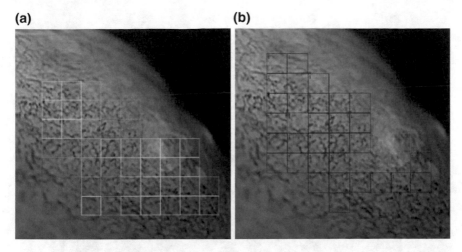

Fig. 2 Indicator grid on endoscopic image of stomach mucosa **a** square areas **b** indication of risk zones using color

training set in order to select interest areas to be examined with different methods including histology. Initial image description in web atlas, created by medical professionals, contained additional data on regularity of mucosa microstructure. No univocal correspondence was determined between the binary regularity criterion, and binary neoplasm criterion, the following model for qualitative assessment of the changes was proposed:

$$K_n = p_1 K_c + p_2 K_r, \tag{3}$$

where K_n is a value for relative coefficient of gastric mucosa change, obtained using neural network; p_1, p_2—weights, K_c is a binary neoplasm criterion, and K_r is a binary criterion of regularity.

Table 1 presents interpretation of the model. Examples of endoscopic image fragments with assessment variants are presented on Fig. 3(a–d).

On the images marked up by medics fragments were cut from selected interest areas using sliding window. Empirically selected window size was 135 × 135 pixel.

Table 1 Neural network-based model for qualitative assessment of mucosal changes

Irregular pattern	Neoplasm	Relative change coefficient, K_n	Color markup	Illustration
Yes	No	0	Green	Fig. 1a
No	No	0.25	Yellow	Fig. 1b
Yes	Yes	0.75	Orange	Fig. 1c
No	Yes	1	Red	Fig. 1d

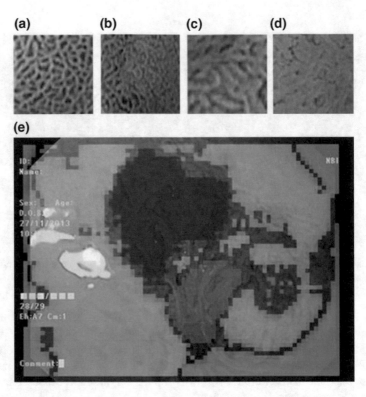

Fig. 3 Examples of image fragments used for training neural network: *a* regular pattern, no neoplasm; *b* irregular pattern, no neoplasm; *c* regular pattern, neoplasm; *d* irregular pattern, neoplasm; *e* endoscopic image of a gastric neoplasm obtained in NBI mode and processed with neural network ($K_n = 1$ *red*; $K_n = 0.75$ *orange*; $K_n = 0.25$ *yellow*; $K_n = 0$ *green*; background is *blue*, distortion and glares are *bright-green*)

For each selected fragment, a value of relative change coefficient K_n was assigned. That resulted in a training set containing more than 30,000 images, equally distributed among four classes corresponding with values of K_n. Neural network was additionally trained to remove glares, recognize background, distorted fragments, and blood. According to cross-validation of neural network use results, the accuracy of assessing K_n value with regard to initial markup was equal to 99%. Example of an image marked up using neural network is presented on Fig. 3. Markup is an overlapping regular grid, with corresponding calculated value of K_n and color marker. For each endoscopic image, the relative shares were calculated for the cells of each color. Descriptor of an endoscopic image is a four-dimensional vector D_4 that represents calculated shares.

When using neural network the accuracy of assessing K_n, obtained using cross-validation with regard to initial markup When using neural network the accuracy of assessing K_n, obtained using cross-validation with regard to initial markup was equal to 93%.

Complex Architecture

Complex architecture is presented on Fig. 4, it is distributed and depends upon user roles. Real-time mode operator (role 1) uses the complex during the endoscopic examination, so he has to use the complete complex functionality of autonomous mode. Hardware of an autonomous module (module 1 on Fig. 4) is based upon the high-capacity graphic card, and includes wide-screen sensor display used to visualize recognition results and hints. Signal from endoscope recorder is transmitted to one of the graphic cards of an autonomous module that interacts with local server (module 2) using https protocol. During the operation of an autonomous module, marked up images are uploaded to local server.

Deferred mode operator (role 2) can change image markup, set values of image attributes, and place image to a training set using web interface. Images placed into training set can be reached by remote operator of a local server (role 3) through the web atlas mode. In order to supplement training set, selected image fragments transmit to a central server (module 3) using https. Training set on a central server is obtained by combining image collections from local servers. Central server operator (role 4) controls image quality.

Neural network training is performed using an external cluster (module 4) in order to not hamper work of a central server operator Neural network training takes several hours, and is regularly performed once in several days. Trained neural network represented by a complex mathematical model is translated from central to local servers, and then—to autonomous modules, in order to revalidate recognition results.

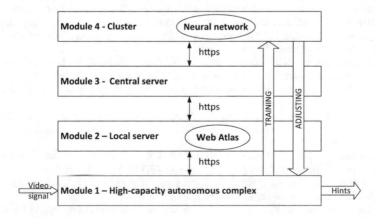

Fig. 4 Complex architecture

Conclusions

The research allowed transformation of visual assessment criteria provided by endoscopists into a set of formal characteristics describing microstructure of gastric and colon mucosa. That was an important step in development of "Smart endoscope"—firmware complex implementing automated endoscopic image analysis and recognition algorithms in real-time mode. Combination of the neoplasm recognition approaches described above allowed obtaining experimental accuracy comparable with accuracy of specialist analysis.

Research results allow hoping that future wide use of CAD systems in endoscopy will allow decreasing analysis subjectivity, improve quality of diagnostics, and cut research cost and time. Use of CADs as pathology classification tool for education and diagnostics will facilitate improving qualification of young specialists, and distribution of expert knowledge that becomes a base for the system intellectual and analytic kernel.

Acknowledgments This work was carried out within the course of a project performed by the SIAMS Company and supported by the Ministry of Education and Science of the Russian Federation. Subsidiary Agreement no. 14.576.21.0018 27. June 2014. Applied research (project) UID RFMEFI57614X0018.

References

1. Buntseva, O.A., Fedorov, E.D., Erendgenova, K.Y.: Delicate features and subtle details for characterization of gastric epithelial neoplasias by high defenition and magnified narrow-band endoscopy. Proceedings of 24th UEG Week 2016, pp. A507–A508. Vienna, Austria (2016)
2. Buntseva, O.A., Galkova, Z.V., Plakhov, R.V.: Modern endoscopic diagnosis of precancerous lesions and early cancers of the stomach and colon using computer decision support systems. Exp. Clin. Gastroenterol. **10**, 88–96 (2014)
3. Ezoe, Y., Muto, M., Horimatsu, T., Minashi, K., Yano, T., Sano, Y., Chiba, T., Ohtsu, A.: Magnifying narrow-band imaging versus magnifying white-light imaging for the differential diagnosis of gastric small depressive lesions: a prospective study. Gastrointest. Endosc. **71**(3), 477–484 (2010)
4. Gadermayr, M., Kogler, H., Karla, M.: Computer-aided texture analysis combined with experts' knowledge: Improving endoscopic celiac disease diagnosis. World J. Gastroenterol. **22**(31), 7124–7134
5. Gadermayr, M., Uhl, A., Vecsei, A.: The effect of endoscopic lens distortion correction on physicians' diagnosis performance. Proceedings of BVM, pp. 174–179 (2014)
6. Gadermayr, M., Uhl, A., Vecsei, A.: Getting one step closer to fully automatized celiac disease diagnosis. Proceedings of IPTA, pp. 13–17 (2014)
7. Hegenbart, S., Uhl, A., Vecsei, A.: Survey on computer aided decision support for diagnosis of celiac disease. Comput. Biol. Med. **65**, 348–358 (2015)
8. Kato, M., Kaise, M., Yonezawa, J., Toyoizumi, H., Yoshimura, N., Yoshida, Y., Kawamura, M., Tajiri, H.: Magnifying endoscopy with narrow-band imaging achieves superior accuracy in the differential diagnosis of superficial gastric lesions identified with white-light endoscopy: a prospective study. Gastrointest. Endosc. **72**(3), 523–529 (2010)

9. Lauwers, G.Y., Carneiro, F., Graham, D.Y.: Gastric carcinoma. In: Bosman, F.T., Carneiro, F., Hruban, R.H., Theise, N.D. (eds.) WHO classification of tumours of the digestive system, 4th edn, pp. 48–58. IARC Press, Lyon (2010)
10. Liedlgruber, M., Uhl, A.: Computer-aided decision support systems for endoscopy in the gastrointestinal tract: a review source of the document. IEEE Rev. Biomed. Eng. **4**, 73–88
11. Omori, T., Kamiya, Y., Tahara, T., Shibata, T., Nakamura, M., Yonemura, J., Okubo, M., Yoshioka, D., Ishizuka, T., Maruyama, N., Kamano, T., Fujita, H., Nakagawa, Y., Nagasaka, M., Iwata, M., Arisawa, T., Hirata, I.: Correlation between magnifying narrow band imaging and histopathology in gastric protruding/or polypoid lesions: a pilot feasibility trial. BMC Gastroenterol. **12**, 17 (2012)
12. Zhang, T.Y., Suen, C.Y.: A fast parallel algorithm for thinning digital patterns. Image Process. Comput. Vision **27**(3), 236–239 (1984)

The Development of Constructive-Technological Decisions on Creation of a Family of Microelectronic Elements on the «Silicon on Insulator» (SOI) Structures to Provide the Ability to Create Sensors of External Influences of a Various Functional Purpose

M.I. Kakoulin, A.V. Leonov, A.A. Malykh, V.N. Mordkovich, A.B. Odnolko and M.I. Pavlyuk

Abstract On the example of the magnetic field sensor shown that the developed sensing element type thin-film SOI MISIM transistor with built-in channel provides the creation of sensors with substantially improved electrical characteristics (magnetic sensitivity, temperature range). The physical model of a sensor is considered and justified the choice of the optimal electrical regimes. It is shown that the operation temperature range of the magnetic sensing element is from LHT up to at least 330 °C. Theoretically predicted the possibility of increasing the temperature limit by 200–300 °C depending on the functional purpose of the sensor. Developed and implemented microelectronic operating voltage stabilizers of sensors and electronic keys operating at temperatures of 230 °C and not less than 300 °C,

---

M.I. Kakoulin · A.V. Leonov · A.A. Malykh · V.N. Mordkovich (✉)
A.B. Odnolko · M.I. Pavlyuk
JSC, ICC Milandr, Moscow, Russia
e-mail: mord36@mail.ru

M.I. Kakoulin
e-mail: kakoulin@ic-design.ru

A.V. Leonov
e-mail: alex25.08@mail.ru

A.A. Malykh
e-mail: malykhanton21@gmail.com

A.B. Odnolko
e-mail: odnolko@miland.ru

M.I. Pavlyuk
e-mail: mikhail@milandr.ru

© The Author(s) 2018
K.V. Anisimov et al. (eds.), *Proceedings of the Scientific-Practical Conference
"Research and Development - 2016"*, https://doi.org/10.1007/978-3-319-62870-7_4

respectively. It is shown that the developed sensor and functional elements provide to create multifunctional multi-channel high-temperature sensor of magnetic field and temperature.

Keywords Magnetic sensor · Hall sensor · Temperature sensor
Misim transistor · Accumulation mode · Depletion mode · SOI technology

Motivation

Lately, the demand for microelectronic sensors that are capable of operating at increased temperatures has substantially elevated in different areas of applications. As an example, automotive, aviation, and space electronics, instrumentation used in the chemical industry, geophysics, oil, and gas industries require sensors that operate in a temperature range of 475–775 °C.

In addition, there is an obvious need in microelectronic sensors that operate in deep cooling conditions down to liquid-helium temperatures (e.g., for control of the characteristics of superconducting magnets in unique physical devices).

It is significant that Si now is the main semiconductor material for nano- and microelectronic devices including sensors manufacturing. Devices based on bulk single-crystal Si are not able to satisfy high-temperature microelectronics as far as cryogenic microelectronics. For example, for Si Hall elements, which are the most-used magnetic sensors, the operating temperature limit does not exceed 150–170 °C [1]. Such upper limit of operating temperature caused by two factors: (i) leakage of the current of the p–n junctions that rapidly increases with the temperature growth and (ii) the thermal generation of exceed electrons and holes in Si, whose concentrations increase with the temperature growth. The conventional method of solving the problem is to apply a wide-band A_3B_5 semiconductors and thin-film multilayer structures for high-temperature sensors based on it, but this solution is much more expensive.

From the other side in case of deep cooling ($T \leq 20$ K), the atoms of donors and acceptors in Si are in the neutral charge state. In this case, the p–n junctions used for separating traditional Hall element from the substrate and the n^+–n junctions used for power supply and for measuring the Hall electromotive force (EMF) disappear and the Hall element actually stops operating.

The purpose of this work is to present our results of design and study of silicon sensors manufactured in the base of the «silicon on insulator» (SOI) technology and to demonstrate that SOI thin-film transistors (TFT) as a sensitive elements of magnetic and temperature sensors provide the temperature range extension for Si sensors as for high temperatures and cryogenic temperature as well. Also discussed the possibility of increasing the operating temperature limit of the microelectronic elements of the service electronics of the sensors.

Object of Study

Sensing element represents the double-gate TF MOSFET with built-in n^+–n–n^+ channel and MISIM field-effect control system integrated with traditional Si Hall element in the same structure [2]. Further, it is denoted as field-effect Hall sensor—FEHS. It's formed in a thin Si of the SOI structure with electron concentration 5×10^{14} cm^{-3}. One of the components of the MISIM field control system is formed by the buried dielectric of the SOI structure (SiO_2) and by the Si substrate with Al metallization. The other part of the control system is traditional for MOSFETs. The additional n^+ contacts for Hall-effect measurements are located on the opposite lateral sides of the n-Si operating layer. The fabricated FEHSes were of different geometry and shapes, here are the results for FEHSes with the geometry of $500 \times 500 \times 0.2$ µm and with both SiO_2 oxides of 350 nm thick.

FEHS can operate in two different modes: in the depletion mode (DM FEHS) near the Si–SiO_2 interfaces (Fig. 1a) and in the accumulation mode (AM FEHS) (Fig. 1b). As one can see from Fig. 1 the main difference of the depletion-mode (DM) FEHS is that there is a peak of the Hall signal in a narrow region of negative gates potentials, that are exceed the module the positive flat-band potential of the

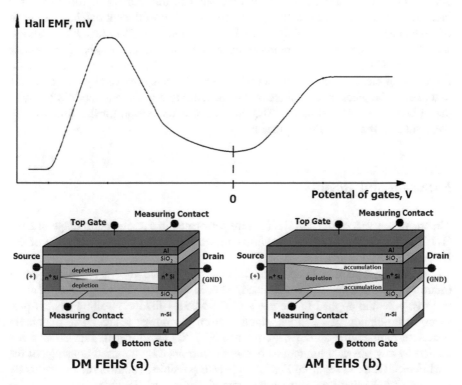

Fig. 1 Typical hall-transfer characteristics of the FEHS and explanation of the design and operation modes: FEHS in depletion mode (a) and FEHS in accumulation mode (b)

Table 1 The calculated dependence of the maximum value of the Hall EMF from the concentration of electrons in the channel of the DM FEHS at constant values of supply voltage (4 V) and magnetic induction (60 mT)

Donor concentration, cm^{-3}	Hall EMF, mV
10^{14}	8
10^{15}	17
10^{16}	50
10^{17}	35

accumulation mode FEHS. The amplitude of the peak characterizing the magnetic sensitivity of the DM FEHS is higher than the maximum Hall signal for the AM FEHS, meanwhile the channel current significantly less. In other words, for the magnetic sensitivity DM FEHS is significantly superior to AM FEHS, and increasing the concentration of donors in the channel of the FEHS to values of the order of 10^{16} cm^{-3}, allowing increasing the amplitude of the peak in several times (Table 1). A further increase in donor concentration leads to a decrease in the mobility of the electrons, i.e., to reduce the magnetic sensitivity of the DM FEHS.

The first disadvantage of the FEHS in DM is the narrow dynamic range of the gate potentials in compare to AM FEHS. Also in the context of this work, the increase of the electron concentration in the channel useful, as it extends the range of operating temperatures of the FEHS. But, unlike the DM, AM FEHS allows achieving the increase of operating temperature with almost no change magnetic sensitivity. This, as will shown further, due to the fact that increasing the concentration of electrons in the channel of the AM FEHS is achieved by increasing the potential of the gates, not a change in the concentration of dopant donor impurity in the Si layer of the SOI structure. The said above is the reason for the discussion in the results of the AM FEHS research.

Experimental Results

The measurements of the AM FEHS were performed in a temperature range of LHT (1.7 K) up to 335 °C. Figure 2a shows the experimentally measured data of the Hall EMF at the near LHT.

Measurement at this temperature is possible despite the fact that the donor impurity in the channel is not ionized. This is achieved due to the fact that the positive potential applied to the gates of the MISIM field controlling system provides accumulation of electrons supplied from the power source. The maximum operating temperature in our experiments (335 °C) as shown in Fig. 2b was not caused by any physical limitations but was determined by the used housings of the AM FEHS. One can see from Fig. 2b, it is also possible to control the magnetically induced signal value by changing the potential of the FEHS gates.

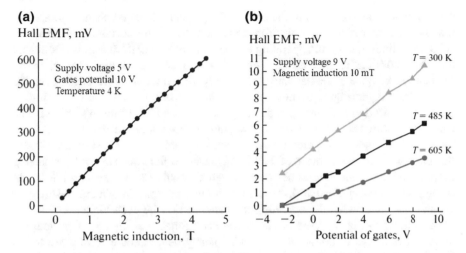

Fig. 2 Experimental results of the AM FEHS measurements: **a** dynamic range at the near LHT; **b** Hall EMF over potential of AM FEHS's gates dependencies

Discussion of the Results

It is known [3] that the use of SOI technology allows to increase the operating temperature of an MOS transistor with an induced channel as compared with analogs made of monocrystalline bulk silicon wafers (MOS transistors with an induced channel are the basis of the CMOS-technology, which is used in the production of the majority of modern silicon and SOI ICs). In a SOI transistor, the areas of the source's and drain's p–n junctions are substantially smaller and, hence, the leakage currents are smaller than those of traditional silicon MOS transistors. In this case, for excess current carriers that occur in the substrate with an increase in the temperature, the buried dielectric layer of the SOI structure acts as a barrier that prevents their penetration into the channel of the transistor. It was already shown in early works on SOI MOS transistors that were published in the 1980s–1990s that the operating temperature limit of these transistors is 200–220 °C. This is tens of degrees higher than that of the silicon analogs. With the development of nanotechnology, not only the length of current-conducting channels of SOI MOS transistors, but also their thickness dramatically decreased. In this regard, the area of the p–n junctions of the source and drain decreased even more, accordingly, the leakage currents decreased and the operating temperature, which reaches approximately 300 °C in modern SOI MOS transistors, increased. The field-effect Hall sensor is a SOI transistor with a built-in channel in which p–n junctions are absent. Thus, the limitation of the operating temperature by the leakage currents is absent. Notable that the downscaling to nano-sizes is not promising for magnetosensitive transistors of an FEHS. This threatens the sensor with a sensitivity loss since the magneto sensitivity of the Hall elements is higher as longer the distance between the Hall contacts is [4]. It is evident that the increase of the concentration of the thermally generated carriers in the area of the partially depleted channel to a value that is close to the average electron concentration in the accumulated area is a substantial limitation on the operating temperature of an FEHS. It follows from here that it is possible to increase the operating temperature of the AM FEHS by increasing the gate potential and accordingly the electron concentration in the accumulated layer. Figure 3 shows the calculated distribution of the concentration of electrons used to estimate the operating temperature limit of the AM FEHS. The calculated estimates of the achievable operating temperature of the AM FEHS by increasing voltages at the gates are summarized in Table 2. These estimates relate to the possibility of using the AM FEHS in magnetic field sensors with an analog output and measurement accuracy of the magnetic field of no worse than 1%. When using such a sensing element in sensors with digital output, in which measurement accuracy is not critical, the operating temperature may be even higher.

According to Fig. 2a, AM FEHS operate comfortably at liquid-helium temperatures when atoms of donors in Si are completely neutral and electrons are fed to the channel of an FEHS by the power supply. It is known that the MOS transistors can operate in these conditions [5]. However, an AM FEHS possesses certain characteristics that facilitate cryogenic measurements. The gate dielectrics in its MISIM system have a sufficiently high positive built-in charge, which allows the formation of accumulated areas near the interfaces even at small supply voltages and gate potentials. From the physical viewpoint, the measurement of the Hall EMF

Fig. 3 Calculated distribution of the concentration of electrons over the Si film cross-section. The *red line* shows the level of intrinsic concentration in Si at the temperature of 575 °C

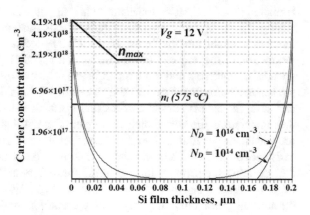

Table 2 The estimated operation temperature limits of the FEHS

Initial concentration of donors N_D, cm^{-3}	Potentials at the FEHS's gates V_g, V	Average concentration of electrons in accumulated areas, cm^{-3}	Estimated operating temperature limit, °C
5×10^{14}	4	2.7×10^{15}	325
	8	1.2×10^{16}	375
	12	2.9×10^{16}	425
	38	2.9×10^{17}	575

at temperatures below 20 K (i.e., in conditions of the neutrality of donor impurity atoms) are of special interest. They allow one to study the influence of surface states near the SiO_2–Si interfaces on the carrier mobility since the contribution of the volumetric carrier mobility in silicon to the Hall signal in this temperature range remains unchanged.

Let us describe the data on the influence of temperature changes in the practically important range (−25 to 325 °C) on a Hall EMF at constant supply voltages, gate potentials, and magnetic induction in the process of measurements (Fig. 2b). According to the figure, the Hall EMF decreases monotonically with the increase in the temperature, which is mainly related to a decrease in the electron mobility [6]. In magnetic sensitive ICs, it is accepted to compensate for this effect by using an additional electron unit as a part of the sensor to decrease the influence of mobility changes on the measured signal [7]. However, for an FEHS-based sensor, this problem can be solved by changing the potentials of the gates that are included in the feedback circuit with Hall contacts [8]. In this case, the temperature coefficient of the dependence of the magnetic sensitivity of an FEHS can be decreased to $\sim 0.02\%/°C$ versus $\sim 0.4\%/°C$, which corresponds to the traditional Hall Si-element [2].

For research of possibility of development of SOI microelectronic functional elements of the sensors, that are operating-capable at the increased temperatures (up to 225 °C according to the technical requirements for the contract) in compare to analogs made of bulk Si were developed and investigated schemes of the power supply stabilizer and the electronic key.

It is shown that the temperature range of the power supply stabilizer can reach 230 °C, and the temperature range of the electronic key can reach at least 300 °C. The choice of these two components resulted from our development of a universal multifunction sensor the magnetic field and temperature with a frequency output [9]. Stable operation of such a sensor with a frequency output requires stabilization of the operating current and the use of electronic keys, switching modes of measurement of the magnetic field and temperature sensor. This development, in particular, showed that the main element of the conversion impact in the frequency —multivibrator, can be made on the basis of similar field-effect transistors including thin-film SOI MOSFETs presented in this work.

Thus it is obvious that a microelectronic sensor with frequency output can be created on the base of thin-film SOI sensing element and thin-film n-channel MISIM transistors, using the developed high-temperature voltage stabilizer and an electronic key. Working temperature of the sensor may be at least 300 °C.

Conclusions

In this work, we proposed and investigated SOI sensing element of sensors of various external factors (magnetic field and temperature), the maximum operating temperature of which 400 °C higher than that for the silicon analogs and significantly exceeds the requirements of the technical specifications.

It is expected that the results obtained will be primarily used in sensory devices for spacecraft.

Acknowledgments Research is carried out (conducted) with the financial support of the state represented by the Ministry of Education and Science of the Russian Federation. Agreement (contract) no. 14.576.21.0064 06 Nov. 2014. Unique project Identifier: RFMEFI57614X0064.

References

1. Бараночников М.Л. Микромагнитоэлектроника. Т. 2. Изд 2-е, доп. М.: ДМК Пресс. 888 с. (с.205–209) (2014)
2. Мордкович В.Н., Бараночников М.Л., Леонов А.В., Мокрушин А.Д., Омельяновская Н.М., Пажин Д.М.: Полевой датчик Холла–новый тип преобразователя магнитного поля. Датчики и системы. **7**, С. 33–37 (2003)
3. Colinge, J.P.: Silicon-on-insulator technology: materials to VLSI, 373 p. (231–236). Springer, Berlin (2004)
4. Popovic, R.S.: Hall effect devices, 2nd edn, 412 p. (238). CRC Press, Florida (2003)
5. Rotondaro, A.L.P., Magnusson, U.K., Claeys, C., Flandre, D., Terao, A., Colinge, J.P.: Evidence of different conduction mechanisms in accumulation-mode p-channel SOI MOSFET's at room and liquid-helium temperatures. IEEE Trans. Elect. Dev. T. **40**(4), 727–732 (1993)
6. Leonov, A.V., Mokrushin, A.D., Omeljanovskaja, N.M.: Features of electron mobility in a thin silicon layer in an insulator-silicon-insulator structure. Semiconductors. T. **46**(4), C. 478–483 (2012)
7. Бараночников М. Л. Микромагнитоэлектроника Т. 1. М.: ДМК Пресс. 544 с. (164–165) (2001)
8. Leonov, A.V., Pavlyuk, M.I.: A stabilizer of micro-and small currents based on a field Hall-effect sensor with autocompensation of the temperature effect. Instrum. Exp. Tech. T. **59** (6), 808–809 (2016)
9. Mordkovich, V.N., Leonov, A.V., Malykh, A.A., Pavluyk, M. I.: Multifunctional sensor with frequency output based on SOI TFT double-gate sensing element. Proceedings of the 2nd International Conference on Sensors and Electronic Instrumental Advances, p. 114. Barcelona, Spain, 22–23 Sept 2016

Thermopile IR Sensor Arrays

V.A. Fedirko, E.A. Fetisov, R.Z. Khafizov, G.A. Rudakov
and A.A. Sigarev

Abstract Thermopile thermo-sensitive element for infrared (IR) sensor array and its optimization is considered. A concept of thermal infrared sensor array based on the micro (nano)-electromechanical system (MNEMS) with nonstationary Seebeck effect is discussed. Infrared absorption of non-stoichiometric silicon nitride thin films has been studied in the region of wavenumbers 500–7500 cm^{-1}. The estimated absorption about 64% is found for 1300 nm layer thickness, which is good enough for thermal sensor.

Keywords Thermal sensor array · Thermopile · Infrared absorption

V.A. Fedirko
National Research University of Electronic Technology "MIET",
Moscow State University of Technology "Stankin" (MGTU "Stankin"),
Moscow, Russia
e-mail: vfed@mail.ru

E.A. Fetisov (✉) · R.Z. Khafizov
National Research University of Electronic Technology "MIET", Moscow, Russia
e-mail: fetisov@inicm.ru

R.Z. Khafizov
e-mail: imagelab@mail.ru

G.A. Rudakov
SMC Technological Centre, Zelenograd, Russia
e-mail: grigory.rudakov@gmail.com

A.A. Sigarev
Moscow Institute of Physics and Technology (State University), Moscow, Russia
e-mail: aasigarev@mail.ru

K.V. Anisimov et al. (eds.), *Proceedings of the Scientific-Practical Conference*
"Research and Development - 2016", https://doi.org/10.1007/978-3-319-62870-7_5

Introduction

Uncooled infrared (IR) imagers [1, P.203; 2, P.66; 3, P.99] have attracted a great deal of interest due to their wide range of practical applications. Their greatest advantage is that they do not need expensive deep cooling systems in contrast to semiconductor quantum infrared photodetectors. In recent years, thermopile IR imager arrays have been actively developed on the base of MEMS technology, which is compatible with the silicon CMOS batch technology [4, P.200; 5, P.239; 6, P.42; 7, P.49]. A sensitive MEMS element comprises an IR high-absorbance dielectric membrane and micro-thermopiles, with the hot contact laying on the membrane and the cold one being in a good heat contact with the substrate. The output voltage of the thermopile, which is generated during the membrane heating (Seebeck effect) is read out by CMOS circuit, which is formed directly on the same chip. Good thermal isolation and low thermal capacity of thermopile-based MEMS elements are the good foundations for the development of uncooled thermal IR-image sensors with high sensitivity, linearity, low power consumption, and high responsivity. Micro-thermocouples are fabricated from polysilicon that ensures sufficiently high thermo-power while their conductivity and thermo-power can be adjusted by the proper doping. Besides, polysilicon is technologically well-proven and widely used material in CMOS technology, which makes it possible to form the sensitive elements and the electronic readout circuit in one chip, to produce efficient imagers and to reduce the production cost.

In this paper, we discuss some aspects of the optimal design and study of MEMS thermopile IR-sensors.

Optimal Design of MEMS Thermopile Element for IR Imager Array

Figure 1 shows schematically a MEMS thermopile element for IR imager array.

The membrane with IR-absorbing layer is suspended on the microcantilevers made of an electro-isolating material with low thermal conductivity. The thermopile is built on the cantilever surface, one contact of the thermopile is heated by the membrane, and the other contact is placed on the silicon substrate with a stable temperature. The output voltage of thermopile is read out by the CMOS integrated

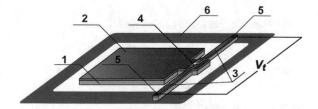

Fig. 1 Principal scheme of the MEMS thermopile element. *1* membrane; *2* absorbing layer; *3* cantilevers with thermopile; *4* hot contact; *5* cold contacts; *6* substrate

circuit, which is formed directly on the chip. The actual structural design can be easily reduced to that arrangement. The membrane and the cantilever beams are formed on a silicon wafer and made, for example, of thermal silica dioxide. SiN_x layer may serve as an IR-absorbing material which provides quite effective IR absorption in the specified wavelength region of 8–14 μm. The thermopile is made of anisotype (p and n-doped) polysilicon buses which are characterized by sufficiently high Seebeck coefficient. The process technology of this MEMS sensor is well compatible with standard CMOS batch technology. Vacuum-sealed package design is supposed to prevent from heat sink through the gas atmosphere.

When designing the IR sensor array, the following key parameters should be taken into account: the number N of sensor elements in the array for the required spatial resolution, frame time τ_f defining the time resolution, and preassigned minimal resolved temperature difference at the object δT_m.

The output characteristic of the thermopile is the open-circuit voltage (output voltage) which for the variable signal on the frequency f is defined by the well-known expression:

$$\Delta V_f = \frac{\alpha \cdot \Delta P_f \tau}{C\sqrt{1 + (2\pi f \tau)^2}}, \tag{1}$$

where α is the Seebeck coefficient of the thermopile, ΔP_f is the corresponding frequency component of the IR emission from the object absorbed by the sensor, C —it's heat capacitance. According to Kotelnikov theorem [8, P.736], the width of the reconstructed spectrum of the variable signal by discrete processing equals a half of the sampling frequency. That restricts the thermal relaxation time τ of the sensitive element by:

$$\tau \leq \tau_f / \pi. \tag{2}$$

Thermal relaxation time of the element is defined by the membrane thermal capacitance and the thermal conductivity of thermopile G_t:

$$\tau \approx C/G_t. \tag{3}$$

The intrinsic noise of a thermopile is defined by Johnson noise of its resistance R that provides the following value of the noise-equivalent temperature difference (NETD) of the object:

$$\text{NETD} = \frac{2C \cdot \sqrt{k_B \text{TBR}}}{\alpha \cdot (\Delta p / \Delta T) \cdot A \cdot \tau} < \delta T_m, \tag{4}$$

where k_B is the Boltzmann's constant, T is the ambient temperature, B is the bandwidth of the readout circuit, A is the area of the absorber, Δp is the excess IR radiation power of black body heated to temperature $T + \Delta T$, absorbed by sensor located in the focal plane of the optical system of the imager. Equations (3) and (4) demonstrate that the lowest NETD of such sensitive MEMS thermopile element is achieved in the structure with single thermopile: serial connection of thermopiles does not result in the increase of the open-circuit voltage of the element because the membrane heating is reduced proportionally due to the increase of total thermal conductivity. At the same time, the total electrical resistance is increasing according to (3), resulting in the increase of NETD.

The number of array elements at a certain size of the chip limits the maximal area of the cell and, respectively, the area S of the thermal sensor itself. The area of the sensor as seen from Fig. 1 is shared between the area of the membrane A and the cantilever area A_c:

$$S = A + A_c; \quad A_c = 2\mu w_t \cdot l, \tag{5}$$

where w_t is the length of one shoulder of the thermopile, l is the length of the cantilever, μ is the "adjacent" coefficient to take the space distances into account.

Thermopile conductivity and its electrical resistance are defined by the formulae, respectively:

$$G_t = \frac{2k_t w_t h_t}{l}, \quad R = \frac{2\rho_t l}{w_t h_t}, \tag{6}$$

where h_t is the thickness of thermopile material (polysilicon), k_t is the thermal conductivity coefficient of the thermopile material, ρ_t is its resistivity (assuming that both shoulders of the thermopile have similar resistance values). Optimization of the structure geometry is then achieved by minimizing the NETD over the variables A and l when:

$$A + 2\mu w_t \cdot l = S, \quad \tau = \frac{cAl}{2k_t w_t h_t} \leq \tau_f/\pi, \tag{7}$$

where c is the thermal capacitance of membrane unit area. That gives [9, P.59; 10, P.125029]:

$$A = \frac{2}{3}S, \quad A_c = \frac{1}{3}S, \quad l = \frac{S}{6\mu w_t}. \tag{8}$$

For a specified τ it can be implemented providing:

$$S \geq 4\sqrt{k_t \mu w_t^2 h_t \tau/c}, \tag{9}$$

that determines the minimal possible area of the sensor for a given τ with all other parameters of the structure being fixed. Vice versa, the maximal thermal relaxation time τ which can be implemented at the specified sensor area S can be expressed as follows:

$$\tau \le \frac{cS^2}{16k_t\mu w_t^2 h_t}. \tag{10}$$

The lowest NETD and the highest open-circuit voltage are reached according to (1) and (3) at the maximal value of thermal relaxation time of the sensor permitted by (2): $\tau = \tau_f/\pi$. However, it can be realized according to (9) providing that:

$$S = S_{\mathrm{opt}} = 3 \cdot \sqrt{2k_t\mu w_t^2 h_t \tau_f/\pi c}, \tag{11}$$

with optimal ratio of the membrane and cantilever areas (8). The corresponding optimized NETD value is as follows:

$$\mathrm{NETD} = \frac{6(k_t w_t h_t)\sqrt{6\mu k_B T r_\square}}{\alpha\eta q j \cdot S_{\mathrm{opt}}^{3/2} \tau_f^{1/2}} = \frac{2 \cdot \sqrt{k_B T \rho_t}}{\alpha\eta q j \tau_f^{5/4}} \cdot (\pi c)^{3/4} \cdot \left(\frac{k_t}{2\mu w_t^2 h_t}\right)^{1/4}. \tag{12}$$

Here we take into account that $(\Delta p/\Delta T) = \eta q j$, where η is the part of incident IR emission absorbed by the membrane in the covered spectral range, q is the optical factor: $q = Ht_o/4$, where H is the lens aperture and t_o is the transmission factor of the lens, $j \approx 2.62$ W/m^2 K for the $8 \div 14$ μm wavelength region; also we assume, in accordance with the above mentioned, that $r_\square = \rho_t/h_t B = 1/2\tau_f$, and we introduce the thermopile sheet resistance .

NETD (12) should satisfy inequality (4), and the optimized area of sensor should not exceed S_p, determined by the specified structural design of the IR imager and the chip size S_c: $S \approx S_p \le S_c/N$. Hence, to achieve the preassigned resolving temperature difference at the object δT_m for the specified structural design and for the frame time τ_f, the parameters of the element should meet the following requirement:

$$\frac{k_t w_t \sqrt{\mu\rho_t h_t}}{\alpha\eta q} < \frac{j\tau_f^{1/2}\delta T_m}{6\sqrt{6k_B T}}\left(\frac{S_c}{N}\right)^{3/2}. \tag{13}$$

Until the size of the array is not very large so that $S_p > S_{\mathrm{opt}}$, the ratio $\chi = A/S$ is restricted by (2): $\tau = \tau_f/\pi$, that results in the expression:

$$\frac{9}{2}\chi(1 - \chi) \cdot (S/S_{\mathrm{opt}})^2 = 1, \tag{14}$$

thus, the optimized ratio χ, evidently, cannot be achieved. Here, as S becomes lower, NETD increases as $\sim S^{-1/2}$. For large-size arrays with $S_p \le S_{\mathrm{opt}}$, the maximal relaxation time according to (10) is $\tau \le \tau_f/\pi$ and $\tau \sim S^2$. The lowest NETD is achieved according to (8) when $\chi = 1/3$, but now the NETD increases much rapidly with the reduction of the element area: NETD $\sim S^{-3/2}$. As a result, the element operational parameters significantly worsen at $S < S_{\mathrm{opt}}$ and that should be taken into account when designing large format imagers arrays with $S_p \le S_{\mathrm{opt}}$.

One can see from (13) that the element technology is faced with higher demands when enlarging the array size, the frame frequency, and the temperature difference resolution. The necessary characteristics can be achieved following Eq. (14) by reducing the width and the thickness of the thermopile, by increasing the absorption capacity of the IR imager and by optimizing the ratio $\sqrt{\rho_t}/\alpha$. When $h_t = 0.1\ \mu$, $w_t = 1\ \mu$, $c = 2\ \text{J/m}^2\ \text{K}$, $\mu = 2.5$, with $\alpha = 300\ \mu\text{V/K}$, $r_\square = 20\ \text{O}$, $\eta = 0.8$, $H = 1$, we find for $\tau_f = 0.05$ s:

$$S_{opt} \approx 1500\ \mu^2, \quad \text{NETD} \approx 8.5\text{мк} \tag{15}$$

That makes it possible to develop, for example, a 128×128 array with 50–60 µm pitch and high-temperature resolution. Note, however, that to register such a signal its pre-amplification directly in the cell should be performed followed by the signal processing in the integrated circuit formed directly on the chip [11, P.1701].

Switchable Thermopile Sensor

A chance to increase the efficiency of thermal imagers with slower response time can be provided by a sensor with switchable thermopile [12, P.596]. In a photosensitive MEMS element of that type, the cold thermojunction is not in a permanent contact with the substrate while the membrane is suspended on additional supporting SiO_2 consoles (are not shown in Fig. 1). The thermopile has cold contacts capable of closing and opening under the effect of elastic and electrostatic forces, the equilibrium contact gap in the open state is supposed to be $\sim 100 \div 500$ nm. The switching process in such a system under the applied voltage is described, e.g., in [2, P.66]. As the thermal conductivity of SiO_2, G_c, is about 30 times lower than that of the polysilicon thermopile, the heat sink from the membrane through the supporting SiO_2 consoles is very small when the thermopile is disconnected from the measurement circuit, while for measuring the output thermo-power the contacts close for a short measuring time τ_r only. The membrane heating temperature (and hence the temperature of hot thermojunctions) in the open cold contact state is then limited by the thermal relaxation time $\tau_0 \approx C/G_c$ which is substantially higher than the relaxation time τ_t when the cold thermojunction is in permanent thermal contact with the substrate. That's why the heating of the membrane may markedly exceed its heating in an ordinary element discussed above. Yet the response time of such an element decreases to the same extent. Nevertheless, providing the same characteristics, the length of the supporting SiO_2 console is much shorter. That allows increasing of the filling factor of an element area S_p with the photosensitive membrane and thus reduces S_p which gives a possibility to increase, in principle, the format of the imager. This is also the way to reduce the time of τ_m of the mechanical switching to $\tau_m \ll \tau_f$. At the same time that makes it possible to shorten the length of the thermopile and thus reduce its intrinsic noise due to the reduction of its resistance providing, of course, that the length of the thermopile in the closed cold contact state ensures the cooling time of the membrane $\tau_t \gg \tau_r$.

Obviously, a nonstationary Seebeck effect is, generally speaking, observed in such an element. The required dynamic of thermal and electrical processes in the sensor array with a switchable thermocouple and SiO_2 consoles is determined by the following hierarchy of characteristic times:

$$\tau_{p,n} < \tau_1 < \tau_m < \tau_r \ll \tau_t \ll \tau_0 \leq \tau_f/\pi. \tag{16}$$

Here $\tau_1 \approx (\rho c l^2 / \pi^2 k_t)$ is the intrinsic thermal relaxation time of the thermopile, and $\tau_{p,n} \sim l^2/D$, ρ being the density, c is the thermal capacitance and k_t is the thermal conductivity of the thermocouple material (polysilicon), l is the length of the thermopile and D is the diffusion coefficient of charge carriers in it. If (16) is realized, the quasi-stationary state of the thermopile is established in the short time τ_1 when the cold contact of the thermopile comes into the contact with the substrate, and the temperature difference $\Delta T \approx \Delta P \tau_0 / C$ at the thermo-junctions is registered by measuring circuit before it diminishes to $\Delta T \approx \Delta P \tau_t / C$.

For an element with the SiO_2 the console length of 100 μ, the width of 2 μ and the thickness of 0.5 μ, with the membrane area of 5000 mm^2 and the thermopile length of $l = 100$ μ, we find: $\tau_0 \approx 1$ s, $\tau_t \approx 0.03$ s, $\tau_1 \approx 5 \times 10^{-7}$ s, $\tau_{p,n} \approx 10^{-7}$ s that satisfy the conditions (17). The measured output voltage of the thermopile is then:

$$\Delta V = \alpha \cdot \Delta T \approx \frac{\alpha \cdot \Delta P}{C} \tau_0 \gg \frac{\alpha \cdot \Delta P}{C} \tau_t. \tag{17}$$

The open-circuit voltage of such an element, as one can see, may significantly exceed the output voltage of an ordinary sensor. However, it is achieved through the corresponding enlarging of the response time of the element which is τ_0 here, so the frame time τ_f can be no less than 3 s.

Therefore, the vacuum IR sensor array with "switchable" thermopiles can be quite effective for the "slow" thermal imaging systems. It may also serve as an alternative to a conventional sensor and quick-response large format systems when a small area of an element limits the length of the thermopile and does not ensure the optimal thermal relaxation time but the technology cannot provide it by reducing its thickness and width. It should be mentioned, however, that the implementation of the proposed elements may face some design, technological and circuit problems.

Study of Non-stoichiometric SiN$_x$ Layers

Silicon nitride is considered to be the possible material for producing of thermo-sensitive elements. With an intensive IR absorption band related to the Si–N stretching vibrations in the 700–1100 cm^{-1} wavenumbers region with the peak at

820–900 cm^{-1} corresponding to the atmospheric 8–14 μm transparency window, SiN$_x$ provides a chance to combine, in principle, the function of supporting membrane of the sensing element and the heat absorber in one material. The PECVD non-stoichiometric SiN$_x$ films possess lower and well-controllable internal stress with good mechanical properties as compared to the stoichiometric Si$_3$N$_4$ films. That enables to use SiN$_x$ films in fabrication of membrane elements and consoles for the IR thermo-sensitive micro-detectors [13, P.58].

We have studied the SiN$_x$ layers with thicknesses from 200 to 1300 nm produced by plasma enhanced chemical vapor deposition (PECVD) on monocrystalline silicon wafers.

The measurements of IR transmittance T and reflectance R spectra in the region of wavenumbers $v = 500$–7500 cm^{-1} were taken on Fourier transform spectrometer "Perkin-Elmer Spectrum 100." The reflectance spectra were measured at the incidence probing IR radiation both on the front surface of the wafer with SiN$_x$ and on the back surface of the wafer without SiN$_x$. Figure 2 shows the spectral functions:

$$A(v) = - \log 10(T(v)/T_0(v)), \tag{18}$$

estimated by the ratio of the transmittance spectra of the spectrometer channel with the sample $T(v)$ and without the sample (an empty spectrometer channel) $T_0(v)$ for the analyzed samples.

To estimate the absorption of SiN$_x$ films in the above IR band region, the reflectance spectra of the samples were analyzed. For this purpose, thick Al layer

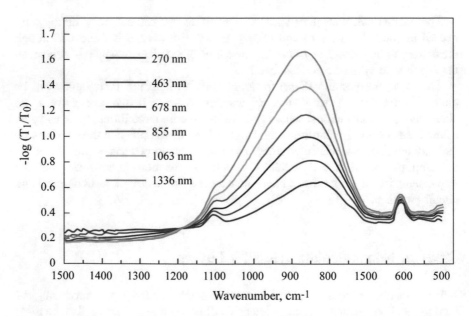

Fig. 2 Spectral functions of absorbance spectra A(v) for SiN$_x$ layers with thicknesses: *1* 270 nm; *2* 463 nm; *3* 678 nm; *4* 855 nm; *5* 1063 nm; *6* 1336 nm

was deposited on the backside of the wafer. The reflectance spectra observed in the $700-1150$ cm^{-1} region ($9-14$ μm) were about $R \sim 35\%$ for the thick SiN$_x$ film (1336 nm) and about $R \sim 55\%$ for thin film (270 nm). The part of absorbed radiation can be estimated as $A \approx 1 - R$, that is equal to $\sim 65\%$ for thick SiN$_x$ film (1336 nm) and to $\sim 45\%$ for the thin SiN$_x$ film (270 nm). Taking into account the double way of IR radiation through the SiN$_x$ film, the one-way absorption A_1 is approximately estimated by the equation $A \approx A_1 + A_1(1 - A_1)$, from that one finds $A_1 \approx 1 - \sqrt{1 - A}$ which gives $A_1 \approx 40\%$ for a thick layer sample and $A_1 \approx 26\%$ for a thin layer sample.

The results allow concluding that the optimal thickness of SiN$_x$ film for effective thermal detection is $1.0 \div 1.5$ μ.

Conclusion

Optimal engineering of a thermopile thermal MEMS element for IR sensor array is considered. Theoretical relations are obtained which optimize the sensor structure to achieve the preassigned characteristics of an imager array such as image format, frame time and temperature difference resolution. Thermal sensor with nonstationary Seebeck effect using "switchable" thermocouple is analyzed. Non-stoichiometric silicon nitride thin films have been studied as an IR absorption layer and are shown to be good enough for use for thermopile thermal MEMS thermosensors.

Acknowledgements Research was carried out with the financial support of the state represented by the Ministry of Education and Science of the Russian Federation. Agreement no. 14.578.21.0009 05 June 2014. Unique Project Identifier: RFMEFI57814X0009 and was partially supported by the Russian Foundation for Basic Research, project No. 15-07-06082-a.

Authors thank L.B. Sharova for critical reading of the manuscript.

References

1. Kruse, P.W., Skatrid, D.D. (eds.): Uncooled Infrared Imaging Arrays and Systems. Semiconductors and Semimetals, 347 p. Academic Press, USA (1997)
2. Fedirko, V.A., Fetisov, E.A., Bespalov, V.A.: Termomechanical nanomembrane IR-imaging sensors. Prikladnaya Fizika (Appl. Phys.). **1**, 66–72 (2010) (in Russian)
3. Fedirko, V.A., Fetisov, E.A., Svidzinsky, K.K.: Performance limit of infra-red thermomechanical imager with optical readout. Univ. J. Appl. Sci. **2**(5), 99–103 (2014)
4. Schaufelbuhl, A., Munch, U., Menfoli, C., Brand, O., Paul, O., Huang, Q., Baltes, H.: 256-pixel CMOS-integrated thermoelectric infrared sensor array. 2001 IEEE MEMS 2001 Conference, pp. 200–203. Interlaken, Switzerland, 21–25 January 2001 (2001)
5. Hirota, M., Nakajima, Y., Saito, M., Satou, F., Uchiyama, M.: 120×90 element thermopile array fabricated with CMOS technology. Proceedings of SPIE Conference, Infrared Technology and Applications XXVIII, vol. 4820. pp. 239–249, 23 January 2003

6. Forg, B., Herrmann, F., Schieferdecker, J., Leneke, W., Schulze, M., Simon, M., Storck, K.: Thermopile sensor array with improved spatial resolution, sensitivity and image quality. Sens. Test Proc., pp. 42–44. Nürnberg, Germany (2011)
7. Khafizov, R.Z.: Infrared Focal Plane Array (FPA) with termopile thermal radiation MEMS sensors. Proceedings of VI All-Russia Science & Technology Conference MES-2014, Extended abstracts, Part II, Institute for Design Problems in Microelectronics of Russian Academy of Science, pp. 49–55. IPPM RAS, Moscow (2015)
8. Kotel'nikov, V.A.: On the transmission capacity of 'ether' and wire in electric communications (1933). Phys. Usp. **49**, 736–744 (2006)
9. Fedirko, V.A., Khafizov, R.Z., Fetisov, E.A.: Optimal design of the MEMS thermopile element for an IR imager array. In: Stempkovsky, A. (ed.) Problems of Perspective Micro- and Nanoelectronic Systems Development—2016, Part IV. pp. 59–64. IPPM RAS, Moscow, 3–7 October 2016 (2016)
10. Xie, J., Lee, C., Wang, M., Liu, Y., Feng, H.: Characterization of heavily doped polysilicon films for CMOS-MEMS thermoelectric power generators. J. Micromech. Microeng. **19**(12), 125029–125036 (2009)
11. Wang, J.-Q., Shen, C.-H.: An offset reduction infrared tracking system with winner-take-all implementation for CMOS thermal microsensor. IEEE Sensors Conference, pp. 1701–1704 (2009)
12. Fetisov, E.A., Fedirko, V.A., Timofeev, A.E.: Study of thermal IR sensor on the base of vacuum micro/nanoelectromechanical system with non-stationary thermocouple's Seebeck effect. Proceedings of the 2016 International Conference on Actual Problems of Electron Devices Engineering, Saratov, vol. 2. pp. 596–601, 22–23 September 2016 (2016) (in Russian)
13. Rudakov, G.A., Sigarev, A.A., Fedirko, V.A., Fetisov, E.A.: Characterization of non-stoichiometric silicon nitride PECVD/ALD films for micro-detectors of IR imager array. Proceedings of 14th International Baltic Conference on Atomic Layer Deposition (BALD-2016), Book of Abstracts, S.-Petersburg, p. 58, 2–4 October 2016 (2016)

Development Signal Processing Integrated Circuit for Position Sensors with High Resolution

G.V. Prokofiev, K.N. Bolshakov and V.G. Stakhin

Abstract The article deals with the problem of creating application-specific integrated circuits transducer signal for position sensors with high resolution. The results of the work on the development of such chips considered various solutions converters angle to code and justify the chosen architecture of the converter on the basis of a digital servo system with interpolation of the input signal. The results of modeling and experimental studies and comparison of developed angle to code converter with other known solutions are described.

Keywords Encoder ASIC · Position sensor · Rotary encoder · Resolvers
Angle sensor

Introduction

Position sensors are widely used in many industries, in particular, this throttle position sensor and electronic power steering in cars, the sensors of the angular position of the rotor brushless motors, position sensors of mobile elements in robotics, position sensors in machine tools and industrial equipment, etc. Such sensors consist of sensitive element and special electronic processing circuit for calculating position code. Sensitive elements of the sensors are used in different principles, including magnetical elements (Hall-effect or magnetoresistive sensors),

G.V. Prokofiev (✉) · K.N. Bolshakov · V.G. Stakhin
VLSI Department, Zelenograd Nanotechnology Center, Moscow, Russia
e-mail: prokofiev@idm-plus.ru

K.N. Bolshakov
e-mail: bolshakov@idm-plus.ru

V.G. Stakhin
e-mail: stakhin@zntc.ru

© The Author(s) 2018   49
K.V. Anisimov et al. (eds.), *Proceedings of the Scientific-Practical Conference*
"Research and Development - 2016", https://doi.org/10.1007/978-3-319-62870-7_6

sine-cosine encoders, sine-cosine resolvers, linear differential transformers (LVDT), or optical sensor systems.

The main trend is the integration of all processing circuitry into a single chip for the purpose of miniaturization of sensors, reducing their costs, and increasing the reliability [1]. Another major trend in the development of signal processing schemes is to increase the resolution of the conversion. For many of actual tasks necessary to provide an angular resolution of one period of sine-cosine signal 15–16 bits and higher.

Also to ensure the greatest breadth of applications, it is necessary to carry out the processing of the signals from all of the most common sine-cosine sensor systems, including resolver, LVDT, sine-cosine magnetic, and optical systems.

Objective

The aim is to develop a single-chip application-specific integrated circuit (ASIC), which provides processing of signals from the position sensors of all major types, with a resolution up to 16 bits per one period of the input sine-cosine signal.

Development of Signal Processing Integrated Circuit

To achieve the stated purpose is required to integrate the single-chip system for generating a drive signal for resolver, LVDT, and specialized signal demodulator with the secondary windings of the transformer, converter angle code, LVDT signal processing unit, and a channel for generating a reference pulse for optical encoders. A key element of such systems is the angle converter code.

Angle to code converters built on two basic principles—tracking loop converter that minimizes the error signal between the input signal and its image in the transmitter memory (table of sine and cosine) [2], and on the basis of direct calculation of the arc tangent of the angle, the most common of which is a system based on CORDIC algorithms [2, 3].

Angle to code converter with direct calculation has fixed the conversion delay is independent of the input signal phase. However, this architecture is sensitive to the quality of the input signal and does not guarantee monotonicity conversion that is necessary for most control systems where position sensors are used for the feedback. To improve the quality of the input signal in direct conversion systems, the provisional carry signal filtering and processing intensive use different algorithms such as Wavelet transformation [4].

Tracking architecture provides guaranteed monotonic conversion, as it represents the code position counter output count pulses from the generator, and the frequency of which depends on the polarity of the error signal between the input signal and the sine and cosine images placed in the converter memory [5]. A tracking system in the tracking mode provides maximum performance, but the time of entering the tracking depends on the current phase of the input signal, and in the worst case, equal to the time needed to iterate over all counter values. For most applications, this feature of the tracking system is not critical.

Tracking converter may be implemented as an analog-to-digital system, and completely in the digital domain. The analog-to-digital implementation of the error signal generated in analog form on the multiplied digital–analog converters (DACs). The maximum resolution of this architecture is around 13 bits, which is primarily due to the limited accuracy of the multiplied DACs because of the mismatch of their elements, as well as with the size of such a high-resolution DACs. As an example, commercial chips based on analog-to-digital tracking system can cause chip iC-NQC [6], as well as a single-chip encoder with a resolution up to 13 bits [7]. Can be realized high-resolution analog-to-digital servo converters based on delta-sigma DAC, but this solution will provide low speed.

For conversion accuracy of 14 bits or more is already applied transformation angle code entirely in the digital domain [8]. This makes it easier to make a preliminary mathematical processing of the input signals to compensate for many of non-idealities of the sensor system.

An analysis of architectural solutions in order to achieve high resolution and the lack of a guaranteed pass code has been selected on the basis of the architecture of the full digital tracking converter.

A key element in determining the accuracy of digital processing is an analog digital converter (ADC). The highest resolution of the ADC has performed on sigma-delta architecture [9]. For converter, a multi-bit sigma-delta modulator of the second order has been developed, made by CIFF architecture (cascade of integrators with feed forward summation) [10]. The modulator converts the input analog signal into a 4-bit sample stream at a frequency of 8 MHz.

Structure of development integrated circuit is shown in Fig. 1.

ASIC consists of internal frequency synthesizer based on direct digital synthesis, it generates sinusoidal signal with programmable frequency and signals amplitude. Sinusoidal signal from synthesizer is used to drive the primary winding of the resolver or LVDT using a delta-sigma DAC.

Detailed conversion signal path for development ASIC is shown in Fig. 2.

ASIC has two conversion channel, includes programmable differential amplifiers PGA1, PGA2 and sigma-delta modulators SDM1, SDM2. The output signal from modulators is fed processing unit provides quadrature demodulation, decimation, and interpolation of the input signal with a resolution of 16 bits and a conversion time of 500 ns. The filter-decimator provides a programmable decimation from 32 to 4096 samples. For the minimum values of decimation, the signal bandwidth is 62.5 kHz. A feature of implemented digital processing system is the interpolation algorithm of the output of ADC samples, allowing to provide a constant value in the

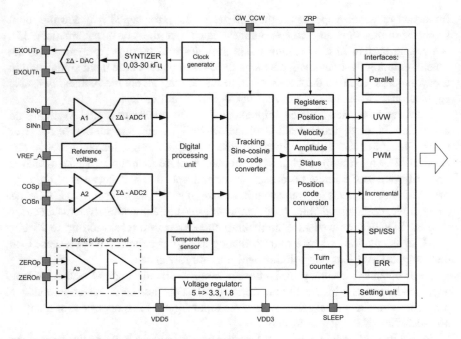

Fig. 1 Structure of development integrated circuit

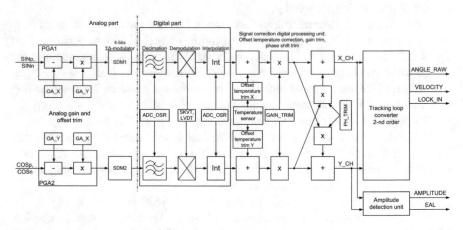

Fig. 2 Conversion signal path of development ASIC

sampling frequency of 2 MHz with 18-bit resolution conversion despite using high decimation of signal values (32–4096). Interpolation is done by filling the intermediate values between the ADC samples with zero values and subsequent processing of the received signal using low pass filter.

Due to the input signal has harmonic type, using interpolation is applicable to the system. This allows the use of ADC with less speed and power consumption, at the same time providing a comparable conversion with converters with significantly more high-speed ADCs.

The filtered and demodulated signal is supplied to the signal correction circuit providing compensation for thermal drift of the offset voltage of input signals by the integrated temperature sensor, independently for each channel, channel gain adjustment, compensation for phase shift between channels.

The corrected signal is supplied to the tracking converter, which converts an input signal into position code with a resolution of 13–16 bits. Next, the code position is adjusted depending on the user settings, counts the number of revolutions, and the combined location code is supplied to the interface circuit.

Structure of tracking loop converter is shown in Fig. 3. Tracking loop converter is minimized error signal *Err* with expression:

$$Err = \sin(\phi) \cdot \cos(Addr) - \cos(\phi) \cdot \sin(Addr)$$

where φ—phase of input signal, *Addr*—position counter code.

Error code integrated at Proportional-Integral controller Pi-reg and connects to code control oscillator CCO (analogous to voltage-control oscillator in analog domain). The CCO depends on the magnitude and polarity of the signal from PI-controller that generates counting pulses UP and DN for the reversive counter. Convert counter code into sine and cosine representations used generator based on CORDIC algorithm in rotation mode with pipelined architecture. Traditionally, such systems for storing sine and cosine values corresponding to counter code using

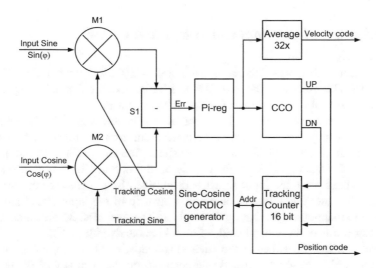

Fig. 3 Structure of designed tracking loop converter

Table 1 Comparison traditional and proposed converted architectures

Parameter	Resolution 16 bit, 0.6 um technology		Resolution 12 bit, 0.18 um technology	
	Traditional solution	Proposed solution	Traditional solution	Proposed solution
	ROM 524.288K	16 bit CORDIC	ROM 24.576K	12 bit CORDIC
Occupied area, mm^2	8.2	1.5	0.09	0.035

a non-volatile memory (ROM), usually keep a quarter of the period, and a signal linking to the full period [5]. This solution provides high-speed conversion, but requires large area on die of ASIC. The proposed solution provides a smaller area occupied by the converter, since it requires to store coefficients slight memory of the 14 values of the binary-weighted arc tangent of the angle. Comparison of occupied area was carried for the traditional architecture using a ROM and proposed solutions. The evaluation was conducted for the same technology, for proposed solution using results from digital synthesis on target standard cell logic library. Comparison result shows in Table 1.

The Table 1 shows that the designed solution can significantly reduce the area of tracking converter.

Velocity calculation used averaged signal from PI-controller.

Amplitude calculation used amplitude detector based on iteration CORDIC converter in vector mode.

Examination of Development Integrated Circuit

Development ASIC was fabricated with 180 nm CMOS technology (X-FAB XH018 CMOS process). Die size 3.6 × 3.6 mm. Nominal clock frequency is 16 MHz. Photo of die of fabricated ASIC is shown in Fig. 4.

For developed ASIC conversion error of sine-cosine signal in the code was calculated. Testing circuit to determine conversion error is shown in Fig. 4. The input circuit model was applied sine-cosine signal frequency of 15 Hz and different amplitudes. Compared to the position code on the output circuit with a reference angle calculated from the input signal is determined by the conversion error.

Figure 5 shows the values of conversion error from the input signal amplitude (in % of the maximum possible signal amplitude) for the ADC 32 decimation mode (providing the maximum bandwidth of 62.5 kHz bandwidth ADC).

It was studied the effect of the filter characteristics of the digital signal processing unit (decimation factor and interpolation) on the accuracy of the development converter. Results are shown in Table 2.

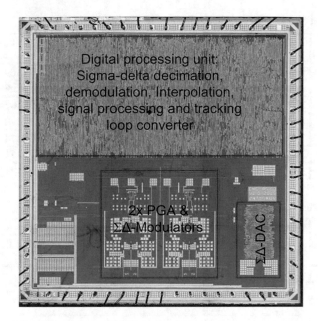

Fig. 4 Photo of die of development ASIC

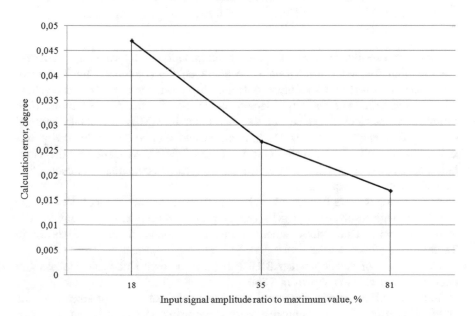

Fig. 5 Conversion error versus amplitude of input signals

Table 2 Effect of filter characteristics of conversion error

Decimation	Interpolation	ADC sampling frequency, at system clock 16 MHz, kHz	Total conversion error, degree
32	none	250	0.0169
64	none	125	0.0209
128	32	2000	0.0167
256	64	2000	0.0170
1024	256	2000	0.0174
2048	512	2000	0.0173
4096	1024	2000	0.0174

Table 3 Conversion error measurement results

Decimation	Interpolation	Conversion error (stationary position), degree	
		Sine-cosine mode	Resolver mode modulation frequency 4.073 kHz
32	0	0.038	–
64	0	0.027	–
128	32	0.022	–
256	64	0.016	–
512	128	0.016	0.016
1024	256	0.011	–
2048	512	0.005	–
4096	1024	0.005	–

Simulation results show that the use of interpolation ADC samples to increase the sampling frequency has virtually no effect on the conversion tracking error transducer. This makes it possible to obtain high-performance converter code angle ADC sigma-delta type for large decimation signal. So for decimation in 4096 samples and at clock frequency of the modulator in 8 MHz we obtain nominal sampling frequency of 1953 Hz, but through the use of interpolation in 1024 points at the filter output we obtain sampling frequency 2 MHz. As shown by the simulation results, the conversion error will not be degraded compared with decimation 32 without using interpolation.

Measurements were conducted with connecting external precision sine-cosine source and with resolver LIR-158. Signal amplitude in both cases was within 90% from maximum value. Measurement results for conversion error are shown in Table 3.

Comparison of development ASIC with others known sine-cosine to position code converter chips is shown in Table 4.

Influence of measuring the ambient temperature has been produced by the conversion error. The measurements were made using an external source of precision sine-cosine signal board with a chip placed in a heat chamber. Measurements were made in the temperature range of −60... + 150 °C. Conversion error over the entire temperature range of samples was 5 or 0.027 degrees.

Table 4 Comparison of development ASIC with others sine-cosine to position code converter chips

Parameter	Development ASIC	AD2S1210, Analog Devices	iC-TW8, iC-Haus	2602PV2AP, NIIEMP	RD-19230, Data Device Corp.
Angle to code converter architecture	Tracking full digital, 2nd order	Tracking full digital, 2nd order	Direct conversion, CORDIC	Tracking, analog–digital, 2nd order	Tracking, analog–digital, 2nd order
Maximum resolution, bit	16	16	16	16	16
Tracking frequency at maximum resolution, Hz	30	125	n/a	2	18
Settling time at angle step 179° for maximum resolution, ms	16	45	n/a	40	50
Conversion error, degree	0.027	0.17	0.08	0.022	0.022
Maximum resolution, bit	48	49	35	270	50
Operating temperature range, °C	−60... + 150	−40... + 125	−40... + 125	−60... + 85	−40... + 85

Conclusions

The research results show the correctness of the chosen chip architecture. According to the measurement results, achieved conversion accuracy is less than 0.03 angular degrees in temperature range −60... + 150 °C at a current consumption of 48 mA and a conversion time of 500 ns. Using the algorithm of interpolation ADC samples allowed for a constant conversion rate does not depend on the value of decimation, without significant degradation of conversion accuracy. Research has shown that the use of interpolation ADC samples in order to increase the sample rate does not affect the conversion error of sine-cosine tracking converter. This makes it possible for the same ADC conversion rate to get a much higher sampling frequency, and thus provide more speed of angle to code converter.

Developed ASIC will create angular and linear position sensors with high resolution. Due to the high degree of integration chip capable of processing the signal from the sensors of different types, which makes it fairly wide range of applications.

Acknowledgements Research are carried out with the financial support of the state represented by the Ministry of Education and Science of the Russian Federation. Agreement no.14.579.21.0059 23. Sept. 2014. Unique project Identifier: RFMEFI57914X0059.

References

1. Sensor Trends 2014.: Trends in future-oriented sensor technologies. AMA Association for sensor technology, (2010)
2. Burke, J., Moynihan, J.F., Unterkofler, K.: Extraction of high resolution position information from sinusoidal encoders. In Proceedings PCIM-Europe 1999, Nuremberg, pp. 217–222
3. Meher, P.K., Valls, J., Juang, T.-B., Sridharan, K., Maharatna, K.: 50 years of CORDIC: Algorithms, architectures, and applications. IEEE transactions on circuits and systems—i: Regular papers, **56**(9), 1893–1907 (2009)
4. Kovtun, D.G., Sinyutin, C.A., Prokofev, G.V., Stakhin, V.G., Obednin, A.A.: Research of digital signal processing methods on the basis of the wavelet transformation for signal processing from a position sensor of the angle-component solvers. Int. J. Appl. Eng. Res. **10** (18), 38808–38811 (2015)
5. Nishimura H.: Tracking loop type digital angle converter and angle/digital converting apparatus. US Patent 7541951 (2009)
6. iC-NQC, 13-bit Sin/D CONVERTER WITH SIGNAL CALIBRATION. iC-Haus GmbH. URL. http://www.ichaus.de/NQC. Accessed 11 Sep 2016
7. iC-MH, 12 BIT ANGULAR HALL ENCODER datasheet. iC-Haus GmbH. URL. http://www.ichaus.de/upload/pdf/Mh_b1es.pdf. Accessed 11 Sep 2016
8. High-Precision Sine/Cosine Interpolation. IC-Haus white paper. URL. http://www.ichaus.de/upload/pdf/WP7en_High-Precision_Interpolation_140124.pdf (2014). Accessed 11 Sep 2016
9. Plassche R.: CMOS Integrated analog to digital and digital to analog converters. Kluwer Academic Publishers (2003)
10. Ziquan, T., Shaojun, Y., Yueming, J., Naiying, D.: The design of a multi-bit quantization sigma-delta modulator. Int. J. Signal Process. Image Process. Pattern Recogni. **6**(5), 265–274 (2013)

Brain-Controlled Biometric Signals Employed to Operate External Technical Devices

**Vasily I. Mironov, Sergey A. Lobov, Innokentiy A. Kastalskiy,
Susanna Y. Gordleeva, Alexey S. Pimashkin, Nadezhda P. Krilova,
Kseniya V. Volkova, Alexey E. Ossadtchi and Victor B. Kazantsev**

Abstract We present a solution to employ brain-controlled biometric signals to operate external technical devices. Such signals include multiple electrode electroencephalographic (EEG) signals, electromyography (EMG) signals reflecting muscle contraction pattern, geometrical pattern of body limb kinematics, and other modalities. Being collected, properly decoded and interpreted, the signals can be used as a control or navigation signals of artificial machines, e.g., technical devices. Our interface solution based on a combination of signals of different modalities is capable to provide composite command and proportional multisite control of different technical devices with theoretically unlimited computational power. The feedback to the operator by visual channel or in virtual reality permits to compensate control errors and provides adaptive control. The control system can be implemented with wearable electronics. Examples of technical devices under the control are presented.

Keywords Neurointerface · Pattern recognition · Electroencephalography
Electromyography · Signal Processing · Biofeedback · Control

Introduction

The development of technologies of human–machine communications using human intellectual functions, e.g., the brain power, to control technical devices (TD) is, certainly, one of the most promising and dynamically progressing directions of scientific and technological development. Such control strategy will theoretical

V.I. Mironov (✉) · S.A. Lobov · I.A. Kastalskiy · S.Y. Gordleeva
A.S. Pimashkin · N.P. Krilova · V.B. Kazantsev
Nizhny Novgorod Neuroscience Centre, National Research Lobachevsky
State University of Nizhny Novgorod, Novgorod, Russia
e-mail: mironov@neuro.nnov.ru

K.V. Volkova · A.E. Ossadtchi
Centre for Cognition and Decision Making, National Research University
Higher School of Economics, Moscow, Russia

K.V. Anisimov et al. (eds.), *Proceedings of the Scientific-Practical Conference
"Research and Development - 2016"*, https://doi.org/10.1007/978-3-319-62870-7_7

have almost unlimited computational power. For example, simple brain-controlled grasping movement of a finger (10^{15} possible combinations of muscle contraction) can be estimated as 10^6 GHz computer equivalent to choose a correct muscle contraction template in real time [11]. Brain mechanisms of such control remain largely unknown. However, being properly collected, detected, and classified brain-controlled signals from the human body can be further utilized for highly efficient, even intelligent control of multiparameter TDs.

This study mainly focused on the development of new methods and technologies used for remote control of TDs respecting to particular applications. The key purpose was the integration of human biometrical, particularly bioelectrical signals, into a control loop. Such an approach involved online collection and interpretation of different modalities signals, particularly, EEG and multisite EMG signals for different TDs and robotic systems. For this purpose, we developed technical solutions associating patterns of human brain and muscular activity with commands of the controlled object, according to a user-defined translation algorithm.

Despite significant achievements in the field of machine learning, quite recently, it was proved that within the existing theories, the creation of a universal algorithm that can adapt to different conditions in a technical control system was theoretically impossible [24]. One of the expected advantages of brain-controlled devices to traditional control electronics could be the possibility of adaptation due to human brain plasticity. Despite the fact that research in neurointerface electronics is currently one of the key world trends, there have not been yet presented commercial devices for TD control where the control signals generation is based on registration, decoding and wireless transmission signals from the human brain and bioelectrical activity of muscles. During the past decade, several groups of researchers and developers have been formed specialized on the development of portable interfaces that can be applied in everyday life. The best known of these are NeuroSky, MindWave, MindBall, and Emotiv EPOC devices. However, these devices have limited functionality in terms of implementation of multisite adaptive control system. Their most significant drawbacks were long response times and relatively low degree of reliability when it comes to classification of bioelectrical activity patterns.

In this paper, we present a solution of EEG and multisite EMG human–machine interface with the user-defined programmable function of translation of sensory signals to motor command. For simplicity, we further call the solution as "SRD -1" abbreviated from "system for registration and decoding."

Materials and Methods

Motor Imagery BCI System

Our brain–computer interface (BCI) designed as a part of the SRD-1 consisted of several elements. Some of them were used traditional EEG data analysis tools, while some other classification algorithms were originally developed. Initially designed for a 32-channel EEG cap (including 10–20 system electrodes and two additional rows over the motor cortex), it can also be used with fewer electrodes, although with some density recommended of their placement over motor cortex. For any given montage, the number of channels can be further optimized with one of the two following procedures. The first one was based on common spatial patterns (CSP) and the second one on spatio-spectral decomposition (SSD) [18].

Physiological basis behind the developed BCI was the phenomenon of sustained desynchronization of the sensory-motor rhythm (SMR) over motor cortex [27]. The designed BCI operated in the so-called asynchronous mode continuously decoding the EEG signals and generating control signal, what permits to a pilot to operate at his/her own place.

The control of any type of BCI required preliminary training, which included both the procedure of BCI pilot obtaining the skill to control the interface and the training of the data processing algorithm with the data from the training recording [15]. This gave us two major fields of potential improvement: first, different practices can be used to improve the learning process of the user. It was shown that certain types of feedback allow more accurate control of the interface (as shown in Fig. 1). Another way was to modify signal processing and data analysis algorithms used to decode commands.

There is a large variety of techniques but the most common sequence of processing steps includes removal of eye artifacts (e.g., with ICA), bandpass filtering, feature vector construction (the method of common spatial patterns, e.g., the CSP, was often used to contrast the data from two classes) and eventually classification. The imple mented BCI partly followed this chain. Signal processing algorithm included scanning for the optimal bands (in the 8–24 Hz range) to distinguish each pair of commands,

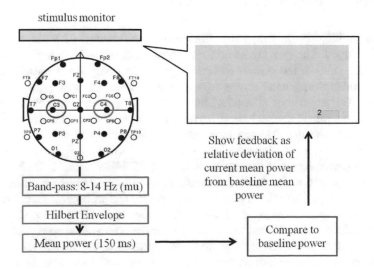

Fig. 1 Neurofeedback signal processing schema

building CSP features for each pair and using regularized linear regression as a classifier. Then the output from linear classifiers for all pairs were processed by multilayer perceptron, which permitted performing multiclass classification.

It should be mentioned that the EEG data analysis was hampered by the signal non-stationarity, high noise, and limited amount of training data, which led to classifier overfitting. As one of the proposed solutions to the overfitting problem in this context, we would like to highlight the method of CSP components choice based on the dipole fit of their topographies. The premise used in this solution was that topographies, which can be accurately described by the field of a dipole (e.g., the value of residual fitting error was under certain threshold), can be considered more plausible from the physiological point of view. Thus, such solution is considered as more reliable in the sense that they reflect the activation of the corresponding regions of the cortex and not the artifacts unrelated to the task.

To perform such analysis, the topography corresponding to each of the found common spatial patterns were fitted with a dipole using the process of minimizing the discrepancy between synthetic data and the topography vector. In practice, for given point in physical space, it involved the calculation of the topographies of three orthogonal dipoles situated at this point and the projection of the CSP topography into the space orthogonal to the space of these topographies. The calculation of the dipole topographies was based on the three-layer spherical volume conduction model of the head [2]. Either each point of the three-dimensional coordinate grid was scanned, or a procedure of non-gradient optimization (e.g., Nelder-Mead Simplex [19]) was used to find the coordinates of the dipole minimizing the residual error.

EMG-Based Neuromuscular Interface

Original multielectrode arrays for simultaneous monitoring of multiple muscles were developed. The first layout contained four pairs of medical AgCl electrodes, which were suitable for surface electromyogram recording of the muscle activity. The electrodes were placed on the flexible tissue which then was worn on the arm (Fig. 2a). The experimental results showed that such array was not suitable for long-term recording (several hours), because the electrodes could easily move. The loose contact with the skin and electrodes motion eventually led to EMG signal distortion.

Then we developed the array consisted of 12 silver coated flexible planar electrodes mounted to the flexible substrate to provide maximum attachment to the body surface (Fig. 2b). The registration was performed in bipolar mode, e.g., the muscle signal was sensed by the pairs of electrodes (6 in total) and one reference electrode which was mounted to the place near the elbow. The array was used to decode gestures generated by the arm. We composed the electrodes to match anatomy of the arm—the diameter of the electrode was 1 cm and interelectrode distance was set to 2 cm.

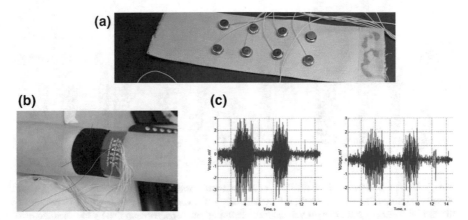

Fig. 2 Multielectrode array for electromyography recording. **a** Array of medical electrodes to record multiple muscles activity. **b** Flexible multielectrode array for arm gesture recording. **c** Signals from 2 different electrodes of the array

The cover material of the electrodes provided stable and long-term recording with amplitude up to several hundreds of microvolts. In particular, such approach can be used in further development of neurointerface for prosthetic limb control in medicine and rehabilitation and commercial interfaces for everyday usage.

Further, registered signals used for classification of nine static hand gestures as motor patterns. During an experiment, users performed two series of nine gestures each, selected in random order. The first series was learning set while the second was testing one. For extraction of the discriminating features from EMG signals root mean square (RMS) was calculated. Each 50 ms RMS feed to a multilayer artificial neural network (ANN) for the feature classification. The standard error backpropagation algorithm was used for learning. Earlier we suggested to use the network of spiking neurons as a feature extractor instead RMS [12]. In particular, we showed that mutual inhibition contrasted EMG signal from different sensors and improved the classification accuracy.

To optimize the ANN performance, we ran the process of gesture recognition on the same set of data and varying the number of layers of the ANN, the number neurons in hidden layers and the learning rate. Figure 3 shows the results. The ANN error dropped significantly between one and two layers and then lightly arose while learning time increased significantly (Fig. 3a). The similar picture one can see with changing the number of neuron in hidden layers (Fig. 3b). Thus, we selected a network with two layers and eight neurons in the hidden layer for further experiments. We also observed that the learning rate 0.01 optimizes both the learning error and the learning time (Fig. 3c). Thus, this value was used in all experimental tests of the interface.

Our software and hardware implement both command control based on pattern classification and proportional control based on muscle effort estimation. We suggested several schemes of combination these strategies [13, 14]. In particular, recognized patterns can be used for choice of movement direction and muscle effort for its speed.

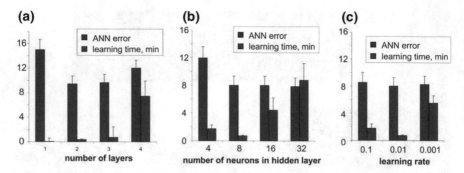

Fig. 3 The performance of the artificial neural network (mean square error and learning time) at classifying nine hand gestures with different number of layers (**a**), different number of neurons in the hidden layer (**b**) and different values of the learning rate (**c**). *Error bars* show the standard error

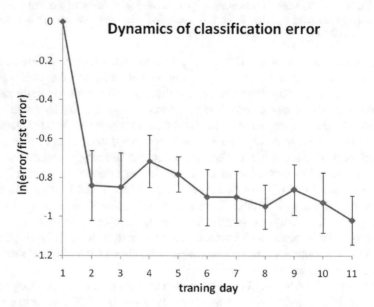

Fig. 4 Evolution of the interface performance due to the training procedure. All data were normalized to the first training day

Note that personal classification accuracy varied significantly. For example, the accuracy of 9 patterns recognition at 10 pilots was 86.5–98.5%. In this regard, we explored the possibility of improving personal performance by training the user. In this investigation, eight pilots showed positive dynamics in the performance after several days of training which included sEMG-feedback and playing a training game with the sEMG-interface. Figure 4 shows this result in term of normalized classification error. The main progress was achieved to the second training day. However, in any case, there should be some short training course before a pilot can effectively operate with the EMG interface.

Results

Motor Imagery BCI System

To estimate the effect of the proposed enhancements of the EEG data analysis algorithm (see Methods), we used pairs of recordings with the identical paradigm. The first one was used as training data and the second one as testing data. Table 1 shows the results for the topography choice procedure. The CSP components choice based on the dipole fitting improved the overall decoding accuracy. There are a number of ways for the further enhancement the decoding quality with this method. First, a simplified common model was used during dipole fitting. Since the accuracy of dipole localization depends substantially on the head model so this method can be significantly improved by using individual head parameters. Second, the heuristics for the optimal threshold choice can be developed (the results given in Table 1 correspond to the threshold of 15%).

Other proposed methods that leverage additional information included the techniques of both CSP filters and neural network weights regularization with the templates calculated for experienced pilots (users who can generate clearer patterns). It can help preventing overfitting and distinguishing informative features.

The designed BCI took its place among the most advanced systems of its kind (see Table 2 for comparison). The unique combination of features permitted to reach high decoding accuracy of multiple commands while maintaining asynchronous paradigm and requiring relatively few electrodes and short training time.

EMG-Based Neuromuscular Interface

The quality of the sEMG data was comparable to the similar ones found in the literature [7, 9, 20, 21, 23, 26]. One of the most important characteristics was the low noise level in the raw signal. Spectral characteristics correspond to the expected values. A significant part of the signal was localized in the range of 20–300 Hz (Fig. 5).

Table 1 Decoding results for the algorithm using all components and the algorithm with the choice of components based on dipole fit

Dataset	Training data size	Decoding accuracy (all components)	Decoding accuracy (choice)
A1	360 s	0.66	0.74
P1	270 s	0.46	0.51
P2	270 s	0.32	0.40
K1	270 s	0.75	0.78
K2	270 s	0.55	0.63

Table 2 Comparison of various motor imagery BCI implementations

Synchronous/asynchronous	Num. of channels	Num. of commands	Training duration	Online feedback	Accuracy (%)	Source	Methods
Synchronous	64	4	6–9 sessions, 240–360 trials total	no	63	[22]	AAR, SVM
Synchronous	64	4	6 sessions, 192 trials each	yes	72	[1]	Bandpass filtering, CSP, regularized LDA
Asynchronous	2	2	4 months of usage of Graz-BCI	n/a	>90	[10]	Decision rule based on power threshold in frequency band
Asynchronous	5	3	A couple of sessions, 3 min (30 trials) each	yes	75	[16]	Local neural classifier [9]
Asynchronous	8	3	~4 sessions, 5 min each	yes	~70	[27]	wavelets, CSFP
Asynchronous	32	4	~6 min	no	81	Designed BCI	Bandpass filtering, CSP, LDA, MLP

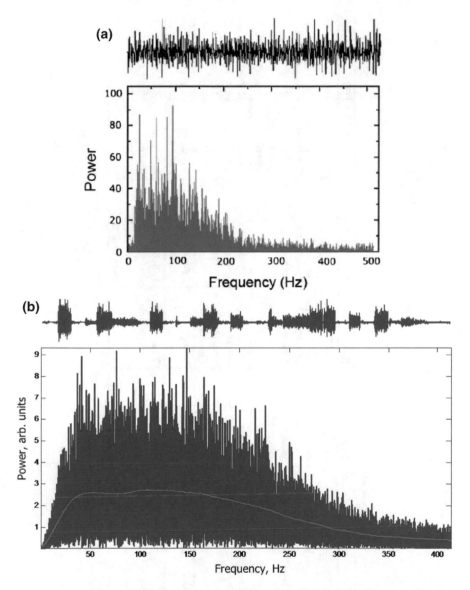

Fig. 5 Power spectrum of sEMG signal **a** figure from [20], **b** SRD-1 based raw data analysis

The accuracy of the pattern classification algorithm was 92% ± 4% in the case of nine gestures and 97 ± 2% in the case of six gestures in the command control mode [12]. Such quite a high accuracy rate was very close to the attainable limit ("error-free") in the development of human–machine interfaces (Table 3).

The complete SRD-1 device consisted of the EEG- and EMG-modules and permitted to control of TDs by both brain intention and muscle effort pattern. The

Table 3 Comparison of various myoelectric control devices

Indicator measured	SRD-1	[4]	[25]	[8]	[5]	[6]	[3]
Recognition accuracy	98.3%	–	96%	>90%	–	~90%	–
Control	Command and proportional control	Consistent proportional control	Motion pattern recognition. Proportional control	Proportional control	Proportional control	Command and proportional control	Motion pattern recognition. Proportional control
Classifier	Hybrid neural network	Linear discriminant analysis	Linear discriminant analysis	Artificial neural network (perceptron)	Artificial neural network (perceptron)	Linear regression	Linear discriminant analysis
Number of gestures/commands	9 gestures	5 gestures	2 DoF, 5 gestures	3 DoF	2 DoF, 4 gestures	2 DoF, 4 gestures	8 gestures
Number of EMG channels/sensors	8 for recording + 1 reference	5	6	7 pairs for each forearm	192-channel electrode array in the monopolar configuration	4 for each type of electrode	12 pairs of bipolar electrodes

(a)   (b)

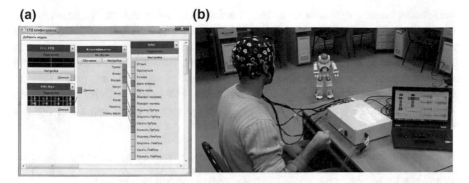

Fig. 6 **a** User interface of the SRD-1 programmable translator, and **b** SRD-1 control of NAO robot by means of the signals of bioelectric muscle and brain activity

list of tested devices included mobile robot LEGO NXT Mindstorms (LEGO, Denmark) [12, 13], humanoid robot NAO (Aldebaran, France), and exoskeleton Ilya Muromets (UNN, Russia) [17]. We used standard SDKs developed by manufacturers of the devices. The connection between the SRD-1 and tested TD was wireless using Bluetooth for LEGO or Wi-Fi for NAO, and for the exoskeleton. If SDK had a support of movement instructions, we sent a macro command directly (e.g., "go forward" in case of NAO and exoskeleton). Otherwise, we implemented required macro command in our software and sent to the device elementary command (e.g., "rotate motor A with speed x%" in case of LEGO). According to our results, it was succeeded to achieve correct control of tested TD by means of the signals of bioelectric activity of the pilot interpreted by the hardware and software system to robotic commands (Fig. 6).

Conclusions

We developed a technical solution of collecting, decoding, and translating multimodal biometric data to control external TDs. Novel algorithms for the classification of human bioelectric activity patterns were developed. In particular, a new approach to the implementation of muscle activity patterns classification was proposed, using a hybrid neural network, thus demonstrating an ability to classify 9 patterns with an accuracy of up to 98.3%. Furthermore, during the study, a possibility to achieve an acceptable (>75%) median accuracy of classification for four motor-imaginary commands, according to EEG activity signals, was demonstrated. Special attention should be paid to the fact that the specified accuracy has been achieved using only 7 electrodes, which is important, from the standpoint of practical use of the developed complex.

The experimental studies of the developed experimental model of the software and hardware complex of recording and decoding system (SRD-1) have been performed. During operational testing, the SRD-1 showed correct operations when controlling real TD (Aldebaran Robotics NAO and an exoskeleton for the lower limbs). We hope that the results of the study will be useful in many areas of human life and activity, including rehabilitation medicine, industrial robotics, and virtual reality systems. We believe that the development of the software and hardware complex designed for interpretation of human bioelectric activity signals will provide a quantum leap in the technical development of these areas and devices expanding human capabilities become a reality.

Acknowledgements The research was carried out with the financial support of the State represented by Ministry of Education and Science of Russian Federation. Agreement (contract) no. 14.581.21.0011 01. Dec 2014. Unique project Identifier: RFMEFI58114X0011.

References

1. Blankertz, B., Müller, K.-R., Curio, G., Vaughan, T.M., Schalk, G., Wolpaw, J.R., Schloegl, A., Neuper, C., Pfurtscheller, G., Hinterberger, T., Schröder, M., Birbaumer, N.: The BCI competition 2003: progress and perspectives in detection and discrimination of EEG single trials. IEEE Trans. Biomed. Eng. **51**(6), 1044–1051 (2004)
2. Dumitru, D., Delisa, J.A.: AAEM minimonograph: volume conduction. Muscle & Nerve. (1991)
3. Earley, E.J., Hargrove, L.J., Kuiken, T.A.: Dual window pattern recognition classifier for improved partial-hand prosthesis control. Frontiers Neurosci. **10**, 58 (2016)
4. Fougner, A., Stavdahl, O., Kyberd, P.J., Losier, Y.G., Parker, P.A.: Control of upper limb prostheses: terminology and proportional myoelectric control—a review. IEEE Trans. Neural Syst. Rehabilitation Eng. **20**, 663–667 (2012)
5. Hahne, J.M., Biessmann, F., Jiang, N., Rehbaum, H., Farina, D., Meinecke, F.C., Muller, K.-R., Parra, L.C.: Linear and nonlinear regression techniques for simultaneous and proportional myoelectric control. IEEE Trans. Neural Sys. Rehabilitation Eng. **22**, 269–279 (2014)
6. Hahne, J.M., Farina, D., Jiang, N., Liebetanz, D.: A novel percutaneous electrode implant for improving robustness in advanced myoelectric control. Frontiers Neurosci. **10**, 114 (2016)
7. Hargrove, L., Englehart, K., Hudgins, B.: A comparison of surface and intramuscular myoelevtric signal classification. IEEE Trans. Biomed. Eng. **54**(5), 847–853 (2007)
8. Jiang, N., Vest-Nielsen, J.L., Muceli, S., Farina, D.: EMG-based simultaneous and proportional estimation of wrist/hand kinematics in uni-lateral trans-radial amputees. J. Neuroengineering Rehabilitation **9**, 92 (2012)
9. Lee, S., Oh, J., Kim, Y., Kwon, M., Kim, J.: Estimation of the upper limb lifting movement under varying weight and movement speed. In: Proceedings of 2011 International Conference on Engineering and Industries (ICEI), Jeju, Korea, 29 November–1 December, pp. 1–6 (2011)
10. Leeb, R., Friedman, D., Mueller-Putz, G.R., Scherer, R., Slater, M., Pfurtscheller, G.: Self-paced (Asynchronous) BCI control of a wheelchair in virtual environments: a case study with a tetraplegic. IEEE Trans. Neural Syst. Rehabilitation Eng. **11**(2), (2003)
11. Llinás, R.R.: I of the vortex: from neurons to self, 320 p. The MIT Press (2001)
12. Lobov, S., Mironov, V., Kastalskiy, I., Kazantsev, V.: A Spiking Neural Network in sEMG Feature Extraction. Sensors **15**(11), 27894–27904 (2015)
13. Lobov, S., Mironov, V., Kastalskiy, I., Kazantsev, V.: Combined use of command-proportional control of external robotic devices based on electromyography signals. Sovremennye Tehnologii v Medicine [Modern technologies in medicine] **7**(4), 30–38 (2015)

14. Lobov, S., Krilova, N., Kastalskiy, I., Kazantsev, V., Makarov, V.A.: A human-computer interface based on electromyography command-proportional control. In: Proceedings of the 4th International Congress on Neurotechnology, Electronics and Informatics (NEUROTECHNIX), vol. 1, pp. 57–64 (2016)
15. Lotte, F., Larrue, F., Muehl, C.: Flaws in current human training protocols for spontaneous brain-computer interfaces: lessons learned from instructional design. Frontiers in Human Neurosci. (2013)
16. Millan, J. del R., Mourino, J.: Asynchronous BCI and local neural classifiers: an overview of the adaptive brain interface project. IEEE Trans. Neural Syst. Rehabilitation Eng. 11(2), (2003)
17. Mironov, V., Lobov, S., Kastalskiy, I., Kazantsev, V.: Myoelectric control system of lower limb exoskeleton for re-training motion deficiencies. Lect Notes Comput. Sci. 9492, 428–435 (2015)
18. Nikulin, V.V., Nolte, G., Curio, G.: A novel method for reliable and fast extraction of neuronal EEG/MEG oscillations on the basis of spatio-spectral decomposition. Neuroimage 55(4), (2011)
19. Olsson D.M., Nelson L.S.: The nelder-mead simplex procedure for function minimization. Technometrics 17(1), (1975)
20. Parker P.A., Englehart K.B., Hudgins B.S.: Control powered upper limb prostheses. In electromyography physiology, engineering, and Noninvasive Applications. IEEE Press Ser. Biomed. Eng. 453–475 (2004)
21. Pistohl, T., Cipriani, C., Jackson, A., Nazarpour, K.: Abstract and proportional myoelectric control for multi-fingered hand prostheses. Ann Biomed. Eng. 41(12), 2687–2698 (2013)
22. Schloegl, A., Lee, F., Bischof, H.: Pfurtscheller, G.: Characterization of four-class motor imagery EEG data for the BCI-competition 2005. J Neural Eng. 2, 14–22 (2005)
23. Tang, Z., Zhang, K., Sun, S., Gao, Z., Zhang, L., Yang, Z.: An upper-limb power-assist exoskeleton using proportional myoelectric control. Sensors (Basel) 14(4), 6677–6694 (2014)
24. Wolpert, D.H., Macready, W.G.: No free lunch theorems for optimization. IEEE Trans Evolutionary Comput. 1(1), 67–82 (1997)
25. Wurth, S.M., Hargrove, L.J.: A real-time comparison between direct control, sequential pattern recognition control and simultaneous pattern recognition control using a Fitts' law style assessment procedure. J. Neuroengineering and Rehabilitation 11, 91 (2014)
26. Yungher, D., Wininger, M.T., Barr, J.B., Craelius, W., Threlkeld, A.J.: Surface muscle pressure as a means of active and passive behavior of muscles during gait. Med. Eng. Phy. 33, 464–471 (2011)
27. Zhao, Q., Zhang, L., Cichocki, A.: EEG-based asynchronous BCI control of a car in 3D virtual reality environments. Chin. Sci. Bull. (2009)

Improving Talent Management with Automated Competence Assessment: Research Summary

N.S. Nikitinsky

Abstract Improvement and automatization of Human Resource Management processes is an aspiring trend in modern business. In this paper, we summarize the results of the research we conducted in order to create a formal Text Mining-based competence assessment model and develop a decision support system for talent management (DSSTM) based on this model. We also conducted several experiments in order to improve the performance of the competence assessment model. The resulting prototype of DSSTM currently undergoes validation in a software development company and shows promising results.

Keywords Competence assessment · Text Mining · Decision support system Word embeddings · Classification

Introduction

Currently, Human Resource Management (HRM) becomes a strategic trend for business. Managers from science-based industries pay special attention to Talent Management, becoming a popular tool in effective organization. As certain parts of HRM can be supported by technology, many enterprises use special software such as HR Information Systems or HR Management Systems.

Competence Management is an old, but still promising area in Talent Management field, consisting of all of a company's formal, organized approaches to ensuring that it has the human talents needed to meet its business goals. Now, there are several Competency Management Systems (CompMS) offering functionality to make the process of competency assessment more formal.

However, many contemporary CompMS propose only limited set of tools, based on traditional competence evaluation methods such as 360 degree-feedback or ordinal professional skills tests. The extensive use of Data Mining and, particularly,

N.S. Nikitinsky (✉)
NAUMEN, Moscow, Russia
e-mail: nnikitinskij@naumen.ru

K.V. Anisimov et al. (eds.), *Proceedings of the Scientific-Practical Conference "Research and Development - 2016"*, https://doi.org/10.1007/978-3-319-62870-7_8

Text Mining can greatly improve speed and quality of competence assessment, making it less human-biased at the same time. This approach should improve business processes and significantly decrease HR expenses. Several researchers and developers work in the field of applying modern Data Science approaches to the field of Competence Management.

Related Work

In our short survey, we will focus on several aspects. First, we will briefly describe attempts to introduce Data Mining into Decision Support Systems (DSS) field. Second, we will turn to contemporary full-cycle HRM systems and outline modern attempts to improve the Talent and Competence Management with Data Mining, including both theoretical models and practical applications.

Currently, several researchers offer promising Data Mining solutions for Decision Support Systems. For example, Dai et al. presented the Mining Environment for Decisions (MinEDec) model that introduced Text Mining technologies into Competitive Intelligence [4]. Bara and Lungu made a further effort to justify the importance of using Data Mining in DSS. In the paper, they described the process of developing DSS using Data Mining techniques and finished with the case study of applying data mining for forecasting the wind power plants' energy production [1].

To support decision making in Talent Management, there exist several multifunctional HRM software products, such as SAP SuccessFactors [17], Oracle Taleo [12], Cornerstone OnDemand [3], and People Fluent, Inc. [14]. These solutions cover the full cycle of employment and all HR processes, including hiring, developing, and retaining the best staff. Nevertheless, their approach to assessment of competences and qualifications is still mainly based on traditional methods.

In the meantime, various approaches for competence assessment have been discussed in academic literature. Some researches try to formalize the process of competence and qualification assessment and create theoretical models for competence assessment.

Thus, Berio and Harzallah in their earlier work discuss the model to manage four essential processes: identification, assessment, acquisition, and usage of the competences [2]. García-Barriocanal et al. developed a generic ontological model, which should facilitate the quantitative competences analysis [5]. Work of Rauffet et al. propose a methodology for assessment of organizational capabilities [16]. Haimovich proposed a mathematical model for assessing qualification efficiency of professionals [6]. However, it suits better for evaluation of qualification of university graduates, not professionals. Hassan et al. proposed theoretical model for an automated competency management system [7].

Researchers also conduct a study on applying standalone Data Mining modules to assist decision-making process in HRM. Ranjan presented the Data Mining approach to discover and extract useful patterns from this large data set to find observable patterns in HR and thus improve the quality of the decision-making process in HRM [15]. Jantan et al. demonstrated how clustering, classification and association rules can be applied to solve modern Talent Management challenges such as detecting employees with similar characteristics, predicting the employee performance and associating the employee's profile to the most appropriate job [8].

DSSTM Concept

The idea to conduct research in the area of Competence Assessment originated from HR managers trying to cope with challenging Talent Management tasks such as searching for the most skillful employees or group of employees or improving speed and quality of employees' competences evaluation. As the result of research, we created a formal model for automating competence assessment and developed a prototype of Decision Support System for Talent Management (DSSTM) based on the model. The main functionality of DSSTM is competence assessment and services built on the top of it.

Currently for competence assessment, the system combines the information from three large sources: employee HR profile, all the text documents produced by employees (e.g., scientific publications, work reports) and results of professional skill tests and other traditional competence assessments methods. To obtain this data, DSSTM uses connectors to the most popular HR software systems, databases, and directory services.

As DSSTM focuses on text document analysis, it obtains texts of documents produced by employees. The system takes into account all the metadata derived from the documents, including document type (report, specification, etc.), co-authors, and results of document evaluation provided by colleagues (such as likes or comments), whereas some data (e.g., information from the employee HR profile) demands relatively simple transformation (e.g., ranking or averaging), preprocessing of documents is a more complex task. Apart from tokenization, stop words removal, and morphological analysis of the contents (i.e., part-of-speech tagging, lemmatization, etc.), we employ word2vec and latent semantic analysis (LSA) for text classification and key terms extraction.

Word2vec is a tool implementing two neural network architectures, which is used for word embeddings analysis. The assumption behind the model is that words located in the similar contexts tend to have semantic closeness (i.e., similar meanings) [11].

LSA is a natural language processing technique, which analyzes relationships between a set of documents and the terms they contain [10]. The assumption behind the algorithm is very similar to that of word2vec. LSA constructs a weighted term-document matrix, where rows represent unique words, and columns represent documents.

Text classification allows DSSTM to define document subject areas such as scientific areas (physics, chemistry, etc.) and then compare vectorized document of certain subject area to a vector of a benchmark document in order to define its quality.

The algorithm for text classification in DSSTM consists of five steps. First, the word2vec model is trained with given text corpus where each text is tagged with certain topic name. Second, the text from each topic in corpus is projected into the word2vec model and transformed into the sum of word vectors, thus we obtain a so-called "topic vector." Third, the input text undergoes the same transformation. Fourth, the system then compares vectorized input text to each topic vector. Fifth, the input text is then tagged with topic names with semantic similarity above the threshold.

For key term extraction, we apply a combination of the LSA approach with a rule-based approach. We first select candidates for key terms from the document with predefined part-of-speech-based rules. Then, we estimate cosine similarity between each candidate-term vector and document vector in the LSA space (thus, we obtain a list of so-called local key terms). After that, we estimate similarity between the document vector and all lemmas in the semantic space (and thus we obtain list of so-called global key terms). Finally, we select top-n key terms (both for global and local key terms) and obtain a list of the most appropriate human-interpretable key terms (including n-grams). The extracted key terms are later used for competence assessment and in other additional modules.

Competence Assessment Model

By a competence, we mean a combination of a skill applied to a certain domain. For instance, the skill "Analytics" applied to a domain of Biology is a competence "Analytics in Biology."

Competence assessment consists of two steps: First step is the identification of competence, when the system checks if an employee has certain competence. Second step is the evaluation of competence, when DSSTM evaluates the competence with help of set of modifiers.

To identify competence, DSSTM uses a rule-based approach. Based on possible features, obtained from profile, text documents or professional skills tests, a user may create rules to identify the presence of competence for an employee. For instance, there may be a rule similar to this: *check for presence of at least 3 documents of type "RnD report" with topic "Physics" AND average semantic similarity of that three documents to the benchmark RnD Physics report must be at least 0.7 AND key terms from that three documents must match benchmark key terms for this competence at least for 70%.*

The result of the competence identification must be a number in order to be used in the process of competence evaluation.

The general formula for competence evaluation thus may be expressed as

$$rate_{comp} = B_{comp} * \left(\frac{1}{3} (B + HR + TXT) \right), \tag{1}$$

where $rate_{comp}$—competence score, B_{comp}—basic score, B—basic parameters modifier score, HR—HR modifier score, and TXT—text modifier score.

We scale the resulting competence score to the conventional scale (say, 1–5, where 1 means junior level of competence and 5—expert compared to other employees with the same competence). Therefore, the results from the formula (1) need to be adjusted accordingly:

$$rate_{comp} = \begin{cases} MaxScale \text{ if } rate_{comp} > MaxScale \\ rate_{comp} \text{ if } rate_{comp} \in Scale \\ MinScale \text{ if } rate_{comp} < MinScale \end{cases}, \tag{2}$$

where $rate_{comp}$—competence score, $MaxScale$—the maximum value for the scale, $MinScale$—the minimum value for the scale, and $Scale$—the scale value within the range.

As the result, an employee gets his competence level evaluated and scaled to a conventional scale. For example, "Analytics in Biology—3."

The formula of competence evaluation (1) is composite and consists of basic score and several modifier scores—Basic parameters modifier, HR modifier, and Text modifier. To calculate every modifier, currently we use more than 10 parameters.

To make sure that all the modifier parameters produce values between 0 and 1, we apply scaling all the modifiers' parameters using well-known Min-Max scaling technique.

Basic score indicates the evidence of the identified competence. It is calculated as ratio of the result of competence identification for certain employee to maximum result of competence identification in a department or whole organization:

$$B_{comp} = \frac{CI_{res}}{CI_{max}}, \tag{3}$$

where B_{comp}—basic score, CI_{res}—the result of competence identification for certain employee, CI_{max}—maximum result of competence identification in a department or whole organization.

Basic parameters modifier takes into account the numerical or categorical parameters from the employee HR profile, such as overall work experience, work experience in the company, amount of KPI achieved, etc.

HR modifier is based on traditional evaluation methods such as 360 degree-feedback, various professional tests, and different surveys. The modifier utilizes scores earned by employee in the tests.

Text modifier evaluates the indirect quality of text documents by estimating text parameters such as readability (calculated via different methods such as Flesh-Kinkaid and SMOG readability indices), uniqueness of lexis, etc.

The general formula to calculate any modifier is the following:

$$MOD = gl * \left(\sum_{i=1}^{n} (D(mod_i) * imp) \right), \tag{4}$$

where D(modi)—adjusted modifier element from 0 to 1 for computational convenience, imp—the weight for each element (by default equal 1), gl—global weight for the element (by default equal 1/n).

The competence assessment algorithm is customizable and allows adding more rules and parameters into the formulas.

Evaluation of separate competences cannot tell CEO or HR manager, how qualified each employee is in general and in comparison to others with same competences and same position in a company. Thus, we also provide evaluation of the *overall qualification level*. Average qualification evaluates the current qualification level of employee. The index represents the average of present competences:

$$Q = \frac{1}{n} \sum_{i=1}^{n} rate_i, \tag{5}$$

where Q—the level of employee qualification and $rate_i$—the current level of certain employee competence.

We also calculate a *Qualification with reference to job requirements*, which evaluates the employee's qualification required for promotion to the next position within the company structure. This parameter compares the current level of employee's qualification with the required level for specific position, which has to be defined in advance.

Professional interest discovery at the same time is intended to define the subject areas an employee is interested in. DSSTM detects two types of interests: global and local. Global interests are determined as *top-n* most frequent subject areas from the employee's text documents. Local interests are determined as *lists of key terms* from clustered personal employee's semantic LSA subspace. For LSA subspace clustering, we employ Clustering by Committee (CBC) algorithm [13]. Lists of key terms are extracted via the above described key terms extraction algorithm.

As a result, we obtain a list of key terms for every local professional interest, which is human-interpretable. Either, we sort the discovered local interests according to their significance (based on occurrence in texts) and relate them to global ones to let CEO or HR managers find out that, for example, certain employee is interested in specific branch of biology.

Application of Assessed Competences

The assessed competences, qualification, and professional interests are used to create a personal DSSTM profile for every employee, perform search for employees and let the employees with similar competences communicate and share knowledge. Furthermore, DSSTM computes competence statistics for a company, which can be used for analysis purposes—for example, to find out strategic competences or the competences that are weakly represented in the company.

DSSTM has several modules built over the competence assessment model, which help the managers in their daily Talent Management tasks:

Information Retrieval System module provides search for employees and task teams based on specified or unknown parameters of competence, qualification, and/or professional interests.

Content recommendation system (*CRS*) is designed to share knowledge between employees by recommending content created by one employee to other employees. The system is based on two approaches: content filtration and rule-based method to fit both employees who produced sufficient amount of documents and newcomers.

Employee recommendation system (*ERS*) is designed to recommend employees to each other, assuming that the employees with close profiles (including competences and qualification) and semantically close documents tend to have common tasks and interests and may help each other in their daily work. The system is based on the same approach as CRS.

Automatic employee catalog was created to group employees with close professional characteristics (including competences) based on all the data about the employees in the system. This catalog is intended to help employers with organization of task team in a bit different way than task-team search.

Additional Experiments

After we created DSSTM prototype, we started to research possibilities of applying some sociological and psychological aspects to the system as a possible parameters or modifiers in order to improve the quality of the competence assessment, since current version of DSSTM lacked such parameters. As first step, we decided to try to work in the Generation Theory framework. As part of the research, we conducted several experiments in order to create a model, which could detect a generation of a person based only on the texts, produced by the person.

Generation Theory is an approach used in management, which is based on specificity of various generations. The key notion of Generation Theory is a generation—a group of people born in a certain period of time who acquired common behavior scenarios, habits, beliefs, and language peculiarities [18].

For the study, we randomly selected in popular social networks (namely, Facebook and Vk.com) 600 Russian-speaking people from one of the five largest

Russian cities, who have been born between 1968 and 1981 (for generation X, excluding borderline years) or between 1989 and 1999 (for generation Millennium, excluding borderline years) and posted on social network more than five large texts created by themselves (no reposts).

We preprocessed all the texts with tokenization, stop words removal and morphological analysis. We extracted key terms from every text, using algorithm for local key terms extraction, which was described previously in the paper. Then, every person's keywords were transformed into vector in word2vec space. We used the pretrained word2vec model, namely, Web corpus from RusVectores [9].

We randomly selected 25% of all person's vectors for every generation. Then, we averaged the vectors, thus obtaining the so-called "generation vector." Therefore, we got Generation X vector and generation Millennium vector. The other 75% of persons from each generation were selected as validation sample. We iteratively compared each vector from validation sample to the Generation X vector and generation Millennium vectors. We related each vector from validation sample to certain generation if it was closer to this "generation vector" than to another.

We conducted 20 iterations, each time randomly selecting the generation samples in order to verify the stability of the approach and obtained average accuracy of classification equal to 0.951 and 0.953 for generation X and Millennium, respectively. Average F1-measure on 20 iterations was at the same time 0.797, which is a decent result.

Results and Discussion

The prototype version of DSSTM has been functioning for 3 months in the software developing company in Russia. The practical implementation of the system demonstrated promising results such as average time to find an employee with required competencies within the company decreased by almost 35% from 2–3 to 0.7–1.5 h. However, we also revealed certain limitations of the model—for instance, the model can detect and estimate mostly technical competencies and it works well only for those employees, who produce at least some text (e.g., programmers, analysts, marketing managers etc.). In addition, it may be a challenging task to create detection conditions for some competences, which are difficult to formalize.

Additional experiments showed us that text-based generation classifier might be utilized as an additional parameter for the DSSTM or other similar system. As a standalone module, the classifier may help HR managers better understand the life values of certain persons who were born in borderline years for two types of generations (e.g., X and Millennium) and may belong to either one or the other generation. Certainly, this classifier model also has limitations, e.g., it should be retrained on newly acquired data often enough, as persons write texts in social networks under influence of various social or political issues, which tend to change in the course of time. Thus, the model built on the texts written recently, will tend to produce worse classification results in future.

Conclusion

Nowadays, automation and formalization of HRM and Talent Management activities such as competence assessment are in high demand for large companies. Several researchers focus on this topic and apply Data Mining techniques to assist decision-making process in HRM.

In this paper, we briefly described the results of our research, which was focused on designing a theoretical framework for competence assessment, its practical implementation, developing additional modules for that implementation, which are useful for business, and conducting experiments in order to find new features for the model, which will be able to improve its quality and business-value.

We created a prototype version of Decision Support System for Talent Management, which was based on a text mining-based theoretical framework for competence assessment. This prototype version has been undergoing validation in a software development company and showing promising results. For a future study on this, we would suggest focusing on improving current limitations of a competence assessment model.

Additional experiments showed that the word2vec-based classifier can be applied to pure text-based detection of generation of a person.

Acknowledgements Research was conducted with the financial support of the state represented by the Ministry of Education and Science of the Russian Federation. Agreement no. 14.579.21.0091 28. Nov 2014. Unique project Identifier: RFMEFI57914X0091.

We would like to acknowledge hard work and commitment from Alexey Korobeynikov and Alexey Nesterenko throughout this study.

References

1. Bara, A., Lungu, I.: Improving decision support systems with data mining. Advances in data mining knowledge discovery and applications (2012)
2. Berio, G., Harzallah, M.: Towards an integrating architecture for competence management. Comput. Ind. **58**(2), 199–209 (2007)
3. Cornerstone. Cornerstone OnDemand official website [Electronic source] URL: https://www. cornerstoneondemand.com/ (date accessed 20.11.2016)
4. Dai, Y., Kakkonen, T.: MinEDec: a decision-support model that combines text-mining technologies with two competitive intelligence analysis methods. Int. J. Comput. Inf. Syst. **3**(1), 165–173 (2011)
5. García-Barriocanal, E., Sicilia, M.A., Sánchez-Alonso, S.: Computing with competencies: modelling organizational capacities. Expert Syst. Appl. **39**(16), 12310–12318 (2012)

6. Haimovitch, I.N.: Metodika integral'noi otsenki effektivnosti kvalifikatsii spetsialista [Methods of integrated performance assessment training]. Sovremennye problemy nauki i obrazovaniya **6**, 1–8 (2013)
7. Hassan, W., et al.: Automated competency management system: an advanced approach to competence management efficiency. Eur. J. Business Manage. **5**(16), 64–74 (2013)
8. Jantan, H., Hamdan, A.R., Othman, Z.A.: Towards applying data mining techniques for talent management. International conference on computer engineering and applications (IACSIT) **2**, 476–481 (2011)
9. Kutuzov, A., Andreev, I.: Texts in, meaning out: neural language models in semantic similarity task for Russian. In: Proceedings of the Dialog 2015 Conference, vol. 2, pp. 133–145 (2015)
10. Landauer, T.K., McNamara, D.S., Dennis, S., Kintsch, W. (eds.).: Handbook of latent semantic analysis. Lawrence Erlbaum Associates, (2007)
11. Mikolov, T., Chen, K., Corrado, G., Dean, J.: Efficient estimation of word representations in vector space. In: Proceedings of workshop at ICLR, (2013)
12. Oracle.: Oracle Taleo Enterprise Cloud Service [Electronic source] URL: http://www.taleo.net/ (date accessed 20.11.2016)
13. Pantel, P.A.: Clustering by committee. University of Alberta, Diss (2003)
14. PeopleFluent.: PeopleFluent HCM Software official website [Electronic source] URL: www.peoplefluent.com/ (date accessed 20.11.2016)
15. Ranjan, J.: Data mining techniques for better decisions in human resource management systems. Int. J. Business Inf. Sys. **3**(5), 464–481 (2008)
16. Rauffet, P., Da Cunha, C., Bernard, A.: A dynamic methodology and associated tools to assess organizational capabilities. Comput. Ind. **65**(1), 158–174 (2014)
17. SAP SuccessFactors official website [Electronic source] URL: https://www.successfactors.com/ (date accessed 20.11.2016)
18. Strauss, W., Howe, N.: Generations. Morrow, New York (1991)

Educational Potential of Quantum Cryptography and Its Experimental Modular Realization

A.K. Fedorov, A.A. Kanapin, V.L. Kurochkin, Yu. V. Kurochkin, A.V. Losev, A.V. Miller, I.O. Pashinskiy, V.E. Rodimin and A.S. Sokolov

Abstract The fundamental principles of quantum mechanics are considered to be hard for understanding by unprepared listeners, many attempts of its popularization turned out to be either difficult to grasp or incorrect. We propose quantum cryptography as a very effective tool for quantum physics introduction as it has the desired property set to intrigue students and outline the basic quantum principles. A modular desktop quantum cryptography setup that can be used for both educational and research purposes is presented. The carried out laboratory and field tests demonstrated usability and reliability of the developed system.

A.K. Fedorov · A.A. Kanapin · V.L. Kurochkin · Yu.V. Kurochkin ·
A.V. Losev · A.V. Miller · I.O. Pashinskiy · V.E. Rodimin (✉) · A.S. Sokolov
Researcher RQC Russian Quantum Center, Moscow, Russia
e-mail: v.rodimin@rqc.ru

A.K. Fedorov
e-mail: akf@rqc.ru

A.A. Kanapin
e-mail: a.kanapin@rqc.ru

V.L. Kurochkin
e-mail: v.kurochkin@rqc.ru

Yu.V. Kurochkin
e-mail: v.kurochkin@rqc.ru

A.V. Losev
e-mail: a.losev@rqc.ru

A.V. Miller
e-mail: avm@rqc.ru

I.O. Pashinskiy
e-mail: i.pashinskiy@rqc.ru

A.S. Sokolov
e-mail: a.sokolov@rqc.ru

© The Author(s) 2018
K.V. Anisimov et al. (eds.), *Proceedings of the Scientific-Practical Conference
"Research and Development - 2016"*, https://doi.org/10.1007/978-3-319-62870-7_9

83

Keywords Quantum cryptography · QKD · Plug & play · Quantum physics popularization · Training workshop

We live in a time of dynamically developing quantum technologies. Over the last decades, an essential breakthrough happened in new applied aspects of quantum physics, such as secure communication networks, sensitive sensors for biomedical imaging, and fundamentally new paradigms of computation. While not wholly agreeing on whether this is the second or third quantum revolution [3, p. 1655; 11, p. 1], the worldwide scientific society points out that the new emerging quantum technologies now promise the next generation of products with exciting and astounding properties that will affect our lives profoundly. In an effort to accelerate the impact of these innovations, many quantum manifests, memorandums, strategies, and other roadmaps were drawn up and adopted in recent years [4, p. 92; 6, 10]. The importance of education of a new generation of technicians, engineers, scientists, and application developers in quantum technologies is emphasized everywhere.

However, we assume that there is arguably very poor awareness of quantum physics amidst non-physicists—even worse than that of GTR. Most of the well-educated people are able to say something about space deformation near a massive object, whereas quantum mechanics does not have a similar key phrase which is easy to recall. It is all the more strange that every day we use the achievements of quantum technologies like semiconductors, but hardly deal with relativistic phenomena. That general ignorance of quantum physics sometimes leads to mishaps, such as the clueless reaction of world media to the recognition by the Russian authorities the prospects of quantum teleportation [7] in June 2016. It was reported by many media sources that Russian scientists were going to develop teleportation, as it was understood in science fiction. All in all, it is important to popularize quantum physics, spreading information about new technologies, achievements, and challenges, creating the right social and regulatory context.

Meanwhile, there are certain difficulties in quantum physics popularization related to the very nature of empirical science. Many phenomena on the quantum scale have no analogs in the common human world so they cannot be expressed via our everyday language. Some spectacular and animated attempts to make an interpretation of quantum mechanics by means of metaphysics or Eastern mysticism are rejected out of hand by the great majority of physicists as distorting reality [2, p. 132]. Besides, most physicists have no clear conception of the interpretation of their most basic theory, quantum mechanics. As a result, a common introduction to quantum mechanics starts with the definition of the mathematical formalism and its interpretation, which is difficult to grasp for an unprepared listener and is not suitable for popularization. An important tool in popularization consists in making scientific results exciting. The best way to intrigue students is to make use of descriptions that are both easy to understand and unusual [2, p. 132].

We consider quantum cryptography to be a very suitable subject for an introductory role to quantum mechanics. Indeed, it allows to grasp the very fundamental principles of quantum physics, such as superposition, non-orthogonal quantum states and so on, actively involving all the necessary concepts to begin with. Second, it is rather easy to understand. According to our experience, it can be basically explained to high school students within half an hour without extraordinary mental efforts. Third, it is mysterious and exciting because, on one hand, it associates with cryptic spy games, and on the other, it guarantees protection based on the laws of nature.

Considering quantum cryptography with respect to the methodological potential, its inspiring aspect deserves to be mentioned. There are several fundamental negative assertions in quantum physics: impossibility of making any measurement without perturbation of the system; the Heisenberg uncertainty principle; no-cloning theorem. Quantum cryptography has a very advantageous look on this background, not prohibiting but allowing absolutely secure cryptographic key distribution.

Let us briefly outline the problems of quantum cryptography and quantum communication in general [10]. Secure communication is important for a wide range of consumers, including enterprises and governments. However, the methods used at present are based on classical cryptography, which can be broken with the completion of a quantum computer. Thus, post-quantum cryptography methods are in need of development, since they cannot be broken by quantum computers. Solutions are already commercially available today, as are tools for quantum random number generation, which are a key foundation in most cryptographic protocols. One of the problems is that they are limited by distance. Existing quantum key distribution devices can operate only over distances up to 300 km. Such a weakness arises from quantum cryptography's own strength—since information cannot be cloned, it also cannot be relayed through conventional repeaters. To reach global distances, trusted nodes or quantum repeaters, perhaps even involving satellites, are needed. Trusted nodes allow for lawful intercept, which is required by many states, whereas quantum repeaters can extend the distance between trusted nodes, taking advantage of multimode quantum memories. The technologies for fully quantum repeater schemes have been tested in laboratory conditions, and are only several years away from reaching the market. Although long-distance transmission is only possible via photons, quantum memory can be realized via trapped ions, atoms in optical resonators, quantum dots, and so on, allowing for storage and processing at repeater nodes. Companies, such as ID Quantique, Toshiba, and British Telecom, are being increasingly involved with quantum communication technologies.

Quantum cryptography is a relatively young science. One can say that it was born in the '70 s as an idea that belonged to Stephen Wiesner who proposed "quantum money," which cannot be forged by copying. In the '80 s it was developed as a first quantum key distribution (QKD) protocol BB84 [1, p. 175] and got its first experimental implementation. Whilst the experiment in the '90 s was considered by a majority of physicists mainly as an expensive toy of philosophical nature, nowadays it has industrial realizations.

We propose a QKD system specialized for student education purposes and scientific research. It has a module structure and can be modified easily. The QKD system is based on the LabVIEW platform and for control signals, National Instruments (NI) FPGA boards are used. We use this platform because of that tool's high popularity in research activity of the academic community, to which our product is aimed for. On the basis of our system, numerous student workshops can be carried out in universities. Due to the quantum basis and optical methods of information transfer, this desktop device suits pretty well for the education of undergraduate physics students. Taking into account challenging tasks to generate, control, modify, and synchronize electric and optical pulses, it is a powerful practical course for electronics engineers and hardware programmers. Involving the necessity to correct errors in sifted quantum keys and privacy amplification, we get a wide field of research for algorithm developers and mathematicians. In sum, a QKD workshop system is an indispensable instrument for general training. All these areas are naturally included in the new promising engineering field—photonics. The photonics industry shows dynamic growth these years, the world photonics market is estimated to be EUR 615 billion by 2020 [8]. The Russian government adopted a roadmap for the photonics development in 2013 [9].

The proposed desktop QKD system "University Apparatus" traditionally consists of two parts Alice and Bob, the former one is in charge of sending quantum signals, the last one—for receiving them. Figure 1 illustrates a design concept of the Alice unit, Bob looks identical. On the right side, one can see a row of 8 red add-on electronic cards with SMA connectors. The laser module, phase, and amplitude modulator drivers are implemented in the form of such cards. These cards can be inserted into the motherboard, situated on bottom of the unit. The main function of the motherboard is to provide commutation between add-on cards and the NI board installed in an ordinary PC (not shown in the figure) and connected to the motherboard via a PCI cable. In addition, the motherboard implements the power supply control and executes a delaying of electric pulses to apply them at the correct time to optic modulators. On the top side of the Bob or Alice unit there is an optic table where fiber optic components can be settled. On the optic table fiber component trays and mating sleeves are shown in the figure. The parts that do not

Fig. 1 Design concept of the Alice's unit in the developed modular QKD setup

require easy access are located under the optic table. Thereby we have a system, which is easily reachable from all sides, where we can reconfigure both the optical scheme and electronic driver cards set. A sufficient amount of additional SMA connectors allows one to connect an oscilloscope for learning the shape of electrical pulses and their position on the timescale. The units are supplied with removable covers and side handles for transportation.

Using a similar modular design, the plug & play QKD scheme has been assembled and tested in different circumstances, see Fig. 2. Among the many possible QKD optical schemes, the plug & play one is characterized by relatively simple configuration and parameters adjustments, so it is ideal to begin the experimental study of quantum cryptography. The plug & play scheme consists of a two-pass auto compensation fiber optic scheme with Mach-Zehnder interferometric phase shift measurements. A detailed description of the plug & play scheme operation principles has been given in many sources, for example see [12, p. 46]. The semiconductor laser LDI-DFB2.5G under control of FPGA board Spartan-6 generates optical pulses 2 ns wide on the standard telecommunication wavelength 1.55 μm with a frequency of 10 MHz. The circulator, beam-splitters, Faraday mirror and phase modulators are quite standard components, we used ones by Thorlabs. The quantum channel and storage line are single mode fibers each 25 km long. For the synchronization of electric and optic pulses in Alice's unit, either a detector, as in the original plug & play system, or a direct coaxial cable leading from Bob to Alice was used. The synchronization coaxial cable may be utilized in an academic QKD system for its simplification and cost reducing. Artix-7 FPGA board is used for the delay of synchronization electric pulses. The single photon detectors are free-running ID Quantique ID230, whose count registers only in 5 ns time windows when a photon can be expected. Use of as short as possible time windows can effectively reduce the quantum bit error rate (QBER) level. Overall control of the electro-optic components is accomplished by NI PCIe-7811R installed in the corresponding PC—Bob or Alice. Key distribution procedure comprises of repetitive sessions of 1000 trains, with 2400 pulses in each train. The raw key is being sifted after each session via the public channel by TCP/IP protocol.

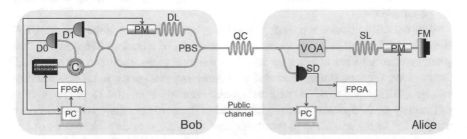

Fig. 2 Schematic diagram of the QKD system Laser—semiconductor laser, *C* circulator, *DL* delay line, *PM* phase modulator, *PBS* polarization beam-splitter, *D0* & *D1* single photon detectors, *FM* Faraday mirror, *SL* storage line, *QC* quantum channel, *VOA* variable optical attenuator, *SD* synchronization detector, *PC* personal computer

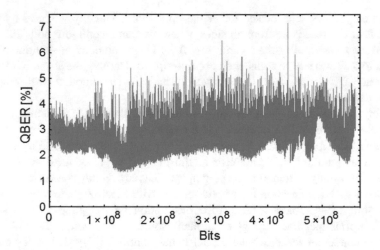

Fig. 3 QBER level as function of distributed key size at 5-day continuous experiment

Figure 3 demonstrates the QBER vs distributed key size, obtained as a result of a 5-day continuous operation in the laboratory, including unsupervised operation on weekends. A truly random 70.2 MB key with 2.6% QBER in average was amassed. Graph plotting is made with QBER averaging for every 2 KB of the distributed key. Because of the system's unstable operation, originating from both software and hardware issues, every several QKD sessions a failure occurred resulting in the very noisy behavior of QBER. After each failure the system restored and continued to work automatically. Smooth changes in QBER are mainly related to the time window adjustment shift because of temperature drift.

Later a stability of the system operation was achieved; some tests outside the laboratory were carried out when the system was on an exhibition. The system demonstrated much less noisy operation there, Fig. 4 shows the corresponding QBER. During the 18 h of intermittent operation, a 4 MB key was distributed with 1.6% average QBER. The abrupt steps of QBER are connected to changes of Alice's optic signal attenuation, the lowest regions correspond to approximately 0.2 photons/pulse.

With the fully functional plug & play scheme based on our modular system, QKD tests were conducted between the two local branches of a bank in Moscow [5]. The quantum channel fiber length was 30.6 km, the total attenuation equal to 11.7 dB. During several hours, it was generated a quantum key at a speed of 1.0 kbit/s and an average QBER equal to 5.7%, with an impulse intensity level of 0.18 photons. With an intensity of 0.35 photons exiting the sender, the corresponding parameters were equal to 1.9 kbit/s and 5.1%. It was calculated the highest achievable speeds of key generation under these conditions to be equal to 1.1 and 2.2 kbit/s, correspondingly.

The software in charge of controlling the system was written in the development environment LabVIEW, created by National Instruments. It is a popular tool used by scientists and others, mostly for data acquisition, instrument control, and

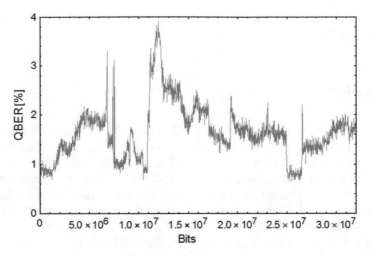

Fig. 4 QBER level as function of distributed key size during field experiment

industrial automation. We consider LabVIEW to be an optimal choice for educational purposes. It is already widely used in universities physical workshops, so it will be familiar to many students. At the same time, those who don't have any experience, can easily get involved, since the programming language is graphical and intuitive. Without diminishing its simplicity, it is also a very powerful tool— many complex tasks usually associated with working with electronics are taken care of by the software itself. The FPGA boards used were chosen because of the simplicity with which they interact with the software. We shouldn't forget to mention that post-processing work can also be carried out in the environment, since it offers all the needed instruments for data analysis.

Understanding how all the signals in the QKD device are related to each other and the electronic drivers that they operate can be a bit of a challenge. In order to solve rather complex problems concerning signal processing and synchronization, students will need to gain a deep understanding of visual dataflow programming, as well as getting used to what stands behind most of the graphical elements used. However, this challenge will allow students to better understand the subtleties of the software as well as the device itself.

Needless to say, the FPGA boards have their drawbacks. For example, the PCI-7811R board does not have the resources required for our device, and thus we had to use a third party FPGA to fulfill that function. The board does not have enough memory to delay information from a train of impulses by values of about 100–1000 μs, which is needed to modulate the impulses by different devices correctly. Another reason for this problem is that at high frequencies the board can only deal with 8-bit integers, which would also mean that even if we could solve the memory problem, we would still be unable to change the delay with sufficiently precise increments. However, in the future we plan to use a more powerful version of the FPGA—PCIe-7841R, which will have the required resources and allow us to implement all the needed functions solely on the board itself, without using any third party FPGA.

Conclusions

Using the presented desktop modular QKD system, one is able to assemble much more complex schemes than the described one. For example, in some realizations an optic modulator is necessary to variate Alice's outgoing optic pulses. The proven possibility to attack such a scheme by means of bright illuminations to dazzle single photon detectors, demands the use of a flare detector on Bob's side for industrial QKD realizations. All these options can be included in our modular platform as it has enough room for specifications. As of this writing, we are finalizing development of our compact single photon detectors that will be positioned under the optic table of Bob unit. As a result, we hope to significantly reduce the cost of the system.

The plug & play scheme implementation we used allows to process optical pulses with a higher repetition frequency than we have done, in the near future we are going to increase the frequency at least up to 40 MHz. However, concerning the transfer rate, the plug & play scheme has certain limitations because the optic pulses pass the quantum channel twice. From this point of view, one-pass optic schemes with polarization encoding appear to be much more prospective, allowing much higher key distribution bitrate. But in one-pass fiber schemes, polarization fluctuation problems arise, which need to be solved. Thus, there is a vast field for further QKD research and development. The presented modular QKD system has all necessary features such as flexibility, usability, and robustness and can be aimed for both education and scientific research.

Acknowledgements Research is carried out with the financial support of the state represented by the Ministry of Education and Science of the Russian Federation. Agreement no. 14.582.21.0009, 14 Oct. 2015. Unique project Identifier: RFMEFI58215X0009.

References

1. Bennett, C.H., Brassard, G.: Quantum cryptography: public key distribution and coin tossing, pp. 175–179 (1984)
2. Dieks, D.: The quantum mechanical worldpicture and its popularization. AntiMatters **1**(1), 131–143 (2007)
3. Dowling, J.P., Milburn, G.J.: Quantum technology: the second quantum revolution. Philos. Trans. Ser. A, Math Phy. Eng. Sci **1809**(361), 1655–1674 (2003)
4. Heracleous, L.: Quantum strategy at apple inc. Organizational Dynamics **2**(42), 92–99 (2013)
5. Kurochkin V.L. et al.: Effect of crosstalk on QBER in QKD in urban telecommunication fiber lines. SPIE EC200 Accepted for publication (2016)

6. National strategy for quantum technologies: a new era for the UK [Electronic source]. https://www.gov.uk/government/news/quantum-technologies-a-new-era-for-the-uk

7. Oliphant R.: Russia aims to develop «teleportation» in 20 years. The Telegraph [Electronic source]. URL: http://www.telegraph.co.uk/news/2016/06/22/russia-aims-to-develop-teleportation-in-20-years

8. Photonics21.org [Electronic source]. http://www.photonics21.org/

9. Photonics roadmap (in Russian) [Electronic source]. http://government.ru/docs/3462/

10. Quantum Manifesto [Electronic source]. http://qurope.eu/manifesto

11. Sowa, A.P., Zagoskin, A.M.: Can we see or compute our way out of plato's cave? J. Appl. Comput. Mathematics. **7**(3), 1–7 (2014)

12. The Physics of Quantum Information Quantum Cryptography, Quantum Teleportation, Quantum Computation ed. D. Bouwmeester, A.K. Ekert, A. Zeilinger, Springer (2000). 315 p

Interactive Visualization of Non-formalized Data Extracted from News Feeds: Approaches and Perspectives

D.A. Kormalev, E.P. Kurshev, A.N. Vinogradov, S.A. Belov and S.V. Paramonov

Abstract In this article, we consider problems related to the task of interactive visualization of data extracted from news feeds, along with possible approaches to its solution. We describe the general concept of visualization taking into account the most recent developments in the field and provide a general description of our approach and the experimental implementation of a visualization system.

Keywords Data visualization · Information extraction · Natural language processing · News feeds

Introduction

The task of information visualization is a long-standing one, however when the information necessary for decision making is presented as non-formalized data (text documents, images, and multimedia), wide usage of such data poses a problem, as most of the existing visualization technologies are oriented at well-structured and normalized information, with an exclusion for some visualization techniques with niche applications.

D.A. Kormalev (✉) · E.P. Kurshev · A.N. Vinogradov
Ailamazyan Program Systems Institute of RAS, Pereslavl-Zalessky, Russia
e-mail: dk@conrad.botik.ru

E.P. Kurshev
e-mail: epk@epk.botik.ru

A.N. Vinogradov
e-mail: andrew@andrew.botik.ru

S.A. Belov · S.V. Paramonov
JSC "CTT Group", Moscow, Russia
e-mail: s.belov@cttgroup.ru

S.V. Paramonov
e-mail: s.paramonov@cttgroup.ru

K.V. Anisimov et al. (eds.), *Proceedings of the Scientific-Practical Conference "Research and Development - 2016"*, https://doi.org/10.1007/978-3-319-62870-7_10

Today's proliferation of digital documents and web expansion prioritize the development of a technology that will enable the use of information from large text collections in the process of decision making. According to various estimates, over 80% of commercial companies' information is presented in the form of text documents. In order to enable the visualization of the information from textual collections, text-to-data technologies are being developed, generally referred to as text mining technologies. These technologies rely on a family of techniques: information retrieval, information extraction, text classification and clustering, topic detection, data unification, etc.

Text-to-data techniques have a number of limitations that constrain the use of visualization methods. In particular, one key limitation is the level of granularity (structuring) of the output information. The granularity level must meet the requirements of the visualization task; however, the existing methods of text mining sometimes fail to provide the necessary level of information granularity (at least without damage to precision and recall). In this case, the degree of structuring is an indicator of how close we get to the concept of "data" (vs. text-based fragments). Another important limitation is the precision of the acquired data. Wrong input information may lead to wrong decisions; therefore the use of text mining technology requires a mechanism to protect against processing errors.

The focus of our study was the effective visualization of data coming from unstructured natural language text collections and feeds. The goal of the project was to develop a technology for interactive visualization of non-formalized heterogeneous data. The intended application of this technology is in decision support systems based on news feed monitoring and analysis. The project scope covered the following subtasks:

(a) develop functional models and algorithms for information object (IO) and link extraction from textual news feeds;
(b) develop functional models and algorithms for preprocessing and visualization of extracted IOs and links taking into consideration their multidimensional nature and heterogeneity;
(c) create an experimental software implementation of the above algorithms;
(d) develop specifications for future development of decision support systems taking advantage of the developed interactive data visualization technology;
(e) develop interface solutions to integrate the developed interactive data visualization technology with other software systems.

A variety of visualization technologies have been developed recently; however, there is no universal visualization method suitable for all decision-making tasks. The vast majority of existing software decision support systems follow some kind of problem oriented approach to visualization. Therefore, the relevance of our study is based on the fact that the task implies the research of visualization methods covering a maximum range of applications in the field of decision support.

The novelty of our research is based on a two-pronged approach to visualization of non-formalized heterogeneous data. The first part is to improve text mining technology, and information extraction technology in particular. The second part is to search for new approaches to semi-structured information visualization, i.e., to develop visualization methods making it possible to work around the flaws of the text mining technology. Both parts are covered in our study. Such a two-way approach facilitates the search for new scientific and engineering solutions and will enhance the functional capabilities of modern decision support systems.

Visualization Concept

The practical aspects of intelligent text analysis (e.g., in application to web feed processing) should be considered in the context of a general applied problem from a certain field: business, administration, research, security, etc. An integral part of the system under consideration is the human expert who is able to synthesize findings from analysis outcomes and make decisions based on them. In this context, a text analysis system becomes an augmented intelligence tool rather than an artificial intelligence (AI) component. The difference is that the former does not replace the human intelligence but enhances its capability to perceive and assess large volumes of data. It is the critical difference between systems that enhance and scale human expertise rather than those that attempt to replicate all of human intelligence.

In the light of intensive development of AI systems, the scope of their application is being considered. The US government published a Request for Information (RFI) in June 2016 [4], asking the leading IT market players to present their views on the use of AI technologies in various social areas.

In response IBM proposed its vision [5] making a case for the critical role of human expertise to enable decision making in complex systems. The company highlights the need to focus on augmented intelligence technology within its own "cognitive computing" concept. Apart from the used technology stack, the key features of the technology are:

- the support for Big Data operations;
- a cognitive interface making it possible to include the human operator "in the loop".

The visual channel is the most natural way of human perception. Even after the IOs and links are extracted from the raw textual data, the resulting dataset is too abstract and incomprehensible for the operator. It is necessary to present this data in a visual form (view) and provide the tools for view interaction and manipulation.

Information visualization is the task of processing abstract data regardless of its nature or source. Input information goes through several stages of transformation producing a set of visual images to be displayed to the user. The visualization system can be described with a universal model covering general data stages and data transformations. In particular, when developing a domain- and application-oriented visualization system, such an approach makes it possible to describe specific system elements detailing their functions and requirements.

A widely used example of such description model is the Information Visualization Data State Reference Model [2]. This model includes the description of data stages and data transformations within an abstract technology allowing for visualization of arbitrary data.

The general structure of the Reference Model is provided in Fig. 1. The Data State Model breaks down each technique into four Data Stages, three types of Data Transformation and four types of Within Stage operators. The visualization data pipeline is broken into four distinct Data Stages: Value, Analytical Abstraction, Visualization Abstraction, and View. Transforming data from one stage to another requires one of the three types of Data Transformation operators: Data Transformation, Visualization Transformation, and Visual Mapping Transformation.

The main output of the natural language text processing is a set of extracted IOs and links. A natural way to present this information is to use graphs. The advanced theoretical base for graph processing is complemented by a broad range of practical

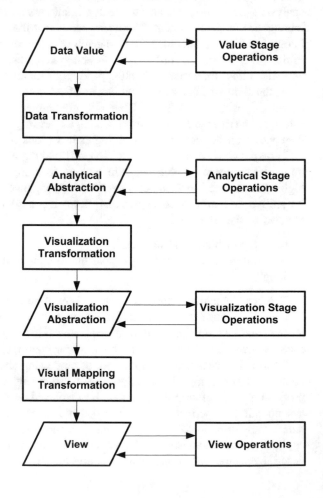

Fig. 1 Information visualization data state reference model

algorithms supporting graphs of different kinds. We can securely assume that natural language processing systems (and other systems with similar information nature and complexity) will use graph representation for analytical abstraction.

Further visual representation of this data may use the "natural" form (direct graph visualization) or other data representation techniques (charts, trees, heat maps, etc.). The choice of visual data representation depends primarily on the type of objects being displayed and on the nature of existing relations between them (e.g., hierarchical vs. non-hierarchical). Therefore, several models of visual representation are to be provided.

The exact combination of object layout and display (rendering) is chosen depending on the number of entities (graph nodes) being visualized, the level of detail, and the volume of metadata.

The characteristics of data representation models for each level, therefore, will be driven by design decisions regarding technology choices.

To make major technology and architecture decisions of a visualization system, one should consider the following questions.

Data access mode. Static data access implies that the abstract analytical data representation can be obtained right after the extraction from the semantic collection, and that it is immutable during further operations with the collection. So, all user queries regarding the visualization of a data set (subgraph) are routed to a static persistent data representation. Dynamic data access implies preparing a visual representation "on the fly," i.e., the queries are routed to the raw data that goes through transformations before being displayed.

The choice of visual representation means. While the attribute set of the IOs in the semantic collection can be extremely wide, we can classify the types of data queries using a certain taxonomy. The choice of visual representation means used in the system is made with regard to the resulting classification.

Requirements to user-level tools. In the vast majority of cases, user actions are associated with dynamic scenarios involving access to additional objects and attributes that are not displayed at the moment.

The above considerations define the set of data transformations that are required to execute the scenario involving either static or dynamic data access, along with UI level tools, graph algorithm set requirements and the functional service architecture to process user queries.

Our Approach and Experimental Implementation

The whole process of interactive visualization of non-formalized heterogeneous data naturally breaks down into two tasks: (1) news feed monitoring and text processing and (2) interactive visualization of extracted IOs and links.

Our experimental software implementation is modular, making it possible to use only the necessary component set depending on the task at hand. The text processing and visualization subsystems may be used independently. They communicate only indirectly, through a fact database supporting arbitrary types of IOs and links, which are modeled indirectly (rather than through a traditional direct schema definition for specific types of entities and relations). There are no prede-fined IOs, links and attribute—the taxonomy can be modified online to support new types of objects.

1. The system connects to a set of news feeds and automatically processes the text of incoming documents according to the extraction task. It should be noted that the extraction task can only be solved when the problem specification and the input text meet certain requirements, i.e., the technology does not involve "real" text understanding, but rather it is a "pattern matcher on steroids." Let us consider this process in more detail.

First, the incoming messages are preprocessed in order to extract the metadata and to obtain the message text for further processing (the incoming messages are not necessarily plain text).

The following linguistic analysis provides domain-independent linguistic information for the text under consideration. Linguistic analysis tools include such means as tokenization, POS tagging, morphology, and syntax analysis. These form the basis for application of information extraction methods.

After the linguistic analysis is done, domain-specific analysis is performed involving pattern matching over linguistic structures and providing additional information about identified fragments of interest. The patterns describe target information and possible contexts in the terms of linguistic features, vocabulary, and semantic categories. Our approach to IO and link extraction mostly relies on deterministic pattern-based algorithms employing finite transducers that operate over interval markup. It has common features with CPSL [1] and later JAPE [3] implementation with a number of enhancements, such as support for new data types and operations, variables, look-ahead constructions, non-greedy quantifiers, posi-tional functions, and a number of implementation-specific optimizations.

Another distinctive feature of the extraction model is that it is supported by a knowledge resource containing both linguistic and domain knowledge used to improve the analysis. The knowledge resource is a specialized shared subsystem that is available to all the extraction modules. It contains domain background information (entity classes, possible relations between them, their attributes, etc.) along with extracted factual information (when the document processing is com-plete the newly extracted content is used to populate the fact database). The knowledge resource also contains domain dictionary information.

When the text processing task is complete, the extracted IOs and links are stored in the fact database for later interactive visualization.

The processing scenario provided above is rather generic and lists only the most notable modules. In fact, the processing pipeline is fully configurable and supports the use of various processing resources through a common API.

2. The visualization subsystem supports a number of visualization scenarios that allow for various ways of graphical data representation. The scenarios include

visualization of IO link structure (using graph layout algorithms), providing aggregated information about IO features (using bar charts) and interactive refining of the view (subgraph) presented to the user. After the subgraph is ready, it can be processed to create a visual image using a range of view options.

The main representation is a graph (including tree-like and network-like structures) with links between OIs. Additional tools that are used to represent aggregated information and interactively refine the view include bar charts and time lines.

The visualization subsystem can be controlled via either a web-based UI (for end users) or a REST API (for integration with external systems). A number of auxiliary technical solutions are provided, including news feed downloader, input document format convertor, and visualization output exporter into common graphic formats.

Conclusions

Building upon theoretical findings (functional models and algorithms) we have developed a suite of scientific and engineering solutions for processing and visualization of data coming from heterogeneous news feeds, and implemented the developed models and algorithms in the form of an experimental software prototype. The prototype performs pattern-based extraction of IOs and links from the input textual data using general linguistic and specific domain knowledge. The extraction results are stored in a fact base for further creation, processing and interactive visualization of a multidimensional link matrix. An API is provided for access from the external systems.

The developed technology and solutions will make a basis for development of decision support systems in the new domains, where previously there were no effective ways to employ the vast arrays of textual information in the decision-making process. Various decision support systems will benefit from the developed technology facilitating the perception of large volumes of complex heterogeneous data and providing a comprehensive picture of the controlled object state. The technology, in general, has significant potential in various industries as it is aimed at improving management quality.

The outcomes of the project can be applied in software systems and solutions for decision support in the public and private sectors. A number of proposed methods and algorithms may be used for a wide range of applications related to automated document analysis, e.g., in the areas of state and corporate governance, business intelligence, defense analytics, marketing, library services and publishing, etc. The results of the study can be used to support further work on such topics as war room information support, open information data mining and social process modeling.

The industrial partner of the study, JSC "CTT Group," and its business partners express interest in the practical application of the results.

Acknowledgments This research was performed under financial support from the state, represented by the Ministry of Education and Science of the Russian Federation. Agreement (contract) no. 14.604.21.0138 dated 06 Nov 2014. Project unique identifier: RFMEFI60414X0138.

References

1. Appelt, D.E.: The common pattern specification language: Technical report/SRI International, Artificial Intelligence Center. (1996).
2. Chi, E.H.: A taxonomy of visualization techniques using the data state reference model. In: Proceedings of the IEEE Symposium on Information Vizualization 2000, INFOVIS '00, Washington, DC, USA, 69–75 2000.
3. Cunningham, H., Maynard, D., Bontcheva, K., Tablan, V.: GATE: a framework and graphical development environment for robust NLP tools and applications. In: Proceedings of the 40th Anniversary Meeting of the Association for Computational Linguistics. Philadelphia, 850–854 (2002)
4. Request for information: preparing for the future of artificial intelligence [Electronic resource] URL. https://www.whitehouse.gov/sites/default/files/whitehouse_files/microsites/ostp/NSTC/preparing_for_the_future_of_ai.pdf
5. Response to request for information preparing for the future of artificial intelligence [Electronic resource] URL. https://www.research.ibm.com/cognitive-computing/ostp/rfi-response.shtml

Development of Pulsed Solid-State Generators of Millimeter and Submillimeter Wavelengths Based on Multilayer GaAs/AlGaAs Heterostructures

V.A. Gergel, N.M. Gorshkova, R.A. Khabibullin, P.P. Maltsev,
V.S. Minkin, S.A. Nikitov, A.Yu. Pavlov, V.V. Pavlovskiy and
A.A. Trofimov

Abstract This paper presents the results of research of electrical characteristics features of multibarrier $Al_xGa_{1-x}As/GaAs$ heterostructures with tunnel-nontransparent potential barriers. Briefly described constructive-technological features fabricated using molecular beam epitaxy. We measured the quasi-static current–voltage characteristics of test items by electric pulses of duration 10^{-6} s and a duty cycle of 103. Characteristics observed with a strong section of the negative differential resistance in the current range of several tens of milliamperes. It is proposed to use this effect for the generation of terahertz electromagnetic radiation. The theoretical interpretation of the observed phenomena on the basis of quasi-hydrodynamic theory of electron drift is briefly stated .

Keywords Terahertz pulse generator · Multi barrier heterostructure
Semiconductor · Microwave radiation

Investigated heterostructure made by molecular beam epitaxy on a conductive substrate of GaAs, is a system of alternating layers of high-alloy ($Nd \sim 10e18$ cm^{-3}) GaAs and barrier layers $Al0.25Ga0.75As$, about 50 nm thick. The first and last barrier layers have a reduced content of Al $x = 0.125$, that

V.A. Gergel · N.M. Gorshkova (✉) · V.S. Minkin · S.A. Nikitov ·
V.V. Pavlovskiy
Institution of Russian Academy of Sciences, Kotelnikov Institute
of Radio Engineering and Electronics of RAS, Moscow, Russia
e-mail: vgergel@mail.ru

S.A. Nikitov
e-mail: nikitov@cplire.ru

R.A. Khabibullin · P.P. Maltsev · A.Yu. Pavlov · A.A. Trofimov
Institution of Russian Academy of Sciences, Institute of Ultra-High Frequency
Semiconductor Electronics of RAS, Moscow, Russia

© The Author(s) 2018
K.V. Anisimov et al. (eds.), *Proceedings of the Scientific-Practical Conference*
"Research and Development - 2016", https://doi.org/10.1007/978-3-319-62870-7_11

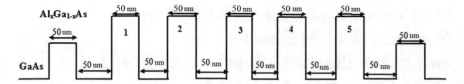

Fig. 1 The energy profile of the conduction band edge of the investigated heterostructures

is, a lower height of energy compared with the other (see the energy diagram on Fig. 1). Then by wet etching on the plate-shaped system quasicylindrical mesa element as a diameter of about 10–15 μm, the side surface of which is isolated CVD SiO_2 deposition. After opening corresponding windows on top of the each element was formed and an ohmic contact of (Au) metallization. The corresponding metal contact has been let down and to the vicinity of the test element base.

Quasi-Static Current–Voltage Characteristics

Quasi-static current–voltage characteristics of test items were measured in pulsed current sequence mode, increasing the amplitude of the duration of 1 μs and a repetition rate of 1 kHz. At the same characteristics, a number of them showed a distinct region of negative differential resistance in the region of relatively high current values <50 mA and on the stress scale Section 1.5–2 B (Fig. 2).

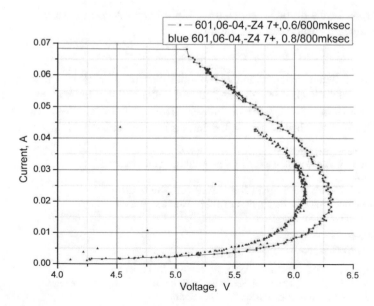

Fig. 2 Experimental quasi-static CVC of test element with NDR

We believe that the observed NDR responds the so-called effect thermo-injective instability, which is a qualitative explanation is that in these heterostructures applied voltage is concentrated mainly in the undoped barrier layers. The adjacent layers of heavily doped GaAs electric field are relatively small. The electrons from the GaAs valleys overcome the corresponding energy barriers by thermo-injection. In this case, the appropriate drift process that ensures the continuity of the current takes place in the so-called non-local energy balance, when the electrons are heated in the undoped barrier layers in which the concentrated high electric field, and cooled in the adjacent heavily doped GaAs layers with a weak electric field. At the same time, a certain portion of the electron heat flux increases with increasing current reaches the next energy barrier and enhances injection of electrons through the barrier into the next GaAs valley. Therefore, to maintain a constant electron current should be reduced electron heating in the previous barrier layer that is provided by a corresponding decrease of incident voltage, which causes the S-shape of the resulting current–voltage characteristics with a strong region of the negative differential resistance. Despite the marked non-locality, the average Joule energy is transformed into a crystal lattice, resulting in a corresponding heating of the entire sample. That is why to reduce the temperature of the sample electrical device is powered by microsecond short current pulses with a repetition frequency of about 1 kHz. For a more convincing interpretation of the results should refer to the so-called quasi-hydrodynamic theory of electron drift (thermal model), which includes a Poisson equation, the equation of continuity of the electron beam (1–2):

$$\frac{d\varphi}{dx} = \frac{q}{\varepsilon\varepsilon_0}(N_d - n) \tag{1}$$

and

$$\frac{\partial j_n}{\partial x} = 0 \tag{2}$$

and the continuity of electron temperature flow (3):

$$\frac{dj_T}{dx} = j_n \frac{d\varphi}{dx} - n\frac{T - T_0}{\tau} \tag{3}$$

in the formulation of Stratton [5, P. 453], and characterized in that in it summands with electrostatic potential φ supplemented summand with zone quasipotential $\varphi_c = -E_c(x)/q$, where $E_c(x)$—the edge of the conduction band, varying with the coordinate in accordance with a change in composition.

The system (1–3) discloses the expression for the flow of the electron density and the electron temperature (4–5):

$$j_n = \mu n \left[\frac{d(\varphi + \varphi_c)}{dx} - (1 - \alpha) \frac{dT}{dx} \right] - \mu T \frac{dn}{dx} \tag{4}$$

and

$$j_T = \left(\frac{5}{2} - \alpha \right) \left(-\mu n T \frac{dT}{dx} + j_n T \right) \tag{5}$$

written on the assumption the power dependence of the kinetic coefficients of mobility and the relaxation time of the electron temperature T (6–7):

$$\mu = \mu_0 \left(\frac{T_0}{T} \right)^{\alpha} \tag{6}$$

$$\tau = \tau_0 \left(\frac{T}{T_0} \right)^{1-\alpha}, \quad 0 < \alpha < 1 \tag{7}$$

Solutions of this system of equations obtained by numerical simulation methods in a certain range of parameters clearly demonstrated S-shaped resulting current–voltage characteristics, indicative of an electric instability considered heterostructures [1, P. 453; 2, P. 1075]. Note that when $\alpha = 1$, the equation of the electron temperature is linear flow continuity in heavily doped GaAs valleys, allowing you to build and relevant analytical solutions for the electron temperature. As expected, diffusion (caused by thermal conductivity) component of the heat flux decreases exponentially in the direction of electron drift in the valley with a decrement decreases with increasing current density, which increases the heat transfer from the previous warming up electrons barrier to the next, making it easy to thermionic bridging and hence reduces the need for heating in the previous barrier, i.e., reduces the voltage drop across it. Note that such an analytical model, [3, P. 481; 4, P. 83], has been widely used in the calculation of the expected characteristics of the test heteroelements, showing a characteristic range of currents and voltages NDR area. Figure 3 shows typical resulting curves obtained under the effect of greatly simplified analytical model.

Previously, we could not consider the structure with uneven barriers and that we strongly limited. In particular, a half-height front barrier empirically necessary to do in order to provide preheating. Therefore, the development of the above model, we abandoned the cyclic boundary conditions and constructed an algorithm sequential solution to the problem of thermionic, starting with the first input of the unit cell comprising a contact layer high-alloy material and the adjacent potential barrier. Optimization was necessary to construct an algorithm that takes into account the uneven barriers and differences valleys (8)–(11). The results of this calculation produce a flow of electrons into the second cell, which is a boundary condition for the character to overcome the electron beam of the second cell and so on. This algorithm allows to consider the structure with different thicknesses of the barrier

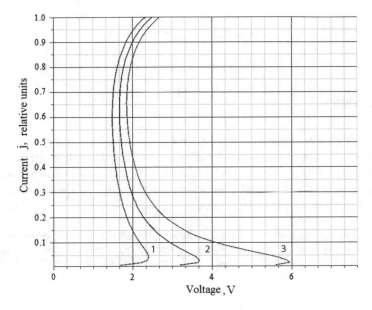

Fig. 3 Typical results CVC modeling in the framework of the analytical model

layers and high valleys. It turns out that the whole relaxation takes at least four barriers, despite the fact that we have to undergo substantial thickness of the layers, subject to the criterion of quality—minimizing heat release. This achieves sufficiently significant value UDF, which provides the ability to generate terahertz radiation at a certain partial minimize stress on the energy barriers, by varying the thickness of the individual layers and the selection of the ratio of the heights of the energy barriers. Compile the above model in the algorithm above formulas (1)–(7) were supplemented by the resulting expression expressing the electron temperature flow in the valleys, allowing to determine the voltage at the next barrier is needed to maintain the value of a given current, bearing in mind that before preceding the first barrier is no barrier, and therefore the adjoining contact layer relatively low electron temperature.

$$T_0 = \frac{1}{1 - \frac{2}{3}j^2} \tag{8}$$

$$V_i^+ = jd^+ T_i \exp\left(\frac{\Delta}{T_i}\right) \tag{9}$$

$$\tilde{V}_N = \sum_{i=0}^{N} V_i e^{-(N-i)Kd_-} \tag{10}$$

$$T_N = T_0\left(1 + \frac{2}{3}j\tilde{V}_N K e^{-Kd_-}\right) \tag{11}$$

Conclusions

Characteristics were observed with a strong section of the negative differential resistance in the current range of several tens of milliamperes. It is proposed to use this effect for the generation of terahertz electromagnetic radiation. The theoretical interpretation of the observed phenomena on the basis of quasi-hydrodynamic theory of electron drift is briefly stated.

Acknowledgments Researches are carried out (conducted) with the financial support of the state represented by the Ministry of Education and Science of the Russian Federation. Agreement (contract) no. 14.607.21.0141 27. October 2015. Unique project Identifier: RFMEFI60715X0141.

References

1. Gergel, V.A., et al.: Semiconductors, **39**, 453–455(2005)
2. Gergel, V.A., et al.: Semiconductors, **39**, 1075–1079 (2005)
3. Gergel, V.A., et al.: Semiconductors, **48**, 481–486 (2014)
4. Gergel, V.A., et al.: Technical Physics, **85**, 83–86 (2015)
5. Stratton, R.: Diffusion of hot and cold electrons in semiconductor barriers. Phys. Rev. **126**(6), 2002 (1962)

Asymmetric Magnetoimpedance in Bimagnetic Multilayered Film Structures

A.S. Antonov and N.A. Buznikov

Abstract The magnetoimpedance (MI) effect in bimagnetic multilayered film is studied theoretically. The multilayer consists of a repetition of the base three-layered film structure having the soft and hard magnetic layers separated by highly conductive non-magnetic spacer. It is shown that the magnetostatic coupling changes the magnetization distribution in the soft magnetic layers and leads to the asymmetry in the field dependence of the impedance. The influence of the number of layers on the asymmetric magnetoimpedance (AMI) is analyzed. The calculated field and frequency dependences of the impedance describe qualitatively the AMI effect observed in bimagnetic multilayers. The results obtained may be used in the development of sensors of the magnetic field.

Keywords Ferromagnetic films · Magnetoimpedance · Magnetostatic coupling Bias field · Sensor applications

Introduction

The magnetoimpedance (MI) implies a change in the impedance of a magnetic conductor with the variation of an external magnetic field. The nature of the MI effect can be explained in the framework of the classical electrodynamics and is related to changes in the permeability and skin depth with the external field.

A.S. Antonov (✉)
Institute for Theoretical and Applied Electrodynamics, Russian Academy of Sciences, Moscow 125412, Russia
e-mail: asantonov@inbox.ru

N.A. Buznikov
Scientific & Research Institute of Natural Gases and Gas Technologies—Gazprom VNIIGAZ, Razvilka, Leninsky District, Moscow Region 142717, Russia
e-mail: n_buznikov@mail.ru

© The Author(s) 2018   107
K.V. Anisimov et al. (eds.), *Proceedings of the Scientific-Practical Conference "Research and Development - 2016"*, https://doi.org/10.1007/978-3-319-62870-7_12

The interest in the MI is supported by the possible use of the effect for the development of sensors of a weak magnetic field (see, for example, [1, 2]). From the point of view of the sensor miniaturization and integration with modern microelectronic technologies, thin-film structures are one of the most attractive materials for sensitive elements of MI sensors.

The linearity and sensitivity for the external magnetic field are the most important parameters in applications of the MI effect. However, for most of the materials, the MI response exhibits nonlinear behavior in the vicinity of zero field, and this is unfavorable for sensor applications. To improve the linear features of the MI response, the asymmetric field dependence of the impedance is promising. The asymmetric magnetoimpedance (AMI) may be obtained by applying the direct bias current or alternating magnetic field to the soft magnetic conductor [3–5]. Note that these methods have some limitations in applications due to an increase in the power consumption.

Another approach of producing the AMI consists in the use of materials with asymmetric static magnetic configuration arising from the magnetostatic interactions or exchange bias. This type of the AMI has been observed initially in field-annealed amorphous ribbons [6, 7]. For film structures, the AMI has been studied extensively in the exchange bias multilayers [8, 9]. It was found that the linear behavior of the MI response can be tuned by modifying the angle between the external field and exchange bias field or by changing the frequency.

The AMI effect has been also observed in three-layered NiFe/Cu/Co film structures [10]. The measured field dependence of the film impedance exhibited asymmetry within a wide frequency range. It was shown that the film structures have biphase magnetic behavior, and the origin of the AMI can be ascribed to the magnetostatic coupling between the soft and hard magnetic layers. The linear behavior of the MI response in bimagnetic films can be tuned by varying the copper spacer thickness and current frequency [10]. A model to describe the AMI effect in three-layered bimagnetic films has been proposed recently [11]. The MI response was found by means of a simultaneous solution of Maxwell equations together with Landau–Lifshitz equation, and the magnetostatic coupling between the hard and soft magnetic layers was described in terms of an effective bias field appearing in the permalloy layer. The model allows one to explain qualitatively the field and frequency dependences of the AMI effect observed in NiFe/Cu/Co film structures.

Note that the field sensitivity of the impedance in the three-layered films is sufficiently low [10]. This fact is related to the small thickness of the film structure, where the skin depth is much larger than the film thickness. To enhance the sensitivity of the AMI, the multilayers consisting of a repetition of the base bimagnetic three-layered film structure have been proposed [12]. It was found that these multilayers exhibit bimagnetic behavior, and the AMI sensitivity increases, which is promising for sensor design. In this work, we extend the approach proposed in [11] to the case of the bimagnetic multilayers and study the effect of the number of layers on the AMI.

Model

Figure 1 shows schematically the studied film structure having the length l and width $w < l$. The structure consists of the soft magnetic layer 1, highly conductive non-magnetic spacer 2, hard magnetic layer 3, and buffer layer 4. The corresponding values of the layer thickness are d_1, d_2, d_3, and d_4. The number of repetitions of the first three layers is denoted as N. The total number of layers equals to $4N - 1$, and the multilayer thickness $D = N(d_1 + d_2 + d_3) + (N - 1)d_4$.

The multilayered film is subjected to the alternating current $I = I_0\exp(-i\omega t)$, and the external magnetic field H_e is parallel to the current. Neglecting edge effects, we assume that the electromagnetic fields depend only on the coordinate perpendicular to the sample plane. Furthermore, we neglect the longitudinal alternating magnetic field arising in the soft magnetic layers due to the cross-magnetization process [13]. In this approximation, the solution of Maxwell equations for the amplitudes of the longitudinal electric field $e_k^{(j)}$ and transverse magnetic field $h_k^{(j)}$ is given by

$$e_k^{(j)} = (c\lambda_k/4\pi\sigma_k)[A_k^{(j)}\cosh(\lambda_k z) + B_k^{(j)}\sinh(\lambda_k z)], \tag{1}$$

$$h_k^{(j)} = A_k^{(j)}\sinh(\lambda_k z) + B_k^{(j)}\cosh(\lambda_k z). \tag{2}$$

Here $k = 1, 2, 3$, and 4 is the layer number; $j = 1, \ldots N$ is the number of the base structure repetition; $A_k^{(j)}$ and $B_k^{(j)}$ are the constants; $\lambda_k = (1 - i)/\delta_k$; $\delta_k = c/(2\pi\omega\sigma_k\mu_k)^{1/2}$; c is the velocity of light; σ_k and μ_k are the conductivity and the transverse permeability of the layer k, respectively. For the non-magnetic layers, $k = 2$ and $k = 4$, the transverse permeability is equal to unity, $\mu_2 = \mu_4 = 1$.

Fig. 1 A sketch of the bimagnetic multilayer geometry and coordinate system used for analysis

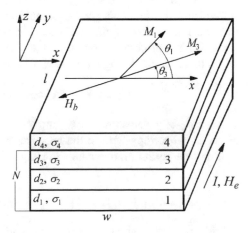

The amplitudes of the electric and magnetic field satisfy the continuity conditions at the layer interfaces [11]. In addition, the amplitude of the magnetic field at the multilayer surfaces, $z = 0$ and $z = D$, is governed by the following excitation conditions:

$$h_1^{(1)}(0) = -2\pi I_0/cw,$$
$$h_3^{(N)}(D) = 2\pi I_0/cw. \tag{3}$$

Thus, the boundary conditions allow one to find the $2(4N - 1)$ constants in Eqs. (1) and (2). The multilayer impedance Z can be expressed in terms of the surface impedance tensor [2]:

$$Z = \frac{\pi l}{cw} \times \left[\frac{e_3^{(N)}(D)}{h_3^{(N)}(D)} - \frac{e_1^{(1)}(0)}{h_1^{(1)}(0)} \right]. \tag{4}$$

Further, it is assumed that the values of the permeability in the magnetic layers are governed by the magnetization rotation only. This approximation is valid at sufficiently high frequencies, when the domain-wall motion is strongly damped [14]. We assume that the magnetic layers have the uniaxial in-plane anisotropy. During the deposition of the multilayer, a constant magnetic field was applied along the short side of the film [12]. Therefore, the direction of the anisotropy axes in the magnetic layers is close to the transverse one. The magnetostatic coupling between the hard and soft magnetic layers induces the effective bias field H_b in the soft magnetic layers. It is assumed that the bias field does not vary over the soft magnetic layer thickness and has the opposite direction with respect to the magnetization in the hard magnetic layers. Note that a similar approach has been used previously to study the AMI effect in amorphous ribbons [15, 16].

The distribution of the magnetization in the magnetic layers can be found by minimizing the free energy. In the hard magnetic layers, the free energy can be presented as a sum of the uniaxial anisotropy energy and Zeeman energy. The minimization procedure results in the following equation for the magnetization angle θ_3 in the hard magnetic layer:

$$H_3 \sin(\theta_3 - \psi_3) \cos(\theta_3 - \psi_3) = H_e \cos \theta_3. \tag{5}$$

Here ψ_3 is the anisotropy axis angle with respect to the transverse direction and H_3 is the anisotropy field in the hard magnetic layers.

For the equilibrium magnetization angle θ_1 in the soft magnetic layers, we obtain [11]

$$H_1 \sin(\theta_1 - \psi_1) \cos(\theta_1 - \psi_1) - H_b \sin(\theta_1 - \theta_3) - H_e \cos \theta_1 = 0. \tag{6}$$

Here ψ_1 is the anisotropy axis angle and $H_1 \ll H_3$ is the anisotropy field in the soft magnetic layers.

It should be noted that there is a difference in magnetic properties between the first soft magnetic layer and the inner soft magnetic layers. This is due to the first soft magnetic layer interacts only with one hard magnetic layer, whereas the other soft magnetic layers are sandwiched between two hard magnetic layers [12]. To take into account this fact, we assume that the bias field in the first soft magnetic layer is two times lower than H_b.

The transverse permeability in the magnetic layers can be found by means of the solution of the linearized Landau–Lifshitz equation. To simplify calculations, we use the so-called electromagnetic approximation [2, 14], where the contribution of the exchange energy is neglected. More rigorous theoretical treatment requires the including of the exchange-conductivity effect in the model [17–19]. However, the contribution of the exchange-conductivity effect to the magnitude of the impedance is relatively low within the high-frequency range studied.

The solution of the linearized Landau–Lifshitz equation leads to the following expression for the transverse permeability in the magnetic layers [11]:

$$\mu_k = 1 + \frac{\gamma 4\pi M_k [\gamma 4\pi M_k + \omega_k - i\kappa\omega] \sin^2 \theta_k}{[\gamma 4\pi M_k + \omega_k - i\kappa\omega][\omega_k - \gamma H_k \sin^2(\theta_k - \psi_k) - i\kappa\omega] - \omega^2}. \quad (7)$$

Here $k = 1$ and $k = 3$ corresponds to the soft and hard magnetic layers, respectively, M_k are the values of the saturation magnetization in the layers, γ is the gyromagnetic constant, and κ is the Gilbert damping parameter. The characteristic frequencies ω_1 and ω_3 can be expressed as

$$\omega_1 = \gamma[H_1 \cos^2(\theta_1 - \psi_1) - H_b \cos(\theta_1 - \theta_3) + H_e \sin \theta_1], \quad (8)$$

$$\omega_3 = \gamma[H_3 \cos^2(\theta_3 - \psi_3) + H_e \sin \theta_3]. \quad (9)$$

Results and Discussion

To investigate the effect of the magnetostatic coupling and number of layers on the MI ratio, we use geometrical parameters of the bimagnetic multilayers studied experimentally [12]. The MI ratio $\Delta Z/Z_0$ is defined as follows:

$$\Delta Z/Z_0 = (Z - Z_0)/Z_0, \quad (10)$$

where Z_0 is the DC resistance of the multilayered film:

$$Z_0 = \frac{l}{w[N(\sigma_1 d_1 + \sigma_2 d_2 + \sigma_3 d_3) + (N-1)\sigma_4 d_4]}. \quad (11)$$

Fig. 2 The multilayer impedance ratio as a function of the external field at $N = 5$ and $f = \omega/2\pi = 0.5$ GHz for different values of H_b. Parameters used for calculations are $d_1 = 25$ nm, $d_2 = 7$ nm, $d_3 = 50$ nm, $d_4 = 10$ nm, $M_1 = 800$ G, $M_3 = 1000$ G, $H_1 = 5$ Oe, $H_3 = 30$ Oe, $\psi_1 = -0.1\pi$, $\psi_3 = 0$, $\sigma_1 = 10^{16}\text{s}^{-1}$, $\sigma_2 = 5 \times 10^{17}\text{s}^{-1}$, $\sigma_3 = 1.5 \times 10^{17}\text{s}^{-1}$, $\sigma_4 = 7 \times 10^{16}\text{s}^{-1}$ and $\kappa = 0.02$

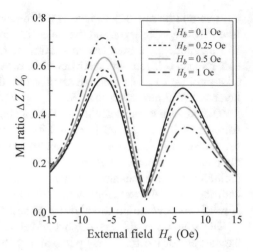

The dependence of the MI ratio of the multilayered film ($N = 5$) on the external field is shown in Fig. 2 for different values of the bias field H_b. In the presence of the bias field, the impedance field dependence shifts with respect to zero field, and there is a difference between the maximum impedance values at the positive and negative fields. The asymmetry growths with the bias field, the negative field peak increases and the positive field peak decreases. The AMI in multilayered films is caused by the fact that the bias field changes the static magnetization distribution and transverse permeability in the soft magnetic layers [11].

Note that the AMI response in multilayers is very sensitive to the anisotropy axis angle in the soft magnetic layers. It was shown for three-layered bimagnetic film structures that the asymmetry in the MI response is maximal at low values of ψ_1 and decreases with an increase of ψ_1 [11]. The asymmetry between the impedance values at the peaks disappears, if the anisotropy in the soft magnetic layer has the transverse direction, $\psi_1 = 0$. The deviation of the easy axis from the transverse direction may be attributed to peculiarities of the bimagnetic multilayer preparation [10].

The variation in the field dependence of the MI ratio with the frequency is presented in Fig. 3. At not very high frequencies, the peak positions in the field dependence of $\Delta Z/Z$ remain nearly the same, and the maximal values of the impedance increase due to the decrease of the skin depth. At frequencies of the order of 1 GHz, the peak positions shift toward higher fields due to the ferromagnetic resonance. In this case, the asymmetry between peaks almost vanishes.

Although the results of modeling describe main features of the AMI effect in bimagnetic multilayered films [12], some experimental data cannot be explained in the framework of the model proposed. It was observed that at frequencies above 1.5 GHz the shape of the impedance field dependence changes. The MI response becomes nonlinear at the low external field, and an additional peak in the impedance field dependence was observed near zero field [12]. At the same time, the

Fig. 3 The multilayer impedance ratio as a function of the external field at $N = 5$ and $H_b = 1$ Oe for different f (GHz). Parameters used for calculations are the same as in Fig. 2

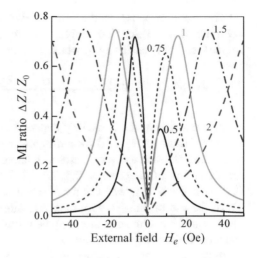

Fig. 4 The multilayer impedance ratio as a function of the external field at $f = 0.5$ GHz and $H_b = 1$ Oe for different N. Parameters used for calculations are the same as in Fig. 2

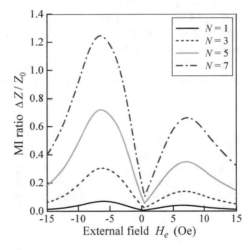

model predicts a monotonic behavior of the film impedance within this field range. The disagreement between theoretical and experimental results may be ascribed to approximations of the model.

The influence of a number of layers on the field dependence of the MI ratio is illustrated in Fig. 4. The MI ratio increases with the number of repetitions N of the base structure, since the skin effect is more pronounced in thick multilayered films. It follows from Fig. 4 that the impedance peak positions shift slightly toward higher fields with an increase of N. However, the shape of the impedance field dependence is not affected by the number of layers. Similar results are obtained for the whole frequency range studied.

The AMI response of the bimagnetic multilayered film shows nearly linear behavior at low fields. To analyze the impedance variation, let us introduce the impedance field sensitivity S, which is defined as follows

$$S = \frac{Z(H_e = -H_1) - Z(H_e = 0)}{H_1 Z_0} \cdot 100\%. \tag{12}$$

Figure 5 shows the frequency dependence of the impedance sensitivity calculated by means of Eq. (12) for a different number of repetitions of the base structure. The field sensitivity increases at relatively low frequencies and attains peak at the frequency of the order of 0.5 GHz. The highest field sensitivity is 1.3%/Oe at 0.55 GHz for $N = 1$, whereas for $N = 7$ it increases up to 22.7%/Oe at 0.45 GHz. Thus, the use of bimagnetic multilayers allows one to raise significantly the AMI effect, and these film structures may be promising for the development of miniature sensors of a weak magnetic field.

In conclusion of this section, note that much attention is paid to the MI effect in multilayered nanostructured films. These films consist of soft magnetic layers separated by thin highly conductive non-magnetic layers. In particular, recently the non-symmetric multilayered films were attracted interest in connection with a development of magnetic biosensors [20]. Although the MI effect in the multilayers was studied experimentally quite well, its theoretical explanation is still missing. The approach presented in this work seems to be useful for the analysis of the MI effect in the non-symmetric multilayered nanostructured films.

Fig. 5 The frequency dependence of the impedance field sensitivity at $H_b = 1$ Oe for different N. Parameters used for calculations are the same as in Fig. 2

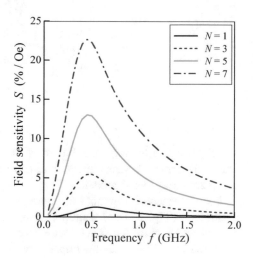

Conclusion

A model to describe the AMI effect in bimagnetic multilayer is proposed. The MI response is calculated by means of a solution of Maxwell equations with Landau–Lifshitz equation. The asymmetry in the field dependence of the impedance is related to the magnetostatic coupling between the magnetic layers. As a result of the magnetostatic coupling, the bias field is induced in the soft magnetic layers, which leads to changes in the static magnetization distribution in the layers. It is demonstrated that the AMI effect increases significantly in bimagnetic multilayers in comparison with the base three-layered film structure. The obtained field and frequency dependences of the impedance are in a qualitative agreement with results of the experimental study of the AMI effect in bimagnetic multilayers [12]. The analysis shows that the bimagnetic multilayers may be attractive sensitive elements for the design of miniature sensors of a weak magnetic field.

Acknowledgements Research are carried out with the financial support of the state represented by the Ministry of Education and Science of the Russian Federation. Contract no. 02.G25.31.0127.

References

1. Knobel, M., Pirota, K.R.: Giant magnetoimpedance: concepts and recent progress. J. Mag. Magn. Mater. **242–245**. Part 1, pp. 33–40 (2002)
2. Phan, M.-H., Peng, H.-X.: Giant magnetoimpedance materials: fundamentals and applications. Prog. Mater. Sci. **53**(2), 323–420 (2008)
3. Kitoh, T., Mohri K., Uchiyama, T.: Asymmetrical magneto-impedance effect in twisted amorphous wires for sensitive magnetic sensors. IEEE Trans. Magn. **31**. N 6, pp. 3137–3139 (1995)
4. Makhnovskiy, D.P., Panina, L.V., Mapps, D.J.: Asymmetrical magnetoimpedance in as-cast CoFeSiB amorphous wires due to ac bias. Appl. Phys. Lett. **77**(1), 121–123 (2000)
5. Panina, L.V., Mohri, K., Makhnovskiy, D.P.: Mechanism of asymmetrical magnetoimpedance in amorphous wires. J. Appl. Phys. **85**(8), 5444–5446 (1999)
6. Kim, C.G., Jang, K.J., Kim, H.C., Yoon, S.S.: Asymmetric giant magnetoimpedance in field-annealed Co-based amorphous ribbon. J. Appl. Phys. **85**(8), 5447–5449 (1999)
7. Kim, C.G., Kim, C.O., Yoon, S.S.: The role of exchange coupling on the giant magnetoimpe-dance of annealed amorphous materials. J. Mag. Magn. Mater. **249**(1–2), 293–299 (2002)
8. García, C., Florez, J.M., Vargas, P., Ross, C.A.: Asymmetrical giant magnetoimpedance in exchange-biased NiFe. Appl. Phys. Lett. **96**. N 23, P. 232501 (3 p.) (2010)
9. da Silva, R.B., Silva, E.F., Mori, T.J.A., Della Pace, R.D., Dutra, R., Corrêa, M.A., Bohn, F., Sommer, R.L.: Improving the sensitivity of asymmetric magnetoimpedance in exchange biased NiFe/IrMn multilayers. J. Mag. Magn. Mater. **394**, 87–91 (2015)
10. Silva, E.F., Gamino, M., de Andrade, A.M.H., Corrêa, M.A., Vázquez, M., Bohn, F.: Tunable asymmetric magnetoimpedance effect in ferromagnetic NiFe/Cu/Co films. Appl. Phys. Lett. **105**. N 10, P. 102409 (5 p.) (2014)
11. Buznikov, N.A., Antonov, A.S.: A model for asymmetric magnetoimpedance effect in multilayered bimagnetic films. J. Mag. Magn. Mater. **420**, 51–55 (2016)

12. Silva, E.F., da Silva, R.B., Gamino, M., de Andrade, A.M.H., Vázquez, M., Corrêa, M.A., Bohn, F.: Asymmetric magnetoimpedance effect in ferromagnetic multilayered biphase films. J. Mag. Magn. Mater. **394**, 260–264 (2015)
13. Makhnovskiy, D.P., Lagar'kov, A.N., Panina, L.V., Mohri, K.: Effect of antisymmetric bias field on magneto-impedance in multilayers with crossed anisotropy. Sens. Actuators A. **81**. N 1–3, pp. 106–110 (2000)
14. Kraus, L.: GMI modeling and material optimization. Sens. Actuators A. **106**(1–3), 187–194 (2003)
15. Buznikov, N.A., Kim, C.G., Kim, C.O., Yoon, S.S.: A model for asymmetric giant magnetoimpedance in field-annealed amorphous ribbons. Appl. Phys. Lett. **85**(16), 3507–3509 (2004)
16. Buznikov, N.A., Kim, C.G., Kim, C.O., Yoon, S.S.: Modeling of asymmetric giant magnetoimpedance in amorphous ribbons with a surface crystalline layer. J. Mag. Magn. Mater. **288**, 130–136 (2005)
17. Kraus, L.: Theory of giant magneto-impedance in the planar conductor with uniaxial magnetic anisotropy. J. Mag. Magn. Mater. **195**(3), 764–778 (1999)
18. Ménard, D., Yelon, A.: Theory of longitudinal magnetoimpedance in wires. J. Appl. Phys. **88** (1), 379–393 (2000)
19. Yelon, A., Ménard, D., Britel, M., Ciureanu, P.: Calculations of giant magnetoimpedance and of ferromagnetic resonance response are rigorously equivalent. Appl. Phys. Lett. **69**(20), 3084–3085 (1996)
20. Kurlyandskaya, G.V., Chlenova, A.A., Fernández, E., Lodewijk, K.J.: FeNi-based flat magnetoimpedance nanostructures with open magnetic flux: New topological approaches. J. Mag. Magn. Mater. **383**, 220–225 (2015)

The Model of the Cybernetic Network and Its Realization on the Cluster of Universal and Graphic Processors

A.E. Krasnov, A.A. Kalachev, E.N. Nadezhdin,
D.N. Nikolskii and D.S. Repin

Abstract The model of the cybernetic network consisting of information and management subnets is offered. For each subnets, its active structural elements and also a way of their association in the general heterogeneous multicoherent network are described. The main functions of active structural elements of such network are considered. For the creation of the model, the object-oriented approach is used. The description of functionalities of main classes intended for the work on clusters of universal and graphic processors is submitted. The approbation of the developed model is executed by a solution of neurobiological tasks.

Keywords Model · Cybernetic network · Information and management subnets
Commutator · Connector · Controller · Object-oriented approach
Neurobiological tasks

Introduction

In recent years, the huge attention is paid to questions of controllability by networks in a classical statement at which the operating signals influence a row, so-called, leading knots, and variables of a vector of a network's condition are connected with

---

A.E. Krasnov · A.A. Kalachev · E.N. Nadezhdin (✉) ·
D.N. Nikolskii · D.S. Repin
State Institute of Information Technologies and Telecommunications, Moscow, Russia
e-mail: e.nadezhdin@informika.ru

A.E. Krasnov
e-mail: a.krasnov@informika.ru

A.A. Kalachev
e-mail: a.kalachev@informika.ru

D.N. Nikolskii
e-mail: d.nikolsky@informika.ru

D.S. Repin
e-mail: r_d_s@informika.ru

© The Author(s) 2018   117
K.V. Anisimov et al. (eds.), *Proceedings of the Scientific-Practical Conference*
"Research and Development - 2016", https://doi.org/10.1007/978-3-319-62870-7_13

its hubs [1–3]. So, for example in [4] influence of attacks on the control of the network is investigated, and it is revealed that the purposeful attack on links of knots is more effective than the stochastic attack to knots. In addition to it in works [5, 6], the new approach in which also dynamics of links is considered which is connected with vectors of a condition of the network.

It seems appropriate to consider the network model with built-in controllers for controlling various connections, as well as its implementation on a cluster-purpose and graphics processors.

The proposed model and its software implementation may be used to solve quite a wide range of tasks of network dynamics, segmentation and recognition signal and image processing, signal processing phased arrays, the simulation of hardware implementations of networks, technologies process control.

The Statement of the Problem

In this paper, we solve the problem of modeling on the functional–structural level the multiple heterogeneous network consisting of two subnets (information and management), formed by a large number of active elements.

The ultimate goal is to realize about a complex network model on a cluster of universal and GPUs to create high-performance, scalable tool for solving various problems of network dynamics, segmentation, data compression, and pattern recognition.

The basic idea of the structural organization of the network, in terms of the introduction of a management subnet, is borrowed from neurobiological studies [6–9].

However, compared with the existing approaches, which describes only the idea of the flow of calcium metabolism and neurotransmitter between synapses and astrocytes, we propose a cybernetic generalization in terms of the interaction between two subnets, one of which is formed by connecting switches and their connectors, and the other is formed by controllers. In the model, each controller polls a few connectors, to give information about their conditions, then adjusts their parameters [10, 11].

The introduction of the controller's subnet allows experimentally examined the relationship issues of governance networks, the formation of a variety network clusters, their influence on the processes of learning networks, as well as, to increase their operational reliability. Using of multiple controllers can also parallelize the control process for networks with a large number of connections ($\geq 10^8$).

Structural Elements of the Network and Its Topology

The proposed network includes following subnets and their structural elements.

Information subnet formed by commutators («Comm», N) and connectors («Conn», $N \times N$).

Commutator (switch) «Comm» has input and output communication ports connected hubs with internal and external inputs and outputs, as well as ports of external control.

Connector «Conn» communicates between switches and has input and output communication ports, as well as ports of external and internal control.

Management subnet formed by controllers («Contr», $M \geq N$).

Controllers «Contr» have input and output communication ports as well as ports of external and internal control. These ports are connected via communication hubs channels.

Setting up the structural elements of a heterogeneous network carried out via the external control trunk—Trunk of an external control (Fig. 1).

The informational subnet is an exchange of information data flows I_{comm} and I_{conn}, convertible switches and connectors thru the corresponding Trunks of switches and connectors.

The input of each switch «Comm» via internal Trunk of switches is connected to the output of the group of connectors «Conn».

The information output of each switch «Comm» is connected via internal Trunks of switches with a group of connectors. For example, the output of the switch $Comm_1$—the information data flow $Icomm_1$ effects on inputs of connectors $Conn_{1,1}, \ldots, Conn_{1, N}$.

In the management subnet, the exchange by streams of S_{conn} and C data via the Trunk1, Trunk2 of internal management is occurred (Fig. 1). Data flow S_{conn} describes the state of connectors and C describes the control by connectors from the controllers.

The exchange by information data flows I_{contr} between all controllers carried out via the internal $Trunk_3$ (Fig. 1).

Since a number of M controllers are much smaller than the number of $N \times N$ connectors, each controller controls a group (set) of connectors. For example, connectors of the group $\{1,1\}, \ldots, \{1, N\}$ ($Conn_{1,1}, \ldots, Conn_{1, N}$) are operated by controller $Contr_1$ which generally performs various controls $C_{1,1}, \ldots, C_{1,N}$. In turn, each m-th controller receives information $S_{\{m\}}$, which characterizes the state of a group $\{m\}$ of connectors connected to it ($m = 1, \ldots, M$).

Functions of the Structural Elements of the Network

Structural elements of the network discussed above have the following active functions.

Commutator or switch («Comm») performs the functions of the spatial and temporal integration of input information signals and generating an output information signal and the storing of parameters of the setting.

To describe the transfer functions of its element as static Boolean expressions, as dynamic equations discussed, for example, [3, 12] can be used. In the latter case, in the n-th switch $Comm_n$ the integration of the dynamic equations system is carried

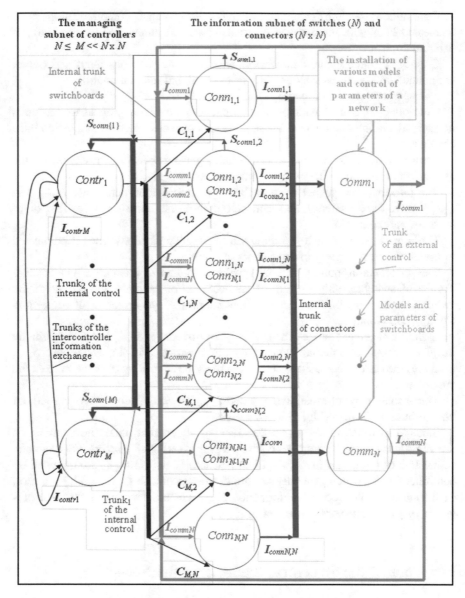

Fig. 1 Functional and structural topology of the cyber network taking into account the exchange of information and control data flows

out. In this case, the specific form of the dynamic equation and its parameters are set via an external control line (Fig. 1).

For the purposes of neurodynamic the condition of formation the value I_{comm} signal output can be set, for example, in the form of: if $I_{comm} \geq I_{thr}$, I_{out}; otherwise $I_{comm} = 0$, where I_{thr}—the signal threshold.

Connector («Conn») performs the functions of the time integration of the input information signal and generating an output signal, formation of the state functions, as well as memory function of parameters from the controller settings.

To describe transfer functions of the element as static Boolean expressions, dynamic equation discussed, for example, in [14] can be used. In the latter case, in k, l-th connector $\text{Conn}_{k,l}$ (k, $l = 1, 2, ..., N \times N$), connecting k-th and l-th switches (see Fig. 1), integration of the dynamic equations was carried out describing the rate of change $I_{\text{conn}}(k,l)$ output information signal and the speed change of the function $S_{\text{conn}}(k,l)$ its state depending on the input information signal $I_{\text{comm}}(k)$ from the kth switch and control signal $C_{\text{cont}}r(m)$ from the m-th controller ($m = 1, ..., M$; $N \leq M \ll N \times N$).

The controller («Contr») sells computing and control functions for a group of connectors.

To describe its operation as static Boolean expressions, dynamic equation is discussed, for example, in [12, 13] can be used. In the latter case, in the m-th controller Contr_m, the integration of dynamic equations is performed.

Input control signal for the controller is the function $S_{\text{conn}}(k, l)$ of the state of the polled k, l-th connector, and the output control signal is the function $C_{\text{contr}}(k,l)$. Communication takes place via a line Trunk of the internal control (Fig. 1).

Each m-th controller produces an output information signal $I_{\text{contr}}(m)$ depending on information signals $I_{\text{contr}}(m-k)$, $I_{\text{contr}}(m + k)$, associated through line Trunk of the communication exchange. The contact for information channels needed for the backup of control functions in case of the failure of any controllers. Using multiple controllers we can also parallelize the process of the control for networks with a large number of links.

For a network with a large number of active elements ($N \geq 10^5$) there are modern technologies based on graphics processors (GPU) and cluster universal processors (CPU).

The following is considered as an object-oriented approach to software implementation of our model on a cluster of general purpose processors and graphics cards.

The Object-Oriented Approach to the Software Implementation of the Model

The library for the modeling of elements of the multiply heterogeneous network of active elements comprises the following groups of classes:

Communicator—contains classes modeling communicators «Comm»: Comms, CommsCPU, CommsGPU, ...
Connector—provides classes that simulate connectors (connections) «Conn»: Conns, ConnsCPU, ConnsGPU, ...
Controller—contains classes that simulate controllers «Contr»: Contrs, ContrCPU, ContrsGPU, ...

Fig. 2 The scheme of
creating specific heirs to run
on a cluster CPU and GPU

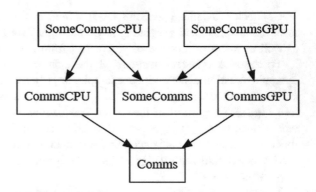

Each class uses software and hardware solutions for the implementation of parallel processing.

The class of Hierarchies is built using object-oriented programming principles. Communicator Groups, Connector, and Controller encapsulate the basic interfaces for their classes.

To create a class containing specific implementation of methods associated with the implementation of specific mathematical models for the specific architecture, the inheritance diagram (Fig. 2) is used.

In this figure, the creation of specific heirs of the Comms class to carry on the CPU and GPU cluster is presented.

The Comms class provides a common interface for all classes, modeling the job of the switch. He is the ancestor for CommCPU and CommGPU classes that contain specialized structures for the data storing for universal and graphics processors.

To develop specific classes that implement specific mathematical models, a base class SomeComms, containing fields and methods necessary to implement the necessary mathematical model is created. Then the classes SomeCommsCPU and SomeCommsGPU, performing calculations on universal and graphics processors are developed.

The polymorphism principle, adopted in object-oriented programming, enables the development of algorithms, working with pointers on base classes, and the network is programmed using pointers on base classes, which in the course of the implementing of various scenarios of the program can be transferred any heirs (Fig. 3).

Figure reflects the cyclical exchange of information (I_{comm}, I_{conn}) and managers (S_{conn}, C_{contr}) data streams.

Each of the classes CommsPerNode, ConnsPerNode, and ContrsPerNode engaged in processing of portion of objects fall on one cluster node. The latter enables the simulation of the work of large networks on a cluster of universal and (or) GPUs.

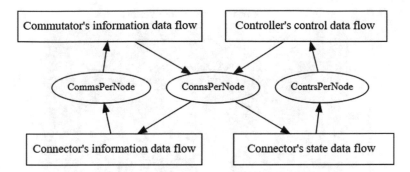

Fig. 3 The scheme of processing streams of data units of software modules

Examples of Solutions of Neurodynamic Problems

As an example, consider following the solution of problems of the modeling the neurodynamic process with the using of the cluster based on Intel Core i5-2300 processor and graphics GM206 CPUs (NVidia Geforce GTX 950).

Experiments with the network of switches. To test the developed object-oriented approach as a switch «Comm» the neuron dynamic equation in the form of E.M. Izhikevich model [14] was chosen, as a connector «Conn» the synapse with free-wheeling and adjustable the amplitude response and the time τ_{conn} of its fall was chosen, as a controller «Contr» the astrocyte with the control function of the amplitude of the time τ_{conn} of the connector signal recession was chosen.

The process of generation of excitation (spikes) of 100 neurons under their relationship in the form of a complete graph and different settings of relaxation times τ_{conn} of synaptic responses was investigated.

Figure 4 shows the activity of neurons, divided into groups (34, 33, 33) with the relaxation times of synaptic responses 0.4, 4, and 40 ms, respectively.

The diagram shows that neurons with long relaxation times of synaptic responses often generate spikes.

At the additional testing, the fully connected network of 1000 neurons was used. The model of E.M. Izhikevich built on the basis of a mathematical model implemented in Matlab package and realized by numerical Euler method was used as the referent set [15]. The results of this model were compared with the results obtained by us on the GPU and CPU (Fig. 5).

In simulation experiments as well as in [15], the alpha rhythm (10 Hz) and gamma wave (40 Hz) were observed. Thereby, the correct operation of the developed model and its implementation, both graphical and universal processors were confirmed.

Experiments on the interaction of subnets switches and controllers. During testing of the developed model, the interaction between the network of connectors Conns and the network of controllers Contr was qualitatively investigated. In the first experiment as a model of switches Comms, the model of a neuron «Integrate

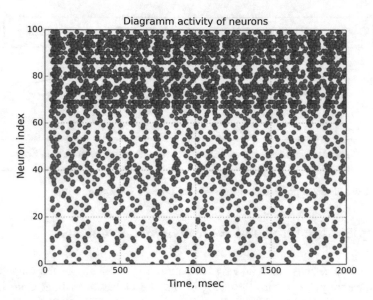

Fig. 4 The activity diagram of three different groups of neurons (0.4, 4 and 40 ms)

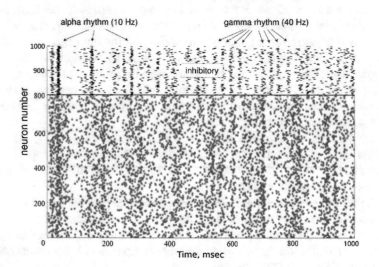

Fig. 5 The activity diagram of fully connected neural network (1000 neurons), implemented on the GPU

and Fire» [15] resistant to interference was taken, as well as models of interaction between the synapse and astrocyte through the sharing of neurotransmitter of works [7–9] were taken.

In the first experiment, switches (neurons) connected to each other through connectors, which are connected to the controller (Fig. 6).

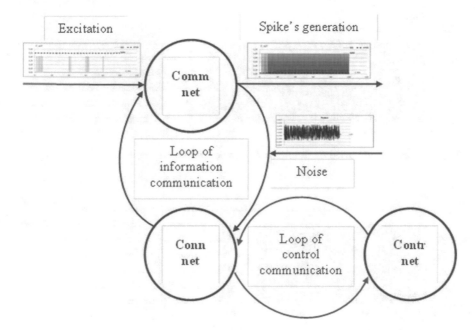

Fig. 6 The experimental scheme of the effect of the controllers network on the functioning of the switches network

On the external information inputs of switches excitation pulses were supplied and to their outputs has been added the additive evenly distributed interference that passed through connectors to internal information inputs of switches.

In the absence of the control loop and under exceeding by exciting pulses, the threshold value switches generated stable spikes on the entire time interval of exciting pulses ($0 \div 100$ ms) regardless of changes of the interference amplitude over a wide range ($0 \div 5$ $\mu A/cm^2$).

After the control loop switch on for one of the links functioning according to regulatory models of neuro- and gliatransmitter exchange [7–9], for some implementations of interference breakdowns generate switch connected to this connection are occurring.

Figures 7 and 8 are examples the effect of different implementations of noise (0.5 $\mu A/cm^2$) at various power of the communication between connectors and controllers.

In a second experiment, the effect of controllers on the connectors in the network with the correct regular topology (complete graph), which is most resistant to variations of relations [4] was studied. As a dynamic model of the switches functioning the model of «Izhikevich» neuron [14, 15] was used. For example, in diagram (Fig. 9a) all weight is the same and equal to 100, and in the diagram (Fig. 9b) for 10% of bonds weights are set equal to 1.

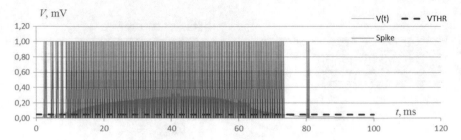

Fig. 7 Responses generated by the switch at a predetermined «rigidity» of the control loop

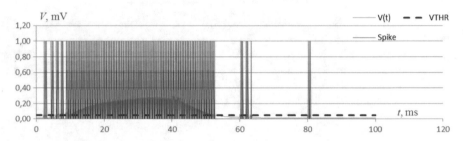

Fig. 8 Responses generated by switches at higher «rigidity» of the control loop

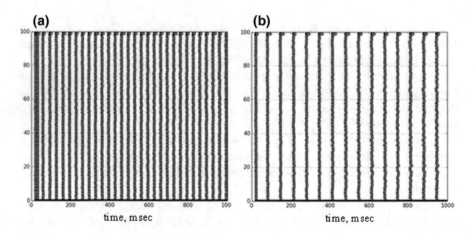

Fig. 9 The activity of neurons at different scales links

Conclusion

Carried out simulations showed that the developed model of the cyber network and object-oriented approach to its software implementation helped to create a very flexible and versatile tool for the solving problems of neurodynamics.

For example, our results of the network simulation coincided with known results of E.M Izhikevich. Also it shows qualitatively the fundamental possibility of the control for the network of commutators by the network of controllers.

The use of GPUs and the CUDA parallel programming technology from NVIDIA shows better performance on large data volumes, compared with the universal processors because of their greater efficiency for the use of computing concepts SIMD (Single Instruction Multiple Data), even without the use of optimization mechanisms.

The investigation of the effect of these mechanisms on the performance of the model under consideration and test it on a modern line of GPU (NVIDIA Pascal) are planned in the course of this work.

In further, the considered model and its advanced software implementation will be used for the experimental study of the stability of the dynamics of complex networks near the critical parameters of their relationship values, handling complex networks.

Acknowledgements Research are carried out with the financial support of the state represented by the Ministry of Education and Science of the Russian Federation. Contract no. 203-15-ЕП.

References

1. Gao, X.-D. et al.: Control efficacy of complex networks. Sci. Rep. **6**, Article number: 28037 (2016)
2. Yang-Yu, L., Jean-Jacques, S., Barabasi, A.-L.: Controllability of complex networks. Nature **473**, 167–173 (2011)
3. Gao, J., Barzel, B., Barabasi, A.-L.: Universal resilience patterns in complex networks. Nature **530**, 307–312 (2016)
4. Pu, C.-L., Pei, W.-J., Michaelsond, A.: Robustness analysis of network controllability. Phys. A **391**, 4420–4425 (2012)
5. Nepusz, T., Vicsek, T.: Controlling edge dynamics in complex networks. Nat. Phys. **8**, 568–573 (2012)
6. Slotine, J.-J., Liu, Y.-Y.: The missing link. Nat. Phys. **9**, 1–2 (2012)
7. Ghanim, U., et al.: Anti-phase calcium oscillations in astrocytes via inositol (1, 4, 5)-trisphosphate regeneration. Cell Calcium **39**, 197–208 (2006)
8. Nazari, S., et al.: A digital implementation of neuron–astrocyte interaction for neuromorphic applications. Neural Netw. **66**, 79–90 (2015)
9. Agulhon, C., et al.: What Is the role of astrocyte calcium in neurophysiology? Neuron. **59**. I. 6, 932–946 (2008)
10. Kalachev, A.A., Krasnov, A.E., Nadezhdin, E.N., Nikolskii, D.N., Repin, D.S.: The hetero-geneous multiply-connected network of active elements. In: In the Collection of

Materials of XIII Inter-national Scientific-Practical Conference «Innovative Information Technologies», pp. 277–280 (2016)

11. Kalachev, A.A., Krasnov, A.E., Nadezhdin, E.N., Nikolskii, D.N., Repin, D.S.: Model of the heterogeneous network for the simulation of neuro dynamic problems. // Modern information technology and IT education. Int. J. **12** (№1), 80–90 (2016)

12. Poznyak, A.S., Sancbez, E.N., Yu W.: Differential neural networks for robust nonlinear control. identification, state estimation and trajectory tracking. Singapore. World Scientific publishing, 422 p (2001)

13. Fu, C.Y.: Closed loop adaptive control of spectrum-producing step using neural networks. US5841651 A (1998)

14. Izhikevich, E.M.: Dynamical systems in neuroscience: the geometry of excitability and bursting. Cambridge. Massachusetts London. England. The MIT Press, 497 p (2007)

15. Izhikevich, E.M.: Simple model of spiking neurons. IEEE Trans. Neural Netw. **14**(6), 1569–1572 (2003)

Autonomous Mobile Robotic System for Coastal Monitoring and Forecasting Marine Natural Disasters

V.V. Belyakov, P.O. Beresnev, D.V. Zeziulin, A.A. Kurkin,
O.E. Kurkina, V.D. Kuzin, V.S. Makarov, P.P. Pronin,
D.Yu. Tyugin and V.I. Filatov

Abstract The paper presents the steps of creating an experimental prototype of an autonomous mobile robot for coastal monitoring and forecasting marine natural disasters. These systems of continuous coastal monitoring are the necessary link in predicting possibilities of developing the resources of the Russian shelf (areas of the Arctic and the Far East). One of the most difficult issues, associated with the creation of the described product, is to ensure the necessary level of mobility in

V.V. Belyakov
Department of Research and Innovation, Laboratory of Modeling of Natural
and Anthropogenic Disasters, Nizhny Novgorod State Technical University
n.a. R.E.Alekseev, Nizhny Novgorod, Russia
e-mail: nauka@nntu.ru

P.O. Beresnev · D.V. Zeziulin · O.E. Kurkina · V.S. Makarov · D.Yu. Tyugin · V.I. Filatov
Laboratory of Modeling of Natural and Anthropogenic Disasters, Nizhny Novgorod State
Technical University n.a. R.E.Alekseev, Nizhny Novgorod, Russia
e-mail: norb132801@gmail.com

D.V. Zeziulin
e-mail: denis.zeziulin@nntu.ru

O.E. Kurkina
e-mail: oksana.kurkina@mail.ru

V.S. Makarov
e-mail: makvl2010@gmail.com

D.Yu. Tyugin
e-mail: dtyugin@gmail.com

V.I. Filatov
e-mail: filatov.v.ngtu@yandex.ru

A.A. Kurkin (✉)
Department of Applied Mathematics, Laboratory of Modeling of Natural and Anthropogenic
Disasters, Nizhny Novgorod State Technical University n.a. R.E.Alekseev, Nizhny
Novgorod, Russia
e-mail: aakurkin@gmail.com

© The Author(s) 2018
K.V. Anisimov et al. (eds.), *Proceedings of the Scientific-Practical Conference
"Research and Development - 2016"*, https://doi.org/10.1007/978-3-319-62870-7_14

inaccessible areas of coastal zones. This problem is solved by development of the chassis of modular design with the possibility to be reequipped with different types of movers (wheeled, tracked, rotary-screw), depending on operating conditions and the physical and mechanical characteristics of the ground surfaces. The presented robotic complex is also equipped with a set of measuring instruments (circular scanning radar, weather station, navigation system, lidars, video cameras), which allows to carry out comprehensive studies of any coastal zone and evaluate the risks and hazards for providing data for engineering simulation of hydraulic systems and structures. The results of experimental investigations of the coastal zone in the south-east of Sakhalin Island, using the developed experimental prototype of the autonomous mobile robot, are given.

Keywords Coastal monitoring · Mobile robotic system · Instrumental data Field observations

Introduction

To carry out coastal zone monitoring, which can be connected with the measurement of the wave climate, ice conditions, the dynamics of spread of pollutants in inaccessible places, is necessary to have reliable means to rapidly take the measurements over a large area. Undoubtedly, the task of evaluating the possible hazards in providing oil and gas production in coastal and offshore fields is extremely important for the Russian Federation.

The use of vehicles and mobile robots with the production of such measurements is very promising [1, p. 2; 2, p. 50; 3,p. 566; 4, p. 89; 5, p. 215; 6, p. 1; 7, p. 7; 8, p. 2]. Radar systems (with some modifications) are applicable as means for carrying out hydrodynamic measurement and assessment of hazards in a coastal zone, determine the velocity and size of drift ice.

The necessity of using radar systems in remote areas requires from the vehicle's structure to be able to adapt to a wide range of operating conditions. A possible solution to this problem is the use of different types of interchangeable movers (wheeled, tracked and rotary-screw) for the expansion of the range of operating conditions.

V.D. Kuzin · P.P. Pronin
Nizhny Novgorod State Technical University n.a. R.E.Alekseev, Nizhny Novgorod, Russia
e-mail: chromium32@mail.ru

P.P. Pronin
e-mail: pavel.pronin2010@yandex.ru

Development of the Autonomous Mobile Robotic System

The project aims to develop a set of scientific and technical solutions in the field of autonomous mobile robotic system (AMRS) for monitoring and forecasting the state of the environment in order to ensure the reliability and safety of hydraulic structures in coastal zones.

In accordance with the intended purpose the following research objectives were formulated:

1. Investigation of physical and mechanical properties of ground surfaces of coastal areas and interaction of different types of movers with terrain, conducting mathematical modeling of the AMRS movement in conditions of coastal zones, selection of the parameters for designing mobile chassis with interchangeable movers (wheeled, tracked, rotary-screw).
2. Development of a list the necessary measurement and research equipment to be installed on AMRS, allowing to monitor coastal zones with the maximum adaptability to the environment.
3. Development of software for the operation of the measuring equipment of the AMRS experimental prototype and its unmanned control system.
4. Development of design documentation, creation of the experimental prototype and conducting experimental tests of the AMRS on the Gulf of Mordvinov (Sea of Okhotsk, Sakhalin Island).

The first step in designing calculations of the AMRS experimental prototype was correct account of special characteristics of ground surfaces of the coastal zones. For this purpose with the support of Special Research Bureau for Automation of Marine Researches (SRB AMR, Sakhalin Region, Yuzhno-Sakhalinsk, Russian Federation) experimental studies of topography and physical and mechanical properties of the coastal areas were conducted by the group of authors. The obtained data were used to develop new statistical models of surfaces of coastal areas to predict the ways of ensuring the efficiency of the mobile robot [9, p. 16; 10, p. 528].

The next step was the implementation of the design calculations and simulation of vehicle-terrain interaction in conditions of the coastal areas. As a result there were selected parameters for designing AMRS's chassis. [11, p. 78; 12, p. 46; 13, p. 940; 14, p. 6].

AMRS uses the navigation equipment of Orient Systems Company. It consists of a high-precision mobile GPS/GLONASS receiver (OC-103), mounted on the chassis, and a base station installed on the ground. The base station (OC-203) transmits the amendments to AMRS's receiver to increase the accuracy of positioning. Navigation equipment is used to obtain the coordinates of the AMRS and bind measured characteristics to a point on the map.

For remote sensing of water surface AMRS uses Omni Directional Radar MRS-1000. The ability to determine the parameters of sea waves by means of the ship's radar is justified by authors in [15, p. 91; 9, p. 13; 16, p. 30].

To monitor the weather conditions of the coastal zones the weather station Vaisala WXT520, which allows to measure the temperature, wind speed and direction, humidity, pressure, precipitation, is used.

For videofixing of waves in addition to data obtained from the radar the AMRS is equipped with a video camera AXIS Q6045-E, which creates a synchronized video stream. This camera is to be also used for AMRS's remote control system.

In order to detect obstacles on the path the AMRS uses laser scanning system, which is a continuously rotating platform with two lidars Sick LMS291Pro.

The board computer Adlink MXE-5400 is set on the AMRS for controlling instrumentation, data collection, data storage and processing. The laptop Panasonic Toughbook CF-31WEUAHM9 is used to remotely connect to AMRS's onboard computer by Wi-Fi, view the data on state of the measuring equipment and send instrumentation commands.

Description of the structure and capabilities of developed software for functioning the measuring equipment of the AMRS experimental prototype and its unmanned control system is presented in [17, p. 6; 10, p. 526].

General views of created experimental prototype are shown in Fig. 1, and the technical characteristics of AMRS's chassis are summarized in Table 1. The modular AMRS's design allows adapting the layout of the chassis depending on the task and modifying its individual units in accordance with the requirements of the end consumer.

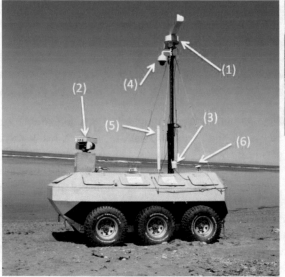

Fig. 1 General view of the AMRS for coastal monitoring: X-band radar (*1*), obstacles detection system (*2*), multi-weather sensor (*3*), high resolution video camera (*4*), long range Wi-Fi antenna (*5*), GNSS antenna (*6*)

Table 1 Technical characteristics of AMRS chassis options

Parameters	Characteristics		
Type of mover	Wheeled	Track	Rotary Screw
Cargo weight, kg	500		
Dimensional (overall) length, mm	3800		
Dimensional (overall) width, mm	2100	2360	30,000
Dimensional (overall) height (in the transport position), mm	3500	3690	3540
Dimensional (overall) height (in the operating position), mm	6400	6590	6440
Parameters of mover	Tire size: 33 × 12, 5–15	Width of the tracks, mm: 400	Diameter of the rotor, mm: 600
Ground clearance, mm	300	490	350
Angle of ascent at full load, deg	30		
Angle of roll (In transport mode), deg	45		
Maximum speed on highway, km/h	45	45	45

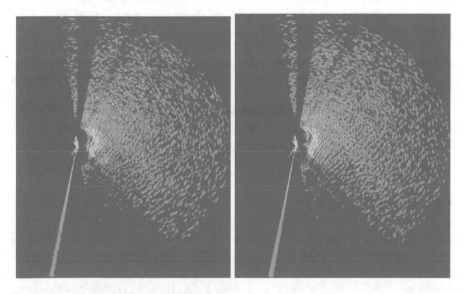

Fig. 2 Screenshots of radar work (Okhotsk Sea)

The AMRS prototype was tested in field conditions in the area of the south-eastern coast of Sakhalin Island (Cape Svobodny) in May–June 2016 with the support of SRB AMR FEB RAS.

Fig. 3 Video from camera during AMRS remote control

During experimental investigations there were conducted measurements of the sea surface and atmosphere: the intensity of the reflection of radio signal from waves of the sea (Fig. 2), air pressure, temperature, relative humidity, wind direction and speed. The results of processing the received dependences between the reflected intensity of the radio signal and distance to the point of the AMRS position are described in [9, p. 14].

When studying the efficiency of the AMRS experimental prototype, the quality of standard maneuvers in the remotely controlled (Fig. 3) and autonomous modes was assessed. The probability of failure and time of equipment operation were also investigated.

Trafficability tests showed that approach of using interchangeable movers allows raising essentially the possibility of the AMRS and significantly expanding the area of its territorial use.

Conclusions

The article describes the main stages of research and approaches to the practical implementation of its results in the creation of the AMRS experimental prototype, used for unmanned coastal monitoring.

Design features of the developed AMRS for specific operating conditions, description of measuring equipment and sensors of unmanned control system have been presented.

In the design, the approach of predicting mobility and determining optimal modes of functioning of the AMRS on preliminarily studied routes of coastal zones has been used.

The results of experimental studies confirm the efficiency of the AMRS for measuring of sea waves, obtaining environmental data, performing typical maneuvers and driving in conditions of coastal zones. The decisive contribution to ensuring the necessary level of AMRS mobility according to terrain characteristics makes the choice of the type of mover.

Thus, the project provides a complete solution to the problem associated with the development of a new kind of AMRSs for coastal monitoring, supplying data for the assessment of hazards and engineering simulation of hydraulic systems and structures.

The results make a significant contribution to the creation methods of unmanned vehicles for monitoring natural objects, emergencies and special operations.

Acknowledgments Research are carried out with the financial support of the state represented by the Ministry of Education and Science of the Russian Federation.Agreement No. 14.574.21.0089 16.Jul 2014Unique project Identifier: RFMEFI57414X0089

References

1. Barber, D.M., Mills, J.P.: Vehicle based waveform laser scanning in a coastal environment. In: The 5th International Symposium on Mobile Mapping Technology, Pradua, Italy, 29–31 May 2007
2. Bio, A., Bastos, L., Granja, H., Pinho, J.L.S., Gonçalves, J.A., Henriques, R., Madeira, S., Magalhães, A., Rodrigues, D.: Methods for coastal monitoring and erosion risk assessment: two Portuguese case studies. J. Integr. Coast. Zone Manag. **15**(1), 47–63 (2015)
3. Didier, D., Bernatchez, P., Boucher-Brossard, G., Lambert, A., Fraser, C., Barnett, R.L., Van-Wierts, S.: Coastal flood assessment based on field debris measurements and wave runup empirical model. J. Mar. Sci. Eng. **3**, 560–590 (2015)
4. Incoul, A., Nuttens, T., De Maeyer, P., Seube, N., Stal, C., Touzé, T., De Wulf, A.: Mobile laser scanning of intertidal zones of beaches using an amphibious vehicle. In: INGEO 2014: 6th International Conference on Engineering Surveying, Prague, Czech Republic, 3–4 April 2014. Slovenská Technická Univerzita v Bratislave. Stavebná Fakulta, pp. 87–92 (2014).
5. Kramer, J., Hunter, G.: Performance of the StreetMapper mobile LiDAR mapping system in "Real World" Projects. Photogrammetric Week '07, pp. 215–225 (2007)
6. Serra, A., Baron, A., Bosch, E., Alamus, A., Kornus, W., Ruiz, A., Talaya, J.: GEOMOBIL: Integration y experiencias de Lidar Terrestre en LB-MMS // Setmana Geomatica. Barcelona. Spain (2005)
7. Ussyshkin V. Mobile laser scanning technology for surveying applications: from data collection to end-products. In: Proceedings of FIG Working Group, Eilat, Israel, 3–8 May 2009
8. Wübbold, F., Hentschel, M., Vousdoukas, M., Wagner, B.: Application of an autonomous robot for the collection of nearshore topographic and hydrodynamic measurements. Coastal Eng. Proc. **1**(33) (2012). doi:10.9753/icce.v33.management.53
9. Kurkin, A.A., Zeziulin, D.V., Makarov, V.S., Zaitsev, A.I., Belyaev, A.M., Beresnev, P.O., Belyakov, V.V., Pelinovsky, E.N., Tyugin, D.Yu. Investigations of coastal areas of the Okhotsk Sea using a ground mobile robot. Ecol. Syst. Devices. **8**, 11–17 (2016)
10. Makarov, V., Kurkin, A., Zeziulin, D., Belyakov, V.: Development of chassis of robotic system for coastal monitoring. In: Proceedings of the 13th European Conference of the International Society for Terrain-Vehicle Systems, Rome, Italy, pp. 524–529 (2015)

11. Kurkin, A., Belyakov, V., Makarov, V., Zeziulin, D., Pelinovsky, E.: Methods of tsunami detection and of post-tsunami surveys. Sci. Tsunami Hazards. **35**(2), 68–83 (2016)
12. Kurkin, A.A., Pelinovsky, E.N., Belyakov, V.V., Makarov, V.S., Zezyulin, D.V.: New trends in tsunami research. Ecol. Syst. Devices. **12**, 40–55 (2014)
13. Kurkin, A., Pelinovsky, E., Tyugin, D., Giniyatullin, A., Kurkina, O., Belyakov, V., Makarov, V., Zeziulin, D., Kuznetsov, K.: Autonomous robotic system for coastal monitoring. In: Proceedings of the 12th International Conference on the Mediterranean Coastal Environment MEDCOAST, V. 2, 933–944 (2015)
14. Zeziulin, D., Beresnev, P., Filatov, V., Makarov, V., Kurkin, A., Belyakov, V.: Development of an unmanned ground vehicle for coastal monitoring. In: Proceedings of the ISTVS 8th Americas Regional Conference, Detroit, MI, 12–14 Sept, p. 75 (2016)
15. Garbatsevith, V.A., Ermoshkin, A.V., Ivanov, I.I., Telegin, V.A.: Use low power marine radar x—band to measure the spatial-temporal characteristics of the ocean wave. Heliogeophys. Res. **13**, 91–96 (2015)
16. Tikhonchuk, E.A., Zaitsev, A.I., Filatov, V.I.: The study of ice drift in Okhotsk sea with radar. Ecol. Syst. Devices. **8**, 29–34 (2016)
17. Belyaev, A.M., Belyakov, V.V., Beresnev, P.O., Kurkin, A.A., Pelinovsky, E.N., Tyugin, D. Yu., Filatov, V.I.: Mobile robotic system for coastal monitoring. Ecol. Syst. Devices **8**, 3–10 (2016)

On Creation of Highly Efficient Micro-Hydraulic Power Plants of Pontoon Modular Design in Conditions of Super-Low Flow Parameters

A.V. Volkov, A.A. Vikhlyantsev, A.A. Druzhinin, A.G. Parygin and A.V. Ryzhenkov

Abstract This paper considers raising efficiency problems and choosing the optimum designed parameters of prospective arrangements for construction of low-head micro-hydraulic power plants for plain relief water bodies, for example, of hydropower plants with a "Kaplan turbine—siphon penstock" power complex studying. The analysis results of the constructive configurations both in floating performance and in stationary placement of micro-hydraulic power plant are presented. This configuration solves the flood accident problems and provides power plants mobility. A criterion for comparison of micro-hydraulic power plant efficiency is developed. It is derived for the optimum solution searching in terms of energy useful utilization by micro-hydraulic power plants hydraulic equipment. The basic approaches are set out for raising efficiency of low-head hydraulic power plants, as well as ways to the optimum selection of parameters of their hydraulic turbines at early stages of design. This is particularly valuable for designers of blade hydraulic machines. The energy efficiency limit of a propeller hydraulic turbine located in a siphon penstock is demonstrated. It is a hydraulic analog (suggested for the first time) of the Betz–Joukowski limit widely used in aerodynamics and design of wind energy plants. New approaches to design flow-power hydraulic units are experimentally supported. Prospects of their application for raising overall efficiency of micro-hydropower plants are substantiated.

A.V. Volkov (✉) · A.A. Vikhlyantsev · A.A. Druzhinin · A.G. Parygin · A.V. Ryzhenkov
NRU "MPEI", Moscow, Russia
e-mail: VolkovAV@mpei.ru

A.A. Vikhlyantsev
e-mail: alexgidro91@mail.ru

A.A. Druzhinin
e-mail: DruzhininAA@mpei.ru

A.G. Parygin
e-mail: parygin_ag@mail.ru

A.V. Ryzhenkov
e-mail: artemrus@inbox.ru

© The Author(s) 2018
K.V. Anisimov et al. (eds.), *Proceedings of the Scientific-Practical Conference*
"Research and Development - 2016", https://doi.org/10.1007/978-3-319-62870-7_15

Keywords Low-pressure micro-hydraulic power plant · Modular design
Floating micro-hydraulic power plant · Small hydropower · Kaplan turbine
Siphon penstock · Hydraulic losses · High rotation frequency · Energy efficiency
Methodology of evaluation · Hydrailic power unit · Standalone power grid
United power grid · Capital costs · Flood safety

Power supply requirements of consumers located on territories remote from central
power supply are of great relevance today. This problem is associated with a range
of acute environmental problems determined by considerable emissions resulting
from traditional power generation facilities operation (thermal, nuclear, diesel
power plants). Therefore, the problems of supplying population with alternative
sources of energy are becoming increasingly important. Among the ways to solving
this problem, the field of hydropower is considered, first and foremost.

The hydropower current state examination and analysis show that today more
than two-third of the hydropower potential from the European Union and the
Russian Federation territories is implemented with large HPPs. The extent of larger
rivers hydropower potential (HP) exploration is not only determined by extensive
hydraulic resources availability, but also the local landscape peculiarities. In particular, the former Soviet republics territories, as well as the eastern and south-
western zones in the Russian Federation are characterized by mountainous regions
and, accordingly, high heads. Regarding the question of HP further exploration, one
should point out that its unexplored part is constituted by smaller water bodies—
plain rivers, various kinds of hydraulic facilities, and retaining/regulating water-
works characterized by low heads and flow low speeds in the channel.

At present, hydraulic units designing which intended for operation in conditions
of exactly such sources, as well as the necessity of developing decentralized power
supply that determined a return to the global trend to development of small
hydropower. It is difficult, and quite probable, impossible to use large hydraulic
units at a small rivers HP exploration; therefore, neither is it possible to apply
well-elaborated methods of design of such hydraulic units hardly suitable for
low-head hydraulic machines.

To solve engineering problems for creating efficient hydraulic machines, it is
necessary:

- to identify the heads and flow rates optimum combinations required for efficient
  operation of low-power energy generation equipment within a micro-hydraulic
  power plant;
- to study the working process proceeding at operation of the same;
- to perform energy characteristics analysis and optimize the geometry of
  hydraulic machines;
- to select perfect design configurations for construction of micro-hydraulic power
  plants.

Today's requirements to the quality of electric energy [1, p. 5] are satisfied the
most conveniently if the hydraulic unit operates at the rotor high rotation frequency,

which is a challenging task in low heads conditions. On the other hand, using low-frequency hydraulic turbine types requires development of efficient low-speed electric generators, or application of multiplicative configurations for their drive. Using multipliers, in its turn, results in a considerable increase in the equipment prime cost comparable to expenses on construction of an entire micro-hydraulic power plant; or (if cheaper multipliers are used) in lower reliability of the micro-hydropower plant and, accordingly, in a considerable rise in operating costs. Therefore, in the view of this paper's authors, application of micro-hydraulic power plants multiplicative arrangements cannot be considered expedient [2, p. 116; 3, p. 10878]. That is why, to solve the problem, we need to develop technological solutions enabling to use a hydraulic machine with a high rotation frequency, or application of low-speed electric generators. In this context, it is proposed to choose the path of developing an efficient hydraulic machine capable to operate with a high rotation frequency of rotor in conditions of low flow heads. Such solution makes it possible to use the simplest, the most affordable, and reliable device—an asynchronous electric machine—as a generator. A review of the types of hydraulic machines and analysis of their operation in conditions of low heads and low flow rates have shown that axial (Kaplan) blade machines make the most suitable type. To reduce the power unit prime cost, to increase its efficiency and to simplify the design as far as possible, a task is considered for creating a hydraulic unit without guide vanes and with the application of a propeller impeller type.

Selection of an energy-efficient design configuration of a micro-hydraulic power plant with Kaplan turbine provides for its being equipped with a device to ensure a sufficiently full conversion of geodetic head into dynamic head, i.e., a considerable increase of the static pressures drop on the hydraulic turbine. A siphon penstock may be used as such device (Fig. 1). Such arrangement enables to locate the

Fig. 1 Design of micro-hydraulic power plant with a siphon penstock

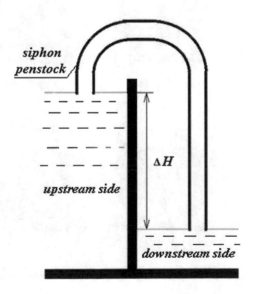

siphon penstock

ΔH

upstream side

downstream side

hydraulic machine in the flow part of the penstock, with its completely immersed into the working fluid. As distinct from open-design configurations, such layout is protected from intense icing of the structure during the cold season. This enables to operate power units all the year round without losing operability and in much broader natural climatic zones. The problems of creating efficient power units in conditions of low heads force to enhance the power of micro-hydraulic power plants through increasing the plant hydraulic turbine characteristic dimensions of, which, however, cannot be considered an efficient solution. As a way out of the situation in place, it is suggested to use the principle of *modularity* of the design, which would enable to use a series of units installed in individual penstocks (the so-called modules) within a single hydraulic power plant. This will increase the power of the micro-hydraulic power plant by making it to have a number of hydraulic units, ensuring prolongation of initial capital investments, gradually "increasing" the power to the required level by staged commissioning of the micro-hydraulic power plant modules, and will make such plants more mobile. In order to analyze the efficiency of various design configurations, estimated studies for choosing the place for installation of a hydraulic turbine in the siphon penstock were performed. Analysis of the working process, energy efficiency. and positive suction head of prospective arrangements, presented in Fig. 2, has shown that the arrangements are very close in terms of the above parameters, provided that the energy efficiency factor, and the efficiency of the turbine (to be considered below)

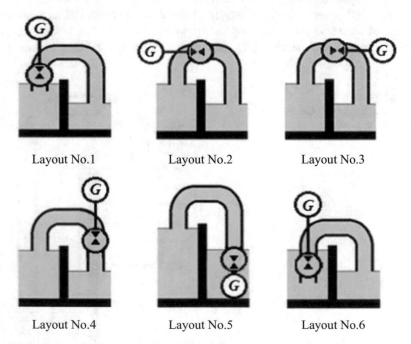

Layout No.1   Layout No.2   Layout No.3

Layout No.4   Layout No.5   Layout No.6

Fig. 2 Arrangement possible options of hydropower unit in the flow part

are identical. However, layout No. 6 has a clear advantage over all the others, first of all, in terms of operation: the hydraulic turbine impeller immersed into the working fluid makes much more convenient process of starting the micro-hydraulic power plant by brief running the hydraulic unit in the pumping mode for filling the flow part of the siphon penstock. This solution totally eliminates the need to use a complex vacuuming system. Moreover, a propeller hydraulic turbine arranged in the siphon penstock flow part enables to proceed from the pumping mode to the turbine mode (the electric energy generation mode) without the direction reversing of the shaft rotation. This simplifies the micro-hydraulic power plant operating and makes it less structurally complex. The advantages of layout No. 6 lead to considerable reduction of costs and the scope of construction work for creating the same and make it the most promising option for practical application. Therefore, further on, the attention will be mostly focused on its theoretical and experimental studies.

The energy efficiency of a hydraulic turbine in a penstock is assessed involving the use of equations for determination of its power, as well as the intensity of the working fluid flow in the penstock [4, p. 16] according to the designed arrangement as shown in Fig. 3.

The intensity of the flow, i.e., the working fluid energy passing through the siphon penstock flow part without hydraulic machine can be presented as follows:

$$N_{s.p.} = \rho g Q_{s.p.} H_{microHPP}, \tag{1}$$

where

$H_{microHPP} = z_{US} - z_{DS}$ is static head, or gross head [5, p. 15] of the flow in Section 1-1,

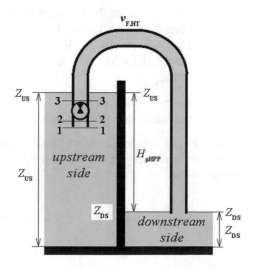

Fig. 3 Designed layout of the "hydraulic turbine—penstock" complex of micro-hydraulic power plant with a siphon penstock

z_{US} and z_{DS}	are elevations of the upstream and downstream levels, respectively,
$Q_{s.p} = F_{s.p}.v_{s.p}.$	is flow rate in the penstock without hydraulic machine,
$F_{s.p.}$	is cross-sectional area of the siphon penstock flow part,
$v_{s.p.}$	is flow rate in the penstock without hydraulic machine.

Below is the equation for power implemented on the hydraulic turbine:

$$N_{ht} = \rho g Q_{ht} H_{theor}^{ht},$$ (2)

where

$H_{theor}^{ht} = H_{disp}^{ht} \eta_{hyd}$	is theoretical head usefully utilized by the hydraulic turbine (Section 3-3),
η_{hyd}	is hydraulic efficiency of hydraulic turbine,
$H_{disp}^{ht} = H_{microHPP} - h_{s.p.}$	is disposable head on the hydraulic turbine (Section 2-2),
$Q_{ht} = Q_{s.p.}$	is flow rate through the hydraulic turbine.

To express the factor in relative units, it is necessary to find the ratio between Eqs. (2) and (1):

$$\frac{N_{ht}}{N_{s.p.}} = \frac{\rho g Q_{ht} H_{theor}^{ht}}{\rho g Q_{s.p.} H_{microHPP}};$$ (3)

With the equation reduced, we have the parameter K_N:

$$K_N = \left(\frac{v_{f.ht}}{v_{s.p.}}\right)\left(\frac{H_{theor}^{ht}}{H_{microHPP}}\right);$$ (4)

where $v_{f.ht}$ is the rate of the flow passing through the hydraulic machine.

The hydraulic turbine located in the penstock creates a considerable hydraulic friction in the same, which results in additional flow energy losses. Therefore, it is necessary to determine losses. This can be done by one of two ways, expressing the losses through:

1. hydraulic loss factor ζ_{ht} or
2. additional head losses h_{ht}^{II} through hydraulic efficiency of the hydraulic turbine,

where it is convenient to use mathematical expressions of these methods as related to the working fluid weight; in this case, these expressions will be as follows (respectively):

$$h_{ht}^{I} = \zeta_{ht} \frac{v_{f.ht}^2}{2};$$ (5)

$$h_{ht}^{II} = g H_{theor}^{ht}(1 - \eta_{hyd}).$$ (6)

As applied to the "hydraulic turbine—penstock" arrangement under examination, the flow energy losses in the hydraulic turbine must be considered at calculation of the flow velocity in the siphon penstock accommodating a hydraulic turbine. With losses in the hydraulic turbine expressed as head losses (according to p. 2), the equation for the flow velocity will be as follows:

$$v_{f.ht} = \varphi \sqrt{2g \left(H_{ht} - H_{theor}^{ht} \left(1 - \eta_{hyd} \right) \right)};$$ (7)

With consideration of $H_{ht} = H_{microHPP} - H_{theor}^{ht}$ and with a *head efficiency factor* introduced as the ratio of theoretical head to working head

$$K_H = \frac{H_{theor}^{ht}}{H_{microHPP}},$$ (8)

the transformed equation will be expressed as Eq. (9):

$$v_{f.ht} = \varphi \sqrt{2g H_{microHPP} \left(1 - K_H \left(2 - \eta_{hyd} \right) \right)};$$ (9)

Now, the ratio of equations for the velocity of flow through the hydraulic turbine (9) to the rate of flow in the water pipeline without the hydraulic machine would give a parameter of the ratio between the flow rates (10):

$$\frac{v_{f.ht}}{v_{s.p.}} = \frac{\varphi \sqrt{2g H_{microHPP} \left(1 - K_H \left(2 - \eta_{hyd} \right) \right)}}{\varphi \sqrt{2g H_{microHPP}}};$$

$$\frac{v_{f.ht}}{v_{s.p.}} = \sqrt{1 - K_H \left(2 - \eta_{hyd} \right)}.$$ (10)

Here Eq. (4) expressed with consideration of (10) gives *energy efficiency factor* K_N and is presented as follows (11):

$$K_N = K_H \sqrt{1 - K_H \left(2 - \eta_{hyd} \right)}.$$ (11)

Similar considerations with *disposable head* (instead of theoretic head) enable to determine the value of the maximum energy efficiency factor unambiguously. Then the expression for K_N will be presented as

$$K_N = K_H \sqrt{1 - K_H},$$ (12)

here, unlike the Eq. (8) K_N is the ratio of the total head H_t to the static head $H_{microHPP}$.

The parameter obtained, expressed in relative units, shows *the share of usefully utilized energy of the flow by a hydraulic turbine located in a penstock*. With this criterion, it is possible to undertake a quantitative comparison of energy efficiency of different micro-hydraulic power plant arrangements comprising a "hydraulic turbine—penstock" complex. Besides, to ensure operation of the micro-hydropower plant in favorable conditions, in terms of energy efficiency (i.e., with the optimum combinations of flow rate, head, and rotation frequency), it is necessary to address the problem of searching for the optimum value of head utilized by the hydraulic turbine usefully. To find its solution, it is important to analyze the behavior of the energy efficiency factor, which can be conveniently presented graphically. Power function K_N in (11) depends on two variables—head efficiency factor K_H and hydraulic efficiency of hydraulic machine η_{hyd}.

Therefore, it would make sense to analyze the behavior of K_N for a series of values K_H in the range $0 \div 1$ for a specific value of η_{hyd}. Below are presented the results of calculation of function $K_N = f(K_H, \eta_{hyd})$ with variation of the parameter K_H within the aforesaid range with an increment 0.1 for a series of hydraulic machines with different hydraulic efficiency in the range $0.5 \div 1$ with the same increment. The graphic result of computations is presented in Fig. 4. With the graphic relationship plotted, it can be seen that energy efficiency factor K_N has a maximum—the optimum value of head efficiency factor K_{Hopt}, which suggests that there exists an optimum value of the ratio $\frac{H_{theor}^{ht}}{H_{microHPP}}$ (according to Eq. (8)), as regards energy efficiency for a particular hydraulic turbine. Parameter K_N, according to (12), is the hydraulic efficiency function only $K_H = f(\eta_{hyd})$ and, as can be seen from the graphic relationship (Fig. 4), it demonstrates the maximum energy efficiency.

Examination of function (11) for extremum enables to determine uniquely the optimum ratio of heads for a perfect hydraulic machine with hydraulic efficiency $\eta_{hyd} = 1$ located in a siphon penstock. It is worth noting that for such hydraulic turbine the value $K_N^{max} = 2/3 \cdot \sqrt{(1 - 2/3)} \approx 0.3849$ is limiting and is similar to the Betz–Joukowski limit used in wind energy engineering [6, p. 15]. This

Fig. 4 Graphical presentation of function $K_N = f(K_H, \eta_\Gamma)$ for a series of η_{hyd}

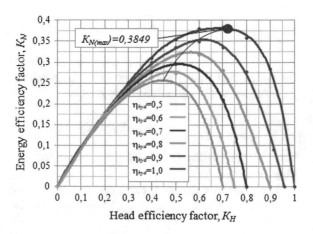

conclusion demonstrates that micro-hydraulic power plant with the "hydraulic turbine—penstock" complex achieves energy efficiency maximum and it occurs if the turbine is designed for the total head is equal to two-third of the disposable head $(H_{opt} = 2/3 H_{microHPP})$.

Experimental studies of the microHPP design configuration with a siphon penstock were performed at NRU "MPEI" involving use a micro-hydropower plant mockup model mounted on an experimental bench (Fig. 5). Power engineering tests enabled to obtain a relation between the maximum power N_{ht}^{max} of hydraulic turbine and the shaft rotation frequency n (Fig. 6).

As can be seen from the curve, the hydraulic unit power would reach its maximum at rotation frequency $n = 1000$ rpm, which is the designed value.

As experimental researches were carried out at different disposable pressure drops on the hydraulic turbine, the results must be reduced to the head design value, i.e., to determine the reduced power N_{11} [7, p. 472]:

Fig. 5 Micro-hydraulic power plant mockup model on NRU "MPEI" experimental bench

Fig. 6 Experimental relations $N_{max} = f(n)$ and $H_{HT} = f(n)$ at N_{max}

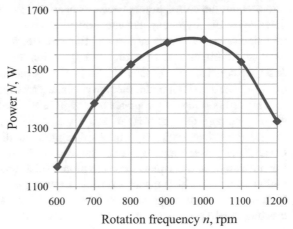

Fig. 7 Experimental relation
$N_{11} = f(n)$

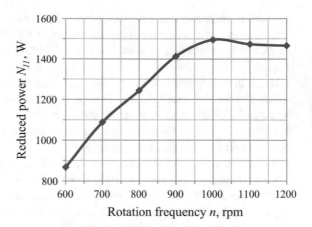

$$N_{11} = N_{max} \cdot \frac{H_{opt}}{H}.\tag{13}$$

For realistic compliance assessment of obtained hydraulic unit experimental characteristics with designed characteristics, according to Eq. (13), the relation $N_{11} = f(n)$ (Fig. 7) was made for the maximum values of measured power. As can be seen from the curve, N_{11}^{max} is obtained at $n = 1000$ rpm and is equal to 1500 W, which is precisely equal to designed characteristics of the hydraulic unit.

Conclusion

The studies presented in the paper enabled to establish analytic relations of the energy useful utilization criterion, which does not only make possible to assess energy efficiency of the micro-hydraulic power plant hydraulic part, but also to take into account the correlation between its head and the working fluid flow rate, which determines the hydraulic turbine power ensuring its designed parameters optimization at *early stages of design*. Elaboration, analysis, and relevant estimates enabled to identify a structural configuration of micro-hydraulic power plant ensuring considerable costs reduction in construction of new and restoration of out-of-operation small hydraulic power plants, mostly, with a siphon penstock used.

The outlined *new approaches to design* small hydropower facilities make possible to engage the potential of small water bodies, which is currently virtually unused (less than 1%), while in some regions small hydropower could make the basis of the power supply system. The results of the research present efficient tools for micro-hydropower plants construction, which are characterized by small inundation of lands, which means exerting a lower load on the ecosystem. Moreover, such power plants require lower capital and operation expenses, contribute to creating conditions encouraging construction and restoration of small hydropower facilities in the Russian Federation and Europe, including introduction of obligations on top-priority connection of mini-hydraulic power plants to decentralized power supply grids, and simplification of obtaining permissions for small hydropower facilities construction [8, p. 4].

As a result of the studies indicate, a modular low-head micro-hydraulic power plant is a power supply efficient source of consumers located in decentralized areas and based on renewable sources of energy.

Acknowledgments Research conducted with the financial support from the state represented by the Ministry of Education and Science of the Russian Federation. Agreement (contract) No. 14.574.21.0076. Unique identification number: RFMEFI57414X0076.

References

1. GOST 32144-2013 Electric energy. Electromagnetic compatibility of technical equipment. Power quality limits in the public power supply systems
2. Parygin, A.G., Volkov, A.V., Ryzhenkov, A.V.: Commentary on the Efficiency of Selected Structural Designs of Low Head Micro Hydraulic Power Plants. Modern Applied Science **9**(4), (2015). doi:10.5539/mas.v9n4p116. URL: http://dx.doi.org/10.5539/mas.v9n4p116
3. Parygin, A.G., Volkov, A.V., Ryzhenkov, A.V., Druzhinin, A.A., Naumov, A.V.: Optimization algorithm of parameters of the low head micro hydraulic power plant at an early design stage. International Journal of Applied Engineering Research **11**(22), 10878-10886 (2016). ISSN 0973-4562. Research India Publications. URL: http://www.ripublication.com
4. Smirnov, I.N.: Hydraulic turbines and pumps. Manual for power engineering and polytechnic higher schools. —M.: Vysshaya Shkola, 400 p (1969)
5. Krivchenko, G.I.: Hydraulic machines: Turbines and pumps. Manual for higher education institutions. —M.: Energiya, 320 p (1978)
6. Okulov, V.L. et al.: Development of optimum rotor theories. —M.: Izhevsk, 120 p (2013)
7. Nikolayenko, Yu.I.: Development of economical hydropower units for low-pressure mini-hydropower plants. Yu.I. Nikolayenko, A.V. Tarasov, G.I. Topazh. News of Samara Research Center of the Russian Academy of Sciences, **12** No.1(2), 472–475 (2010)
8. Dobrokhotov, V.I., Kargiyev, V.M. et al.: Renewable Energy bulletin. —M.: "Intersolarcenter", 21 p May 2005

Development of Scientific and Technical Solutions to Create Hybrid Power Source Based on Solid Oxide Fuel Cells and Power Storage System for Responsible Consumers

A.I. Chivenkov, E.V. Kryukov, A.B. Loskutov and E.N. Sosnina

Abstract This article deals with hybrid power source (HPS) based on solid oxide fuel cells (SOFC) developed by the authors. HPS includes generating, power storage, integration, and active-adaptive control systems. Generating system includes modular electrochemical generator based on SOFC and reformer. Power storage system consists of capacitive storage and accumulation batteries. Accumulation batteries are created of alkaline nickel–cadmium batteries with improved energy characteristics. The base of integration system is current distribution converter. For the active-adaptive control system realization, the algorithm of HPS functioning has been designed. The conducted research of HPS experimental prototype characteristics allowed to confirm the efficiency of scientific and technical solutions.

Keywords Smart hybrid power source · Solid oxide fuel cell · Capacitive storage Nickel–Cadmium battery · Actively adaptive control system

A.I. Chivenkov (✉) · E.V. Kryukov · E.N. Sosnina
Nizhny Novgorod State Technical University n.a. R.E. Alekseev,
Nizhny Novgorod, Russia
e-mail: chyvenkov@mail.ru

E.V. Kryukov
e-mail: e.kryukov@rambler.ru

E.N. Sosnina
e-mail: sosnina@nntu.nnov.ru

A.B. Loskutov
Head of Department, Nizhny Novgorod State Technical University
n.a. R.E. Alekseev, Nizhny Novgorod, Russia
e-mail: loskutov@nntu.nnov.ru

© The Author(s) 2018
K.V. Anisimov et al. (eds.), *Proceedings of the Scientific-Practical Conference "Research and Development - 2016"*, https://doi.org/10.1007/978-3-319-62870-7_16

Introduction

A great attention of different international scientific groups is paid to the development of power installations (PI) based on fuel cells (FC) and their application in power supply systems [1, 2]. Japanese, American, and German scientists have achieved significant success in this research. Thus, German company Siemens Westinghouse developed and tested PI with power from 5 to 300 kW [3]. Company General Electric tested compact electrochemical generator with power from 1 to 6 kW [4]. Company Cummins-SOFCo developed PI based on FC pilot prototype with power 1 kW, which had continuously worked for 2000 h using natural gas [5].

The operation of the FC is based on the principle of direct conversion of chemical into electrical energy [6]. The main advantages of FC installations are high efficiency that can reach 85% considering heat recovery, and environmental friendliness. Due to the lack of direct fuel chemical contact with the oxidant, the amount of harmful emissions is almost 100 times less when compared with conventional power installations [7].

However, widespread application of FC in power supply systems is constrained by a number of problems [8]. There is high cost of PI based on FC due to the complexity of its manufacturing technology and the high cost of the materials and high temperature of FC operation, whereby it takes a long time to start the PI and to achieve the optimum operating condition. Low FC maneuverability makes use of this PI inefficient at the irregular daily schedule of electric group load of consumers.

The problem of improving FC efficiency can be resolved by applying them with other energy sources. One of solutions to this problem is to create a hybrid power source (HPS) on the basis of fuel cells and power storage system. The experimental prototype of such a source is developed at the Nizhny Novgorod State Technical University n.a. R.E. Alekseev (NNSTU) [9]. Solid oxide fuel cells (SOFC) are used as the main energy source; accumulation batteries (AB) are used as an additional energy source and capacitive storage (CS) is applied to cover peak loads.

This approach is used by other researchers [10]. The novelty of the suggested solution is that developed HPS allows to realize active-adaptive PI modes changing with the constant generating power of electrochemical generator based on SOFC and dynamic changing of load power which had not been considered before. Developed PI has improved maneuverability, reduced fuel consumption, and increased service life of SOFC.

This article contains scientific and technical solutions to develop of HPS based on SOFC and HPS prototype research results.

Scientific and Technical Solutions to Hybrid Power Source Development

Developed HPS includes generating system, power storage system, integration system and an active- adaptive control system [11].

Generating system with electrochemical generator based on SOFC provides a continuous operation of HPS in basic part of schedule of consumers electrical load. Electrochemical generator has modular execution. High temperature parts are combined in one block: SOFC stack, reformer, afterburning-heating system. SOFC stack consists of planar elements. Reformer is a matrix convertor with high efficiency of hydrocarbon conversion into synthesis gas.

Power storage system consists of capacitive storage and accumulation batteries (AB). If the load decreases lower specified value the electrical energy accumulation from the network happens in AB. In case of load excess of given power, AB gives stored electrical energy into the network. Capacitive storage is used for peak load supply.

The realization of AB is done based on nickel–cadmium accumulators with sintered-type electrodes (sintered-type NCA). This type of accumulators is produced by the German company « Hoppecke » and used by high-speed trains in Russia [12]. However, producing such an accumulators complicated technological operations are applied that is why AB are very expensive. The technology of sintered-type NCA producing is developed by NNSTU scientists which let us to reduce electrodes cost saving high specific electrical characteristics and make the manufacturing more ecological friendly. The specific cost of such an accumulators is 1.5–2 times lower than analogs with the same energetic characteristics able to work in wide range of charge and discharge current densities and temperatures.

The producing technology of positive electrode does not contain chemical metallization, the negative electrode is made by pressing and does not contain metalized foundation. Such technologies are not used nowadays.

Integration system provides generating and power storage systems connection and HPS currents redistribution. The HPS integration system basis is current distribution converter (CDC) between its elements [13]. Schematic diagram of the CDC, developed by the authors, is shown in Fig. 1.

CDC includes three DC/DC upconverters. DC/DC converter 1 converts the voltage of 50 V generated by SOFC to 400 V, that is rated for the load and the capacitive storage, DC/DC converter 2 to voltage of 100 V, that is, operating

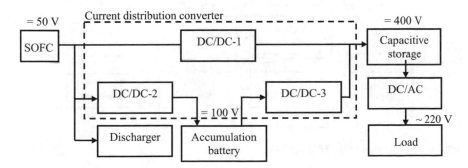

Fig. 1 Schematic diagram of current distribution converter

voltage for the accumulation batteries. DC/DC converter 3 is used for AB and capacitive storage voltage matching.

Active-adaptive control system realizes monitoring of HPS and load electrical parameters and also regulating and control of current distribution between the HPS elements and load depending on consumers power. Control system works according to algorithm which is shown in Fig. 2.

HPS experimental prototype has created for verification of developed solutions efficiency.

The basic criterion of the algorithm is to maintain the voltage on the capacitors in a predetermined range (minimum—350 V, maximum—450 V). The minimum level of voltage matches the ability to ensure the efficiency of the DC/AC inverter—stabilization of the output AC voltage of 220 V, 50 Hz, and the implementation of consumers power supply in the power range from 0 to nominal value.

Under the condition of constant power generation and load, the regulation of the current AB values and capacitors is carried out at the expense of the three structural elements: DC/DC converter 1, DC/DC converter 3 and AB state diagnostic unit.

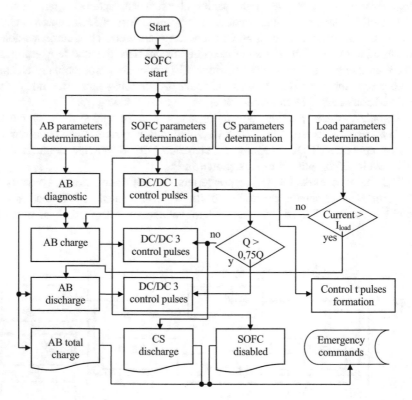

Fig. 2 An algorithm of the HPS operation

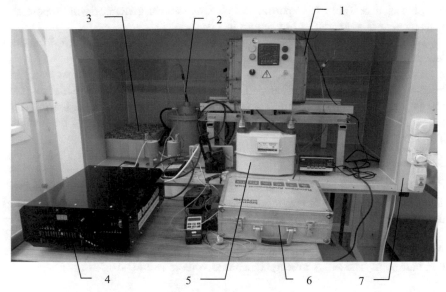

1 – thermobox with SOFC stack; 2 – matrix converter; 3 – NCA battery; 4 – current distribution converter; 5 – gas flow meter; 6 – gas analyzer; 7 – fume hood

Fig. 3 HPS experimental prototype

Hybrid Power Source Experimental Prototype

Experimental prototype includes four modules: generation system module, power storage system module, current distribution converter, and active-adaptive control system. HPS experimental prototype photo is shown in Fig. 3.

Maximum HPS experimental prototype output power is 1/5 kVA, output voltage —220 V, 50 Hz.

Following tests have conducted on experimental prototype: electrochemical generator electrical characteristics tests; NCA battery charge and discharge characteristics test; current distribution converter electrical characteristics tests.

Results and Discussion

1. *Generating system research*

Electrochemical generator electrical characteristics research has been conducted with hydrogen consumption 4 l/min and air consumption 15 l/min. Research results are presented in Table 1. Experimental electrochemical generator volt-ampere and watt-ampere characteristics are shown in Fig. 4.

The results showed that there is significant drop of voltage with the working time less than 10000 h.

Assessment of emergency developing possibility while the PI is working has been conducted during the research. Theoretical analysis and experimental research results showed that HPS start and stop are the most dangerous modes. During start, heating speed exceeding may cause uneven heating and destruction of the element. The similar situation is possible when cooling.

2. *Power storage system research*

NCA charge and discharge characteristics experimental research has conducted. The time dependencies of current and voltage were obtained at different charge and discharge modes. The results are shown in Figs. 5, 6, 7, and 8.

The results analysis showed that the discharge time is about 6 h with current 2.5 A (Fig. 5). Battery voltage is stable for about 5.5 h, there is its rapid decline in the last half hour. The NCA average discharge voltage is 14.0–14.5 V.

Table 1 Electrochemical generator research results with hydrogen consumption 4 l/min and air consumption 15 l/min

Output power, W	Output voltage, V	Output current, A	Temperature, °C
86	26	3,31	750
185	24,5	7,55	748
340	20,2	16,83	749
514	16	32,13	748

Fig. 4 Experimental electrochemical generator volt-ampere and watt-ampere characteristics 1, 2— characteristics after series of experiments, 3, 4—original characteristics

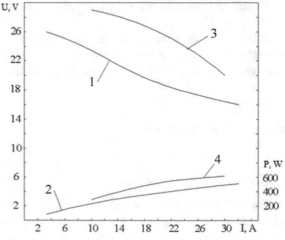

1, 2 – characteristics after series of experiments, 3, 4 – original characteristics

Fig. 5 NCA discharge
voltage changing with
discharge current 2.5 A

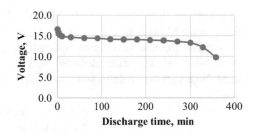

Fig. 6 NCA discharge
voltage changing with
discharge current 20.0 A

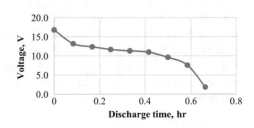

Fig. 7 NCA charge voltage
changing with charge current
2.5 A

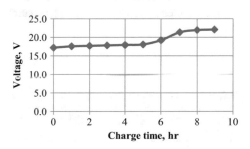

Fig. 8 NCA charge current
changing with charge voltage
19.5 V

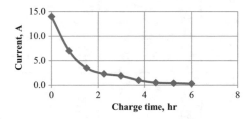

Increase of the discharge current reduces the discharge time to eight times. The dynamics of the operating voltage during AB discharge with discharge current 20.0 A is shown in Fig. 6.

To provide NCA high discharge characteristics, its charge can be carried out in two ways: either by constant current or by constant voltage.

The charge at a 2.5 A constant current was carried out to 120% of rated capacity. AB charging curve can be divided into two sections (Fig. 7). Initially, main reactions are the formation of higher oxides of nickel in the positive electrode and the metal cadmium in the negative when the AB voltage is 16.5-18.0 V. Further significant increase of voltage is due to the occurrence of oxygen liberation side reactions of nickel oxide electrode and hydrogen on the cadmium electrode. The final discharge AB voltage is up to 22.0 V at chosen charge mode. Charge time is 8–10 h.

The charge of the battery can be accelerated by applying a constant voltage mode. The charge carried out with voltage 19.5 V.

At the initial period of the charge, current can reach values of 13.0–15.0 A (Fig. 8). Then there is its rapid reduction. After 1.5 h there is some stabilization of the current, but after 3 h the amount of current is reduced again and it is stabilized at the level of 0.2–0.3 A. Charge time is 4–5 h.

Results analysis showed that DC source output power should be not less than 50.0 W when charging by a constant current. Battery charge by constant voltage requires the use of semiconductor devices with an output power of up to 300.0–500.0 W.

3. *Integration system research*

DC/DC-1, DC/DC-2 и DC/DC-3 converters as part of HPS current distribution converter electrical characteristics research was conducted. Results are shown in Table 2.

Table 2 Converters electrical characteristics research results

№	Input voltage, V	Input current, A	Output voltage, V	Active load, W
DC/DC-1				
1	50	4,9	406	120
2	80	3,8	402	
3	50	19,4	394	860
4	80	14	394	
5	50	34,4	390	1680
6	80	22	394	
DC/DC-2				
1	50	2,4	100	60
2	80	2,2	100,5	
3	50	6,2	99	220
4	80	4,9	99	
5	50	2,4	105	–
6	80	2,4	106	
DC/DC-3				
1	100	3,5	402	120
2		10,6	397	860
3		18,5	397	1680

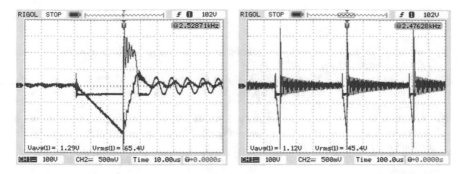

Fig. 9 DC/DC Converter 1power transistor voltage (*blue* line) and the resistive current sensor voltage (*red* line). 50 V input voltage and 120 W load power

As an example, Fig. 9 shows the waveforms of voltages at the power transistor of the DC/DC Converter 1 and the resistive current sensor with the input voltage of 50 V and load power 120 W.

Result's analyses showed that when input voltage changes the transistor open state time and the transistor closed state time also changes (converter frequency). The transistor open state time for DC/DC-1, DC/DC-3 at 50 V is 28 µs, closed state —400 µs, at 80 V—18 and 430 µs, at 100 V—14 and 480 µs.

When load changes the transistor closed state time changes (converter frequency), transistor open state time is almost unchanged.

Conclusions

Conducted HPS experimental prototype research allowed to prove the efficiency and effectiveness of developed scientific and technical solutions.

The accepted solutions allow to stabilize the SOFC generated power under varying schedule of the consumer electric load. This mode increases the duration of the SOFC operation at maximum utilization rate (energy consumption per generated power unit).

Developed HPS may be implemented as a mobile version, and stationary. Mobile version is designed to cover the deficit of electric energy in areas with rapidly developing infrastructure and load capacity exceeding the capacity of existing power networks. Stationary HPS based on SOFC and storage system would provide reliable and qualitative power supply for remote consumers of electrical energy.Research is carried out with the financial support of the state represented by the Ministry of Education and Science of the Russian Federation. Agreement no. 14.577.21.0073 05.Jun 2014. Unique project Identifier: RFMEFI57714X0073.

References

1. Korovin N.V.: Fuel cells and electrochemical power plants. Power Engineering Institute publishing house, Moscow (2005). 280 p
2. Laosiripojanaa, N., Wiyaratnb, W., Kiatkittipongc, W., Arpornwichanopd, A., Soottitantawatd, A., Assabumrungratd, S.: Reviews on solid oxide fuel cell technology. Eng J **13**(1), 65–83 (2009)
3. SOFC tests and demonstrations: summary. Siemens AG. [Online]. URL. http://www.power generation.siemens.com/products-solutions-services/products-package/fuel-cells/demonstrations/demonstrations-summary
4. GE Fuel Cells. [Online]. URL. http://www.hydrogen.energy.gov/pdfs/htac_nov14_5_wellington.pdf
5. Cummins Power Generation. 10kWe SOFC Power System Commercialization Program. [Online]. URL. https://www.netl.doe.gov/File%20Library/Events/2003/seca/Daniel-Norrick.pdf
6. Somov S.I. Status of researches, developments and practical applications of solid oxide fuel cells. Proc. Solid oxide fuel cells and power installations based on them Conf (2010)
7. Kiselev I.V.: Energy efficiency improving of solid oxide fuel cells and study their application to low power consumers energy supply. Russian federal nuclear center all-russian research institute of experimental physics (2013)
8. Jain, S., Jiang, J., Huang, X., Stevandic, S.: Modeling of fuel-cell-based power supply system for grid interface. IEEE Trans Ind Appl **48**(4), 1142–1153 (2012)
9. Loskutov A.B., Sosnina E.N. (and etc): The development of scientific and technical solutions to build hybrid power source based on SOFC and storage system for responsible consumers. Nizhny Novgorod State Technical University, Nizhny Novgorod, Research Rep. GR 114101670042. June 2014
10. Izadian A., Famouri P.: Low cost high efficiency converter for solid oxide fuel cell residential application. Proc. American Science and Technology Conference (2007)
11. Loskutov A.B., Sosnina E.N., Chivenkov A.I., Kryukov E.V.: The development of hybrid power source based on SOFC for distant electricity consumers' power supply. Proc. 2015 IEEE Innovative Smart Grid Technologies - Asia (ISGT ASIA) Conf, pp. 1–6 (2015)
12. Hoppecke batterien [Online]. URL. http://www.hoppecke.com/
13. Chivenkov A.I., Grebenshchikov V.I., Antropov A.P., Mikhailichenko E.A.: Enhanced features of the voltage inverter of renewable energy sources and industrial network systems. Engineering Herald of Don. 24(1) (2013)

Automated Control Unit of Power Flow in Intellectual Electricity Distribution Network

M.G. Astashev, D.I. Panfilov, P.A. Rashitov, A.N. Rozhkov and D.A. Seregin

Abstract Results of the development of scientific and technical decisions on creation automated control unit power flow in the intellectual electricity distribution network with microprocessor control system based on modern technologies of design of power electronics devices and digital control systems are presented. Goals and objectives of the study are indicated. Relevance, novelty and practical significance of the work are shown. Properties and opportunities created in the framework of the project equipment are listed.

Keywords Power system · Transport capacity flows · Operating modes
Transformers · Models · Control system

The concept of smart grid with active-adaptive network requires implementation in power grid wide class of control devices for managing power traffic.

Automated control unit power flow (ACUPF) designed to manage energy flow in an active-adaptive electric smart grid networks. There are more than 10 places in Unified National Electric Grid of Russia where implementation of ACUPF is cost effective.

As a basis for ACUPF building, the project proposed to use the phase-shift device (PSD) with a semiconductor switch. High-speed reliable PSD with a

M.G. Astashev · D.I. Panfilov (✉) · A.N. Rozhkov
K. A. Krug of JSC "ENIN", Moscow, Russia
e-mail: Panfilov@eninnet.ru

M.G. Astashev
e-mail: Astashev@eninnet.ru

A.N. Rozhkov
e-mail: RozhkovAN@eninnet.ru

P.A. Rashitov · D.A. Seregin
Department of Industrial Electronics, National Research University "Moscow Power Engineering Institute", Moskva, Russia
e-mail: Rashitov333@mail.ru

D.A. Seregin
e-mail: SereginDA@mpei.ru

© The Author(s) 2018
K.V. Anisimov et al. (eds.), *Proceedings of the Scientific-Practical Conference "Research and Development - 2016"*, https://doi.org/10.1007/978-3-319-62870-7_17

semiconductor switch is one of the most technically and economically effective tools for power management in active-adaptive electric networks. PSD with semiconductor switch is not evaluated and implemented so far. The widespread prototypes of PSD are phase-shift transformers with mechanical control devices under the load, have performance with hundreds of times less than the proposed solution in the project. Proposed PSD solution characterized by the original topology and control algorithms of its semiconductor converter, which provide enhanced ACUPF reliability and functionality. The proposed PSD solution, as opposed to the alternative, ensures equality voltage modules at its inputs and outputs, greatly expands the possible range of controlled phase-shift angles and improves its dynamic characteristics.

The aim of the project is ACUPF development with microprocessor control system based on of power electronics devices and digital control systems. ACUPF designed for intellectualization of electric power transmission and distribution process in active-adaptive electric networks and provide high levels of reliability and efficiency of energy transportation and control via power transmission lines.

Major tasks solved in the framework of the project:

- Review of scientific literature and patent search;
- Topology justification of ACUPF;
- Simulation models development of power line with ACUPF location, ACUPF and control system development;
- ACUPF control algorithms development;
- Physical model for the study and debugging ACUPF design;
- Physical model and experimental sample of ACUPF design and manufacturing;
- Physical model and experimental sample of ACUPF evaluation.

The project is the next step for practical realization of active-adaptive power grid. The proposed solution allows to create a world-class technology for the development and implementation of high-speed devices for energy management in power grid with power semiconductor devices and digital control systems.

One of the features of the project is to use Russian components base of power electrical engineering, electronics, and software, which, in turn, contributes to the technological independence of the country in the field of electricity.

In the framework of the project, ACUPF experimental sample was created in conjunction with the physical model of a three-phase power line of 10 kV with a variable phase-shift at the ends of the line. They are unique and can be used for design and evaluation of different ACUPF structures.

The main project results are:

- New methods of ACUPF control;
- Simulation models of ACUPF and power line;
- Proposed approach for simulation and evaluation of ACUPF performance in its specific location in power grid;
- A physical model including experimental sample of ACUPF and power system model (Fig. 1). The model is intended for research and testing of received

(a)

(b)

Fig. 1 a Scheme of the physical model and the experimental sample of ACUPF; **b** Power transformers of physical model and ACUPF; **c** The phase of the semiconductor switch of ACUPF; **d** Compartment of ACUPF control system

(c)   **(d)**

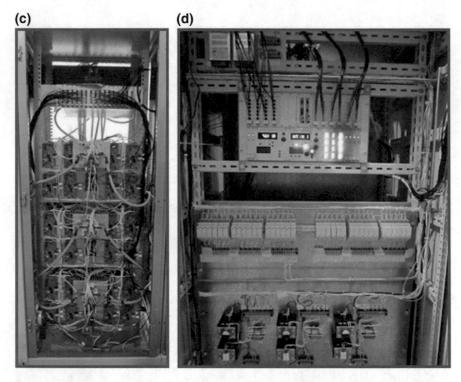

Fig. 1 (continued)

scientific and technical solutions for the intellectualization of electric power transmission and distribution process in the intelligent electrical networks and allows debugging of ACUPF work in various modes;

- Automatic mode control block structure of ACUPF control system that allows adapting to the parameters of the power system operation mode;
- Methods and control algorithms of semiconductor converter as part of ACUPF and algorithms functioning of ACUPF mode automatic control unit;
- Software and hardware of ACUPF digital control system;
- Experimental confirmation of the basic theoretical positions developed in the project.

Participants of the project: one Doctor of Engineering Sciences, four Candidate of Engineering Sciences, as well as the Research Officers and engineers of JSC "ENIN", National Research University "Moscow Power Engineering Institute" and LLC "Togliatti Transformer."

The results of the research: defended one PhD thesis, one dissertation presented to the defense in December 2016, received two patents [1, 2], published eight scientific articles in Russia and abroad [3–10].

Conclusions

1. The effectiveness of automated control unit power flow (ACUPF) with semiconductor switch for power grid modes control had been demonstrated.
2. New algorithms, hardware, and software of ACUPF control system had been developed.
3. ACUPF simulation models had been developed and evaluated.
4. Physical models of power system and ACUPF have been manufactured.
5. Experiments obtained on physical model of power system and ACUPF confirmed the main projects proposals: methods of semiconductor switch control, models of power line, PSD, control system, algorithms, software, and hardware of control system.
6. The guidelines for ACUPF design for its specific location in power grid had been designed.Research are carried out) with the financial support of the Ministry of Education and Science of the Russian Federation. Agreement (contract) no. 14.579.21.0045 26 Aug 2014. Unique project Identifier: RFMEFI57914X0045

References

1. Panfilov D.I., Novikov A.A., Astashev M.G., Novikov M.A · The secondary power source. Patent RU №2601419 Issued 01.15.2015 г. (In Russian)
2. Remizevich T.V., Rashitov P.A., Panfilov D.I., Astashev M.G., Novikov M.A., Fedorova M. I., Verbitskaya A.A.: The control system phased switching windings of the shunt transformer of phase shift device. Patent RU №154310 Issued 23.07.2015 г. (In Russian)
3. Astashev M.G., Novikov M.A., Panfilov D.I., Rashitov P.A., Remizevich T.V., Fedorova M. I.: Regarding electrical transmission line's operating modes with controllable phase shifters. Proceedings of the Russian Academy of Sciences. Power Engineering Journal. vol. 1, pp. 1–9 (2016). (In Russian)
4. Astashev M.G., Novikov M.A., Panfilov D.I., Rashitov P.A., Remizevich T.V., Fedorova M. I.: Unbalance mode of traffic flow power control schemes in the intellectual electricity distribution network. Proceedings of the Russian Academy of Sciences. Power Engineering Journal, vol.4, pp. 45–53 (2015). (In Russian)
5. Astashev M.G., Novikov M.A., Panfilov D.I., Rashitov P.A., Fedorova M.I.: A simplified analytical model for the study of unbalance work phase rotation device with a thyristor switch. Proceedings of the Russian Academy of Sciences. Power Engineering Journal, vol.1, pp. 91–104 (2015). (In Russian)
6. Astashev, M.G., Novikov, M.A., Panfilov, D.I., Rashitov, P.A., Fedorova, M.I.: Simplified analytical model for open-phase operating mode of thyristor-controlled phase angle regulator. Therm. Eng. **62**(13), 928–937 (2015)
7. Voronin P.A., Rashitov P.A., Astashev M.G.: Single-purpose circuitry model thyristor. Electrical engineering Journal, vol.9, pp. 21–30 (2015). (In Russian)
8. Panfilov D.I., Astashev M.G., Rashitov P.A., Rozhkov A.N.: Analysis of key switching methods of AC thyristor bridge. Proceedings of the Russian Academy of Sciences. Power Engineering Journal, vol.4, pp. 148–159 (2014). (In Russian)

9. Panfilov D.I., Rozhkov A.N., Astashev M.G.: Controlled Phase Shifters Model for Power Grid Operating Modes Calculation. In: IEEE 16 International Conference on Environment and Electrical Engineering. Conference Proceeding, Florence, Italy, 7–10 June 2016, pp 118–122

10. Rozhkov A.N., Astashev M.G., Rashitov P.A.: The Influence of Control Methods of AC Thyristor Bridge on its Switching Modes. In: 17th International Conference of Young Specialists on Micro/Nanotechnologies and Electron Devices. Proceedings. Erlagol, Altai, 30 June–4 July 2016, pp. 544–548

The Partial Replacement of Diesel Fuel in Hot Water Boiler with Syngas Obtained by Thermal Conversion of Wood Waste

O.M. Larina, V.A. Lavrenov and V.M. Zaitchenko

Abstract This paper presents experimental results on the use of syngas produced from waste wood by two-stage pyrolytic conversion method, in the implementation of heating system on a basis of hot water boiler. The method of two-stage pyrolytic biomass processing, combining the waste wood pyrolysis, and subsequent heterogeneous cracking of volatile pyrolysis products in charcoal bed, provides a high degree of energy conversion of raw material into syngas with a lower calorific value of 10–11 MJ/m^3. The possibility of partial replacement of diesel fuel in hot water boiler with syngas was shown.

Keywords Biomass · Gasification · Heating systems · Hot water boiler
Pyrolysis · Syngas · Thermal conversion · Wood waste

Introduction

Development of technologies that allow efficient use of wood waste for energy purposes is an important task in terms of rational use of natural resources. During processing of wood, only 28% of the original weight of wood turns into a lumber, the rest becomes waste. An alternative to the direct combustion of waste wood is processing into gas suitable for use as fuel for the boilers of the existing heating systems. Methods of thermal conversion of the wood biomass into gas can be divided into two main types: gasification and pyrolysis.

O.M. Larina (✉) · V.A. Lavrenov · V.M. Zaitchenko
Joint Institute for High Temperatures of the Russian Academy of Sciences,
Moscow, Russia
e-mail: olga.m.larina@gmail.com

V.A. Lavrenov
e-mail: v.a.lavrenov@gmail.com

V.M. Zaitchenko
e-mail: zaitch@oivtran.ru

© The Author(s) 2018
K.V. Anisimov et al. (eds.), *Proceedings of the Scientific-Practical Conference
"Research and Development - 2016"*, https://doi.org/10.1007/978-3-319-62870-7_18

165

Gasification is a partial oxidation process to yield a syngas, the main combustible components are carbon monoxide, hydrogen and methane. It also contains large amount of ballast gases: nitrogen (air gasification), carbon dioxide and water vapor. Furthermore, syngas contains various impurities, such as tars and particles of ash and carbonaceous substance [3]. Air, oxygen, steam, or mixtures thereof may be used as the oxidant in the gasification process. Syngas obtained by air gasification has a lower calorific value of not greater than 6 MJ/m^3 [7]. This gas can be burned in boilers.

Pyrolysis is the thermal decomposition of the raw material without oxidant access. The products of the pyrolysis are a gas mixture (consisting mainly of H_2, CO, CO_2, CH_4, C_nH_m, and N_2), liquid fraction (mixture of water and pyroligneous liquor) and solid carbon residue. The gas mixtures produced from biomass, have a lower calorific value of 20 MJ/m^3 [4]. The ratio of the masses of liquid and gaseous products is about 1.5, and most significantly depends on the heating rate [5]. The main disadvantages of pyrolysis from the viewpoint of obtaining gas mixtures are relatively low specific gas yield which does not exceed 0.3–0.4 m^3 per 1 kg of raw material and high carbon dioxide content (up to 30% vol.). This causes a low efficiency of energy conversion of the feedstock into gaseous products: the ratio of the energy content of the pyrolysis gas to the calorific value of the feedstock does not exceed 0.3.

Increasing the degree of the raw material conversion can be achieved by processing of the liquid fraction in the gas. There are catalytic [6] and noncatalytic methods [5]. In this paper, to produce gas from waste wood used a method similar to that proposed for the processing of wood chips in [8] and then studied in detail in [1, 2]. It is based on the cracking of the pyrolysis products formed during the heating of raw material, in the bed of porous carbon residue maintained at a fixed temperature of about 1000°C. This scheme has been adopted as the basis for creating a pilot plant, allowing to obtain a syngas with enhanced characteristics (more than 90% vol. of H_2 and CO, the lower calorific value of about 11 MJ/m^3, the almost complete absence of tar in the gas). Obtained syngas can be effectively used as a substitute for diesel fuel in existing boilers.

The Experimental Technique

Experiments on the waste wood processing into syngas and its further use in boiler were carried out on the area of Energonezavisimost Ltd. (city of Nizhny Novgorod).

The heating system was used floor cast iron boiler "RIELLO RTT 93" with nominal thermal power of 100 kW with the diesel fuel burner "CUENOD NC12H101". The heaters "KEV-36T3W2" set in the heated rooms were used as a heat load of the heating system. Syngas produced in the thermochemical conversion module was accumulated in the elastic polymeric gas holder, from which it was fed through a pressure regulator into the special co-combustion nozzle of burner (Fig. 1).

Fig. 1 Combustion head of the fuel-oil burner with a nozzle for co-combustion of gaseous fuels

Fig. 2 Scheme of the
measurement system of the
boiler and burner parameters:
G1—syngas volume meter;
P1—the diesel fuel pressure
meter; T1—temperature of
the combustion products; T2
—temperature of water

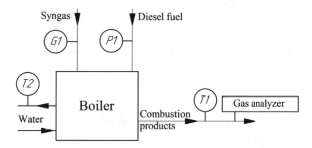

Scheme of the measurement system of the boiler and burner parameters is shown
in Fig. 2. The measurement system of the boiler and burner allows to determine the
following parameters:

- the diesel fuel consumption (calculates based on the diesel fuel pressure);
- the syngas consumption (gas flow meter "SGMN-1 M-G6");
- the oxygen volumetric content (O_2) in combustion products (gas analyzer
  "Askon-02");
- the combustion products temperature (chromel-alumel thermocouple (type K),
  connected to thermometer "Aktakom ATT-2004").

Each experiment included two main stages. The first stage is obtaining of syngas
and its accumulation in the elastic polymer gas holder. At the second stage,
obtained syngas is fed to the burner. For testing of the synthesis gas and diesel fuel
co-combustion five operation modes have been selected. Three operation modes on
the diesel fuel only (modes 1, 2, and 3) and two operation modes of partial substitution of diesel fuel with syngas (modes 4 and 5). The first mode is intended to
work at the nominal diesel fuel consumption and the base air flap setting.

In all other modes, the air flap setting remained unchanged from the nominal mode 1. In the second mode, the diesel fuel consumption decreased to reduce the thermal power of the burner by about 10% and in the third mode—by 20%. In the fourth and fifth operation modes the diesel fuel consumption corresponds to the second and third operation modes, but the syngas consumption was adjusted so that the thermal power of the boiler reaches value corresponding to the nominal mode 1.

Results and Discussion

The feedstock and syngas characteristics, obtained in elemental analyzer "Elementar Vario Macro Cube", gas flow analyzer "MRU Vario Plus Industrial «Syngas»" (O_2, H_2, CO, CO_2, C_nH_m, N_2) and gas chromatograph "Chromos GH-1000" (hydrocarbons) are shown in Table 1.

The measurement results of the combustion products parameters and the calculated values of power and efficiency of the boiler in the five operation modes are shown in Table 2.

Table 1 The feedstock and syngas characteristics

Parameter	Dimension	Value
The feedstock type		Pine shavings
The feedstock consumption	kg/h	5.0
The feedstock composition (on the wet /dry /dry ashless basis)		
Moisture content	% mas.	8.80/0/0
Ash content	% mas.	0.48/0.53/0
The elemental composition:		
Carbon (C)	% mas.	47.68/52.28/52.56
Hydrogen (H)	% mas.	5.54/6.07/6.10
Oxygen (O)—as a residual	% mas.	37.43/41.04/41.26
Nitrogen (N)	% mas.	0.05/0.06/0.06
Sulfur (S)	% mas.	0.02/0.02/0.02
Volatiles at 700°C	% mas.	84.5/83.0/−
The syngas volume	m^3	6.5
The syngas specific volume	m^3/kg	1.3
Chemical composition of dry syngas		
Hydrogen (H_2)	% vol.	49.2
Carbon monoxide (CO)		40.8
Carbon dioxide (CO_2)		5.0
Nitrogen (N_2)		1.8
Oxygen (O_2)		0.0
Hydrocarbons (C_nH_m), among them:		3.2

(continued)

Table 1 (continued)

Parameter	Dimension	Value
– methane (CH_4)	% vol. of C_nH_m	88.9
– ethane (C_2H_6)		1.1
– этен (C_2H_4)		1.0
– propane (C_3H_8)		4.7
– пропен (C_3H_6)		0.3
– i-butane (C_4H_{10})		1.2
– n-butane (C_4H_{10})		0.9
– i-pentane (C_5H_{12})		0.4
– n-pentane (C_5H_{12})		1.5

Table 2 Key parameters of the burner and boiler in the different operation modes

Parameter	Dimension	The operation mode				
		1	2	3	4	5
The diesel fuel consumption	kg/h	8.76	7.92	6.98	7.92	6.98
The syngas consumption	m³/h	0	0	0	3.6	7.4
	kg/h	0	0	0	2.29	4.72
The substitution degree	%	0	0	0	10.7	21.9
Parameters of the combustion products (boiler outlet):						
Temperature	°C	297	292	286	299	303
Oxygen concentration (O_2)	% vol.	3.2	4.74	6.20	3.12	3.08
Excess air ratio	–	1.20	1.32	1.46	1.19	1.19
Boiler parameters:						
Thermal power	kW	90.4	80.9	70.5	90.7	90.4
Efficiency	%	87.2	86.3	85.3	87.2	87.1

The presented data show that the boiler efficiency decreases in modes 2 and 3 compared to nominal mode 1 because of the excessive dilution of the combustion products with air. The boiler thermal power value recovers in modes 4 and 5 due to the syngas supply. The efficiency value in operation modes 4 and 5 is equal to the corresponding value in the nominal operation mode 1, which indicates that the partial replacement of the diesel fuel not deteriorated the combustion conditions. These results confirm the possibility of replacing the diesel fuel in existing hot water boilers to syngas obtained by two-stage thermal conversion of waste wood.

Conclusions

In this paper, the results of the syngas application obtained by the method of two-stage pyrolytic biomass conversion, combining the feedstock pyrolysis, and subsequent heterogeneous cracking of volatile pyrolysis products in charcoal bed.

The possibility of partial substitution of the diesel fuel with syngas obtained from wood waste by two-stage pyrolytic conversion method is shown. It is experimentally confirmed that the substitution of 10.7 and 21.9% of the diesel fuel with syngas obtained from waste wood occurs without reducing the thermal efficiency of the boiler.Research are carried out with the financial support of the state represented by the Ministry of Education and Science of the Russian Federation. Agreement no. 14.607.21.0073 20.Oct 2014. Unique project Identifier: RFMEFI60714X0073.

References

1. Батенин В.М., Бессмертных А.В., Зайченко В.М. и др. Термические методы переработки древесины и торфа в энергетических целях. Теплоэнергетика. 2010. № 11. С. 36–42
2. Батенин В.М., Зайченко В.М., Косов В.Ф и др. Пиролитическая конверсия биомассы в газообразное топливо. Доклады Академии наук. 2012. Том 446. №2. С. 179–182
3. Гелетуха Г.Г., Железная Т.А. Обзор технологий газификации биомассы. Экотехнологии и ресурсосбережение. 1998. № 2. С. 21–29
4. Зайченко В.М., Качалов В.В., Лавренов В.А. и др. Двухстадийная термическая конверсия древесной биомассы в синтез-газ. Экология и промышленность России. 2016. Том 20. № 11. С. 4–9
5. Кислицын А.Н. Пиролиз древесины: химизм, кинетика, продукты, новые процессы. М.: Лесная промышленность, 1990. 313 с
6. Dayton D.: A review of the literature on catalytic biomass tar destruction. Milestone Completion Report NREL/TP-510-32815. National Renewable Energy Laboratory, USA, 2002. 33 p
7. Kosov V.V., Kosov V.F., Maikov I.L. et al. In: High calorific gas mixture produced by pyrolysis of wood and peat: proc. of 17th European Biomass Conference and Exhibition, 29 June–3 July 2009, Hamburg, Germany. P. 1085–1088
8. Snehalatha K. Chembukulam, Arunkumar S. Dandge, Narasimhan L. Kovilur et al.: Smokeless fuel from carbonized sawdust. Chem. Prod. Res. Dev. 20, 714–719 (1981)

The Experimental Research on Independent Starting and Autonomous Operation of HDTB Considered as a Basic Block of AES Based on Supercritical Hydrothermal Destruction

A.D. Vedenin, V.S. Grigoryev, Ya.P. Lobatchevskiy, A.I. Nikolaev, G.S. Savelyev and A.V. Strelets

Abstract This article deals with an independent starting system (ISS) of the hydrothermal destruction test bed (HDTB) as a part of the layout of the autonomous energy system (AES) based on environmentally friendly technology of supercritical hydrothermal destruction (SCHD) of organic waste and fuels to supply electric and heat energy to small-scale distributed energy facilities. Independent starting system and HDTB tests which may be considered as AES basic block while using liquefied hydrocarbon gas (LHG) as a fuel for AGP-30 gas-piston power plant (GPPP) included into ISS have been described in the paper. Formation of combustible gas as a result of hydrothermal destruction of aqueous mixtures of organic waste and fuels has been shown experimentally. The produced combustible gas may have rather high net calorific value (NCV), which could define its suitability to be used to provide autonomous operation of HDTB.

A.D. Vedenin · V.S. Grigoryev · Ya.P. Lobatchevskiy ·
G.S. Savelyev · A.V. Strelets (✉)
Scientific Research Institute of Agricultural Mechanization (FGBNU VIM),
Moscow, Russia
e-mail: streletsav@ya.ru

A.D. Vedenin
e-mail: advedenin@yandex.ru

V.S. Grigoryev
e-mail: 1117731@mail.ru

Ya.P. Lobatchevskiy
e-mail: Lobachevsky@ya.ru

G.S. Savelyev
e-mail: ovlov2012@yandex.ru

A.I. Nikolaev
Dmitry Mendeleev University of Chemical Technology of Russia, Moscow, Russia
e-mail: aeshilus@gmail.com

© The Author(s) 2018
K.V. Anisimov et al. (eds.), *Proceedings of the Scientific-Practical Conference
"Research and Development - 2016"*, https://doi.org/10.1007/978-3-319-62870-7_19

Keywords Autonomous energy · Renewable energy sources · Supercritical technologies · Hydrothermal destruction · Gas-Piston power plant

Introduction

The development of renewable energy sources (RES) may remain a key task for Russian power-engineering in respect that around 15% of population is considered to consume energy from small-scale distributed energy facilities [1] while inhabiting two-thirds of the Russian territory. The importance of the decentralized energy development designated under the Energy Strategy of Russia for the Period up to 2035 Decree and aimed at "maximizing the cost effective use of domestic fuel and energy resources, developing cost effective decentralized and individual heat supply systems..." with regard to the facilities of housing and community amenities (HCA) as well as of agricultural and industrial production may be increasing due to the continual widening of an environmental damage scope caused by their organic wastes. It stands to reason that a considerable thought has been given to environmental safety by the society and the authorities, so the 2017 year is announced as a Year of Environment in Russia by the President of the Russia Federation (RF). The official plan of events for the Year of Environment approved by the RF government includes a special section dealing with waste disposal problems.

Despite the remarkable technological progress in developing power plants based on RES, the investigations on the development of AES to produce energy during organic waste and low grade fuels recycling may be at the stage of the experimental data accumulation as well as evaluation of physicochemical properties and the elucidation of the main trends during the process of destruction along with the facility design development [2].

Thus, the advantages of using power plants based on SCHD could be rather a significant factor for industries involving the production of organic waste and fuels (e.g., lignin-containing waste, animal and vegetable waste, wood waste) in hard-to-reach and remote areas, being a feedstock for them.

The technology of supercritical hydrothermal destruction or, as it is also called, supercritical water oxidation (SCWO) is based on the use of supercritical water (temperature and pressure are higher than 374.15 °C and 220.64 bar, respectively). Under such conditions, not only may water obtain properties different from those of liquid and vapor phases such as high density, low viscosity, and substantial diffusion, but also a change in its nature (from polar to almost nonpolar substance) may occur. This enables water to dissolve various organic matters including those forming solid coal-like matters and tars. Minimum demands on feedstock humidity may be imposed by conducting the reaction in an aqueous medium. A number of investigations [3] on supercritical water gasification (SCWG) of various types of biomass and methods for realization of biomass are known, though the application problems of the product in terms of ensuring autonomous operation of the energy system have not yet been considered.

Being a variety of gasification of conventional fuels and biomass of different origin [4], hydrothermal pyrolysis may be used solely to produce gas fuel having high content of hydrogen, methane and other high-energy ingredients. As a result of various investigations on hydrothermal gasification, almost any formation of carbonaceous matters may be observed. External heat supply may be required while conducting hydrothermal pyrolysis as the pyrolysis process is endothermic, which may correspond to HDTB allothermal regime.

At the same time, hydrothermal oxidation may imply addition of the oxidizer to supercritical water and allow obtaining the complete conversion of organic matter to water and carbon dioxide [5–7]. Formation of incombustible gas–vapor mixture may occur at rather appreciable heat evolution, which may ensure the conducting of the process without constant external heat supply and correspond to HDTB autothermal regime [8]. Thus, HDTB operation may open up alternative possibilities for the energy production due to the use of gas-piston power plant (GPPP) in allothermal regime, and it would be advisable to use steam microturbine in autothermal regime.

Previous investigations as well as an accumulated experience have shown the capability of the power plants based on SCHD to recycle a wide range of organic waste including pesticides and other persistent organic pollutants (POPs) [7, 9]. However, to provide usability of the SCHD technology at various power plants, the mobility as well as the autonomy of the processing plants. The presence of power supply source independent of external network may be one of the methods to provide the autonomy of energy systems, for example, in terms of energy supply. Thus, a need has been recognized for the development of HDTB independent starting and autonomous operation. An experimental research on the ISS including GPPP has been carried out in the context of the investigation, while liquefied hydrocarbon gas (LHG) has been used as a fuel.

Methods

The independent starting system of HDTB (Fig. 1) based on AGP-30 gas-piston power plant (rated capacity of 30 kW and fuel consumption of 15 nm^3/h) equipped with a control system may include the following:

- LHG gas cylinders (E);
- accumulator battery (ACB);
- gas filter with a shut-off valve (FSV);
- gas reducer-evaporator (RE);
- gas pressure and flow sensor (GFS);
- shut-off needle valve block (VB);
- electric water pump (WP);
- electric direct-flow water heater (DWH).

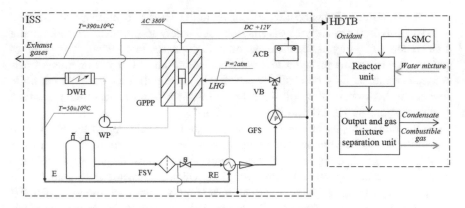

Fig. 1 A schematic of HDTB independent starting system

The use of LHG (propane/butane mix [10]) as a fuel for AGP-30 may be determined by its broad accessibility and relative cheapness as well as the possibility to be transported and stored in household gas cylinders.

The experimental research has been carried out in compliance with test agenda and procedure developed within the context of the second stage of the applied research and exploratory development.

The principle of ISS operation may be the following. To provide constant gas-phase fuel feeding to the gas fuel system of GPPP, the water heater should be brought into operation before starting GPPP. Circulation of refrigerating fluid through the water heater could be obtained by GPPP motor pump powered in turn by the accumulator battery. On opening the shut-off valve, gas after additional cleaning in the filter may flow from the cylinder to the reducer-evaporator where it may transform into gas phase with the required pressure. Then gas may flow through the gas pressure and flow sensor and the valve block and enter the input of the AGP-30 gas fuel system (Fig. 2).

The operator should perform starting algorithm procedures stipulated by the instruction, and gas may be fed from the cylinder as well as the power plant may be started in a manual mode. After starting, the engine may automatically turn into an operative mode with unvarying operating parameters of the power plant, consequently through low idle and nominal speed condition. After GPPP reaches full operation, HDTB could be connected to it for experimental verification of the possibility of independent starting and autonomous operation. The reactor could be heated up within 1 hour while water could be fed to reach the point of operating parameters of T = 450 ÷ 550 °C, P = 23 MPa due to the energy produced by the power plant.

To conduct SCHD reaction, 10% (by mass) aqueous mixtures of organic waste and fuels divided into 3 kg portions have been prepared at the developed process equipment for aqueous mixture preparation which may be the part of AES: 1 is rape oil emulsion; 2 is crushed buckwheat hull suspension; 3 is whey solution. The aqueous mixtures have been fed to the reactor which may be a key element of the

Fig. 2 ISS and HDTB during test

Table 1 Parameters monitored during test

Parameter	Unit of measurement	Value
Energy consumption in allothermal and autothermal regimes	kW	16.5/7.8
Gas fuel consumption of HDTB in allothermal/autothermal regime	nm^3/h	12/8.1
Gas fuel pressure before gas pressure and flow sensor input	atm	1.95–2.2
Operating pressure in SCHD reactor	MPa	23–27
The amount of organic fuel fed to the reactor (for each portion: emulsion, suspension, solution)	l	2.8

test bed by a high-pressure proportioning pump in a periodic mode (100 ml per one in-feed). SCHD reaction has been carried out in allothermal as well as autothermal regime within 60 s, while 100 ml of 50% (by mass) hydrogen peroxide solution has been fed in autothermal regime in parallel. During HDTB independent starting and operation in various modes, the value monitoring of several parameters such as LHG consumption by GPPP; energy input; discharge of water–organic mixtures (Table 1) stipulated by the program and testing methods have been carried out, and the parameters have been fixed with high-pressure pump running.

The gas–vapor mixture produced in the course of SCHD reaction has been separated into fluid and gas phases in a heat exchanger. Gas phase has been accumulated in a gasometer included in HDTB, whereupon the samples have been taken to study the phase.

Results and Discussion

Since high calorific gas may be a reaction product of hydrothermal pyrolysis of aqueous mixtures of organic waste and fuels, then gases sampled in allothermal regime of HDTB autonomous operation could be used to determine gas composition as well as to evaluate its usability as a fuel for GPPP.

Initially, the experimental verification of the produced gas combustibility has been carried out by means of the Bunsen burner (Figure 3).

Then, to analyze the combustible gas, samples have been taken into gas pipette of gas sampling and storing instrument. The samples composition analysis of the gas produced in the course of hydrothermal destruction of fuel and waste samples has been carried out by chromatographic method in a laboratory of the Common Use Centre at D. I. Mendeleev University of Chemical technology of Russia. The results of the sample analysis have shown that the produced gas may have the compositional resemblance to synthesis gas and net calorific value (Table 2) calculated by the computational method (State Standard 22667-82) with the following formula:

$$Q = \sum_{i=1}^{n} Q_i \cdot C_i$$

where Q_i is gas combustion heat of the ith component in gas and C_i is proportion of the ith component in gas.

Fig. 3 The experimental verification of SCHD reaction product

Table 2 The results of gas sample analysis produced during tests on HDTB independent starting

No	Combustible gas (CG)	Concentration of the components								Q_{net} (MJ/m^3)
		CO	CO$_2$	H$_2$	CH$_4$	C$_2$H$_6$	C$_2$H$_4$	C$_3$H$_8$	C$_3$H$_6$	
1	Synthesis gas (for a comparison)	15–18	30–32	38–40	9–11	–	–	–	–	11–12
2	CG (during whey destruction)	0.17	22.33	34.92	14.33	2.24	0.87	0.28	0.73	11.11
3	CG (during rape oil destruction)	2.87	25.96	19.6	14.27	9.52	10.86	–	–	20.08
4	CG (during crushed buckwheat hull destruction)	1.57	20.11	23.84	19.63	9.48	8.57	1.61	4.29	22.39

The results of gas samples analysis and net calorific value may bear witness to the possibility of using the combustible gas as a fuel for GPPP.

Thus, plans have been made to develop the obtained results at the next stage of the applied research and the experimental research. The work on optimization of the heat energy efficiency of independent starting and autonomous operation of the test bed would be done while conducting an experimental research on modernized HDTB. To achieve this, equipping ISS with the process equipment preparing the combustible gas to be used as a fuel for GPPP as well as selecting efficient parameters of SCHD reaction and methods of the reaction product quality improvement have been planned.

Conclusion

The independent starting system of HDTB which may be considered as a basic block of AES based on supercritical hydrothermal destruction of organic waste and fuels to produce electric energy has been developed and tested. The experimental research has verified the possibility of independent starting and autonomous operation of the test bed for various types of aqueous mixtures. The qualitative and quantitative composition of the combustible gas produced during HDTB test may provide the possibility to use it as a fuel for GPPP. The test has verified the efficiency of AES blocks (process equipment for preparing aqueous mixtures, independent starting system, and hydrothermal destruction test bed). The promising directions for follow-up studies and its realization have been identified.

Acknowledgement The study has been financially supported by The Ministry of Education and Science of the Russian Federation, Grant Agreement No. 14.607.21.0126 dated October 27, 2015. Unique identifier: RFMEFI60715X0126.

References

1. Popel, O.S.: Renewable energy sources in present and state-of-the-art energy. Russ. Chem. J. **52**(6), 96–106 (2008) (in Russ.)
2. Mazalov, Y., Pustovgar, A., Grigorev, V., Vedenin, A., Adamtsevich, A.: Technology for hydrothermal destruction of organic fuel materials. Appl. Mech. Mater. **752**, 873–877 (2015).
3. Yakaboylu, O., Harinck, J., Smit, K.G., de Jong, W.: Supercritical water gasification of biomass: a literature and technology overview. Energies. **8**(2), 859–894 (2015)
4. Efimov, N.N., Fedorova, N.V., Mirgorodskiy, A.I., Kolomiytseva, A.M.: Gasification of Organic Waste and Fuels. Advantages Present Nat. Sci. **1**, 15–21 (in Russ.) (2007)
5. Psarov, S.A.: Thermal effects of organic matters oxidation in supercritical water. Author's abstract of master's dissertation, p. 24. Novosibirsk (in Russ.) (2006)
6. Fediayeva, O.N.: Transformations of Low Grade Fuels in Supercritical, p. 256. Ph.D. dissertation, Novosibirsk (in Russ.) (2014)
7. Astakhova, L.V., Grigoryev, V.S., Mazalov, Yu.A., Nizovtsev, V.E.: High-toxic substances disposal. Prod. Ecol. **8**, 66–70 (in Russ.) (2011)
8. Mazalov, Yu.A., Bersh, A.V., Merenov, A.V. and others.: Production of heat and electric energy by organic waste combustion in aqueous media. Energy Saving and Energy Supply in Agriculture. Proceedings of the 6th International Scientific Conference on Applied Sciences and Engineering (2008). Part 1. Problems of Energy Supply and Energy Saving. 390–395, (in Russ.)
9. Mazalov, Yu.A., Svitsov, A.A.: The use of SCWO for recycling organic waste. Water Mag. 1 (77), 36–38, (in Russ.) (2014)
10. State Standard 5542–2014.: Natural combustible gases for commercial and domestic use. Specification. Standartinform, Moscow (2014), (in Russ.)

Development of a Multifunctional All-Terrain Vehicle Equipped with Intelligent Wheel-Drive System for Providing Increased Level of Energy Efficiency and Improved Fuel Economy

V.V. Belyakov, P.O. Beresnev, D.V. Zeziulin, A.A. Kurkin, V.S. Makarov and V.I. Filatov

Abstract The use of hydrostatic transmission driveline as a part of multi-wheeled all-terrain vehicles on ultralow pressure tires allows achieving the efficient power distribution for driving wheels depending on the conditions of vehicle-terrain interaction. This provides a significant increase in average speeds of vehicle movement in difficult road conditions (the maximum possible traction force is implemented by automatic maintaining the required level of wheel slip). At the same time, the minimum energy losses of wheel-soil interaction (improved fuel economy) and environmental safety of wheels, when operating on soft soil and vegetation, are provided. Installing the hydrostatic transmission allows to optimize layout scheme at the expense of the free selection of placement of transmission units. Application of integrated combination of active drive wheels, tires ultralow pressure, independent suspension system, the ability to control all the wheels according to any given algorithm (ability to control turning by all wheels) when

V.V. Belyakov · P.O. Beresnev · D.V. Zeziulin · A.A. Kurkin (✉) ·
V.S. Makarov · V.I. Filatov
Nizhny Novgorod State Technical University n.a. R.E.Alekseev,
Nizhny Novgorod, Russia
e-mail: aakurkin@gmail.com

V.V. Belyakov
e-mail: nauka@nntu.ru

P.O. Beresnev
e-mail: norb132801@gmail.com

D.V. Zeziulin
e-mail: denis.zeziulin@nntu.ru

V.S. Makarov
e-mail: makvl2010@gmail.com

V.I. Filatov
e-mail: filatov.v.ngtu@yandex.ru

© The Author(s) 2018
K.V. Anisimov et al. (eds.), *Proceedings of the Scientific-Practical Conference
"Research and Development - 2016"*, https://doi.org/10.1007/978-3-319-62870-7_20

179

driving on soft soil allows to effectively carry out the transportation operation with preservation of the ecology of the soil cover.

Keywords Multifunctional all-terrain vehicle · Hydrostatic transmission driveline Efficiency · Fuel consumption

Introduction

In the market of modern all-terrain vehicles (ATV) there has been formed quite developed segment of vehicles on ultralow pressure tires. The maximum ground pressure of these vehicles is much lower than that of known types of wheeled and tracked vehicles. Furthermore, the method of rotation at the expense of steered wheels makes the minimal detrimental impact on the soil and makes possible the preservation of soil and vegetation cover even when turning with a small radius.

It should be emphasized that the ATV tires on ultralow pressure tires with wheel arrangement 8×8 are inherently potential competitors of tracked vehicles. They exert significantly less harmful impact on the soil cover, while providing a similar carrying capacity. The most well-known vehicles are the following: Shaman [1], Tundra [2], Trom [3], Xpen [4], Staratel [5], Strannik [6].

However, for movement on difficult terrain in case of sudden change in traffic conditions to ensure the efficiency of vehicle functioning is only possible by timely alterations of basic operation modes of separate units and systems of the vehicle, then most of the models in this class of ATV do not possess.

The current trend of ATV developers around the world is the development of intelligent control systems, capable of providing pre-emptive response effect on the various situations, which arise while driving, and significantly improving the performance characteristics of vehicles by introducing so-called adaptive mobility (agility).

By present time in studies as the most effective means of improving the mobility and efficiency of vehicles there is noted the use of «flexible intelligent» transmissions, which can be implemented on the basis of electromechanical or hydrostatic drives [7, p. 17].

«Intelligent» driveline provides an adjustable vehicle behavior, in particular, the multi-axle vehicle, operating in off-road conditions. The advantage of this transmission is in the ability to provide fast and accurate response to changes in force and kinematic parameters of motion. The best object of automatic control systems among all currently existing types of continuously variable transmission is the hydrostatic transmission because of the very low inertia and high «rigidity» of drive [8, p. 182; 9, p. 17]

Development of the Multifunctional All-Terrain Vehicle

The purpose of the project is obtainment of significant scientific and technological results, which help to start creating high-performance multifunctional all-terrain vehicles (MATV), equipped with intelligent wheel-drive system for providing increased level of energy efficiency and improved fuel economy.

The purpose requires solving the following tasks:

1. Analysis of the layout schemes and designs of 8 × 8 vehicles and the existing power distribution strategies for driving wheels;
2. Development of mathematical model of MATV movement in off-road conditions and conducting simulation with the definition of rational parameters and obtaining initial data for designing;
3. Development of mathematical model of intelligent wheel-drive system functioning and control algorithms, identification of settings of the hydrostatic transmission, ensuring the achievement of the required energy efficiency on terrain;
4. Development of design documentation, manufacturing of an experimental prototype of MATV with subsequent adaptation of control laws of the wheel-drive system;
5. Conducting research tests of the experimental prototype under different operating conditions, and identifying ways to improve the design.

The use of hydrostatic transmissions combining individually regulated power actuator of each wheel with automatic control system allows achieving the efficient torque distribution for driving wheels depending on the conditions of vehicle-terrain interaction. The best-known vehicle with the hydrostatic transmission is three-axle vehicle «Gidrohod-49061», designed by the Central research and development automobile and engine institute «NAMI». From the results of evaluating demand for the generated MATV it must be concluded that competitive vehicle must relate to a lesser weight category [10, p. 947; 11, p. 517], have independent suspension, the control system of all wheels turning and ultralow pressure tires.

Selection of rational parameters of the MATV and obtaining initial data for designing was carried out according to the results of simulation modeling of the MATV motion in off-road conditions and analysis of statistical dependencies between power characteristics, capacity, average ground pressure and total mass of vehicles of this class [12, p. 4].

When selecting chassis parameters there was used optimization method based on the choice of design parameters contributing to ensure the minimum energy consumption for MATV moving on terrain. The installation of hydrostatic transmission driveline ensures the optimal arrangement of vehicle aggregates (due to more the free choice of placement of transmission units).

The most well-known strategies of power distribution for driving wheels of multiaxial vehicles, including vehicles with electro-mechanical systems and hydrostatic transmission, were analyzed by authors in [13, p. 123; 14, p. 159]. Restrictions in the issue of their technical realization were identified.

The studies have revealed that the most effective scheme for the MATV is hydro-differential wheel-drive of sides (Fig. 1). The ability to adjust the torque on the sides can improve throughput and influence maneuverability. There are possibilities of oversteering and skid steering, and there is no circulation of parasitic power between wheels of one side. Adjusting the torque on each of wheels based on readings of inertial navigation system, angular rate sensors and pressure sensors in hydraulic lines. The change of working volumes of hydraulic motors as required allows implementing maximum adhesion and minimal resistance to movement at each wheel when driving on typical soils of terrain.

Development, debugging, and evaluation of the efficiency of different algorithms of torque distribution in accordance with the condition of maximum traction force and minimum energy consumption for transportation were conducted using simulation model of MATV's wheel-drive system.

Parameters, characteristics, and adjustments of the hydrostatic transmission, ensuring achievement of the desired energy efficiency on terrain, were determined.

During carrying out virtual tests the following parameters of the MATV motion were determined: angular accelerations, slippage, and torque at the wheels; velocities and accelerations of MATV's center of mass; ground reaction forces

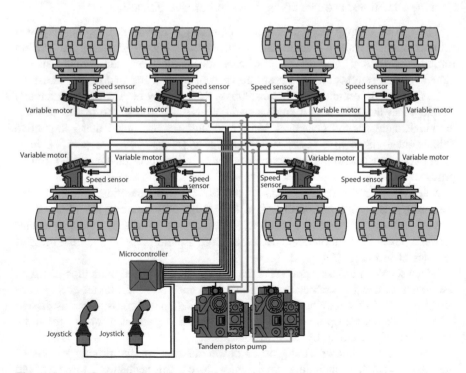

Fig. 1 Structural transmission scheme

implemented by wheels. Realization of control of MATV motion in the simulation model was carried out by using PID-controller.

Parameters of hydraulic aggregates are calculated from the following expressions:

pump:

$$T_H = p_w q_H (2\pi \eta_H)^{-1}; Q_H = q_H \omega_H \eta_{V_H},$$

motor:

$$T_M = p_w q_M \eta_M (2\pi)^{-1}; Q_M = q_M \omega_M \eta_{V_M}^{-1},$$

where T_H, T_M—torques on the shaft of the pump and motor; Q_H, Q_M—maximum output of the pump and motor, respectively; p_w—pressure-drop of the working fluid in lines of hydraulic machines; q_H, q_M—working volume of the pump and motor; η_{V_H}, η_{V_M}—volumetric efficiency of the pump and motor, respectively, ω_H, ω_M—angular velocities of the pump and motor.

The distribution of torque for shafts of motors approximately corresponds to the ratio of their working volumes [15, p. 336]:

$$T_{M_1} : T_{M_2} : T_{M_3} : \ldots : T_{M_n} \approx q_{M_1} : q_{M_2} : q_{M_3}; \ldots : q_{M_n}.$$

Angular velocities of hydraulic motors are calculated from the dependence [15, p. 336]:

$$Q_H \approx q_{M_1} \omega_{M_1} + q_{M_2} \omega_{M_2} + q_{M_3} \omega_{M_3} + \ldots + q_{M_n} \omega_{M_n}$$

Applied to the drive wheel torque from the hydraulic pump is spent to overcome rolling resistance, wheel acceleration and the implementation of traction. The general equation of wheel dynamics has the form:

$$I_K \cdot \dot{\omega}_K = T_M i_{p.M} \eta_{p.M} - T_{co\Pi p},$$

where $T_{co\Pi p}$—drag torque of wheel rotation (load); I_K—moment of inertia of the wheel, $\dot{\omega}_K$—angular acceleration of the wheel, $i_{p.M}$—final drive gear ratio; $\eta_{p.M}$—efficiency of the final drive.

The torque of resistance to wheel rotation is determined by the rolling resistance of the wheel $T(R_z)$ and torque that is created by the tangential component of the force of wheel–soil interaction $T(R_x)$:

$$T_{co\Pi p} = T(R_z) + T(R_x).$$

The model of work of the hydrostatic transmission driveline during the motion of a multiaxial vehicle is described in more detail in [16, p. 61].

To select rational options of the hydrostatic transmission control system the following algorithms were analyzed [17, p. 224; 18, p. 79; 19, p. 3; 20, p. 41]:

- individual slip control for wheels of a side with a certain linear velocity of the center of mass of MATV's chassis (with the iterative selection of the optimum value of slippage);
- torque control algorithm for wheels of a side based on the known minimum wheel angular velocity («high-threshold» control with limitation of the angular acceleration of the wheels);
- torque control algorithm based on the average wheel angular velocity of a side.

Virtual tests of MATV moving on the ground with the changing characteristics of resistance and adhesion are the most revealing, since they reflect the real character of the vehicle–terrain interaction. Examples of changes in the instantaneous fuel consumption in cases of using different algorithms are shown in Fig. 2.

According to the results of numerical modeling there was carried out an assessment of MATV energy efficiency and fuel economy, which allowed drawing conclusions about the effectiveness of the wheel-drive system control algorithms. When moving on « mixed » surface the developed control algorithms alongside with the radical increase of trafficability allow to increase the efficiency of the MATV movement up to 10% and reducing fuel consumption up to 18% depending on the selected regulation algorithm.

General views of 3D model and experimental prototype of produced MATV are presented in Fig. 3, and technical characteristics are given in Table 1. In making

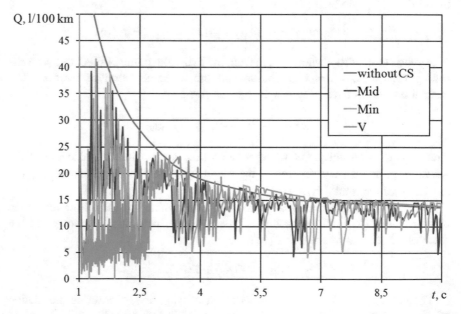

Fig. 2 Graphs of fuel consumption changes at various regulation laws: «without CS»—without control system, «Mid»—algorithm based on the average wheel angular velocity, «Min»—«High-threshold» control, «V»—individual slip control

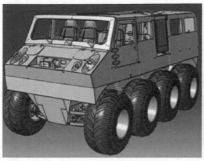

Fig. 3 General views of 3D model and experimental prototype of the produced MATV

Table 1 MATV technical characteristics

Parameters	Characteristics
Wheel arrangement	8 × 8
Number of seats (including the driver)	13
Body	amphibious (floating)
Curb weight, kg	4500
Cargo weight, kg	3500
Dimensional (overall) length/width, mm	5650/2800
Dimensional (overall) height, mm	2800
Wheelbase, mm	4350
Track width, mm	2100
Ground clearance, mm	550
Power, kW (hp) for rpm	110 (150) for 2400
Torque, N·m for rpm	490 for 1500–1700
Driveline	hydrostatic
Hydraulic pumps	2 axial piston pumps with adjustable volume. Working volume of 125 cm^3
Hydraulic motors	8 axial piston motors with adjustable volume. Working volume of 107 cm^3
Steering System	Turning of each wheel/Skid steering system
Suspension system	Independent suspension
Tires	ultralow pressure tires (dimension 52,2 × 25,5–24)
Maximum speed on highway/water, km/h	60 (5)
Minimum turning radius, m	15
Angle of ascent at full load, deg	30

Fig. 4 Fragments of field trials

design and technological solutions the approach, that provides the highest unification of components and assemblies, as well as the use of parts, assemblies and
units of domestic production in the MATV design, was used. [21, p. 41; 22, p. 24].

The central stage of the procedure for creating a workable technology is to
conduct field tests of an experimental prototype (Fig. 4) to identify ways of the
design optimization. Conducting experimental researches allowed to evaluate the
efficiency of the MATV; reliability of operation in matters of trafficability, stability
of motion, and maintain performance of its units in difficult climatic conditions;
determine the values of parameters and indicators of performance properties of the
designed chassis.

MATV has the ability to control all its wheels on any given algorithm and
achieve rational angles of rotation of the wheels. This technical result is achieved by
the fact that hydraulic cylinders for turning control are used in the MATV for each
wheel (but not only one for each axle as in analogs).

Conclusions

The research results are technical solutions for creating high-performance multifunctional all-terrain vehicles on low pressure tires with hydrostatic transmission
driveline and the possibility to control turning by all wheels.

Technical implementation of the proposed solutions was carried out taking into
account the results of the analysis of layout schemes and designs of 8 × 8 ATV, as
well as the existing power distribution strategies for driving wheels of multiaxial
vehicles. Structural, functional and hydraulic circuits of the MATV have been
developed.

Conducting simulation of the MATV movement with the intelligent wheel-drive
system in off-road conditions has allowed to select rational parameters of the
chassis, ensures the development and virtual testing of torque distribution laws for
driving wheels according to conditions of their interaction with the terrain, providing minimum energy consumption for transportation. Forecasting MATV

performance with hydrostatic transmission control system has been organized on the base of study of chassis dynamics when driving on the ground surface with variable characteristics.

Preliminary design has been developed and the MATV experimental prototype has been manufactured. Torque distribution algorithms for driving wheels, designed accompanying imitation modeling, have been adapted for hydrostatic drive control system of the MATV experimental prototype.

Research testing of the MATV experimental prototype has been carried out, specifications and values of operational properties indicators of the designed chassis have been determined. Finalization of the MATV prototype has been performed according to test results.

Acknowledgements Research are carried out with the financial support of the state represented by the Ministry of Education and Science of the Russian Federation. Agreement no. 14.574.21.0107 8.Sep 2014 Unique project Identifier: RFMEFI57414X0107

References

1. AVTOROS, All terrain vehicle 8x8 Shaman URL. http://avtoros.info/en/node/240. Accessed 16 May 2016
2. Group firms SKARN URL. http://www.skarn.ru. Accessed 29 April 2016
3. Off-road vehicles 8x8 Trom URL. http://trom8x8.ru. Accessed 01 March 2016
4. All-Terrain Vehicle « Xpen » URL. http://xpen.komi-nao.ru/. Accessed 11 May 2016
5. All-terrain amphibious vehicles of Russia URL. http://вездеходы-амфибии.рф. Accessed 05 May 2016
6. All-Terrain Vehicle « Strannik » URL. http://atvstrannik.ru. Accessed 15 March 2016
7. Pliev, I.A.: Control algorithms of the power delivered to the wheels of four-wheel drive. J. Automotive Eng. **3**(74), 16–18 (2012)
8. Gorelov, V.A.: Forecasting of the characteristics of the curvilinear motion of all-wheel-drive vehicle with the steering 1-0-3 at various laws of the control wheels of the rear axle: Cand. diss. p. 200. Moscow, MGTU im. N.E. Baumana, (2008)
9. Chizhov, D.A.: Development of complex techniques to improve energy-wheel drive wheeled vehicle: Cand. tehn. sci. diss. p. 146. Moscow (2012)
10. Belyakov, V., Kurkin, A., Makarov, V., Zeziulin, D.: Multifunctional vehicle for coastal areas. Proceedings of the 12th International Conference on the Mediterranean Coastal Environment, MEDCOAST. **2**, 945–951 (2015)
11. Zeziulin, D., Makarov, V., Belyaev, A., Belyakov, V.: Development of multi-wheeled all-terrain vehicle with hydrostatic transmission driveline. In: 13th European Conference of the International Society for Terrain-Vehicle Systems, p. 517–523, Rome, Italy, 21–23 October 2015
12. Barakhtanov, L.V., Beliakov, V.V., Zeziulin, D.V., Makarov, V.S., Manianin, S.E., Tropin, S. L.: Substantiation of rational design of cross country transport vehicle with 8 × 8 axle arrangement. Russ. Eng. Res. **6**, 3–5 (2015)
13. Serebrennyi, I.V.: Increased support cross-wheel drive vehicle through the rational distribution of power at the wheels. Cand. tech. sci. diss. p. 161. Moscow (2009)
14. Ushnurtsev, S.V., Keller, A.V., Usikov, V.Yu.: Control method for power distribution between driving wheels of car of universal purpose on mutual deviation of kinematic and power factors. Omsk Sci. Bull. 1–107 (107) (2012)

15. Shukhman, S.B., Solov'ev, V.I., Prochko, E.I.: Theory of power drive all-terrain vehicles wheels. – M.: Agrobiznest-sentr publ., (2007). p. 336
16. Kurmaev, R.Kh.: Method of increasing the efficiency of all-wheel drive multi-axis machine with hydrostatic transmission through the use of corrective algorithms. Cand. tech. sci. diss. p. 229. Moscow (2009)
17. Lepeshkin, A.V.: Methods of creating an "intelligent" automatic adaptive transmission control multidrive wheeled vehicle. Izvestiya MGTU MAMI. 2012. V. 1. **2**(14), 222–228
18. Gorelov, V.A., Maslennikov, L.A., Tropin, S.L.: Forecasting performance of curvilinear motion in multi wheeled vehicles for different all-wheel steering laws. Sci. Educ. Bauman MSTU. **5**. Available at: http://technomag.edu.ru/doc/403845.html (2012), Accessed 10 April 2016
19. Gorelov, V.A., Kotiev, G.O., Miroshnichenko, A.V.: Synthesis of control traction motor for individual drive wheeled vehicle. Sci. Educ. Bauman MSTU. **12**. Available at: http://www.technomag.edu.ru/en/doc/282533.html (2011), Accessed 20 Aug 2012
20. Gorelov, V.A., Kotiev, G.O., Miroshnichenko, A.V.: Individual drive control algorithm wheel propulsion vehicles. Herald of the Bauman Moscow State Technical University. Series Mechanical Engineering. Special issue Power and transport engineering. 2011. p. 39–58
21. Beliakov, V.V., Zeziulin, D.V., Makarov, V.S., Kurkin, A.A.: Development of a multi-axle all-terrain vehicle with hydrostatic transmission. Proceedings of Higher Educational Institutions. Machine Building. **10**(679), 39–48 (2016)
22. Beliakov, V.V., Beresnev, P.O., Filatov, V.I., Shapkina, Yu.V.: Multifunctional vehicle for coastal zones. Ecol. Syst. Dev. **8**, 23–28 (2016)

Development and Implementation of an Integrated Approach to Improving the Operating Cycle and Design of an Energy-Efficient Forced Diesel Engine

K.V. Gavrilov, V.G. Kamaltdinov, N.A. Khozeniuk, E.A. Lazarev and Y.V. Rozhdestvensky

Abstract The methods and means of improving the operating cycle and design of forced diesel, its fuel efficiency, and the limitation of the thermomechanical loading were considered.

Keywords Specific power · Mixture formation · Fuel injection
Combustion procedure parameters · Fuel efficiency · Diesel design elements

Introduction

The aim of this study is to increase the competitiveness of possible by intensification of scientific research and finding of new technical solutions for improving characteristics of a diesel.

This fully applies to the developed six-cylinder V-shaped diesel engine, with a cylinder diameter of 150 mm and a piston stroke of 160 mm (further diesel), with

K.V. Gavrilov (✉) · V.G. Kamaltdinov · N.A. Khozeniuk · E.A. Lazarev ·
Y.V. Rozhdestvensky
South Ural State University, Chelyabinsk, Russia
e-mail: gavrilovkv@susu.ru

V.G. Kamaltdinov
e-mail: kamaltdinovvg@susu.ru

N.A. Khozeniuk
e-mail: khozeniukna@susu.ru

E.A. Lazarev
e-mail: lazarevea@susu.ru

Y.V. Rozhdestvensky
e-mail: 79080716674@yandex.ru

© The Author(s) 2018
K.V. Anisimov et al. (eds.), *Proceedings of the Scientific-Practical Conference "Research and Development - 2016"*, https://doi.org/10.1007/978-3-319-62870-7_21

turbocharging and intercooling, intended for application in ground transportation vehicles.

Earlier exploratory research into advanced models identified a number of problems, the most important of which require an integrated approach to further improve the operating cycle [9, p. 2; 1, p. 2; 8, pp. 3–15] and design using a mathematical simulation of the main processes in the inside-cylinder space, in the gas path components, the hydrodynamic phenomena in the fuel supply, and the cooling and lubrication systems of diesel.

An integrated approach is based on the following key assumptions:

- common design, structural, and functional technical solutions for creating a forced modification diesel;
- the availability of technical solutions for computational and theoretical control with subsequent adjustment based on the results of tests of a diesel engine;
- compliance with the requirements of the unification of technical solutions design, low metal content, adaptability in production, and economic feasibility;
- the use of mechanisms and systems of the new generation of structural and operational materials to provide design improvement;
- management of the energy-efficient diesel engine by using modern software and computing complexes;
- including a modern microprocessor element base in the structure of control of the executive power and hydromechanical high-speed devices.

The aim of the study is to develop technical solutions to production based on the Russian components of an energy-efficient forced diesel and a specific capacity of at least 35 kW/l for ground transport vehicles.

To achieve these goals, the following problems need to be solved.

1. Adaptation and development of existing and new mathematical models, methods of calculation and technical solutions for the elements of forced diesel engines, including

- improving the operating cycle of diesel;
- providing the required thermodynamic state of the gas environment and the conditions of heat and gas exchange in the inside-cylinder space in the basic processes of the operating cycle;
- providing high turbine supercharging with intercooling;
- intensification of the process of fuel injection with electronic control,
- improvement of inside-cylinder configuration space of the combustion chamber and the gas distribution phases;
- evaluation and improvement of the strength characteristics of the piston, connecting rod, crankshaft, and crankcase of a diesel engine with a high maximum pressure of gas in the cylinder;
- determination of tribological characteristics to enhance the bearing capacity of the crank and turbocharger rotor bearings and the performance of the "piston-cylinder liner" system;

- evaluation and improvement of the strength characteristics of elements of the fuel injector for high-pressure injection.
2. Experimental evaluation of the main technical solutions for improving the systems and diesel units.
3. Creating a prototype of energy-efficient forced diesel with domestic component parts.
4. Bench testing of the developed prototype of the diesel engine.

Literature Review

The basis of the operating cycle of a diesel with direct injection is enough to put into practice widely approved diesel engine volumetric mixture formation and combustion method, implemented in the undivided combustion chamber, located at the bottom of the piston.

The geometric compression ratios are 13.5 … 14 units. It is necessary to choose the amount of compression performed on the basis of compromise and due to the need to limit the maximum pressure of the gas in the cylinder and ensure a reliable start-up at low temperatures.

The piston of a forced diesel is bimetallic with oil gallery cooling and a special profile of the skirt. The complex design of the piston is due to the high thermo-mechanical loading of elements of inside-cylinder space [12, p. 798]. The undivided combustion chamber has a small heat-sensitive surface, which together with a small whirlwind ratio causes a reduction in the heat transfer to the piston and cooling system.

The individual cylinder heads include four valves, the combined inlet and outlet channels with a minimum of hydraulic losses. Liquid cooling cavities in the heads require an individual unit separate from the coolant supply.

The fuel injection equipment includes a common rail accumulating system with electronically controlled fuel injection. The inclined fuel nozzle with a central location and indoor spraying has eight to ten sprayer holes [7, p. 4].

Diesel has a high turbocharged deep intercooling. According to the analysis of flow and working characteristics, we selected a diesel engine with a turbocharger with a one-stage compressor.

Particular attention is paid to improving the efficiency of mixing and complete fuel burn, taking into account the limitations imposed by the thermomechanical stressed parts [12, p. 798]. To achieve the specific power of diesel at least 35^{-2} kW/l is necessary to solve the following problems:

- providing charge air options, guaranteeing the mixture with an air–fuel ratio of not less than 1.8…2.0;
- intensification of injection and uniform distribution of fuel in the combustion chamber to achieve the best homogeneity of the mixture;

– providing timely and complete combustion of fuel with the least possible duration to increase the efficiency of the combustion process.

Methods

The objectives of improving the operating cycle of diesel were solved using well-known and new methods:

– intensification injection and uniform distribution of fuel in the combustion chamber was achieved using a fuel supply apparatus with a high-pressure injection (160…200 MPa) and an optimum number of holes in the sprayer [7, p. 4];
– ensuring timely and complete combustion of fuel with the least possible duration was achieved in the implementation of the above-mentioned methods, combined with increased use of air charge in the combustion chamber and electronically controlled fuel injection;
– reduction of thermal and mechanical loading of components and parts of inside-cylinder space of a diesel engine was achieved by the thermal emission intensity in the initial stage of the combustion process and minimizing the thermal losses;
– cylinder group reliability is achieved by using composite bimetallic pistons with a special profiling forming the skirt undivided combustion chamber;
– the required intensity of heat transfer and heat exchange in the cooling system of regulation is achieved by a dual feed of the coolant in the cylinder head and block.

Increased efficiency of the turbocharger in the diesel is provided by using a single-stage high-pressure compressor with a two-level vane diffuser and the turbine vane nozzle assembly is completed with a turbocharger with a small rotor inertia and short gas path system. As an intermediate charge, air coolers use air–liquid heat exchangers with minimal hydraulic resistances for air and fluid paths.

The use in a diesel fuel supply accumulator-type apparatus with high-pressure injection (160…200 MPa) and an optimum number (till 10) sprayer atomizer holes reduced the duration of fuel injection to 1.5 ms at a sufficiently uniform distribution of fuel in the combustion chamber [7, p. 4]. This agrees with the results of other authors [2, p. 302–303; 11, p. 192].

Use of the cylinder heads of the two inlet channels with profiled boundary contours and minimal flow resistance on the operating speed of the engine provided increased efficiency in filling the cylinder with a fresh charge of up to 0.95. At the same time, we considered the dimensions (depth, the maximum diameter of the piston head profile) of the undivided combustion chamber in the piston and the specific recommendations of the leading manufacturers of diesel engines for the relevant cylinder diameter, and the location and inclination of the fuel injector in the cylinder head.

Electronic fuel injection control allowed a biphasic injection fuel injection advance angle adjustment depending on the rotational speed of the crankshaft of a diesel engine and the load. As a result, the maximum pressure and the speed of its rise in the cylinder and the combustion temperature were reduced. This effect reduces the heat transfer from the gases to the surrounding elements of inside-cylinder space and temperature level of details, the stress in the dangerous area, loads on the main and connecting rod bearings, and diesel vibroacoustic activity. Reduced heat loss also contributes to increasing the excess air ratio.

Theoretical and Experimental Evaluation of the Parameters of the Operating Cycle and Characteristics of the Diesel

At the initial stage of creating a diesel engine (the design stage), using computational studies we established indicators of the operating cycle for expected characteristics of the fuel burn. For these purposes, we used the method of synthesis of the operating cycle [5, pp. 874–875; 6, pp. 489–492], which was developed in South Ural State University.

An indicator diagram of the pressure P and the temperature T of gases is shown in Fig. 1. Analysis of the indicator diagram shows that the indicators and parameters of the combustion process are as follows: maximum gas pressure of 18.4 MPa, the maximum rate of rise of the pressure of 0.38 MPa/deg., the maximum cycle temperature is 1850 K, the start of combustion timing angle of 5 deg. before TDC, process time combustion 135 deg., and maximum speed fuel burn 0.022 deg^{-1}.

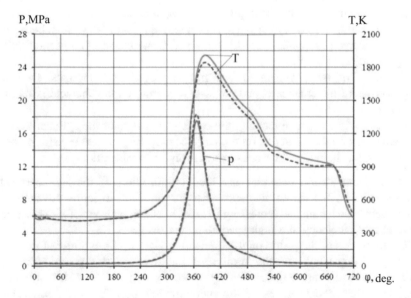

Fig. 1 The indicator diagram of the pressure P and the temperature T of gases in the cylinder of the diesel—single-stage supercharging and two-stage supercharging

Using computer modeling we determined and optimized the following:

- the basic processes of the operating cycle (fuel supply, mixture formation, combustion) and parameters of heat transfer, based on mathematical models of the fuel jet mixing, chemical kinetics, and combustion of fuel nonreacting turbulent gas flow in the combustion chamber with the determination of temperature fields of the combustion chamber surfaces by the finite element method;
- the combustion chamber configuration, based on a mathematical model of gas dynamics reacting turbulent two-phase flow in a three-dimensional setting with a simultaneous selection of the boost parameters, design injector nozzle, and fuel law;
- geometry inlet cylinder head channels, based on a mathematical model of the dynamics nonreacting turbulent gas flow with high Mach numbers in the three-dimensional setting and minimization of areas with the greatest dissipation of turbulent kinetic energy;
- gas distribution phases, based on mathematical models of mass and energy balances of gas in the combustion chamber, intake and exhaust manifolds, one-dimensional flow in the channels of the gas path and submodels of elements of the turbocharged system and charge air;
- boost parameters and characteristics of aggregates with intercooling using a synthesis of operating cycle of diesel;
- dynamics and lubrication of a complex loaded tribosystem with reciprocating and rotational movements of the elements, solution of the boundary conditions for hydrodynamic pressure using multigrid algorithms for bearings with complex geometry of the lubricating layer, taking into account the lubricant sources on the friction surfaces, micro- and macrogeometry, non-Newtonian behavior of the lubricants [10, pp. 22–221; 13, pp. 46–47], and the solution of multi-criteria optimization problems;
- the effect of misalignment of bearings and crankshaft journals, the elastic properties of the crankcase, and crankshaft bearing on the characteristics of the diesel;
- thermal and mechanical loading, the stress–strain state of the cylinder head and crankcase of a diesel engine on the basis of modeling of heat transfer processes in the solid, liquid, and gaseous systems under consideration.

The implementation activities included optimization of the mix parameters characteristics of the gas-turbine supercharging and fuel injection advance angle and reducing the length and increasing the fuel injection pressure. This provides the required fuel efficiency at a high specific power of diesel.

The dynamics of the piston on the lubricating layer in the cylinder of the engine are largely dependent on the profile of the skirt of the piston. The main hydromechanical characteristics (HMC) of the "piston–cylinder" tribosystem are $h_{\min}(\tau)$—instantaneous values of the minimum oil film thickness; $p_{\max}(\tau)$—instantaneous values of the maximal hydrodynamic pressure; $h_{\min}^*$—average value of $h_{\min}(\tau)$; $p_{\max}^*$—average value of $p_{\max}(\tau)$; $N(\tau)$ and N^*—instantaneous and average power loss of friction; Q^*—the average flow rate of oil in the direction of the combustion chamber; and T^*_{eff}—the average effective temperature of the lubricating layer. To assess the influence of the design parameters on the HMC of the interface "piston–cylinder" of diesel we performed parametric studies for the maximum torque mode.

In accordance with the calculation method given in [3, p. 249; 4, p. 519] we define the profiles of piston skirt in cold and hot conditions, and a profile was built up in the form of a parabola approximation (Fig. 2).

Table presents the results of calculation of the HMC of the "piston–cylinder" tribosystem for initial and recommended interface design before and after optimization. Additionally, we analyzed a non-symmetrical design of the skirt of the piston (see Table 1). It is clear from these results that the optimization of the geometric parameters of the original design improves HMC by 7%.

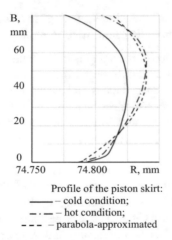

Profile of the piston skirt:
———— – cold condition;
— · — – hot condition;
- - - – parabola-approximated

Fig. 2 Profiles of the piston skirt

Table 1 The results of calculation of the HMC of the "piston–cylinder" tribosystem for initial and recommended interface design before and after optimization

Interface design	N^*, W	Q^*, cm³/s	inf h_{min}, μm	h^*_{min}, μm	supp$_{max}$, MPa	p^*_{max}, MPa
Initial	463.96	76.48	36.13	124.68	5.533	1.293
Optimized with symmetrical skirt	455.89	69.90	21.46	127.65	5.966	1.364
Optimized with non-symmetrical skirt	450.23	68.88	36.41	129.65	5.690	1.277

Discussion

The results given above can help to develop integrated technologies with the following features:

- optimization of the parameters of units of boost at a mass flow of air to 1 kg/s and a high degree of pressure increase with deep intercooling according to the criteria of fuel efficiency and thermomechanical loading of the diesel;
- surface profiling of the undivided combustion chamber and a piston skirt forming a bimetallic composite piston, with longitudinal contours of inlet and exhaust channels in the cylinder head to improve gas exchange in the cylinder;
- optimization of the fuel accumulating injection system for the fuel pressure to 200 MPa, and cyclic feed to 0.4 g per cycle;
- improved elements and cooling structure for reducing heat transfer to the coolant;
- improved reliability of the diesel, taking into account the high thermal mechanical loading, providing fluid friction mode in the basic tribounits.

The practical significance of the study is to implement an integrated approach to designing energy-efficient forced diesel on the basis of a differentiated analysis of the functioning of its basic mechanisms and systems. Suggested recommendations were checked using the new motorless bench [7, p. 2–3]. The original design of the bench allows us to register the development of fuel torches at up to 40.000 frames per second (Fig. 3) and can be recommended for the study of the different fuel systems.

Motor testing of diesel on the motor bench HORIBA-SCHENCK DT-2100-1 confirmed the results of the preliminary analysis of the effectiveness of implemented measures to improve the operating cycle. Suggested a new technical solution in the design of diesel engines can be recommended for the practical design of forced diesel engines based on existing and developed technologies for their practical implementation.

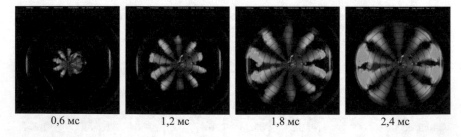

| 0,6 мс | 1,2 мс | 1,8 мс | 2,4 мс |

Fig. 3 The development of fuel torches in a constant volume chamber (sprayer 10×0.27 mm; $p_{t.p.} = 165$ MPa; $\tau_{y.inj.} = 1.5$ ms)

Comparison of the results of the study with the results of similar work indicates that the achievement of a given level of diesel by competitor companies is provided at a substantially greater weight and size with a similar level of fuel efficiency.

Conclusions

A brief analysis of current trends in the development of diesel engines and theoretical aspects of their implementation show that reserves are not exhausted in this direction. It is necessary to combine known and new methods and means of improving the operating cycle with advanced solutions in the design of the basic mechanisms and systems. As a result, we were able to achieve the required technical level. As a result, we were able to achieve the required technical level of the designed diesel.

Acknowledgements This research was carried out with the financial support of the Ministry of Education and Science of the Russian Federation for the implementation of applied research studies, lot 2014-14-579-0109. Agreement (contract) no. 14.577.21.0102, September 16, 2014. Unique identifier of Applied Scientific Research Project RFMEFI57714X0102.

References

1. Dober, G., Guerrassi, N., Karimi, K.: Mixture preparation and combustion analysis, a key activity for future trends in diesel fuel injection equipment. SIA Diesel Powertrain International Conference, Rouen, p. 10. http://delphi.com/images/news/2012/Delphi-Combustion-Technical-PaperSIA-Diesel-Rouen-2012.pdf/ (June 2012). 10p
2. Eagle, W.E., Morris, S.B., Wooldridge, M.S.: High-speed imaging of transient diesel spray behavior during high pressure injection of a multi-hole fuel injector. Fuel **116**, 299–309 (2014)
3. Goritskiy, Y., Ismailova, Y., Gavrilov, K., Rozhdestvensky, Y., Doikin, A.: A Numerical Model for Mechanical Interaction of Rough Surfaces of the "Piston-Cylinder Liner" Tribosystem. FME Trans. **43**, 249–253 (2015)
4. Goritskiy, Y., Gavrilov, K., Rozhdestvensky, Y., Doikin, A.L: A numerical model of mechanical interaction between rough surfaces of tribosystem of the high forced diesel engine. Procedia Engi. **129**, 518–525 (2015)
5. Kamaltdinov, V.G., Lysov, I.O., Nikiforov, S.S.: Diesel Engine operating cycle optimization with simulation of combustion process by double-Wiehe function. Procedia Eng. **129**, 873–878 (2015)
6. Kamaltdinov, V.G., Markov, V.A., Lysov, I.O.: Modeling the combustion process of a powerful diesel engine. Procedia Eng. **129**, 488–494 (2015)
7. Kamaltdinov, V.G., Rozhdestvensky, Y.V., Lysov, I.O., Popov, A.Y., Nikiforov, S.S.: Experimental investigations of the effects of electric control impulse on injection characteristics of common rail type injector. Indian J. Sci. Technol. **9**(42) (2016). doi: 10.17485/ijst/2016/v9i42/104225. 4p

8. Lazarev, E.A.: Basic principles, methods and means of improving the efficiency of the combustion process to improve the technical level of tractor diesel: a monograph. Publishing Center of SUSU, Chelyabinsk (2010). 289p

9. Mozer, Franc K.: Trends and solutions in the development of commercial diesel. Proceedings of the International Scientific Conference "Turbocharging of automobile and tractor engines", p. 6–12, Protvino, 24–25 June 2009

10. Mukchortov, I., Zadorozhnaya, E., Levanov, I., Pochkaylo, K.: The influence of poly-molecular adsorption on the rheological behaviour of lubricating oil in a thin layer. FME Trans. **43**, 218–222 (2015)

11. Postrioti, L., Buitoni, G., Pesce, F., Ciaravino, C.: Zeuch method-based injection rate analysis of a common-rail system operated with advanced injection strategies. Fuel. **128**, 188–198 (2014)

12. Lu, Y., Zhang, X., Xiang, P., Dong, D.: Analysis of thermal temperature fields and thermal stress under steady temperature field of diesel engine piston. Appl. Therm. Eng. **113**, 796–812 (2015)

13. Zadorozhnaya, E.A.: Solving a thermohydrodynamic lubrication problem for complex loaded sliding bearings with allowance for rheological behavior of lubricating fluid. J. Machinery Manufacture Reliability **44**(1), 46–56 (2015)

The Development of the New Type Universal Collective Survival Craft with Unmanned Control Function for Evacuation of Personnel in Emergency Situations of Natural and Technogenic Character on the Arctic Shelf

I.A. Vasilyev, R.A. Dorofeev, J.V. Korushova, A.A. Koshurina
and M.S. Krasheninnikov

Abstract The article presents the concept and the process of development of the new type universal collective survival craft. The article shows the arrangement and the interior of the vehicle and describes solutions for the main systems—chassis, life support systems, control systems, and auto piloting, showing decisions in the passenger compartment. In addition, authors describe the development process included in strength calculations of technical solutions. In conclusion, the article gives a conclusion on possible areas of use of the developed solutions.

Keywords Rotor-screw vehicles · Arctic · Rescue operations · Strength calculations Autopilot system computer vision · Robotics · Life boat · All-terrain vehicles

I.A. Vasilyev (✉)
Laboratory of Autonomous Systems and Adaptive Control of Robotic Systems, The Russian State Scientific Center for Robotics and Technical Cybernetics, Sankt-Peterburg, Russia
e-mail: vas@rtc.ru

R.A. Dorofeev · J.V. Korushova · M.S. Krasheninnikov
Nizhny Novgorod State Technical University Named After R.E. Alekseev,
Nizhny Novgorod, Russia
e-mail: blonde-o@yandex.ru

J.V. Korushova
e-mail: korushova.julia@mail.ru

M.S. Krasheninnikov
e-mail: maxim.krasheninnikov@mail.ru

A.A. Koshurina
ESC "Transport", Nizhny Novgorod State Technical University Named After R.E. Alekseev,
Nizhny Novgorod, Russia
e-mail: allakoshurina@yandex.ru

The Arctic is a special region, which in addition to the wealth of natural resources, characterized by difficult environment conditions: prolonged low temperatures, strong erosion effects of seawater, storms, strong gusty wind, poor visibility conditions, and others. The task of the Arctic region development is complicated by considerable remoteness from the nearest human habitat areas and a low level of infrastructure.

The rapid development of the Arctic region is an important issue for improving economic potential of Russia. Currently, the development of the Arctic region is associated with the development of oil and gas fields of the continental shelf of North Sea (Barents; Pechora and Okhotsk). The problem of development of a transport net has the highest priority for the region. The special role in this process is given to safety. The transport must perform complex functional tasks and conform to highest standards of reliability and safety.

Statistical analysis of accidents on offshore platforms, for the period from 1965 to 2016, confirms the ineffectiveness of conventional rescue facilities in the northern territories. The main restrictions are as follows:

- air rescue equipment (helicopters and small aircrafts) is limited by poor flight conditions, which are typical for the Arctic region;
- rescue equipments such as rafts and lifeboats are not able to move inter broken ice (about 9 months per year);
- submarines, in addition to the very high prices, have difficulties with mounting under the platforms (which have an extensive network of pipes under water surface). Also, it requires a large number of units of submarines for rescue of all personnel platforms (up to 250–350 persons).

The growth of victims number of accidents on the platforms is connected with the increase in jobs in the northern seas, in the absence of an effective rescue equipment. An average of 67% of staff dies at the time of the accident in the northern seas. For example, on December 18, 2011, 53 personnel members out of 67 died on Sakhalin Island during transportation of the drilling platform "Kola".

Scientists and engineers from Russia, Norway, Canada, and the United States are trying to solve the problem of the efficient rescue operations in the Arctic. The solution of the problem consists in the development of the new type of rescue equipment, which is the amphibious all-terrain vehicle. According to government estimates, 1000 units of rescue equipment may be required by 2020 year. This amount constitutes 7–10% of the total number of vehicles required for the Far North and the Arctic.

Currently, several all-terrain vehicles (wheeled, tracked, rotary-screw) are proposed for the task. The list includes hovercraft Arctica, all-terrain vehicles ARKTOS, Vityaz', A.R.C., and others. However, one of the key problems is the choice of construction materials and protection methods of icing and extreme overheating due to the vehicle passage through a burning oil spill (which can happen in accidents). Regulatory documents (SOLAS 74; GOST R 52638-2006;

Rules of the Russian Maritime Register of Shipping and the International Life-Saving Appliance Code) contain strict requirements of fire protection for rescue vehicles. The design of the rescue vehicle should allow being in an area of burning oil at least for 8 min. The vehicle immediately returns to the zone of low temperatures after overcoming of the area of burning oil. The temperature changing can cause high thermal stresses, leading to damaging of the material of the hull.

Currently, the team of the scientific and educational center "Transport" of Nizhny Novgorod State Technical University has developed the first experimental model of the universal rescue vehicle. This prototype is intended for testing of technical solutions for reduction of human casualties in accidents on offshore oil platforms. The rotary-screw mover was taken as a base for the rescue vehicles concept Figs. 1 and 2.

Particular attention was paid to the design of functional equipment and ergonomics. The psychological aspect has a great importance, because a person in a

Fig. 1 The visualization of the universal rescue vehicle appearance

Fig. 2 The universal rescue vehicle moving on water

stressful situation should make quick and right decisions. That was the main reason for the development of intuitive and friendly inner space of the vehicle. The basic sample of the rescue vehicle would belong to the light class rescue vehicles and be designed for 14 people and 2 crew persons, but also it can accommodate six more people if necessary. The protection of people inside the vehicle from overloads appearing during driving on difficult terrain (icefields, wormwood, movement in broken ice, and others) is the essential task. That is why every person in the vehicle is located on an individual seat (Fig. 3) with protection from mechanical vibrations. The seats have the possibility of conversion to a sleeping berth for a case of long rescue operation. Thanks to individually adjustable backrest, footrest, headrest, and lumbar support, the chair can be customized for each person.

The dimensions and the amount of the seats make a decisive contribution to the forming of the passenger compartment, the layout, and appearance of all survival crafts (Figs. 4 and 5).

The vehicle will have the following dimensions: length—11.5 m, width—5 m, and height—4.5 m. Total weight is 13 tons, of which up to 3 tons is load capacity of the vehicle. The layout of the universal rescue vehicle is shown in Fig. 6.

In accordance with the above scheme of the vehicle, the pilot cabin is located in the nose, then the passenger cabin, bathroom and utility room, and the room for the power plant. Other technical systems are located under the passenger compartment floor.

The universal rescue vehicle is developed for autonomous work for 3 days. The maintenance of comfortable conditions in the vehicle during the operation is achieved through the following sections: 1. the compartment-containing toilet, shower, sink, and cabinet for clothes (Fig. 7); 2. the room with basic supplies of food and medicine, as well as a mobile kitchen for cooking or heating food (Fig. 8).

Fig. 3 The individual seats

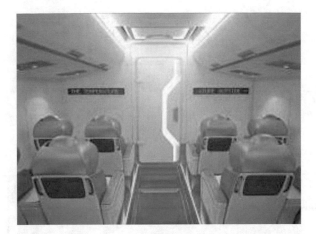

Fig. 4 The passenger compartment (the front view)

Fig. 5 The passenger compartment (the rear view)

The vehicle is equipped with life support systems to protect from the major risk factors associated with the possibility of overcoming the burning oil stains, long stay in low-temperature conditions at high humidity, and others. The main systems are as follows: firefighting, autonomous air supply, de-icing, climate control, autonomous power supply, sound insulation, vibration isolation, and lighting. The most important systems are duplicated.

The issue of permeability and high driving quality is of paramount importance for the Universal rescue vehicle. The movement on difficult terrain (ice fields, broken ice, hummocks, and others.) sets the task of creating a unique multi-function universal chassis system. This system was developed in the course of the project. The vehicle has a suspension system, which adapts its own characteristics to the

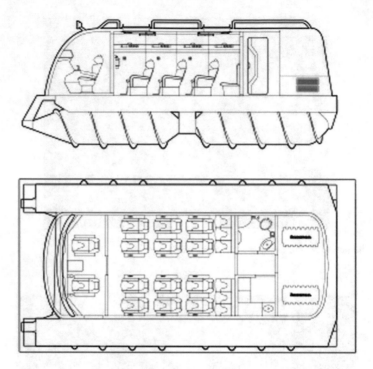

Fig. 6 The layout of the universal rescue vehicle

Fig. 7 The restroom

type of support base. In accordance with the data of the work [2, p. 328] the vehicle must have the ability to climb from water to ice barrier at the height of 2 m. For this task, the chassis system has the opportunity to change the position of basic elements [1, p. 1266]. The significant suspension travel can significantly reduce the dynamic loads on the passengers and crew.

Fig. 8 The utility area

Rotary-screw propeller will experience cyclic loads. One of the characteristic modes of the load is the process of the entrance from water to ice. Performed strength calculation has confirmed the correctness of design decisions [2, p. 5606]. The stress level does not exceed 250 MPa. It corresponds to the zone of elastic deformation for the mover's material. Figure 9 shows the stress–strain state of the rotary-screw mover during the entrance from water to ice.

Arctic operating conditions set special requirements to structural materials for engineering. The choice of materials is of particular importance for the Arctic rescue vehicle. Materials must conform to difficult climatic conditions and must not allow degradation of properties in significant temperature ranges. The combination of all of the desired characteristics is made possible by a combination of modern materials. The use of sandwich panels for the construction of the hull allows a low specific gravity to achieve the required performance for stiffness and strength of the hull.

The autopilot system was created for the pilot workload reduction. The system provides quick deciding in a disasters situation, as well as in a complex and changing terrain. The system provides the following kinds of control: simple, alert, tracking (semi-automatic), adaptive, and navigation by GPS ("movement along a predetermined path"). The system is based on a vision system that combines sensors with different physical natures: video cameras, range finders, radar, sonar, and others. Complex vision system provides spatial orientation in difficult visibility conditions.

The composition of the technical equipment of computer vision system is a hardware–software complex "improved vision" (Enhanced Vision Systems, EVS). EVS system-generated random image information appears below the pilot in real time on the multi-function remote control as a panoramic image of the situation behind the cabin. The autopilot system and the supervision system were carried out jointly with the Russian State Scientific Center for Robotics and Technical Cybernetics (RTC).

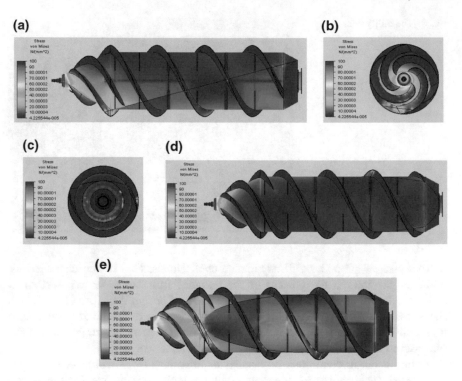

Fig. 9 The strain–stress (the scale 1 x 1): **a** *left*; **b** *front*; **c** *rear*; **d** *top*; **e** *bottom*

Currently, the experimental model of the universal rescue vehicle was built (Fig. 10). The model has the following characteristics: length 5.5 m, width 2.6 m, height 2.4 m, Curb weight/gross 3500/4500 kg, diesel engine 186 kW, hydraulic transmission, and the maximum speed in water 12 km/h and in the snow 23 km/h.

The sample is equipped with the original integrated onboard control system with the function of the autopilot, including navigation systems and computer vision. At the end of 2016, complex tests of the vehicle are planned. The main goal of the tests is the estimation of driving performance on different surfaces, including open water.

Fig. 10 The experimental model of the universal rescue vehicle

Conclusions

The level of the developed technical solutions allows their applying in other areas of potential use of rotary-screw vehicles. It was achieved, thanks to the complexity of the assigned task and the depth of study. The universal multi-function chassis might be efficiently used in the following areas: rescuing people, securing the Northern Sea Route, environmental monitoring, execution of works in shallow water, elimination of oil spills, pipeline construction, and geological prospecting. Patents protect results of intellectual activities of the project team.

Acknowledgments Research is carried out (conducted) with the financial support of the state represented by the Ministry of Education and Science of the Russian Federation. Agreement (contract) no. 14.577.21.0105, September 16, 2014. Unique project Identifier: RFMEFI57714X0105.

References

1. Koshurina, A.A., Krasheninnikov, M.S., Dorofeev, R.A.: Strength calculation and analysis of equalizer beam embodiments for the operated equalizing beam suspension of the Universal rotor-screw rescue vehicle for the arctic. Procedia Eng. **150**, 1263–1269 (2016)
2. Skutin, A.A., Naumov, A.K., Kubyshkin, N.V.: Evaluation of morphometric characteristics of the icebergs in the Barents Sea with the full data for modeling the drift // Sbornik trudov 10-j Mezhdunarodnoj konferencii po osvoeniju resursov nefti i gaza Rossijskoj Arktiki i kontinental'nogo shel'fa stran SNG "RAO/CIS Offshore-2011", SPB (2011)
3. Koshurina, A., Blokhin, A., Krasheninnikov, M., Dorofeev, R.: The strain-stress state of the rotor during entrance of the rotor-screw vehicle from water to ice. ARPN J. Eng. Appl. Sci. **9**, 5603–5607 (2016)

Development of Active Safety Software of Road Freight Transport, Aimed at Improving Inter-City Road Safety, Based on Stereo Vision Technologies and Road Scene Analysis

V.E. Prun, V.V. Postnikov, R.N. Sadekov and D.L. Sholomov

Abstract The article considers the active safety system of road freight transport. The stereoscopic computer vision is the core of the system. The article also describes the major algorithms of active safety and the accuracy characteristics of algorithms' application.

Keywords Advanced driver assistance systems · Computer vision
Stereo vision · Road scene analysis · Road scene segmentation
Road-shape recognition · Object detection · Pedestrian detection
Vehicle detection · GPGPU

Introduction

It is widely known that the major causes of road accidents (about 85%) are the mistakes and non-compliance with the traffic laws. Speeding, crossing into oncoming traffic, inattentive driving, sleeping during the driving, and other causes are the most frequent driving mistakes. One of the ways to improve the driving safety is to deploy active safety systems, aimed at timely notification of a driver of

V.E. Prun (✉) · V.V. Postnikov · D.L. Sholomov
JCS "Cognitive", Moscow, Russia
e-mail: v.prun@cognitive.ru

V.V. Postnikov
e-mail: pvv@cognitive.ru

D.L. Sholomov
e-mail: d.sholomov@cognitive.ru

R.N. Sadekov
MEI "Institute of Engineering Physics", Moscow, Russia
e-mail: sadekovlar@mail.ru

hazardous situations. Such systems are known as advanced driver-assistant system (ADAS) [1].

The potential of such systems is being intensively developed, primarily due to the technological equipment of the vehicles and the development of intelligent algorithms for the analysis of road scenarios. Many automobile manufacturers have already introduced for additional car features to facilitate the process of driving and to make it more secure and more comfortable. The laws enable the progress in the deployment of such systems, so that the introduction of active safety systems will be mandatory for new models of trucks from 1 Nov 2018 (EU Directive 3472012). Hence, the systems and their functional capabilities will be further evolving, which emphasizes the rationale for the research.

Related Works

Many automobile manufacturers as well as representatives of the research community are currently working in the field of development of 'advanced driver assistance systems'. The existing projects may differ in the techniques used: high-performance camera [2], radar, camera, and ultrasonic sensor [3], so in a list of solvable problems, e.g., speed bump detect [4], pedestrian detect [5], and vehicle detect [6].

At the Russian national level the 'advanced driver assistance systems' and the robotic transport systems are best developed by 'KB Avrora' [7] and the Central research and development automobile and engine institute NAMI [8].

The results presented in this paper are new both in terms of the techniques and the algorithms used, so they will be of great interest and used for both national and international developers.

General Description

The objective of the project is to create advanced driver assistance systems that provide the truck driver with efficient information on dangerous traffic situations according to the stereoscopic vision systems.

The introduction of active safety systems will increase the level of road safety, will reduce the number of road accidents, and will reduce the cost of ownership of the vehicle.

The following hardware configuration was used to implement the project. A laptop with Intel core i7 processor, 8 GB of RAM, and a video card GTX960 m (2 GB VRAM) served as an estimator for the stereo system. Two USB cameras (IDS) are combined with stereo rod with a 1.57 m base. The cameras are synchronized via the synchronization cable in a master–slave mode. The degree of mal-synchronization is less than 5 ms. Geolocation sensor and strapdown inertial navigation system Xsens with integrated GPS antenna are used. Additional integration with sensors is also possible. Lidar is mounted on the front bumper and scans in a parallel to the road surface. Front radar is with narrow and mean angles, and infra-red range measuring system.

C++ with the library Open CV 3.1 [9] was the main development language. The complex is cross platform. The software is widely employing the parallel computing on multicore processors GPU with CUDA technology to provide the performance of ~ 10 frames per second.

Due to some limitation in scope, the paper deals only with the following most in-demand driver assistance systems: lane departure warning system, pedestrian protection system, forward collision warning system, and traffic sign recognition system.

Lane Departure Warning System

Lane departure warning system can work on roads with clearly visible lane markers at the speeds exceeding 50 km/h. The following sequential steps constitute the lane detection algorithm: identifying the road lanes, determining the parameters of lane markers, and selecting of road lanes.

First, the image I_t is filtered aimed to identify the lines which are road markers. The procedure is accomplished by using the difference of gaussians. This procedure converts the original color image (Fig. 1a) to a grayscale one.

Second, the fast Hough transform is used to identify the road lanes. The Hough transform is a numeric technique used in detecting the lines in an image. The fast Hough transform makes it possible to determine the common vanishing point of the lines in the image.

Third, the lines detected are checked to be parallel and further these lines are selected. The information from inertial sensors is employed at this stage, which

Fig. 1 The results of lane edges detection on the highway

allows to identify the horizon in the image (Fig. 1), as well as the number of road lanes.

Then, one can determine the dependence of the brightness of pixels on the length of the horizon line. The dependence has several extrema, which correspond to abscissas of the intersection points of the lines detected with the horizon. Considering the data on the number of the road lanes, we select the brightest intersection points, which are further referred to as road marking lines.

This software was tested on a set of the 5-s-long video clips. Actual vehicle departure off the road was visually determined in a case when the wheel crossed 10 cm or more over the road marking line (in compliance with the European standard [10]). The total number of the lane-changing video clips equaled 122. The tests of the proposed algorithm software showed that the true positive index (TP) was 83% while the number of false negatives (FN) was 1.2%.

The LDWS on a truck is active only when the lane changing is not supported by the turn signal.

Pedestrian Protection System

The pedestrian protection system is active all the time since the turn on of the system. The pedestrian protection algorithm, which is the core of the system, is built on the decision trees with broad attribute space [11]. The LUV transformation, the Gradient magnitude, and the Histogram of gradient are used to estimate attributes.

By the LUV transformation, we mean color space CIE L*u*v*, which is calculated from an image in RGB format, according to the known empirical formulas given in [12]. This transformation allows to perform image norming, to allocate the most effective signal and to even the outliers. The gradient is calculated in each point of the image using a formula of difference along two directions by applying a mask.

The histogram of oriented gradients (HOG) is formed as a sum of gradient module in the directions. The histogram consists of a number of cells. The size of each cell is 6 x 6 pixels.

Therefore, the attributive description of X pedestrians is represented by a vector comprising three LUV transformation parameters, one M, and six HOG parameters.

The method of decision trees with a depth of two was used as the classification method. The attributive information and its threshold value are located in the nodes of trees (obtained in the training stage) which is used to classify the measurements into two categories—with or without a pedestrian ($Y = \{-1, 1\}$) in the image, respectively.

The AdaBoost algorithm is employed to train the classifier. Further two samples are developed for the learning procedure: positive sample with images of pedestrians and negative one with other images. The 90 x 36 pixel-size images with pedestrians in different weather conditions (winter, summer, rain) were used as reference images. The training samples can be mathematically represented as pairs

$\{\mathbf{X}_1, \mathbf{Y}_1\},\ldots,\{\mathbf{X}_m, \mathbf{Y}_m\}$, where m is a number of pairs of values used for the training purposes. The classifier F: $F(\mathbf{X}) \rightarrow \mathbf{Y}$ is involved in the process of training the AdaBoost algorithm.

To detect pedestrians in the image we use previously trained classifier, and the detector receives a fragment of the original 90×36 size image. The search for pedestrians throughout the image, as well as the removal of constraints on the size of the pedestrians, is carried out by the use of sliding window and image scaling technologies. The reduction of the first and second kind errors is carried out by the use of the information on the scene geometry. The results of the pedestrian detection are exemplified in the figure (Fig. 2).

The detector application in the image is accelerated by techniques [11]. The results of testing for field data have proved that the TP probability of the system is 78.7%, and the FN is 9.4%.

The driver's alert of a potential hazardous proximity with a pedestrian is based on the calculation of time T_m (time to collision) before the collision and comparing it with a certain threshold, based on the equation

$$T_m = -Z/V = \frac{\Delta t}{S-1} > t_{add},\tag{1}$$

where Z is the relative distance between the objects, V is the relative speed, $S = w_{t-1}/w_t$, $w_{t,t-1}$ is the width of the rectangular area in the image describing the pedestrian at time instants t and $t-1$, respectively, Δt is the time interval, t_{add} is the tolerance when exceeded, and an 'alert' message is activated.

Equation (1) is true at cruise. Given the acceleration, which is a frequent case because of heavy braking, Eq. (1) may be as follows:

$$T = T_m \cdot \frac{1 - \sqrt{1 + 2 \cdot C}}{C},\tag{2}$$

Fig. 2 Pedestrians, detected in the image (marked with rectangles), and the line of the local horizon (thickened line)

where $C = 1 + \frac{T_m^t - T_m^{t-1}}{\Delta t}$.

Forward Collision Warning System

The system is implemented by the use of a hybrid detector. The system operation is carried out at speeds exceeding 5 km /h. The proposed algorithm for pedestrian detection, which is trained on samples containing vehicles, is used as a first detector; the AdaBoost with Haar-like features [Haar-like features] is the second method of detection.

Haar-like features consist of rectangular areas within which the intensities of the pixels of the original image (in grayscale) are summed up with weights $\{-1, 1\}$ for the black- and white-colored rectangles, respectively.

Examples of Haar-like features for detecting vehicles moving behind are presented below (Fig. 3). The first figure focuses on detection of the vehicle bumper, and the second and third figures show the side arches of the vehicle.

The AdaBoost algorithm employs decision trees with a depth of one as the weak classifiers. The detectors are simultaneously run, and each of them gives solutions with a probability from 0 to 100%. The probability is statistically calculated, by counting the number of matches of the objects found in the original image, taken at different scales and by moving the sliding window. The objects found by the first and the second detectors are summed and further filtered in the final stage with the information on the scene geometry. Upon the application of filtering, we evaluate the possibility of finding the object's height h at a given point of the scene, by using information from a stereo system; the object is excluded in a case when the operation is not possible.

Haar-like featured detectors proved to be good at detecting far-away objects. It works on a wide scale pyramid and detects objects ranging in size from 30 to 300 pixels. The average time of such a detector in the FULLHD image equals 50 ms. The realtime performance of object detectors is obtained by using geometric

Fig. 3 Examples of Haar-like features used in the system

Fig. 4 Vehicles detected in the image (marked with *rectangles*) and the line of the local horizon (*dark thickened line*)

filtration technique, which is often reffered to as foveal vision module [13, 15]. The accuracy of detection is increased using image spatial normalization—forward rectification [14]. The quality of the detector operation is TP 86, 9% while the number of false negatives FN is 7% (Fig. 4).

Traffic Sign Recognition System

This system is designed to alert the driver of violations associated with the non-fulfillment of the traffic signs requirements.

A few detectors, each designed for the localization of particular category signs, are employed to detect road signs. The detectors used in the system are divided into two categories: first, 'bitmap' detectors, which calculate features on the fragment bitmap, and second, 'vector', which identify related components and analyze the latter by their shape and color distribution. The above division into categories is rather relative because the two types of detectors respond to both color and shape, but show their specific character.

Bitmap detectors use the Viola-Jones method, and the training is performed on a special color channel, calculated for a specific family of marks. For example, the channel $I = k_R R - k_G G - k_B B$ is used for limiting and warning signs, while the yellow-frame signs are calculated with $I = k_G G - k_B B$, and k_R, k_G, k_B are real coefficients. The Viola-Jones method uses features that are very similar to Haar-like features. Within the work done, we considered the features, aimed at recognition of rectangular, diamond-shaped, triangular, and circular forms. The system uses four bitmap detectors aimed at the detection of each class of marks and colors such as 'bitmap detector of rectangular blue marks'. These detectors are useful in a range of distances from 5 to 50 m on a wide-angle camera.

Vector detectors are used for an image, which is four times diminished to ensure the search speed. This detector is designed primarily for searching the signs at short distances (up to 30 m). The vector detector builds color masks, based on the allocation of the required color in the HSV color space; then, it carries out morphology operations, defines related components, and selects objects in the image. It later filters components by selecting smooth-edged signs on the basis of the matrix filters and accepts the samples. The vector detector filters by finding the objects satisfying the predetermined geometrical shapes (circle, rectangle, triangle, and diamond) with dynamic programming methods.

The information on regions of interest intended to search for signs in places of their probable location can speed up the detectors' performance.

The vector detector is complemented by OCR module (Optical character recognition) for some groups of signs (such as 'speed limit'). This module also uses the analysis of related components; it removes decoys by calculating the width and the height ratios of the objects found, and it compiles and recognizes text fragments in the alphabet, which is characteristic for the analyzed subset of signs. The algorithms used were proposed by OCR Cognitive Cuneiform, Cognitive Forms, developed by the JSC 'Cognitive'.

The hypothesis, which includes the information on the detected region of an image, on a presumed sign group to which the latter belongs and on a priority, is a result of the detecting unit. The highest priority is given to the hypothesis, supported by both detectors. Coincidence is determined by calculating the degree of overlap of regions and comparing it with a threshold.

Neural network approach is further used to classify road signs. Neural network classifier is based on a fully related neural network with two hidden layers, trained to operate with large amounts of data. A three-channel color image of a road sign in the space of BGR ($16*16*3 = 768$ neurons for the first level of networks) is fed at the input of the network. The output layer contains 57 neurons, corresponding to the desired classes of signs.

We used 50 thousand signs, marked on the video frames to train the network. Real and synthesized images of signs should be used while training the second level of networks. The synthesized images were generated taking into account possible

Fig. 5 Traffic sign recognition system (sign 'pedestrian crossing', the probability of detection = 1.0 for the *right* sign and = 0.97 for the sign on the *left*)

projective distortions, rotations, displacements, and other noises, typical of real images. The training sample set of networks included 'failure' classes, which do not contain fragments of road signs. Without taking into account the work of detectors, an independent assessment of the accuracy of the network was 99%. An example of the traffic sign recognition system is shown in the figure (Fig. 5).

The results of the complex testing of the whole sign processing subsystem (which includes a set of detectors and classifiers) for the advanced driver assistance system showed the following characteristics of TP, which is 80–85.4% and FN ~ 6.2–10% (excluding stereo matching procedure).

Conclusions

The paper considers the active safety system of road freight transport. The main element of the software complex is a stereo camera. The paper describes the operation of the core systems of the software complex. There have been developed training and testing systems with the image database as a result of the project execution. The performance of all systems involved is of a frequency not lower than 10 Hz.

Acknowledgments The research is carried out with the financial support of the state represented by the Ministry of Education and Science of the Russian Federation. Agreement no. 14.582.21.0002, September 29, 2014. Unique project Identifier: RFMEFI58214X0002.

References

1. Mobilyeye/Main page (Online resource). Available from http://www.mobileye.com. Accessed on: 02.07.2016
2. Mihir, M., Swami, P.: High performance front camera ADAS applications on TI's TDA3X Platform. In: IEEE 22nd International Conference on High Performance Computing, pp. 1–8 (2015)
3. Khan, J., Elxsi, T.: Using ADAS sensors in implementation of novel automotive features for increased safety and guidance. In: 3rd International Conference on Signal Processing and Integrated Networks, pp. 753–758 (2016)
4. Devapriya, W., Nelson, C. Kennedy Babu, Srihari, T.: Advance driver assistance system (ADAS)—speed bump detection. In: International Conference on Computational Intelligence and Computing Research, pp. 1–6 (2015)
5. Geronimo, D., Lopez, A.M., Sappa, A.D.: Survey of pedestrian detection for advanced driver assistance systems. IEEE Trans. Pattern Anal. Mach. Intell., 1239–1258 (2010)
6. Lee, K.J., Kim, C., Lee, K.-R.: A 502-GOPS and 0.984-mW dual-mode intelligent ADAS SoC with real-time semiglobal matching and intention prediction for smart automotive black box system. IEEE J. Solid-State Circuits, 1–12 (2016)
7. KB 'Avrora'/Main page (Online resource). Available from http://www.kb-avrora.ru/projects/avtopilot-na-baze-gazel-biznes.html. Accessed on: 02.07.2016

8. Central Research and Development Automobile and Engine Institute NAMI Project Lada Kalina/Main page (Online resource). Available from http://nami.ru/projects/drone. Accessed on: 02.07.2016
9. Open CV/Main page (Online resource). Available from http://opencv.org/. Accessed on: 02.07.2016
10. EUR-Lex. Acessto European Union Law/Main page (Online resource). Available from http://eur-lex.europa.eu/legal-content/EN/TXT/?uri=CELEX%3A32012R0351. Accessed on: 02.07.2016
11. Dollar, P., Perona, P.: The Fastest Pedestrian Detector in the West (2010)
12. CIELUV/Main page (Online resource). Available from https://en.wikipedia.org/wiki/CIELUV. Accessed on: 02.07.2016
13. Postnikov, V., Krohina, D., Prun, V.: Road shape recognition based on scene self-similarity. Proc. SPIE Vol. Vol., 9445 (2015)
14. Prun, V., Polevoy, D., Postnikov, V.: Forward rectification—spatial image normalization for a video from a forward facing vehicle camera. In: Ninth International Conference on Machine Vision, International Society for Optics and Photonics, pp. 103410W–103410W (2017)
15. Prun, V. et al.: Geometric filtration of classification-based object detectors in realtime road scene recognition systems. In: Eighth International Conference on Machine Vision (ICMV 2015), 9875p (2015)

Analysis of the Stress State in Steel Components Using Portable X-Ray Diffraction

S.A. Nikulin, S.L. Shitkin, A.B. Rozhnov, S.O. Rogachev
and T.A. Nechaykina

Abstract Comparative measurements have been performed on 20GL steel and grade 2 steel samples put under load in the elastic area in a three-point bending test to evaluate stress using stationary and new developed portable X-ray diffractometer. The results obtained showed that there was a good concurrence in the stress values measured by the stationary and portable diffractometers to within an accuracy of 10% (adjusted for calculation and measurement errors). The experimental results confirm the promising outlook for the practical application of developed portable X-ray diffractometer.

Keywords Railway transport components · X-ray diffraction · Residual stress Portable X-ray diffractometer

Introduction

The X-ray technique occupies a special place among existing non-destructive testing methods of measuring mechanical stresses in metals and alloys such as ultrasonic [1, p. 750], magnetic [2, p. 620; 3, p. 9], laser-interferometric [4, p. 45], acoustic [4, p. 46] and other methods, because it is the only direct non-destructive

S.A. Nikulin · A.B. Rozhnov (✉) · S.O. Rogachev · T.A. Nechaykina
National Science and Technology University MISiS, Moscow, Russia
e-mail: rojnov@nm.ru

S.A. Nikulin
e-mail: nikulin@misis.ru

S.O. Rogachev
e-mail: csaap@mail.ru

T.A. Nechaykina
e-mail: nechaykinata@gmail.com

S.L. Shitkin
Joint Stock Company Railway Research Institute, Moscow, Russia
e-mail: shitkins@yandex.ru

© The Author(s) 2018
K.V. Anisimov et al. (eds.), *Proceedings of the Scientific-Practical Conference*
"Research and Development - 2016", https://doi.org/10.1007/978-3-319-62870-7_24

method of residual stress evaluation as it offers direct measurement of the crystal lattice deformation by displacement of diffraction peaks [5, p. 367; 6, p. 12; 7, p. 201]. Although it achieves stress measurements with high accuracy, the X-ray method requires a special preparation of the surface of a measured specimen. The measurement tool which the technique relies on is the X-ray quantas having wavelength which is commensurate with the measured spacing in the crystal lattice of a metallic structure, and the reflection angle from the crystal lattice planes is strongly correlated with the wavelength of X-ray radiation and the lattice spacing in a crystalline sample and is governed by the diffraction equation. The X-ray method to measure surface stresses which is used in the diffractometer is based on the registration of diffraction peaks which shift relative to their unstressed position on the surface of a specimen.

In order to ensure radiation safety, modern diffractometers use "soft X-rays" with a quantum energy not greater that 20–30 keV which allow residual stress measurements through the thickness of a thin surface layer equal to the penetration depth of X-rays (not exceeding a few tens of micrometers for steel [8, p. 102]). This feature allows measurement of only surface stresses using an X-ray method.

Following stationary X-ray diffractometers used for laboratory measurements, portable X-ray diffractometers that appeared a few years ago can be taken out into field for measurements of stresses in a structural material or component. This is partly due to the adoption of a new solution built on the application of a non-focusing X-ray optics technique for the parallel primary beam (The Debye-Scherrer Scheme) which makes not only the parallel surfaces of a specimen, but all crystal lattice planes involved in the diffraction process [9, p. 317, 5. p. 73].

Introduction of new methods, technologies, and diagnostic equipment often requires special tuning or ad hoc adjustments, which may result in a number of comparative verification tests to be conducted. This paper describes the results obtained from verification testing of an originally developed portable X-ray diffractometer onto steel samples cut out from railway transport components.

Materials for Analysis

Two types of steel materials were used in this study: 20GL steel (used for solebars of a bogie frames) and grade 2 steel under Russian Standart 10791 (used for railway wheels). The chemical composition of the two types of steel is shown in Table 1. Samples having a length of 60 mm, width of 10 mm and thickness of 1 mm were cut out by the method of electro-erosive cutting from full-scale railway transport components. The samples were studied in the normalized condition, but some samples were additionally annealed at 600 °C with 1 h exposure and slow cooling (with furnace).

Table 1 The chemical composition of the steel samples cut out from railway transport parts

Steel	Chemical composition, mass%								
	C	Mn	Si	Cr	Ni	Cu	P	S	Other
20GL	0.19	1.33	0.46	0.21	0.31	0.25	0.016	0.006	Al = 0.03
Grade 2	0.61	0.76	0.27	0.29	0.12	0.27	0.012	0.011	V = 0.07

Method of Analysis

Preparation of the surface. The X-ray measurement site of each sample surface was subjected to grinding followed by polishing and electrolytic etching to remove work hardening which may be present on the component surface as a result of both the manufacturing and operational processes and cutting out test samples from the components.

The samples were manually grinded and polished with a portable polishing machine Accupoll by starting with P 400 lower grade grit sandpaper and ending with P 2500 higher grit sandpaper. The surface received its final polish by means of a piece of velvet with abrasive suspension Masterprep (0.05 microns) applied onto it. The polishing process took 2–3 min and resulted in the polished surface having a mirror bright finish. To remove the work-hardened layer, electrolytic etching using a 4% HNO_3 solution in ethanol to a total depth of 200 µm was carried out by use of an electrolytic etcher Polimat 2. Each sample was prepared for the study in the middle of its surface (Fig. 1).

Mechanical loading. The tensile stress was created by placing the samples under load by the three-point bending method in a special loading device (Fig. 2).

Stress was created by applying a load to the middle of the sample by driving a screw. Surface stress measurements were made at a point opposite to the point of load application located on the surface of the deformed sample. The value of stress created was set by turning the loading screw a set number of times. Preliminary calculations showed that a significant surface elastic tensile stress which would not

Fig. 1 The appearance of a sample to measure surface stresses

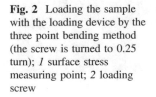

Fig. 2 Loading the sample with the loading device by the three point bending method (the screw is turned to 0.25 turn); *1* surface stress measuring point; *2* loading screw

exceed the yield strength of material could be created by taking 0.5–1 full turns on the screw. This is the way each sample was tested under the same loading to determine surface stresses.

Stress Measurements. Stress measurements were performed by means of a mock-up of the portable X-ray diffractometer originally designed and a serial stationary X-ray diffractometer Rigaku SmartLab. When preparing a sample for measurements it was put under pressure by taking turns on the loading device screw, so that the sample experienced stress in its central part. The screw was turned 0.5, 0.75, and 1.0 full turns.

The designed portable X-ray diffractometer included the following main units: a position-sensitive detector; two X-ray tubes combined with collimation devices; a goniometer; a special tripod for mounting the goniometer and its location relative to a measured component; laser tuning and alignment devices; a power supply unit and electronics interfaced with a personal computer. The design of the tripod and goniometer allows the Bragg plane where X-ray tubes and detector located to be tilted forward and backward at 30° in the frontal direction (Fig. 3).

The measurements were performed using the method of slope. Diffraction peaks were recorded using CrKα radiation for multiple locations of the X-ray tube (angle values = 0°, 10°, 20°, 30°) from diffraction plane 211. Then the locations of the "centers of gravity" of the diffraction peaks were determined after cutting out the background by means of special software "StressControl" and auxiliary applications.

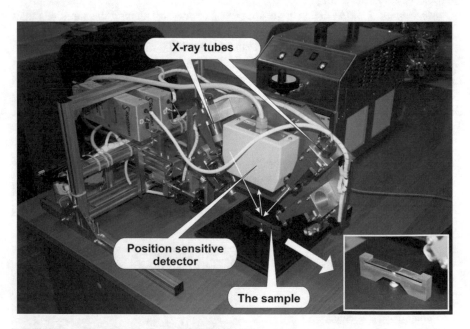

Fig. 3 The sample placed under portable X-ray diffractometer

A linear relationship between relative deformation ε and angle shift of diffraction peak allowed using generalized Hook's Law, to determine surface stress σ in the direction normal to the tilting axis of a Bragg plane on the specimen surface through a tangent of the slope of the straight line $\varepsilon = k \times \sin^2 \psi + b$, where $\sigma = k \times E / (1 + \mu)$, where E, μ—X-ray elastic constants (elasticity modulus, Poisson ratio), b—unspecified coefficient which determines straight line position at $\psi = 0°$.

To verify the test results, additional measurements were made on the same samples held under the same loading conditions using the multipurpose X-ray diffractometer Rigaku SmartLab (Fig. 4) and with application of CuKα radiation and a graphite monochromator. Measurements of stress were conducted by the "$\sin^2 \psi$" method in the "pseudo" slope of the sample, i.e., by the method of rotations. A diffraction plane was used (310) to analyze macro-stresses. Diffraction peaks were registered by use of a parallel X-ray beam to reduce the effects of tilt defocus. The sample was tilted (turned) by $\psi = 0°$; $30°$; $45°$ and $60°$.

After this, analogous to measurements by portable X-ray device, data were processed with software the stress magnitude was calculated by

- determining the position of the centers of gravity of the diffraction peaks (2θ value) after the background has been cut out, where θ—diffraction angle;
- calculating the relative deformation of the sample ε caused by loading applied by formula:

$$\varepsilon = \text{ctg}\theta_0(\theta_0 - \theta_\psi), \tag{1}$$

Fig. 4 The sample placed in the loading device shown in the chamber of the X-ray diffractometer Rigaku SmartLab with designated angles

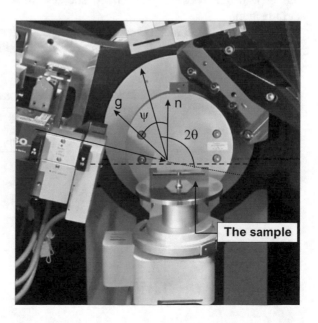

where θ_0 is the initial angle of diffraction. This corresponds to a specimen in unloaded condition; θ_ψ is the offset angle of diffraction (formed by deformation under slope of a Bragg plane to the angle);

- using the $\ll \sin^2\psi \gg$ method to build a graphical dependence between the relative deformation ε of the crystal lattice and $\sin^2\psi$ in the direct form $\varepsilon = k \sin^2\psi + b$ by the method of least squares. The magnitude of the residual stress in the direction on the specimen surface coinciding with the Bragg tilt plane (normal to tilting axis of a Bragg plane) was determined on the basis of the linear relationship built. The magnitude of the stresses is proportional to the tangent of the slope of the linear relationship established (2):

$$\sigma = k \frac{E}{1 + \mu},\qquad(2)$$

where E and μ are Young's modulus and Poisson's ratio for this material correspondingly;

σ is the stress in a given direction.

Relative error of stress value under all measurements was determined taking into account a scatter of diffraction peak locations for different ψ angles.

Analysis Results

The results of the stress measurements are shown in Table 2, and Fig. 5 illustrates the dependence between the surface stresses measured and the magnitude of deformation.

Table 2 and Fig. 5 make is evident that the surface tensile stresses measured on all the samples by use of both diffractometers increase linearly with further increase

Table 2 Tensile stress measurements results obtained on the steel samples

Steel	Heat treatment	Number of screw turns, n	Stress obtained by the stationary diffractometer, MPa	Stress obtained by the portable diffractometer, MPa
Grade 2	Annealing	0.00	–	0
	Normalization	0.00	+37	0
		0.50	+147	+240
		0.75	+213	+280
		1.00	+279	+330
20GL	Annealing	0.00	–	0
	Normalization	0.00	+40	+20
		0.50	+181	+220
		0.75	+254	+280

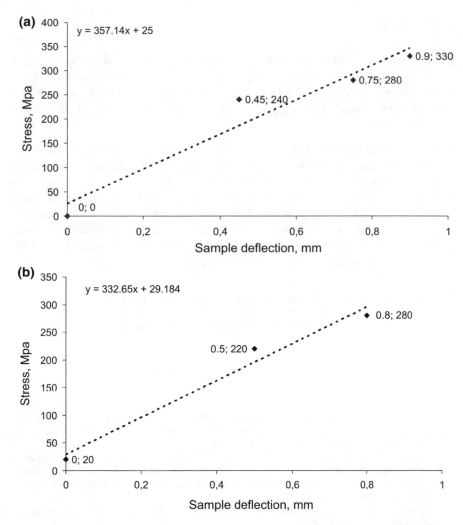

Fig. 5 Relation between stresses and deflection determined on the samples in the three point binding: **a** grade 2 steel sample, GOST 10791, after normalization; **b** the 20GL steel sample after normalization

of each sample deflection. For the annealed samples, as expected, the value of residual stresses is equal to zero (under zero deflection).

Table 2 shows good coincidence of the results obtained by the portable and stationary diffractometers, in particular when the screw is turned to 1 turn (maximum tensile stress).

When the screw is less 1 turn out (0.5), which results in lower tensile stress, there is a greater discrepancy in the results which can be explained by errors in the calculation of diffraction peaks with a slight (weak) shift of the diffraction peak relative to its position in the unstrained condition.

Conclusions

1. By displaying zero values stress measurements performed on all the annealed samples proved the correctness of the angular setting of the portable X-ray diffractometer and pointed out the possibility of using such samples to conduct a preliminary check on the angular setting and efficiency of the portable X-ray diffractometer.
2. Tensile stress measurements performed by means of the portable and stationary X-ray diffractometers showed a linear increase in tensile stress with increasing degree of deformation by bending in all the samples.
3. The comparison of the surface stress measurements performed on the steel samples by use of the portable X-ray diffractometer with the measurement results obtained by the stationary X-ray diffractometer and calculations made showed a good concurrence in the stress values to within an accuracy of 10% (adjusted for calculation and measurement errors).

Acknowledgements Researches are carried out with the financial support of the state represented by the Ministry of Education and Science of the Russian Federation. Agreement no. 14.578.21.0020 06.June 2014. Unique Project Identifier: RFMEFI57814X0020.

References

1. Anisimov, V.A., Katorgin, B.I., Kutsenko, A.N., others.: Acoustical Tensometry. In: Non-Destructive Control. Directory. Under the general editorship of Kluev V.V., 8 vols. V. 4, 736 p, book 1, Mashinostroenie, Moscow (2006)
2. Kluev, V.V.: Magnetic Control Methods. In: Non-Destructive Control. Directory in 8 volumes under the editorship of Kluev V.V., corresponding member of the Russian Academy of Sciences, vol. 6, 848 p, Book 1/ Kluev V.V. Muzhitsky V.F., Gorkunov E.S., others, Mashinostroenie, Moscow (2006)
3. Pankovsky, Y.P.: Hardware realization of certain magnetic non-destructive control methods. World Meas. (5), 9–12 (2005)
4. Prilutsky, M.A.: Methods of Assessment of Welded Metal Structure Strain-Stress State. News of High Schools, Mashinostroenie. (2008), N. 4, pp. 45–50
5. Kraus, I., Ganev, N.: X-ray analysis of the inhomogeneous stress state. In: Snyder, R., Fiala, J., Bunge, H.-J. (eds.) Defect and Microstructure Analysis by Diffraction, pp. 367–401. Oxford University Press (1999)
6. Noyan, I.C., Cohen, J.B.: Residual Stress Measurements by Diffraction and Interpretation. Springer, NY (1987)
7. Vasiliev, D.M.: Diffraction Methods for Structure Analysis, 502 p. The Saint-Petersburg State Technical University, Saint-Petersburg (1998)
8. Sosnin, F.R.: Radiation control. In: Non-Destructive Control. Directory in 8 volumes under the editorship of Kluev V.V., corresponding member of the Russian Academy of Sciences, vol. 1, 560 pp. Book 2. Mashinostroenie, Moscow (2003)

9. Enisherlova, K.L., Lutzau, A.V., Temper, E.M., Gorjachev, V.G., Tkacheva, T.M.: Combined investigation of silicon-on-insulator structures. In: SILICON 2010, 12 Scientific and Business Conference Roznov pod Radhostem, Czech Republic, 2–5 Nov 2010, pp. 315–331
10. Luttsau, A.V., Krymko, M.M., Enisherlova, K.L., Temper, E.M., Razgulyaev, I.I.: Analysis of heterostructures by the one-crystal x-ray diffraction method. In: Electronics Engineering Materials, pp. 72–78. (2012)

The VLSI High-Level Synthesis for Building Onboard Spacecraft Control Systems

O.V. Nepomnyashchiy, I.V. Ryjenko, V.V. Shaydurov,
N.Y. Sirotinina and A.I. Postnikov

Abstract Using small spacecrafts for a wide range of research and applied purposes is one of the major trends in the aerospace field. Modular-network architectures implemented on the "system-on-chip" hardware platform provide required characteristics of onboard control systems. Selecting this system architecture significantly increases demands on very large-scale integration (VLSI) design efficiency and project solution quality. In this paper, we propose a new approach to VLSI high-level synthesis based on a functional-flow parallel computing model. The modified VLSI design flow uses a functional-flow parallel programming language Pythagoras, which allows describing a VLSI operation algorithm with the maximal degree of parallelism. An offered intermediate representation of VLSI architecture in the form of a control-flow graph and a data-flow graph provides an opportunity for synthesizing circuits and verifying projects on the stage of a formal description, without returning to previous hierarchical levels of the project. A set of software tools supporting new design process is developed. The proposed technology is successfully tested on the example of a digital signal processing function. Further, this technology is suggested for use in the synthesis of onboard control system components for small spacecrafts.

O.V. Nepomnyashchiy (✉) · I.V. Ryjenko · V.V. Shaydurov ·
N.Y. Sirotinina · A.I. Postnikov
SibFU—Siberian Federal University, Krasnoyarsk, Russia
e-mail: 2955005@gmail.com

I.V. Ryjenko
e-mail: rodgi.krs@gmail.com

V.V. Shaydurov
e-mail: shidurov@icm.krasn.ru

N.Y. Sirotinina
e-mail: nsirotinina@sfu-kras.ru

A.I. Postnikov
e-mail: alpost@mail.ru

© The Author(s) 2018   229
K.V. Anisimov et al. (eds.), *Proceedings of the Scientific-Practical Conference*
"Research and Development - 2016", https://doi.org/10.1007/978-3-319-62870-7_25

Keywords Small spacecraft · Control systems · Parallel computing
Data flow · Functional programming · Integrated circuit · Algorithm
Formal verification · High-level synthesis

Introduction

Using small spacecrafts for a wide range of scientific and applied problems is one of
the major trends in the aerospace field. This is due to a number of reasons:
development time reduction, relatively low cost, universal reconfigurable platform.
An onboard control system is the main element in the spacecraft control. Currently
accepted approaches to building onboard spacecraft control systems are based on a
centralized hierarchical architecture. This approach does not meet modern
requirements for onboard control systems. Modular-network onboard control system architectures have significant advantages over centralized. However, their
implementation requires changing conceptual approaches to designing onboard
control systems. Technological risks of introducing a qualitatively new architecture
can be reduced using the latest achievements in the field of Russian microelectronics, as well as implementing new approaches to designing VLSI, which would
reduce development time while increasing the reliability of onboard equipment.

Studying the subject area shows that the "system-on-chip" (SoC) technology
allows implementing a functionally complete module set for the small spacecraft
onboard control system on the base of a single VLSI chip. Benefits of using the
SoCs as a hardware platform are the following:

- reducing the cost of VLSI production, especially in case of medium- and
  small-batch production,
- reducing the time-to-market cycle of new product,
- supporting flexible configurability of a system in accordance with requirements
  of the particular project,
- providing high product reliability through testing regular platform structure that
  performs by the manufacturer,
- debugging support,
- reducing power consumption.

The principle of extra-integration based on SoC technology provides an
opportunity to optimize the internal structure and to reduce redundancy, which is
typical of systems using universal components. This optimization determines the
high economic efficiency of project decisions, both through direct savings (reducing
the number of component on the printed board, board area and so forth), as well as
by indirect savings (low energy consumption, high reliability, productivity, minimal
amount of hardware debugging, etc.).

Problem Statement

Selecting the SoC as a hardware platform for small spacecraft onboard control systems significantly increases demands on VLSI development process efficiency and project solution quality. Modern computer-aided design tools and environments, despite the impressive achievements, have a number of significant problems. First, we should mention the global problem of a gap between a number of gates that could be implemented on a single chip and a number of items that can feasibly be designed and verified in a reasonable amount of time. For complex projects, verification and testing come to the fore. The end-to-end verification is important because late error detecting always turns into significant additional time and financial costs. The share of time spent on project verification comes up to 60–80% of the total development time [1, p. 51].

The analysis of the current problem state shows that designing complex single-chip systems requires a new, architecture-independent approach. On the one hand, it should provide separating an initial algorithm representation from a target hardware architecture; on the other hand, it should make possible translating the initial algorithm representation into the RTL-level with simultaneous verification. In this connection, the task of finding such a method for describing parallelism of initial algorithms that further could be "compressed" in accordance with resource-specified constraints of a target hardware platform is still relevant.

Related Works

Attempts to solve these problems are mainly aimed at eliminating the semantic gap between high-level- and low-level descriptions of the system being developed. Analyzing the current state of affairs in integrated circuit design flows, we outline the following ways of improving the initial representation of a VLSI operating algorithm:

1. Expanding Hardware Description Languages (HDL) with special constructions for high-level behavioral description.
2. Creating VLSI synthesis tools on the base of an existing high-level programming language, which is used in software development.
3. Creating a new high-level hardware description languages by way of synthesizing of HDL and high-level programming languages.

An example of the first approach is SystemVerilog [2, p. 7]. This hybrid language is formed from HDL Verilog by introducing high-level constructions for behavioral VLSI description.

The second approach is called High-Level Synthesis (HLS) [3, p. 72] supported by Xilinx company. This method is based on introducing a number of constructions into the basic high-level programming language for describing integrated circuit topologies.

The last option is implemented in languages such as HandelC [4, p. 18]. Also are known a number of projects on the use of functional programming languages for the

VLSI synthesis (Lava, Hume, Erlang) [5, p. 244; 6, p. 93; 7, p. 192]. The use of a functional language for this purpose is the most promising, since the algorithm description in a functional language is most suitable for a parallelization.

The algorithm representation in the form of a data-flow graph provides the most complete description of natural parallelism of an algorithm. The works of Dongarra [8, p. 36; 4, p. 3] convincingly demonstrate that solution of portability problem for parallel programming lies in changing programming paradigm. This paradigm [9, p. 6] must meet the following requirements:

- the absence of an imperative control on calculation process (control on the base of data readiness),
- the data-flow model,
- massive parallelism.

An effective implementation of the data-flow graph for VLSI, in particular for the FPGA platform, is discussed in [10, p. 6; 11, p. 12]. However, described models use explicit control of computation, are oriented on the limited application area (digital signal processing), do not support the portability, and are not used as a high-level languages for VLSI synthesis [10, p. 67].

Proposed Approach

The review of the subject area shows that existing methods do not meet modern demands. VLSI circuit design needs a new architecture-independent approach, which should provide an opportunity for automatic translating the initial algorithm representation from top to bottom with simultaneous verification and independence of the initial algorithm representation from the target hardware architecture. Another important requirement is the opportunity for initially describing the VLSI operation algorithm with maximum parallelism, which would be "reduced" during design in view of resource constraints of a target hardware platform.

The generalized scheme of data processing in a VLSI circuit contains a data signal set and a control signal set. The control signal set includes data ready signals. Each block of the circuit can process data independently and in parallel with other under condition of availability of data confirmed by the data ready signal. If an analogy with parallel programming languages, such a model corresponds to the data-flow model. Thus, effective solutions in this direction can be found using a functional parallel programming paradigm.

One of such models is the functional-flow model of parallel computing. Functional-flow parallel (FFP) programming language Pythagoras [12, p. 71] is designed on the basis of this paradigm. The language provides the initial description of a task with the maximum parallelism.

Portability of a parallel algorithm between different architectures in this case is achieved by a parallelism convolution.

The proposed technology of VLSI architecture-independent design is based on the FFP paradigm [12, p. 71; 13, p. 35]. It complements ESL VLSI design methodology.

Unlike existing HLS methods, the proposed methodology is based on the FFP description of VLSI functioning algorithm with the use of the Pythagoras language. The language allows a designer to create the architectural-independent single-chip system description in the form of parallel program with the maximal parallelism [12, p. 71]. Then, the initial algorithm is translated into data-flow and control-flow graphs. At the next step, HDL-graph is formed based on the mentioned graphs. As a result, the designer obtains a set of architectural solutions in terms of hardware description language with a given degree of parallelism [13, p. 35].

It is expected that the proposed approach provides an opportunity for effective conversions of a high-level representation, formal verification, and functional testing with maximum coverage of VLSI architecture on initial design stages. In addition, it simplifies Design Space Exploration (DSE) process. The main advantages of the methodology are the effective VLSI representation at higher levels of abstraction in ESL design, and as a consequence, the optimal architecture solutions at the gate level.

The authors propose HLS design flow for the VLSI synthesis based on FFP approach. The flow implies the following additional design stages:

The development of the VLSI operation algorithm with the use of FFP programming language. At this stage, the initial algorithm represents in architecture-independent high-level form with maximum degree of parallelism.

The formal verification, optimization and debugging of the FFP code. This stage includes the formal debugging of the VLSI operation algorithm according to the design task without binding to a target platform.

The synthesis of intermediate VLSI representation the in form of data-flow and control-flow graphs. In addition to the graph synthesis, this step provides the intermediate representation of argument types and constants in accordance with the initial program code and constrains.

Data typing and graph optimization. Data typing is performed according to the data type specifications obtained on the previous step.

The intermediate synthesis of data-flow and control-flow graphs. This stage is implemented in cases of transition to the corresponding phase of the conventional HLS flow or combination cycles of conventional and FFP synthesis.

The design space exploration. This stage aims at searching the optimal implementation of the VLSI operating algorithm taking into account the given hardware platform constraints. The problem is solved by reducing and transforming the algorithm parallelism in accordance with the resource constraints. The initial maximal-parallel representation of the algorithm provides the maximal coverage of the solution space and automated search of the optimal one.

The start and final stages: the problem statement, the formation of technical task and specification package, the synthesis of circuit and control signals, the VLSI synthesis on the register-gate level are fully comply with the conventional HLS flow [13, p. 35; 14, p. 239].

Tool Support for the Proposed High-Level Synthesis Flow

The initial description of the algorithm implemented on a VLSI is performed using an integrated development environment of FFP programming language Pythagoras. The Pythagoras language compiler provides the conversion of FFP functions into a set of reversive data-flow graphs (RDFG), describing the information dependences between operations. An additional utility generates a control-flow graph (CFG) using RDFG as input. CFG defines the order of the RDFG vertices execution. This approach provides a wide variety of computing control strategies, from serial to data flow. Selection of concrete computing control strategy defines by the developer. RDFG and CFG together form a program, which can be performed by Pythagoras interpreter for checking its correctness prior to RTL synthesis.

The developer can use additional tools to perform the following functions:

- RDFG and CFG optimization for improving FFP program effectiveness,
- debugging and code analysis at run time, thus providing error searching and tracing,
- formal verification.

These tools allow the developer to debug and test the initial algorithm prior to synthesis VLSI architecture.

A synthesizer transforms the intermediate FFP algorithm presented in the form of RDFG and CFG in accordance with the proposed HDL synthesis algorithm. At the first stage, it synthesizes the graph representation including type declaration and calculation, constant calculation. At the next stage, the synthesizer deletes the calculated graph vertices. The result of this stage is the optimized (minimized) graph representation. The final step includes generating input and output ports and a name (register) table, connecting input ports with registers and generating HDL description for the developed module in interpretation mode.

Results

Consider implementation of a typical digital signal filtering function—multiply accumulate. The function was created and debugged using the developed toolkit. The function implements multiplication of two argument pairs and summation of the multiplication results. The initial function description in FFP language is given below (Fig. 1).

The initial function translates into data-flow graph (Fig. 2).

The next stage is argument typing. The function receives as arguments two input lists containing two values each: ((int.16, int.16), (uint.15, uint.15)). This example uses two 16-bit and two 15-bit unsigned numbers.

The following phase is optimization. Calculating types, calculating, and removing optimizable vertices are executed in course of traversing graph in a

MultSum << funcdef Param {

  A <<Param:1;

  B <<Param:2;

  ((A:1,B:1):*,(A:2,B:2):*):+>>return;

  }

Fig. 1 The initial function description in FFP language Pythagoras

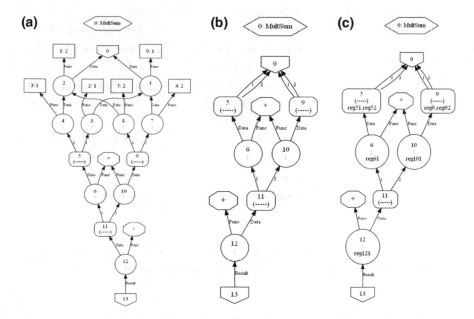

Fig. 2 **a** Initial reversive data-flow graph, **b** optimized data-flow graph, **c** typed HDL-graph

"symbolic interpretation" mode. Figure 2b shows the result of calculating and removing graph vertices.

At the next stage, a name table is generated in the following order. First, registers for all vertices related to the vertex of argument are generated and connected with input ports. Registers for vertices related to interpretation operation are generated at the next step. Then registers for vertices related to results are generated and connected with output ports. At last the name table is generated for the entire module as a whole (Fig. 3c)

The vertices of the graph (Fig. 3c) contain generated registers (table function names).

HDL description is synthesized on the basis of the optimized data-flow graph. Automatically synthesized HDL description in the Verilog language for this example is shown in Fig. 3.

Fig. 3 The function description in Verilog language

```
module MultSum                  begin : input_process
(                                   reg51 <= in1;
 input clk,                          reg91 <= in2;
 input rst,                          reg52 <= in4;
 input [15:0] in1,                   reg92 <= in5;
 input [15:0] in2,                   end
 input signed [14:0] in4,        always @(posedgeclk) begin
 input signed [14:0] in5,            reg101 <= reg91 * reg92;
 output reg signed [31:0] out1   end
);                               always @(posedgeclk) begin
reg signed [14:0] reg92;             reg61 <= reg51 * reg52;
reg signed [30:0] reg101;        end
reg signed [31:0] reg121;        always @(posedgeclk) begin
reg signed [14:0] reg52;             reg121 <= reg61 + reg101;
reg signed [30:0] reg61;         end
reg [15:0] reg91;                always @(posedgeclk) begin
reg [15:0] reg51;                    out1 <= reg121;
begin                            end
always @(posedgeclk)             endmodule
```

Note that the result of the synthesis presents the so-called "pure HDL", i.e., the code without binding to a target platform. This code can be implemented based on any VLSI of any manufacturer, which meets project resource constraints.

Conclusion

A new methodology of VLSI synthesis on the base of FFP paradigm is proposed. This methodology extends the conventional VLSI HLS flow. Unlike existing HLS methods, the proposed methodology allows us to develop an initial description of the VLSI algorithm with the maximal parallelism. In addition, a high-level project description is translated into RTL automatically. These particular specificities coupled with architectural independence should enable a developer to perform formal verification, effective testing, debugging, and optimization. We assume that all mentioned advantages should improve effectiveness, reliability, and quality of developed projects and decrease time expenditures.

For implementing the proposed approach, the modified HLS design flow and appropriate tool support is developed. A series of test projects on creating graph models, functional verification, and debugging, as well as generating HDL description with the following synthesizing RTL representation has been executed. Testing confirms the suitability of the proposed technology for SoC design. Further, this technology is suggested for use in the synthesis of onboard control system components.

Acknowledgements Researches are carried out with the financial support of the state represented by the Ministry of Education and Science of the Russian Federation. Agreement (contract) no. 14.578.21.0021 05.Jun 2014. Unique project Identifier: RFMEFI57814X0021.

References

1. Losev, V., Lumps And., Putr M. Institut of design of devices and systems (Cadence university): Quality, stability, future. Electronic components. **4**, S.49–54 (2008)
2. IEEE Std 1800-2012: IEEE standard for systemverilog-unified hardware design, specification, and verification language, 1275 p (2013)
3. Vivado design suite user guide. High-level synthesis. URL: http://www.xilinx.com/support/documentation/sw_manuals/xilinx2014_4/ug902-vivado-high-level-synthesis.pdf
4. Handel-C Language Reference Manual, 348 p. Celoxica Limited (2005)
5. Alves, C.: Erlang inspired Hardware. In: International Conference on Field Programmable Logic and Applications, pp. 244–246. Milano (2010)
6. James-Roxby, P.: Lava and J Bits. From HDL to bit stream in seconds, pp. 91–100. In: 9th IEEE Symposium Field-Programmable Custom Computing Machines (2001)
7. Sérot J., Michaelson, G.: Compiling hume down to gates. In: Draft Proceedings of 11th International Symposium on Trends in Functional Programming, pp. 191–226. Madrid (2011)
8. Bosilca, G., Bouteiller, A., Danalis, A., Faverge, M., Herault, T., Dongarra, J.: PaRSEC: Exploiting heterogencity to enhance scalability. IEEE Comput. Sci. Eng. **15**(6), 36–45 (2013)
9. Bosilca, G., Bouteiller, A., Danalis, A., Dongarra, J.: Dataflow-based task execution through PaRSEC for HPC. In: IEEE 8th International Symposium on Embedded Multicore/ManycoreSoCs, pp. 1–52. Aizu-Wakamatsu, Japan (2014)
10. Ab Rahman, AAHB.: Optimizing dataflow programs for hardware synthesis. Ph.D. thesis, p. 170. EPFL, Lausanne (2014)
11. Johansen, J.: Design and Implementation of a Novel Dataflow Model and an Intermediate Representation Language for High Level Synthesis on Field Programmable Gate Arrays [Электронный ресурс]. URL http://projekter.aau.dk/projekter/files/71270246/main.pdf (датаобращения: 20.11.2016)
12. Legalov, A.I.: The functional language for creation of architecture-independent parallel programs. Comput. Technol **1**(10), 71–89 (2005)
13. Nepomnyashchy, O.V., Shaydurov, V.V., Legalov, A.I., Ryzhenko, I.N.: The technology of architecture-independent, high-level synthesis for VLSI, vol. 57, No. 3, pp. 35–39. Reports of Academy of Science of Russian Federation. Novosibirsk, NGTU (2014)
14. Nepomnyashchy, O., Legalov, A., Tyapkin, V., Ryzhenko, I., Shaydurov, V.: Methods and algorithms for a high-level synthesis of the very-large-scale integration. WSEAS Trans Comput **15**(Art. #22), 239–247 (2016)

A Concept of Robotic System with Force-Controlled Manipulators for On-Orbit Servicing Spacecraft

I. Dalyaev, V. Titov and I. Shardyko

Abstract To enhance longevity and safety of unmanned spacecraft, especially on high orbits including geostationary ones, it is thought suitable to exploit on-orbit servicing (OOS) spacecraft (servicers) equipped with manipulation system. The employment of the manipulator is by the requirement for fine manipulation in tasks such as refueling, equipment replacement, orbit correction and maintenance tasks. The key feature of a manipulator to fulfill this list is the force-torque feedback. This paper describes a general concept of the manipulator to carry out the on-orbit servicing tasks along with the common and kinematic description of a ground prototype. A brief look at graphical user interface designed for testing purposes is represented. A ground validation facility is described and the description of series of performed tests is included along with conclusions.

Keywords On-orbit servicing · Space robotic systems · Supervised autonomy
Ground testing · Validation · Control system · Force-torque control
Robot software

Introduction

Taking into account the expanding exploration and exploitation of space, it is important to have facilities capable of conducting maintenance of the spacecraft being currently used and coping with such situations as failures or fuel depletion in that spacecraft, further addressed as client. Namely, this facility should be able to capture the client or dock with it, to replace the failed unit, to refuel the object,

I. Dalyaev · V. Titov · I. Shardyko (✉)
RTC, Saint-Petersburg, Russia
e-mail: i.shardyko@rtc.ru

I. Dalyaev
e-mail: igor@rtc.ru

V. Titov
e-mail: vtitov@rtc.ru

© The Author(s) 2018   239
K.V. Anisimov et al. (eds.), *Proceedings of the Scientific-Practical Conference*
"Research and Development - 2016", https://doi.org/10.1007/978-3-319-62870-7_26

to perform required maintenance tasks and to drive the client away from its orbit if it cannot continue functioning [1]. This goal can be achieved by a space system consisting of host spacecraft and robotic system designed to execute specified tasks.

A Servicing Spacecraft Concept

A number of projects and prototypes is known in the research area, which have been developed throughout the world in recent years. Those that stand out include "Orbital Express" experiment carried out by NASA in 2007 [2], project DEOS by DLR, Germany [3], and a prolonged experiment previously related as FREND with its follower being now developed by DARPA known as Phoenix [4]. The latter is shown on Fig. 1 (*left*).

This paper presents a concept of spacecraft that was designed for the tasks given above, which is depicted on Fig. 1 (*right*). It comprises a host spacecraft, a manipulator, a tool kit, a payload unit and a vision system unit. An approach with only one manipulator was chosen after the appropriate research due to its less weight and cost whereas providing essential functionality [5, 6].

Functional and Technical Description of the Manipulation System

Considering the servicer's required functionality, a list of tasks for manipulation system can be drawn, including [7]:

Fig. 1 DARPA phoenix project and spacecraft for on-orbit servicing (Geometric model)

- grasping (and releasing) objects;
- transporting objects;
- inserting objects into their receptacles;
- executing tool-oriented manipulations: cutting, drilling, screwing, etc. (operations).

Consequently, the manipulator meets a number of requirements resulting from the list. First of all, it has an end-effector that enables the diversity of supported operations by connecting with different tools, including a gripper and a tool for each operation. Next, the manipulator has an appropriate 7-DOF arm kinematic structure to allow the full working area coverage and the ability to exchange tools, wherein the excessiveness kinematics provide the essential dexterity. Each DOF is represented by a revolute joint. Besides, the arm produces sufficient pull and push force on the end-effector either with or without the tooltip.

There are some external factors that affect the manipulator design. When using the servicer on high orbits such as geostationary ones, or in lunar or Martian space, there will be significant signal delays leading to instability in manipulator behaviour if an operator controls it from Earth. Also, the high cost of both the assumed mission and the client requires an ultimate level of safety while performing servicing tasks. As a result, the control system of manipulator has supervised autonomy mode, when an operator needs only to appoint a task and the local control system should perform and verify all required actions. Nevertheless, an operator still has the ability to take over control and stop any action at the arbitrary moment of time in case of anomalies or undesired behaviour. The force control subsystem is implemented as well to make the local control system capable of tracking contact tasks, adapting motion trajectories accordingly and stopping motion if contact forces or any joint torques exceed allowed values.

Each of the manipulator joints is a mechatronic unit, which consist of a brushless DC-motor, a compound gear of planetary and harmonic drive parts, motor position sensor, joint position sensor and joint torque sensor. A motor position sensor also acts as a joint speed sensor. High reduction ratio provides for high forces and torques required to perform contact tasks while the speed is not of great importance in context of servicing operations. Finally, each joint contains a group of housekeeping sensors, ensuring temperature monitoring and overcurrent protection.

Manipulator Control System

Manipulator control system provides a number of motion types to conform with the manipulator task list:

- rigid moving along the required trajectory in Cartesian space;
- moving with required velocity vector in Cartesian space;
- applying required force-torque vector (FT-control) in Cartesian space;

- hybrid motion combining moving along some surface with applying force against it;
- moving along the required trajectory in Cartesian space with impedance control.

Mentioned Cartesian space may be represented by a Cartesian coordinate system, either attached to the manipulator base or to its end-effector, or taught to manipulator by an operator. To set the orientation the system of Euler angles ZYX has been chosen [8, p. 11].

To gain assurance that the control system will successfully work on real prototype, an experimental model of manipulator has been manufactured and a test software has been written. A picture of experimental model along with its kinematics is shown in Fig. 2, screenshots of a software interface window are depicted in Fig. 3. The left and right parts of Fig. 3 show two different kind of input capabilities. In the interface on the left an operator can only enter data for performing a one-coordinate motion while on the right it is possible for an operator to set motion requirements on any number of coordinates up to six.

Fig. 2 Experimental model of manipulator and its kinematics

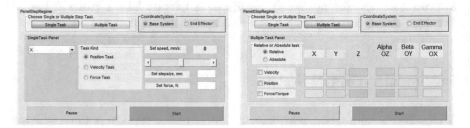

Fig. 3 Test software interface

The last three motion types belong to force control methods, which are crucial for autonomous system in undetermined environment ensuring its safety. A research has been conducted to formulate corresponding control laws and test them on kinematic and dynamic models of the manipulator to verify the laws so that they can be implemented in the software.

Experimental Model Testing on the Ground Validation Facility

A series of experiments has been conducted to test the mentioned control laws. Experiments were held on the ground validation facility (testground) also designed and manufactured in this project. The testground, depicted in Fig. 4, consists of several areas, which are [9]: "Smooth Trajectory" (2), "Rough Trajectory" (3), "Sockets" (4), "Dynamometer" (5), "Screw Connections" (6), "Scales" (7). The given numeration is in accord with Fig. 4, number 1 stands for testground base.

"Trajectory" areas have been designed to examine the ability of control system to cope with obstacles unknown beforehand. The task was to perform a spline trajectory connecting the points located so that there was a part of a solid object on the straight line between them. Moving in accordance with the impedance control law, the manipulator arm was able to make the motion due to a virtual stiffness that allowed necessary deviations from desired trajectory while virtual torques and forces remained moderate. When it was impossible to execute the command

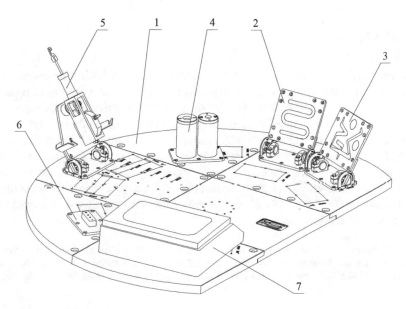

Fig. 4 Ground validation facility

without exceeding the limit FT-values, the arm was stopped to wait for further commands.

The idea of "Sockets" area was to simulate the process of linking two mutual parts (peg-in-hole operation). A case of particular interest here was when the relative position of the parts was not determined strictly, i.e. there was a linear or angular displacement between axes of these parts. Again the impedance control was in action and the range of permitted displacements was found for which the task had been successfully executed. The defined range was admitted as sufficient to meet the manipulator requirements. In the cases when a displacement was out of range, the FT-feedback values exceeded the limits and the system stopped itself to wait for the response of an operator.

As "Dynamometer" area speaks for itself, its objective was to apply force in the chosen direction. The same goes to "Screw Connections" area, where the main criterion was the equality between the task and actual tightening torque. In the course of these experiments the FT-control was applied to the manipulator motion. Tests showed the conformity of actual values to the appropriate tasks.

The purpose of "Scales" area was to examine the behavior of the arm when moving along the scales surface, applying to this surface a force of the required degree. The description clearly states that the task conforms to a hybrid velocity-force motion type specification. The experiments included a series of moves along the scales surface with different velocity and force tasks. Actual velocity was evaluated by the vector of joint speeds and actual force was measured by the scales. The deviations from the required values were within 10% for force and 5% for linear velocity.

Conclusions

This paper presents the concept of a servicing spacecraft with a force-controlled manipulation system along with an experimental model and a facility designed to tune and test the control system. The list of tasks was described for both spacecraft and manipulator and a set of control laws was formed. A number of experiments were held with the model on the testground, successful results of which verified the applied control laws. Future work will include more thorough design of a servicer mechanical system and selection of electronic components suitable for the target environment. Conversely, special attention will be paid to the enhancing of manipulator performance related to increasing motion smoothness and expanding the number of feasible tasks.

Acknowledgements Research are carried out (conducted) with the financial support of the state represented by the Ministry of Education and Science of the Russian Federation. Agreement (contract) no. 14.578.21.0046 16.Sept 2014 Unique project Identifier: RFMEFI57814X0046

References

1. Разработка технического (проектного) облика робототехнической системы с очувствленными по усилению манипуляторами в составе сервисного космического аппарата: отчёт о ПНИ (2 этап - промеж.) / ЦНИИ РТК. – 2015. – 137 с
2. On-Orbit Mission Updates [Электронный ресурс]. URL http://archive.darpa.mil/orbitalexpress/mission_updates.html (дата обращения: 10.09.2016)
3. Reintsema, D., Thaeter, J., Rathke, A., Naumann, W., Rank, P., Sommer, J.: DEOS—The German robotics approach to secure and de-orbit malfunctioned satellites from low earth orbits. In: Robotics and Automation in Space (i-SAIRAS), Sapporo, Japan (2010)
4. Henshaw, C.G.: The DARPA phoenix spacecraft servicing program: Overview and plans for risk reduction. Naval Center for Space Technology. In: The 12th International Symposium on Artificial Intelligence, Robotics and Automation in Space, Montreal, Canada (2014)
5. Технический облик средств робототехнического обеспечения сервисного спутника, предназначенного для продления сроков активного существования космических аппаратов/Даляев. И.Ю, Шардыко И.В., Кузнецова Е.М. // Экстремальная робототехника: Труды международной научно-технической конференции. – СПб: Изд-во « Политехника-сервис » , 2015. – С. 1 181 – 186
6. Lopota, A., Dalyaev, I., Shardyko, I., Kuznetcova, E., Belezyakov, I.: Means of robotic support for on-orbit servicing. In: B. Katalinic (ed.) Proceedings of the 26th DAAAM International Symposium, pp. 0865–0870, Published by DAAAM International, ISBN 978-3-902734-07-5, ISSN 1726-9679, Vienna, Austria
7. Разработка технического (проектного) облика робототехнической системы с очувствленными по усилению манипуляторами в составе сервисного космического аппарата: отчёт о ПНИ (1 этап – промеж.) / ЦНИИ РТК. – 2014. – 202 с
8. Siciliano, B., Khatib, O.: Handbook of Robotics, 1611 p. Springer, Berlin (2008)
9. Разработка технического (проектного) облика робототехнической системы с очувствленными по усилению манипуляторами в составе сервисного космического аппарата: отчёт о ПНИ (4 этап – промеж.) / ЦНИИ РТК. – 2016. – 127 с

Development of Microlinear Piezo-Drives for Spacecraft Actuators

A.V. Azin, S.V. Rikkonen, S.V. Ponomarev and A.M. Khramtsov

Abstract The article describes the development and experimental studies of microlinear piezo-drive for control reflective surface devices of large-sized transformable spacecraft antenna reflectors. Research target—experimental investigation of the microlinear piezo-drive to determine stable oscillatory system operating modes which would include improved energy conversion parameters. The following characteristics are briefly presented: test stand construction-design description, identification of oscillatory system resonant and actual frequencies under inertia load. A series of experiments have been conducted for both different preliminary tensions and inertia mass values.

Keywords Microlinear piezo-drive · PZT stack · Oscillatory system
Frequency characteristics · Test stand · Durability · Reliability

The problem, to reduce the spacecraft (SC) system mass-dimension parameters, is especially acute in the space field. One current solution is to replace electro-mechanical drives in different SC units for piezo-drives which, in its turn, would decrease the mass-dimension parameters n-fold times [1, p. 196; 2, p. 103; 3, p. 160].

Control reflective surface devices of large-sized transformable spacecraft antenna reflectors involve a number of technical specifications: mass, dimensions, thrust force, vibration displacement, operation stability, durability, reliability in space environment.

A.V. Azin (✉) · S.V. Rikkonen · S.V. Ponomarev · A.M. Khramtsov
National Research Tomsk State University, Tomsk, Russia
e-mail: antonazin@niipmm.tsu.ru

S.V. Rikkonen
e-mail: rikk2@yandex.ru

S.V. Ponomarev
e-mail: psv@niipmm.tsu.ru

A.M. Khramtsov
e-mail: khramtsov.home@gmail.com

© The Author(s) 2018
K.V. Anisimov et al. (eds.), *Proceedings of the Scientific-Practical Conference*
"Research and Development - 2016", https://doi.org/10.1007/978-3-319-62870-7_27

Such piezo-drives embracing the above-mentioned TS do not exist abroad or in Russia. It is mostly due to the absence of a reasonable understanding of microlinear piezo-drive operation, no calculation methods and design for MLPD as a single oscillatory system [1, p. 196; 4, p. 86; 5, p. 1].

The problem targets involve the following:

1. development of a new design conception for microlinear piezo-drives (MLPD) of spacecraft actuators, including aerospace industry;
2. development of scientific and methodological design documents of microlinear piezo-drives for SC actuators;
3. development of static and dynamic problem-solving methods in designing spacecraft MLPD actuators with world standard mass-dimensional energy indicators;
4. development of MLPD test unit and energy source with required q-band based on numerical simulation;
5. development of experimental research program and methods for MLPD of SC actuators;
6. development of experimental test stand design of MLPD to estimate the performance;
7. development and validation of technical requirements for new MLPD design.

One-Dimensional Modeling of MLPD Performance

MLPD packet mode is a mechanical oscillatory system operating in alternate mode and regenerating electrical energy into mechanical energy load. This system should be explored as a single oscillatory system including all component elements. All system elements should operate in the following integrated mode—"conversion-generation-consumption" energy. In this case, all these elements are operating synchronous, in coherent mode transferring active power from energy source to consumer. Under these conditions, high-efficient energy conversion and oscillatory system efficiency could be observed.

Electric analogy method is applied to select the MLPD design, element material and SC operation mode protype-conversion of complex oscillatory mechanical systems in electrical circuit and further transformation based on Kirchhoff laws. This conversion is illustrated in Fig. 1: mechanical MLPD system based on APM-2-7 and SC circuit scheme replacement. Solving the resulting equivalent electrical scheme, one can get a SC frequency response, resonating frequency, force on load, and vibration displacement. One-dimensional model calculation results are in Fig. 5 [1, p. 200; 4, p. 91; 5, p. 5].

Finite element modeling is applied to test one-dimensional mathematical simulation of MLPD and material selection for its design.

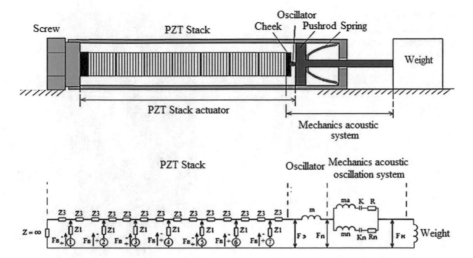

Fig. 1 Electrical circuit scheme of MLPD replacement with 7-laminar MLPD, mechanic acoustic system and load

Three-Dimensional MLPD Modeling

Reliable MLPD operation depends on element material selection. Stress–strain state of the MLPD structure under operational loads is analyzed on the basis of calculations.

MLPD includes stack, screw, PZT Stack actuator, cheek with oscillator, pushrod and seal. Finite-element MLPD model is illustrated in Fig. 2. Maximum stress intensity zone in the MLPD reinforcement construction is the oscillator and the pushrod contact. According to the optimal mechanical and acoustic parameters the following element materials were selected: steel, aluminum, and plexiglass. MLPD reinforcement model involves the following material combinations (sequence of material element: cheek-oscillator-pushrod, seal—rubber used in all cases): steel-steel-plexiglass; aluminum-aluminum-plexiglass; aluminum-aluminum-steel; all aluminum; all steel.

Based on numerical calculations steel-steel-plexiglass materials were selected due to quality and reliability characteristics. Amplitude-frequency characteristics of pushrod vibration displacement based on calculations results are shown in Fig. 5.

Based on the mathematical approach, the designer can select suitable material for each MLPD element according to its design.

Experimental Research

According to modeling results MLPD test unit was designed. MLPD is illustrated in Fig. 3 including piezo-drive elements and assembled MLPD. To determine the force on the load in MPLD, there is a piezoelectric power indicator.

Fig. 2 Finite-element model.
1 PZT stack, *2* Cheek,
3 Oscillator, *4* Pushrod,
5 Spring

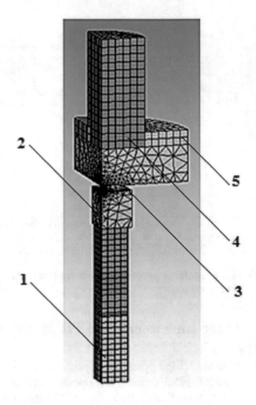

Fig. 3 MLPD based on APM
2-7: *1* Pushrod, *2* MLPD
jacket, *3* Force pickup lead,
4 PZT stack pickup lead,
5 PZT stack APM-2-7,
6 Force sensing unit

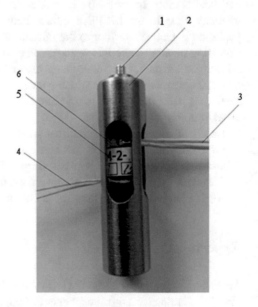

A test stand was designed to obtain research data on the MLPD modes. The test stand is used to investigate MLPD operation modes (Fig. 4a). MLPD modes depend on the following characteristics: SC type and PZT Stack power; vibration acceleration (mass load/mass weight); PZT Stack pretension force; PZT Stack tension; PZT Stack current; frequency response. During MLPD tests, operation modes were recorded as electrical signals: PZT Stack tension, PZT Stack current, PZT Stack force indicator signal, vibration acceleration (mass load/mass weight). During pretesting one more characteristic—pretension force was determined.

The test results showed the following: frequency characteristics of load acceleration, recalculation of defined vibration displacement load characteristics, and frequency parameters of force on load. According to SC frequency characteristics, operating oscillating frequencies were defined.

Varying SC characteristics and frequency responses on the test stand it is possible to investigate the numerous operation modes of this system. Based on the results it was concluded that there is an optimal frequency response for this load, where maximum force on load, maximum vibration displacement load, and maximum power on load. Figure 4 presents the test stand. Description of the following scheme: on the base (5) via adjusting screw (4) the PZT Stack (6) is installed through force sensor (7) which bottoms on the pushrod (2), the pushrod (2) via elasticity (8) acts on the inertia load/weight load (1). Acceleration inertia load (1) is measured by acceleration sensor (9). Adjustment screw (4) provides the actuator preload force.

MLPD power is produced from the multichannel power supply with adjustable frequency and voltage. Figure 4b shows the overall view of the designed workplace for MLPD operation modes.

When applying AC voltage of different frequencies, the actuator generates disturbance force through the oscillator-pusher acting on the inertia load, causing its

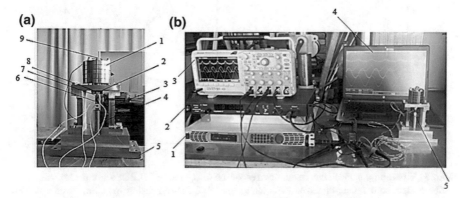

Fig. 4 MLPD operation modes studying workplace: **a** test stand: *1* Inertia mass load (weight load), *2* Pushrod, *3* Supports, *4* Adjusting screw, *5* Base, *6* PZT stack APM-2-7, *7* Force sensor, *8* Elasticity, *9* AR 1019 accelerometer; **b** overall view: *1* DC power supply, *2* AC multi-channel, variable frequency, AC voltage, *3* Electronic oscilloscope, *4* PC information archive, *5* Test stand

vibration at predetermined frequency. Based on the vibration displacement magnitude, magnitude of inertia load is determined, frequency characteristics are based
on the experimental results. Actuator current is maintained at 0.5 A, whereas power
supply voltage was decreased proportionally to the actuator capacitance. The
experiment was conducted at weight mass $M_w = 3$ kg and preload strength
$F_0 = 240$ H.

Frequency response analysis obtained by mathematical modeling of
one-dimensional and three-dimensional models showed good agreement with the
frequency responses (curves 1 and 2, Fig. 5a). Experimental frequency response
also showed an oscillating frequency close to the theoretically calculated characteristics X, μm.

When inertia load changes, SC oscillating frequency decreases to the increase of
inertia mass (Fig. 6a).

MLPD temperature stability tests show that frequency response values do not
change with the temperature increase. When the temperature approaches the limit
values of the oscillating frequency of the passport, MLPD moves towards higher
frequencies (curves 1, 2, 3 Fig. 5b), which is in agreement with [6, p. 41; 7, p. 272;
8, p. 3; 9, p. 1].

An important point is the actuator preload force in operating MLPD. It is noted
that pre-tensioning is the reason of actuator pre-deformation "smoothing" the
malfunction of its production and assembly and providing a "single mechanism"
packet for SC operation mode. When the preload force increases, the power to the
load increases (Fig. 6b).

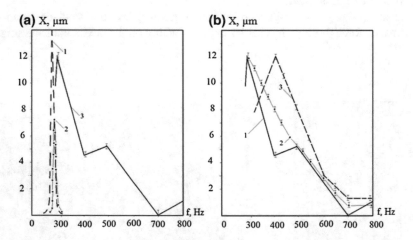

Fig. 5 Vibration displacement frequency responses of operating MLPD: **a** Frequency characteristics comparison: *1* One-dimension modeling results, *2* 3D-dimension modeling results, *3* Test
results; **b** Vibration displacement frequency test responses during temperature stability test:
1 temperature 21 °C, *2* temperature 50 °C, *3* temperature 80 °C

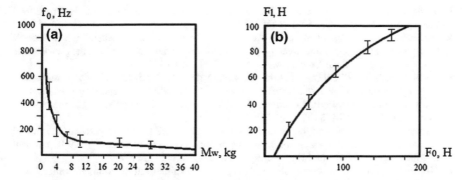

Fig. 6 Test dependences: **a** MLPD operating oscillation f_0 and weight mass M_w; **b** F_1 force on the load at oscillating frequency of system to F_0 pre-load force

Conclusions

1. Newly developed approach for identifying oscillatory frequency system for MLPD;
2. Updated mathematical model to calculate different designed MLPD modes;
3. Experimental data are in good agreement with the results of numerical experiments for three-dimensional and one-dimensional mathematical models, as SC with inertia load is an oscillatory system, then based on the calculation results corresponding MLPD configuration with desired range of oscillating frequency can be selected;
4. Experimentally, pretension is the reason of the actuator pre-deformation, smoothing the malfunction of its production and assembly and providing a "single mechanism" packet for SC operation mode. When the preload force increases, the power to the load increases;
5. Designed multi-channel power supply provides synch and coherent operation modes of several actuators.

Acknowledgement Researches are carried out with the financial support of the state represented by the Ministry of Education and Science of the Russian Federation. Agreement no. 14.578.21.0060 23 sept. 2014. Unique project Identifier: RFMEFI57814X0060.

References

1. Ponomarev, S.V., Rikkonen, S.V., Azin, A.V.: Issledovanie jelektromehanicheskih processov v p'ezojelektricheskoj sisteme [Study of electromechanical processes in a piezoelectric system]. News of higher educational institutions. Physics **2**, 196–202 (2014)
2. Park, S.: Single vibration mode standing wave tubular piezoelectric ultrasonic motor: a thesis ... for the degree of Master of Applied Science, 136 p, Ryerson University, Toronto (2011)

3. Wang, Z., Li, Y., Cao, Y.: Active shape adjustment of cable net structures with PZT actuators. Aerosp. Sci. Technol. **26**, 160–168 (2013)
4. Ponomarev, S.V., Rikkonen, S.V, Azin, A.V.: Modelirovanie kolebatel'nyh processov p'ezojelektricheskogo preobrazovatelja [Simulation of oscillatory processes of the piezoelectric transducer]. Tomsk State Univ. J. Math. Mech. **2**(34), 86–95 (2015)
5. Ponomarev, S.V., Rikkonen, S., Azin, A., Karavatskiy, A., Maritskiy, N., Ponomarev, S.A. The applicability of acoustic emission method to modeling the endurance of metallic construction elements. In: IOP Conference Series: Materials Science and Engineering, vol. 71, Tomsk (2015). doi:10.1088/1757-899X/71/1/012056
6. Davoudi, S.: Effect of temperature and thermal cycles on PZT ceramic performance in fuel injector applications: a thesis ... for the degree of Master of Applied Science, p. 99 Department of Mechanical and Industrial Engineering University, Toronto (2012)
7. Henderson, D.A.: Novel piezo motor enables positive displacement microfluidic pump, vol. 1. pp. 272–275. NSTI Nanotech, Santa Clara (2007)
8. Henderson, D.A.: Simple ceramic motor. In: 10-th International Conference on New Actuators, vol. 1. pp. 1–4. Bremen (2006)
9. Henderson, D.A., Sheryl, L.: Piezoelectric motors move miniaturization forward, pp. 1–2. Electronic products, New York (2006)

Design of Dynamic Scale Model of Long Endurance Unmanned Aerial Vehicle

V.S. Fedotov, A.V. Gomzin and I.I. Salavatov

Abstract Dynamic scale model of aircraft is a model established in accordance with the laws of dynamic similitude and used in the research of flutter and other aeroelastic facts. A result obtained under laboratory conditions was transferred to full-scale aircraft with conversion.

Keywords Unmanned aerial vehicle · Dynamic scale model · Long endurance unmanned aerial vehicle · Flying laboratory · Aviation research
Scaling

The scope of work carries out in the AO NPO OKB M.P. Simonova in order to manufacture a dynamic scale model of long endurance unmanned aerial vehicle (UAV). As part of these activities scaling factors were determined by geometric, aerodynamics, weight, inertial scaling factors and engine-propeller power plant indicators, structural construction of dynamic scale model was designed, designed and manufactured tooling, manufactured, assembled, and tested a dynamic scale model.

In the process of creating a new prototype, we have to carry out extensive research of its aerodynamic performance, stability, controllability at the different flight modes, with different center-of-gravity. Most of these issues can be resolved by examining the model of the aircraft in an air tunnel, but the final testing and debugging is performed on a plane built. Under this system of research, defects can be detected very late, to remove them usually requires debugging and design of the second version of the aircraft. All these make the design of the aircraft much more expensive, delays the launch of a series and it leads to system aging, when all spent money does not give the desired impact. Conducting pre-flight study on dynamic

V.S. Fedotov · A.V. Gomzin · I.I. Salavatov (✉)
AO NPO OKB M.P. Simonova, Kazan, Russia
e-mail: ilmir52@gmail.com

V.S. Fedotov
e-mail: info@okbsimonova.ru

A.V. Gomzin
e-mail: info@okbsimonova.ru

© The Author(s) 2018   255
K.V. Anisimov et al. (eds.), *Proceedings of the Scientific-Practical Conference
"Research and Development - 2016"*, https://doi.org/10.1007/978-3-319-62870-7_28

scale flying model brings some benefit. A significant part of the future prototype defects can be detected and eliminated in advance. Due to the high cost of production of UAV dynamic scale model flight research carried out in real flight conditions allows to approach the full-scale tests on the UAV with greater confidence.

Purpose and field of application

- A flying laboratory for testing control algorithms similar algorithms of full-scale unmanned aerial vehicle;
- Simulation and testing on dynamic scale model on various emergency power rating UAV kind.

Dynamic similarities in relation to the UAV are as follows:

- Similarity of geometrical characteristics;
- Similarity aerodynamic characteristics;
- Similarity rotor-propeller group;
- Weighing likeness;
- Similarity on the inertial characteristics.

The first step in the creation of the dynamic scale model is to determine the design definition. Be aware that small elements create manufacturing complexity and it is necessary to make the structural analysis, to make adjustments. Dynamic scale model has the same weight and stiffness distribution and provides a semblance of aerodynamic effects.

The work on the selection of scaling factors is performed in the developed design definition. All geometrical dimensions should be scaled with the scale factor equal to 1/5.66.

When this model is made of original material it will be stiffer. That means the deformations, stresses, twisting, and bending angels are proportional to scale.

Scaling Coefficients Calculations

In the process of calculation were used formulas. Some formulas are derived from open sources, while others were withdrawn by yourself.

Geometric characteristics

- The wingspread of UAV—m;
- Choose wingspread of dynamic scale model—n;
- Scaling factor: $scale = \frac{m}{n}$;
- Length factor: $f_{length} = scale$;
- Thickness coefficients of the compound material: $f_{compoundthick} = scale$

If the thickness of the compound material becomes smaller than the thickness of the used material, a single layer of material is applied in manufacturing.

Aerodynamic characteristics

- Surface factor of dynamic scale model: $f_s = \text{scale}^2$
- Volume of dynamic scale model: $f_v = \text{scale}^3$
- Wing loading factor: $f_{\text{pressure}} = \frac{f_m}{f_s} = \text{scale}$
- Flight speed factor: $f_{\text{speed}} = \sqrt{f_p} = \sqrt{\text{scale}}$
- Deflection angle: $f_{\text{defl.angl}} = \frac{f_{\text{force}} f_{\text{length}}^2}{f_I} = \frac{\text{scale}^5}{\text{scale}^5} = 1$
- Coefficient Reynolds number: $f_{\Re} = f_{\text{speed}} \cdot f_{\text{length}} = \text{scale}^{1.5}$

After manufacture and assembly carried out center-of-gravity positions dynamic scale model similar centering of full-scale UAVs within a predetermined range of the arithmetic average of the chord.

Characteristics of the rotor-propeller group

- Engine power: $P = P_1 \cdot f_{\text{power}}$
- Propeller diameter: $D = D_1 \cdot f_{\text{length}}$
- Propeller speed: $\text{RPM} = \text{RPM}_1 \cdot f_{\text{RPM}}$
- Propeller tip speed: $\omega = \text{RPM}_1 \cdot f_{\text{RPM}} \cdot \pi \cdot D_1 \cdot f_{\text{length}} = \text{RPM}_1 \cdot \pi \cdot D$
- Ratio speed: $f_{\text{RPM}} = \frac{f_{\text{speed}}}{f_{\text{length}}} = \frac{\sqrt{\text{scale}}}{\text{scale}} = \text{scale}^{-0.5}$
- The relative pace of the screw: $f_\lambda = \frac{f_{\text{speed}}}{f_{\text{RPM}} \cdot f_{\text{length}}} = \frac{\sqrt{\text{scale}}}{\text{scale}^{-0.5} \cdot \text{scale}} = 1$
- The weight of power plant: $W_{\text{pp}} = W_{\text{pp1}} \cdot f_m$

Weighing likeness

- Weight: $f_m \text{scale}^3$

Inertial characteristics

- The moment of inertia factor $f_I = f_{\text{comp.thick}} \cdot f_{\text{length}}^3 = \text{scale}^4$

Analysis of the values carried out after the calculations determined the areas of strength and weakness. There are such cases, when the calculated thickness of cover degenerate and model becomes non-complying. In such cases, it is necessary to ignore the laws of scaling and seek solutions to meet the strength requirements.

Design the Plane

Design of the aircraft was carried out with the use of modern computer-aided design (CAD) system and support of the product lifecycle: Siemens NX, ANSYS, Teamcenter. These programs allow you to conveniently work with dynamic scale model, identify inconsistencies in the design and make the necessary changes in order to minimize costs, definition performance, create working construction documentation.

Dependency presented above was tested in the creation dynamic scale model. Their consistency was tested at the characteristic element of the aircraft structure—cowl of nacelle.

The initial data for the production of a cowl of nacelle is the airfoil that is obtained from the Central Hydrodynamic Institute Zhukovsky (TsAGI). Solid model elements are created with the calculated thickness of cover in computer-aided design system by according scaled airfoil (Fig. 1).

Composite materials that have good characteristics by design and technological parameters were selected as the basic structural material of airframe of dynamic scale model, because modern constructions are required to have a high strength with minimal weight and overstability, robustness, durability, stiffness, manufacturing and maintainability, aerodynamic performance.

Different technologies are applied during the production of details of composite materials which are different in complexity, cost, and equipment. The selection of a technology is conditioned to the volume of production, the degree of preparation, economic evaluation of production efficiency.

AO NPO OKB M.P. Simonova is equipped with modern mechanical processing equipment. Methods of the vacuum and autoclave molding of large composite components are controlled and widely used there. It is necessary to have a forming tool for the creating construction elements.

Forming tool is created from the same material as the cowl of nacelle in order to get the same expansion coefficient during heating at the forming process. Forming tool consists of two elements: the top and bottom parts. The basic elements of molding tool of the aerodynamic contour should be designed with considering the production technology. Molding tool of nacelle of cowl made of glass fiber which is more economically efficient and simple. Forming tools of composite material are lighter, they can retain the shape and size, have a high mechanical properties and low contraction. It enables their manufacture with narrow tolerances.

Forming tool has shapes form, so it required doing master-model. Master-model is made on CNC production center from Medium Density Fiberboard (Fig. 2). Tool for molding hood of nacelle was made by the master-model. It is necessary to create the electronic model for the production of the master-model on CNC production

Fig. 1 Cowl of nacelle.
a solid model, **b** exploded view

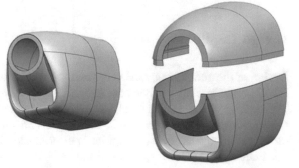

Fig. 2 Master model nacelle of hood

center. The electronic model is created in the program Siemens NX. There are many modules for creating processing programs for production center.

On the basis of the enterprise master-models and forming tool for molding structural elements are made (Fig. 3).

All parts, including the lightly stress loaded thin-walled cover and the elements of the load carrying structure are made of composite materials (Fig. 4). In manufacture used modern components such as reinforcing materials from in glass and carbon fiber.

The main aircraft performances are shown in Table 1, the overall dimensions in Fig. 5.

At the moment, dynamic scale model of UAV long duration flight creates on AO NPO OKB M.P. Simonova and undergoing flight tests.

Fig. 3 Forming tool. **a** bottom matrix, **b** top matrix)

Fig. 4 Hood of nacelle

Table 1 Aircraft
performances

Gross weight	40 kg
Payload mass	10 kg
Flight endurance	5 h
Flight altitude envelope	50–4000 m
Radio reach	50 km
Max length	100 m

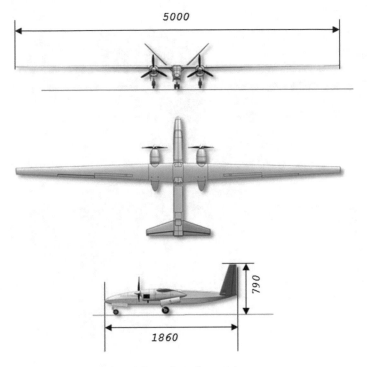

Fig. 5 Dimensional characteristics of dynamic scale model

Conclusions

Using dynamic scale model in the development of new aircraft can significantly reduce the time for design of production. This is achieved by carrying out pre-flight test at critical control modes. These post-flight values from dynamic scale model are transferred to the full-scale plane by conversion factors. In addition to the use of such a dynamic model for aviation research, this development provides a great practical advance in the production of small UAVs. This class of unmanned aerial vehicles has proven itself in the field of monitoring the area and infrastructure.

Acknowledgements *Researches are carried out (conducted) with financial support of the state represented by the Ministry of Education and Science of the Russian Federation. Agreement (contract) no. 02.G25.31.0122 14.Aug.2014*

References

1. Cheranovskii, O.R.: Dreams and reality or the way to science (2010)

Features of the Development of Regional Transport Models

P.V. Loginov, A.N. Zatsepin and V.A. Pavlov

Abstract Improvement of the Russian legislation in the field of strategic and territorial planning administrative units of the Russian Federation repeatedly increased the demand for specialists in the analysis and forecasting of social and economic development of the state authorities and local government entities, in the modern tools of transport modeling. The spectrum of regional facilities in their size and in their administrative status multivariable tasks for problem solving with traffic. In this condition, important thing is the development of building technology of transport models that have universal possibilities for different types of the researching area. Currently, the best developed methods are methods of transport models of cities and agglomerations. Meanwhile, also important thing is using transport modeling as a tool for the planning and development of federal and regional transport systems. In the development of regional transportation model should be taken into attention, the specificity of the transport system functioning in regional level, which does not allow to use the methodological apparatus intended for the development of urban models. The proposed methodological approaches can be used in the development and updating of the regional transport models, helping to reduce the cost of the process and being effectively applied in the researching areas on any administrative level.

Keywords Transport modeling · Regional transport model · Matrix of correspondence

P.V. Loginov (✉) · A.N. Zatsepin · V.A. Pavlov
"Kvantex", Moscow, Russia
e-mail: office@kvantex.ru

A.N. Zatsepin
e-mail: 6234914@gmail.com

V.A. Pavlov
e-mail: office@kvantex.ru

© The Author(s) 2018   263
K.V. Anisimov et al. (eds.), *Proceedings of the Scientific-Practical Conference
"Research and Development - 2016"*, https://doi.org/10.1007/978-3-319-62870-7_29

Introduction

Currently around the world there is a tendency of development of the regional transport systems. This process is accompanied by an increase in economic activity of the population, the redistribution of traffic flows, the formation and the growth of urban centers, etc. So, these processes significantly affect on the transport mobility of the population, analysis reveal the main directions of development of economy and improvement citizen's standard of living. For implementation of the most efficient and less labor-intensive analytic, we need to develop new methodologies and tools to assess the processes that accompany the development and changes in transportation systems. Planning the development of transport systems in the Russian Federation is carried out on three levels: federal, regional, and local. If the methodology of the transport models of cities and agglomerations sufficiently developed by now [1, 2], the development of regional models is an actual scientific problem.

The authors had a task to develop the concept of information technology of macromodel transport systems building for typical Russian region and modeling methods of existing and forecasted traffic flows.

Currently, the technology of building information models in different levels of transport systems has a similar typology. But these models look like a unique product of scientific researching in the annex to the specific object of research.

Creating of a generic technology for building the transportation system model of region requires certain restrictions in raw data and the results of further processing and calibration. The main objective of the developed technologies is the creation of sufficiently reliable information macromodel of regional transport system in the existing information and resource constraints. This model should allow forecasting of the transport system development. Increasing of individual objects' reliability or areas of the region's transport system and the creation of multimodal logistics facilities are the tasks of the next level, and their solution depends on the uniqueness of a particular area.

Transport modeling is a modern tool for decision-making in the systems of transport planning on different levels. The demand of this tool in the development of transport systems is stable and has a high level. In recent years, the need of transport models development repeatedly increased. It is caused by the introduction of new requirements for the development of projects and schemes of the organization of traffic in the territories of municipalities of the Russian Federation, which establishes mandatory construction of transport models in the design process. Unfortunately, this tool is expensive: the development of the transport model—is large scientific and research work, which requires the collection and processing of significant volumes of raw data. The using of developed transport models require special knowledge and training, that makes them difficult to use. Nowadays a lot of the administrations of municipalities in the Russian Federation do not have this tool and do not use it for solving problems of transport planning.

This situation is a serious problem that requires a comprehensive approach to solution. One of the costly stages in the development of the transport model is the stage of collecting raw data. Foreign colleagues from developed countries can do it easier because of a permanent monitoring of the transport system and periodic sociological research usually conducted in their countries [3]. The necessary data are available partly like scattered and not systematic in large number of administrations of the Moscow Area in the Russian Federation.

Thus, one of the urgent tasks of the work is searching for alternative sources of data.

This work covers the following aspects of the authors in the framework of applied research:

- Development of methodological approaches to the construction of a regional transport model of the Russian Federation typical region;
- Searching for the original data sources that reduce the cost of construction of regional models and the development of methods of pre-treatment.

Defining the Concept of Choosing the Optimal Set of the Main Components of Information Technology of Macro Models of Typical Russian Region

The authors selected standard algorithm that include following components:

- baseline data of transport supply and demand;
- model of transport demand. A modified classic four-step method of calculation used for creating the model. The advantages of the chosen model are accurately describing all the stages of transport demand formation with low requirements for computing power of personal computers;
- model of the transport supply;
- distribution model of transport demand for offer.

For creation of these components, the authors have developed a hardware-software complex of transport region system modeling (PAK MTSR). This complex will automate the collecting and processing information of traffic flows process, save and process the basic data of the characteristics of the roads; the proposed modeling technology of transport supply and demand can be implemented with the help of specialized software.

The complex also has a high mobility and computing power. It is important in region conditions. In addition, to ensure the adequacy of the generated transport model and its relevance, the creating of the typical Russian region information macromodel technology includes these procedures

- calibration of the regional transport models;
- evaluation of the quality of the final transport model version with predetermined criteria;
- updating of the model according to source data.

Model calibrating result is conformity of raw data and results of modeling. Difference of approaches to building macromodel of a typical Russian region:

- creation of a model with "from global to local" method with using federal strategies for the development of transport systems;
- initialization of supply and demand models level with using specifics of the region;
- using of mobile PAK MTSR equipped with sensors measuring vehicle flows.

Transport macromodel technology conception has been developed as a result of science researching that described in the current work.

The Features of the Calculation of Transport Demand Within the Regional Model

The main differences in approach to the modeling regional transport system are identified on the phase of the modeling transport demand. For models of the cities and agglomerations, the most intensive are some labor correspondences. Herewith a pendulum character of this kind of correspondence makes great relevance of transport models, which are developed for morning or evening peak period.

In the scale of the region (subject of the federation), the most intensive become business travels and freights and their distribution of time makes daily allowance transport model more actual. It should be noted, that the labor correspondences also shall be calculated in the consisting of the regional model, but their temper is perfectly expressed by localization in the districts of big localities, that result into causing point loads on the regional transport system at all. The specific layer of the demand makes significant load on the regional transport system that was made by recreational correspondences. The specificity includes both the distribution of time (the most intensive periods are weekends and holidays and also on the vacation time) and the susceptibility to the influence of external factors, for example weather. The other layers of demand, that are connected with a rides to the medical institutions, for example, makes quite low load and cannot be considered in the modeling the regional transport model. So business, labor and freights correspondences should be considered first within the regional transport model (in the regions of big localities).

The using of this information gives the most accurate results in the case of the transport of zoning on the borders of MO, which is quite acceptable, and sometimes even excessively when building a regional transportation model.

To calculate the matrix of correspondences for the business and cargo segments, the authors propose the use of as input data of the tax reporting, presented in free access at the website of the Federal tax service of the Russian Federation [4]. All provided reports each year aktualisierte and formed within the context of municipalities. The use of this information gives the most accurate results in the case of the transport of zoning on the borders of MO, which is quite acceptable, and sometimes even excessively when building a regional transportation model.

On the basis of the tax on property of organizations (NIO), about the tax base for the unified tax on imputed income for certain types of activities (UTII), the tax base for the tax paid by taxpayers in connection with the application simplified tax system (STS) can be obtained by estimating the number of economic entities in the context of municipalities. To assess the composition of the truck fleet can be used information about the tax base and structure of charges for the transportation tax (TN).

Accordingly, the calculation of the matrices of freight and business correspondence is based on the conclusion that their number is on average proportional to the number of business entities and the allocation is based on a "gravity" principle.

The calculation of the matrices of labor correspondence methodologically refined [1] and requires the collection of data on employment and working population in the context of the transport areas, in the present embodiment, in the context of MO. The greatest difficulty causes the job assessment, because the authorities are at best accounting for large- and medium-sized enterprises.

To estimate the number of jobs in the context of MO, the use of report 5-NDFL the Federal tax service of the Russian Federation—"the Report on tax base and structure of charges to tax on income of physical persons, withheld by tax agents" [4]. It contains data on the number of individuals paying the tax to incomes of physical persons (NDFL), which corresponds to organized workplaces.

Thus, the use of tax statements as the source data allows to calculate the matrix of correspondences for all major segments of demand within the regional transport model.

Optimization of Organization Methods of Developing and Updating Regional Transportation Models

Regional road network consists of roads of Federal, regional or intermunicipal and local value. Technical condition of the road network in most areas beyond the regulatory framework and is subject to annual random fluctuations in the spring. The volumes of necessary repair work within the region do not allow to neglect this factor in transport modeling. Thus for developing and updating regional transportation models require the rapid collection of information about the condition of the road network.

In the absence of monitoring data of the intensities of traffic flows on major regional highways, this task is necessary to solve the developer transport model. Across the region it is necessary to conduct a substantial amount of measurements of the intensities of traffic flows, it is required to consider possible variations of values depending on weekday or weekend and time of year.For solving the mentioned problems, we propose the use of the above specialized hardware-software

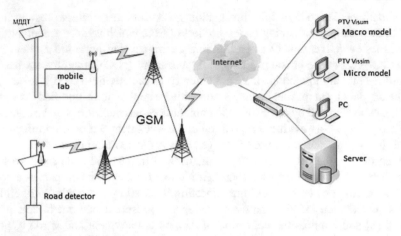

Fig. 1 Structural scheme of PAK MTSR

complex simulation of the transport system of the region. Figure 1 shows a block diagram of the PAK MTSR.

PAK MTCP consists of a mobile complex which is intended for carrying out measurements, and stationary complex, which is used for data processing, development and calibration of traffic models.

The mobile complex is based on a mobile laboratory, for example, KP-514CMΠ produced by "Spetsdortekhnika" (Saratov, RF) [5], which collects data on the characteristics and condition of the road network. Measuring the intensities of traffic flows in the composition of the mobile complex includes the sensors of traffic. Stationary sensors traffic (SDT), for example models of Hand-Wave-A-01 [6], it is proposed to use for long-term monitoring of traffic flows. For the operational refinement of data measurements of the intensities of traffic flows using mobile sensors, traffic (MDDT), for example, ThermiCam ETH 10-7042 [7].

As you collect the mobile set of input data is transmitted to the server PAK MTSR, where a database of transport modeling. In the process of developing a regional transport model to the work of the mobile complex adjustments can be entered. For example, if essential differences of settlement and natural (real) part of road network have been noted, goal is set to mobile complex operator to do addition measurements.

Application PAK MTSR for tasks collect baseline data in order to build and update regional transportation models to improve simulation quality and reduce development costs. These advantages are achieved through the use of the developed transport model, the most complete and current source of data, and optimization procedures for obtaining them.

Conclusions

Today the Russian Federation significantly stands behind developed countries in transport system development; therefore this issue is one of the top-priority for the state. Optimal transport system development on the federal, regional, and field levels is possible with the usage of up-to-date instruments of transport modeling in making management decisions.

In this work, methodological approaches to the development of the regional transport models are suggested, their usage allows to optimize the process of source data acquisition as it is the most cost-based stage during the development. The resulting reduction in expenditure for the development of the regional transport model will facilitate a wider usage of the instrument for the transport system development on the regional and federal levels.

Gratitude

The studies were conducted in accordance with the Agreement No. 14.588.21.0001 from the 26 September 26, 2014. With the Ministry of education and science of the Russian Federation on granting subsidies in the execution of works on the theme "creating technologies of construction of information models of transport systems in the model region of Russia, including the development of algorithm for calculations of the efficiency from exploitation of toll and other roads of Federal, regional and local level, on the basis of the program complex world level PTV Vision®VISUM/VISSIM (Germany)" together with "A+S Consult GmbH Forschung und Entwicklung" (Dresden, Germany) in the framework of applied research in priority areas with the participation of research organizations of member countries of the EU in the framework of activity 2.2 of the Federal target program "Research and development on priority directions of development of scientific-technological complex of Russia for 2014–2020. Unique identifier of applied research (project) RFMEFI58814X0001.

Acknowlegements The studies were conducted with the financial support of the state, represented by the Ministry of Education and Science of the Russian Federation № 14.588.21.0001 Agreement of 26 September 2014 Unique identifier for Applied Scientific Research (project): RFMEFI58814X0001.

References

1. Aliev, A.S., Strelnikov, A.I., Shvetsov, V.I.,Shershevskaya, Yu.z.: Simulation of traffic flow in a major city with an application to the Moscow agglomeration. Automatic. Telemech. **11**, 113–125 (2005)
2. Safronov, E. A.: Transport system of cities and regions: a training manual. pp. 12–25. Publishing house ASV (2005). ISBN 5-93093-345-6

3. TUD—Forschungsprojekt "Mobilität in Städten—SrV" [Electronic resource]—Mode of access: https://tu-dresden.de/die_tu_dresden/fakultaeten/vkw/ivs/srv/. Free. The title. Screen
4. Stat reporting [Electronic resource]—Mode of access: https://www.nalog.ru/rn62/related_activities/statistics_and_analytics/forms/. Free. The title. Screen
5. Comprehensive road laboratory "Track" [Electronic resource]—Mode of access: http://sdtech.ru/store/lab/trassa/trassa.html. Free. The title. Screen
6. Hand-Wave-A-01. ДСW01П [Electronic resource]—Mode of access: http://strelka-wave.ru/product/arrow-wave-a-01-dcw01p.html. Free. The title. Screen
7. ThermiCam Specifications [Electronic resource]—Mode of access: http://www.flir.ru/cs/display/?id=61844. Free. The title. Screen

Part II
NanoScience and NanoTechnology

The Influence of AlGaN Barrier-Layer Thickness on the GaN HEMT Parameters for Space Applications

A.G. Gudkov, V.D. Shashurin, V.N. Vyuginov, V.G. Tikhomirov,
S.I. Vidyakin, S.V. Agasieva, E.N. Gorlacheva and S.V. Chizhikov

Abstract The results of simulation of field-effect microwave high-electron-mobility transistors (HEMTs) based on GaN/AlN/AlGaN heterostructures are presented. The research allowed to determine the optimal thickness of the AlGaN barrier layer for achieving high microwave capacity implementation.

Keywords Gan · Hemt · Heterostructure · Thickness · Space application
Numerical simulation

A.G. Gudkov (✉) · V.D. Shashurin · S.I. Vidyakin · S.V. Agasieva · S.V. Chizhikov
Department of Instrumentation Technology, Bauman Moscow State
Technical University, Moscow, Russia
e-mail: profgudkov@gmail.com

V.D. Shashurin
e-mail: schashurin@bmmstu.ru

S.I. Vidyakin
e-mail: bmsturl@gmail.com

S.V. Agasieva
e-mail: s.agasieva@mail.ru

S.V. Chizhikov
e-mail: chigikov95@mail.ru

V.N. Vyuginov
Svetlana-Elektronpribor Ltd, Saint Petersburg, Russia
e-mail: vvyuginov@yandex.ru

V.G. Tikhomirov
Department of Radiotechnics and Electronics, SPbETU «LETI», Saint Petersburg, Russia
e-mail: vv11111@yandex.ru

E.N. Gorlacheva
Department of Industrial Logistic, Bauman Moscow State Technical University,
Moscow, Russia
e-mail: gorlacheva@yandex.ru

© The Author(s) 2018   273
K.V. Anisimov et al. (eds.), *Proceedings of the Scientific-Practical Conference*
"Research and Development - 2016", https://doi.org/10.1007/978-3-319-62870-7_30

GaN transistors allow to significantly extend the capabilities of microwave devices. Electron mobility in combination with a high-electron density in the region of the 2D electron gas makes it possible to implement high current densities in the transistor-channel cross-section and high gain. However, optimization of heterostructure transistors still remains a complicated and expensive procedure [1–3]. The results of numerical simulation and calculation for the AlGaN barrier layer of high-electron-mobility transistors (HEMTs) are shown in this paper.

We have chosen typical heterostructure GaN HEMT transistor as the object of model. Briefly, it consists of a substrate, made of sapphire or silicon carbide SiC, a buffer layer of undoped GaN (thickness of several microns), a barrier layer of AlGaN (thickness of 10 nm) and a passivating silicon nitride layer SiN, which may be absent in a simplified case. This structure is represented on the Fig. 1.

According to the recommendations from the working [4], we have optimized the used mathematical models for the analysis of our heterostructures working. We have chosen two-dimensional hydrodynamical mathematical model, that often is used in many industrial simulation systems [5, 6], which, combining with the original physical models of the behavior of electrons in dielectrics and semiconductors, gave the good results. The same approaches are described in many workings [7–9]. Choosing the model of electron transporting, we took account balance between speed of calculations and sufficient accuracy. Nowadays it is clear that calculations of the drift—diffusion model cannot satisfy the demands of practice in the calculations of submicron transistors [10–16]. Accounting of the results by the Monte Carlo method in conjunction with the existing hydrodynamic model is the best choice in many cases, in our opinion. In general, we solve the system of three differential equations in partial derivatives. The Poisson's equation, describing the field, the current continuity and electron energy balance equation are solved self-consistently. This system of equations is supplemented by the specific equations for material medias, for example, the mobility of the charge carriers, the electron density, etc. The most important advantage is of complex analyze opportunity of ionization processes, defect formatting, and electron transport in the active area of the transistors.

Fig. 1 The typical diagram of heterostructure for producing of GaN HEMT

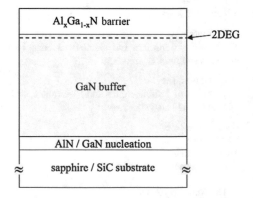

Effects of ballistic and quasi-ballistic movement of electrons in strong hetero-geneous electric fields become significant with reducing of semiconductor devices working area length to 30–300 nm [17–19]. The sizes of structures lead to fun-damental changes in the physics of the devices when designing devices for working in conditions of radiation. It happens because characteristic spatial scales of vari-ation of the electric field are compared with relaxation lengths of the energy, with electron impulse and with electron mean free path, the characteristic dimensions of devices workspace are comparable with the distance between the areas of defects, also the characteristic dimensions of devices workspace are comparable with the sizes of the defects. In this case analysis of radiation resistance supposes a using of the two-dimensional approach and considering of a number of new effects, con-nected with the heating of the electron gas under radiation exposure and scattering of carriers on the radiation defects. To analyze the radiation exposure on the sub-micron semiconductor devices was used quasihydrodynamic method of the charge carriers movement describing.

Studying effect of a heterostructure barrier layer on transistor's static charac-teristics in an equipment of space application, should consider a potential effect of specific external factors such as intensity of ionizing radiation on conductivity of two-dimensional electron gas and a number of the other GaN heterostructure parameters and investigated field-effect transistor, including the effect of metal-lization contacts topology [12, 20, 21].

Because of great amount of works with suggestion to drop heterostructure layers with Si, it is important to take into account the results of experimental investigations of the conductivity of GaN epitaxial structures, doped with a Si, concentration 1×10^{18} cm^{-3} and 1.8×10^{18} cm^{-3}. Usually, electrical measurements are carried out by van der Pauw method in a wide temperature range 40–300 K. The effect of electron irradiation with energy of 1 meV in a dose range 10^{14}–10^{16} cm^{-2} on the conductivity of such structures is interesting. Today it is known that in the range of low temperatures, 40 K conductivity with low activation energy E2, depending on the initial level of doping is observed. In the sample with a lower concentration of 1×10^{18} cm^{-3}, the activation energy is greater and it is equal to 0.08 meV. Half of activation energy E2 = 0.04 meV is observed in a sample with a higher concen-tration, equal to 1.8×10^{18} cm^{-3}. This behavior is typical for strongly doped semiconductors with conductivity in the impurity band. In the range of high tem-peratures 77–300 K extrinsic conductivity with activation energy E1, having a small dependence on the impurity concentration is observed. It is possible to connect the value of this energy that is equal to 2.5 meV with the transition of carriers from the impurity conduction band to the lower edge of the GaN con-duction band. Conducted researches showed that the conductivity of strong doped epitaxial GaN structures in the temperature range 40–300 K is characterized with two activation energies, E1 and E2. The conductivity activation energy E2 can be associated with conductivity of impurity band within which the transitions of carriers with localized states of the impurity band in the states above conduction band edge. Quantity of this energy has strong dependence on the level of doping and decreases with increasing of dopant concentration. The activation energy E1 characterizes the transitions of carriers from the impurity band to the conduction band edge of the GaN crystal and is almost independent on the concentration of silicon impurities. The main static electrical characteristics of the material turned out to be resistant to radiation exposure with electrons, but decreasing of the charge carrier mobility in the area of two-dimensional electron gas channel forming gives grounds to refuse doping heterostructure for using in the space application equipment and to achieve high concentration of electrons using piezodoping effect layers of AlGaN/GaN.

A simplified scheme of two-dimensional section of GaN HEMT transistors for simulation is shown in Fig. 2.

In the first stage of the study, the numerical models [2, 22, 23] were adapted to the specific features of the configuration and fabrication technology of actual device structures. It is well known that the parameters of the undoped AlGaN barrier layer located near the two-dimensional electron channel have a significant effect on the characteristics of a HEMT. Versions of the HEMT heterostructure configuration with AlGaN barrier-layer thicknesses within 10–25 nm range and a fixed Al mole fraction of 25% were calculated, and ways of its optimization were studied.

The calculations revealed the strong effect of the barrier-layer thickness on the transconductance (Gm) (Fig. 3).

Fig. 2 Simplified scheme of two-dimensional cross-section of AlGaN / GaN HEMT transistors for simulation of accounting charges on the interfaces

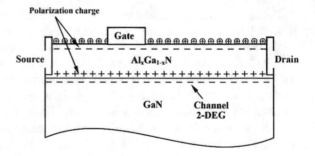

Fig. 3 Calculated dependences of the transconductance (Gm) and gate potential with different AlGaN barrier layer thickness

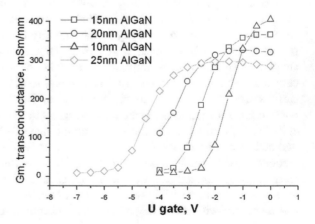

It is easily seen that the drain current significantly increases as the AlGaN barrier-layer thickness increases from 10 to 25 nm. At the same time, a decrease in the barrier-layer thickness leads to a significant increase in the transconductance. However, a decrease in the drain current in the region of small thicknesses of the AlGaN barrier layer means that it is impossible to achieve a sufficiently high power density in a transistor produced from such structure even regardless of current-collapse effects. At the same time, the fabrication of a thin barrier layer is very attractive for improving the high-frequency characteristics of the transistors and reducing short channel effects at a sub micrometer gate length. Considering the above, for purposes of clarity, we can build on the one graph the dependence of the drain current on the thickness of the barrier layer (Fig. 4) and dependence of the outer slope on the thickness of the barrier layer. It is clear that the trends have different directions and it needs to find the balance between current density, received from the device and its reinforcing properties.

Choosing thickness of barrier layer should to consider the other effect, specific only for GaN HEMT heterostructures. Especially it can be important in the design of onboard equipment of space application. Numerical calculations show the presence of strong electric field domain under the gate, more precisely, at the edge of the gate stock. Using the typical operating conditions of the GaN heterostructure transistor the drain voltage is in the range of 30–60 V. Also, electric field intensity distribution is strongly heterogeneously. In the area at the age of the gate stock, the field intensity can reach many megavolts per centimeter. However, this inevitably leads to the appearance inverse piezoelectric effect in the thin barrier layer that can cause significant mechanical stresses at the surface and in the bulk structure [4, 24]. The appearance of the inverse piezoelectric effect is especially dangerous in a strong electric field. Even with very good heat removal from the transistors when

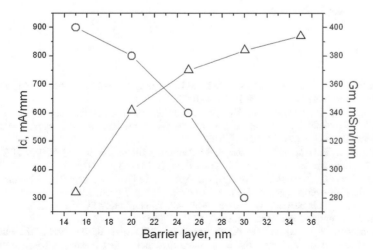

Fig. 4 Calculated dependences of the transistor drain current (Id) at a drain potential of 6 V and the I–V characteristic transconductance (Gm) on the AlGaN barrier layer thickness

the temperature in GaN HEMT channel does not exceed tens of degrees Celsius, it is possible mechanical damage to the heterostructure because of the existence of the inverse piezoelectric effect. In the more favorable case, it can lead to the significant and unexplained by conventional factors (electrical degradation of the structure, external heating, self-heating, the effect of structural defects, the effect of plating defects contacts and so forth) reducing of the equipment life based on the powerful GaN electronics, or what is even more dangerous, to the sudden failures of such equipment, especially in space technologies.

We plan to investigate the described effect in subsequent works described effect in details, but we already can give some important recommendations to the equipment designers. It is important pre-mathematical modeling of specific processes in the GaN heterostructure layers, in this case, imposes the limitations on the thickness of the layers and the necessity to inject additional field electrode in the GaN HEMT construction.

Conclusion

It was shown a strong dependence of the transistor's drain current (Id) and the drain current (Id)—gate voltage (Ug) characteristic transconductance (Gm) on the thickness of the AlGaN barrier layer as a result of the study by numerical simulation.

Acknowledgment Research is carried out (conducted) with the financial support of the state represented by the Ministry of Education and Science of the Russian Federation. Agreement (contract) no. 14.574.21.0116 12. November 2014. Unique project identifier: RFMEFI57414XO116.

References

1. Faraclas, E., Anwar, A.: AlGaN/GaN HEMTs: experiment and simulation of DC characteristics. Solid-State Electron. **50**(6), 1051–1056 (2006)
2. Tikhomirov, V., Parnes, Ya., Vyuginov, V.: Optimization of the parameters of HEMT GaN/AlN/AlGaN heterostructures for microwave transistors using numerical simulation. Semiconductors. **50**(2), 244–248 (2016)
3. Vitanov, S., Palankovski, V., Murad, S.: Predictive simulation of AlGaN/GaN HEMTs. In: IEEE compound semiconductor IC symposium, p. 131–134 (2007)
4. Ambacher, O., Majewski, J., Miskys, C.: Pyroelectric properties of Al(In)GaN/GaN hetero- and quantum-well structures. J. Phys. Condens. Matter. **14**(13), 3399–3434 (2002)
5. Vitanov, S., Palankovski, V.: Normally-off AlGaN/GaN HEMTs with InGaN cap layer: A simulation study. Solid-State Elect. **52**, 1791 (2008)
6. Tikhomirov, V. G., Maleev, N. A., Kuzmenkov, A. G.: Study of the effect of the gate region parameters on static characteristics of microwave field-effect transistors based on pseudomorphic AlGaAs-InGaAs-GaAs heterostructures. Semiconductors. **45**, 1352 (2011)

7. Tang, Z., Huang, S., Tang, X., Li, B., Chen, K.J.: Influence of AlN passivation on dynamic ON-resistance and electric field distribution in high-voltage AlGaN/GaN-on-Si HEMTs. IEEE Trans. Elect. Dev. **35**(7), 723–725 (2014)
8. Xu, Z., Wang, J., Liu, Y.: Applications of photorefractive materials in information storage, processing and communication. IEEE Elect. Dev Lett. **34**, 855 (2013)
9. Yang S., Huang, S., Chen, H.: A positive threshold voltage can be obtained by decreasing the barrier thickness which builds up the foundation for enhancement mode MOSHEMT devices. IEEE Elect. Device Lett. **33**, 979–981 (2012)
10. Huq, H., Islam, S.: AlGaN/GaN self-aligned MODFET with metal oxide gate for millimeter wave application. Microelectron. J. **37**(7), 579–582 (2006)
11. Chang, Y., Zhang, Y., Zhang, Y., Tong, K.: A thermal model for static current characteristics of AlGaN/GaN high electron mobility transistors including self-heating effect. J Appl Phys. **99**(4), 044501(5) (2006)
12. Grasser, T., Ting-Wei, T., Kosina, H., Selberherr, S.: A review of hydrodynamic and energy-transport models for semiconductor device simulation. Proc. IEEE **91**(2), 251–274 (2003)
13. Arulkumaran, S., Ng, G., Liu, Z., Lee, C.: High temperature power performance of AlGaN/GaN high-electron-mobility transistors on high-resistivity silicon. Appl Phys Lett. **91**(8), 083516(3) (2007)
14. Brannick, A., Zakhleniuk, N.A., Ridley, B.K., Shealy, J.R., Schaff, W.J., Eastman, L.F.: Influence of field plate on the transient operation of the AlGaN/GaN HEMT. IEEE Elect. Device Lett. **30**, 436–438 (2009)
15. Chang, Y., Tong, K., Surya, C.: Numerical simulation of current/voltage characteristics of AlGaN/GaN HEMTs at high temperatures. Semicond. Sci. Technol. **20**(2), 188–192 (2005)
16. Huque, M., Eliza, S., Rahman, T., Huq, H., Islam, S.: Temperature dependent analytical model for current–voltage characteristics of AlGaN/GaN power HEMT. Solid-State Elect. **53**(3), 341–348 (2009)
17. Luo, B., Allums, J.K., Johnson, W., et al.: Effect of proton radiation on DC and RF performance of AlGaN/GaN HEMTs. Appl. Phys. Lett. **79**(14), 2196 (2001)
18. Kyung-ah Son, Anna Liao, Gerald Lung et al.: GaN-based high temperature and radiation-hard electronics for harsh environments. Micro—and Nanotechnology Sensors, Systems, and Applications II (2010)
19. Ming-Lan, Zhang, Xiao-Liang, Wang, Ling, Xiao Hong-, et al.: Neutron irradiation effect in two-dimensional electron gas of AlGaN/GaN heterostructures. Chin. Phys. Lett. **25**(3), 1045–1048 (2008)
20. Polyakov, V., Schwierz, F.: Influence of electron mobility modeling on DC I-V characteristics of WZ-GaN MESFET. IEEE Trans. Elect. Dev. **48**, 512–516 (2001)
21. Vurgaftman, I., Meyer, J., Ram-Mohan, L.: Band parameters for III–V compound semiconductors and their alloys. J. Appl. Phys. **89**(11), 5815–5875 (2001)
22. Vitanov, S., et al.: High-temperature modeling of AlGaN/GaN HEMTs. Solid-State Elect. **54**, pp. 1105–1112 (2010)
23. Ambacher, O., Foutz, B., Smart, J., Shealy, J., Weimann, N., Chu, K., et al.: Two-dimensional electron gases induced by spontaneous and piezoelectric polarization in undoped and doped AlGaN/GaN heterostructures. J. Appl. Phys. **87**, 334–344 (2000)
24. Yan, W., Zhang, R., Xiu, X., Xie, Z., Han, P., Jiang, R. et al.: Temperature dependence of the pyroelectric coefficient and the spontaneous polarization of AlN. Appl. Phys. Lett. **90**(21), 212102(3) (2007)

Application of Volume-Surface Hardening by High-Speed Water Flow for Improving Static and Cyclic Strength of Large-Scale Castings from Low-Carbon Steel

S.A. Nikulin, A.B. Rozhnov, T.A. Nechaykina, V.I. Anikeenko,
V.Yu. Turilina and S.O. Rogachev

Abstract The article reveals the possibility of application of volume-surface hardening (VSH) by high-speed water flow for increasing static and cyclic strength of solebars of freight bogies manufactured from low-carbon cast steel 20 GL type. Formation of gradient structures with variable strength and ductility after VSH to the depth of 5–8 mm from the surface provides improved strength by 1.6–2.1 times compared to its normalized state and increases fracture resistance of solebars under static and cyclic loadings.

Keywords Hardening · Water flow · Low-carbon steel · Castings
Solebars of freight bogies

S.A. Nikulin (✉) · A.B. Rozhnov · T.A. Nechaykina · V.I. Anikeenko ·
V.Yu. Turilina · S.O. Rogachev
National Science and Technology University MISiS, Moscow, Russia
e-mail: nikulin@misis.ru

A.B. Rozhnov
e-mail: rojnov@nm.ru

T.A. Nechaykina
e-mail: nechaykinata@gmail.com

V.I. Anikeenko
e-mail: viki.zabo@gmail.com

V.Yu. Turilina
e-mail: veronikat@rambler.ru

S.O. Rogachev
e-mail: csaap@mail.ru

© The Author(s) 2018   281
K.V. Anisimov et al. (eds.), *Proceedings of the Scientific-Practical Conference*
"Research and Development - 2016", https://doi.org/10.1007/978-3-319-62870-7_31

Introduction

Nowadays, improvement of fracture resistance of solebars of freight bogies aimed at ensuring their reliable operation is regarded to be an actual issue. Solebars of freight bogies are casting units of complex configuration manufactured from cast low-carbon steel 20 GL type. As for the heat treatment for the solebars of freight bogies, there is envisaged normalization or normalization with annealing. As a result of this type of heat treatment, solebars of freight bogies are characterized by relatively low strength properties (tensile strength approx. 550 MPa, yield strength approx. 360 MPa); and the internal residual tensile stresses on the surface might reach the level up to minus 50 MPa. This is the cause of their high sensitivity to stress concentrators and low cyclic life while being operated. The most effective way to eliminate the harmful influence of stress concentrators and to prolong the service life is their strengthening by volume-surface hardening method (VSH) by high-speed water flow (spray-quenching) [1, p. 544]. While applying controlled hardening by high-speed water flow, unlike traditional hardening in water or oil, heat transfer process can be significantly intensified [3, p. 257; 4, p. 201; 5, p. 4170]. This type of treatment generates a strength gradient in the product section and provides internal residual compressive stresses within its surface layers [1, p. 544]. Meanwhile VSH technology allows improving strength properties of solebars of freight bogies without changing the material and manufacturing technology of the castings.

Within the framework of this paper, we assume to investigate the impact of VSH by high-speed water flow on the microstructure and mechanical properties under static and cyclic loading of large-sized castings solebars of freight bogies as well as on their fragments manufactured from low-carbon cast steel. Previously we have already investigated the fracture mechanisms of solebar fragments [2, p. 31].

Materials and Methods of Investigation

Chemical composition of steel 20 GL of the solebars of freight bogies having been studied, percent (mass): 0.19 C; 1.18 Mn; 0.38 Si; 0.24 Cu; 0.14 Cr; 0.12 Ni; 0.06 Al; 0.018 S; 0.022 P. Fragments for study and testing (with thickness up to 20 mm) were cut from the solebars of freight bogies belonging to one consignment. Layout of the cutting is shown in Fig. 1.

Fig. 1 The solebar and cut fragments before the test

There has been conducted a comparative analysis of the structure and mechanical properties of the solebars of freight bogies in normalized state (excerpt for 2 h at 930–960 °C, cooling in air) and after VSH effected at 890–1000 °C with cooling to 20–30 °C by high-speed water flow speeds of 800 °C/sec.

Microstructure was studied applying Axiovert 40 MAT (Carl Zeiss) optical microscope and HITACHI TM-1000 scanning microscope operated in the mode of secondary electrons. Quantitative metallographic analysis was conducted applying Thixomet software environment.

Vickers microhardness was measured by Micromet 5101 (Buehler) microhardness tester and digital camera Mitron MTV-1 62W1P under load 1 H applied within 10 s.

Mechanical tensile test of proportional flat samples with thickness of 3.5 mm have been carried out on Instron 150 LX universal testing machine with 1 mm/min loading rate at room temperature. Tensile testing samples have been cut out of the solebar fragments after the VSH, close to hardened surface.

Three-point bending tests of solebar fragments have been carried out at room temperature with a loading rate 25 kN/min on the EBMC-200 PU test machine. The test has been finished when the first macro-cracks have been detected and identified by acoustic emission (AE) measurements. Alongside, there have been identified critical (before the advent of the crack) strain and residual deflection of top wall fragment after removal of the load.

Vertical loading three-point bending type cycle test of full-scaled solebars of freight bogies has been carried out on facility with electrohydraulic loading device equipped with hydropulser with maximum effort at 100 ton-force and applying specially designed loading supports (Fig. 2). Tests have been carried out in asymmetric loading mode with loading frequency 5 Hz. Basic distance between

Fig. 2 Cyclic testing of the solebar

supports has been 1850 mm. Cyclic loading has been carrying out until fracture of solebars of freight bogies at base load cycles equal to 2×10^6.

Results and Discussion

Microstructure. Microstructure analysis outcomes demonstrated that after normalization the fragments from 20GL steel have a ferritic-pearlitic structure with ferrite grains sized 25 ± 5 μm and perlite colonies with plates of cementite having thickness of 0.1–0.3 μm and interlamellar distance of 0.3–0.7 μm (Fig. 3a, d).

VSH leads to creation of gradient structure in the fragment. As for the surface layer of the sample after VSH (at a distance of 2–3 mm from the surface), there has been identified mixed martensite-like structure (Fig. 3b, e). There are troosto-martensite segment with excess ferrite on grain boundaries. On the depth from the surface layer of the fragment after VSH, the shaped structure representing a ferrite-cementite composite with interlamellar distance of no more than 0.2 μm, maintaining orientation of fine-needled martensite is being formed. As for the middle part of the sample fragment, there has been identified mixed structure (ferritic-cementite composite (sorbite) with small percentage of allocations of different morphology ferrite on grain boundaries) ensuring high toughness of the core (Fig. 3c, f).

Microhardness. After the normalization process has been carried out, microhardness along the whole section of fragments achieved 150–200 HV. After VSH there exists a hardened area with microhardness of 400–480 HV to the depth of 5–8 mm. In the central area of the fragment microhardness achieved 300–380 HV. Thus, the application of VSH leads to improvement of hardness at subsurface layers

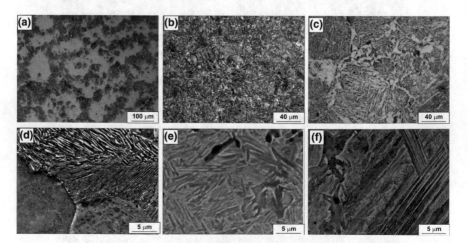

Fig. 3 The microstructure of the samples after normalization (**a, d**) and after VSH in the surface layer (**b, e**) and in the core of the sample (**c, f**)

in the solebars of freight bogies to the depth up to 5–8 mm by 2.0–2.5 times compared to normalized state.

Static and cyclic strength. Comparison of mean values of mechanical properties of samples from hardened surface layer of fragments of solebars of freight bogies after VSH and the samples from fragments of solebars of freight bogies after normalization demonstrates that VSH ensures improved tensile strength by 1.6–1.8 times (up to 965 MPa) and yield strength by 1.9–2.1 times (up to 710 MPa) compared with normalized status provided that relative elongation is not below 10%.

Three-point bending tests of the solebars of freight bogies fragments demonstrated that VSH affects the kinetics of deformation defects accumulation. Layer cracks appeared in hardened layer do not propagate until the maximum load has been applied and a main crack emerging further impedes within ductile core. When loading a three-point bending of the fragments after the normalization the main crack emerges even before the peak load, and its propagation takes place continuously over time until the final failure of the fragment. The loss of bearing capacity of fragments after normalization occurs under load of 1.5–1.7 times lower than for fragments after VSH.

Resistance to fatigue failures of the solebars of freight bogies, after VSH is by 20% higher than for the solebars of freight bogies in a normalized state. The value of the fatigue resistance safety factor of the solebars of freight bogies after VSH is 2.4, compared with a value of 2.1 for normalized solebars of freight bogies.

Conclusions

1. There has been demonstrated improvement of strength properties of large-scale products (thickness up to 20 mm) manufactured from low-carbon cast steel 20 GL, e.g., solebars of freight bogies when applying volume-surface hardening (VSH) by high-speed flow of water.
2. VSH leads to creation of gradient structure along the fragment wall thickness: hardened surface layer thickness of 5–8 mm with martensite-like structure and toughness core with sorbite and ferrite structure.
3. The VSH leads to increased tensile strength by 1.6–1.8 times and yield strength by 1.9–2.1 times in surface layers of the solebars of freight bogies compared to the ones in normalized state.
4. Solebars of freight bogies thermal hardening by VSH method provides their improved resistance to fracture under cyclic loading.

Acknowledgments Researches are carried out with the financial support of the state represented by the Ministry of Education and Science of the Russian Federation. Agreement no. 14.581.21.0009 03. Oct 2014. Unique Project Identifier: RFMEFI58114X0009.

References

1. Fedin, V.M., Borts, A.I.: Volume-surface hardening of freight bogie springs from steels with reduced and controlled hardenability. Metal Sci. Heat Treat. **51**(11–12), 544–552 (2009)
2. Nikulin S.A, Khanzhin V.G., Oguenko V.N., Nikitin A.V., Rozhnov A.B., Turilina V.Yu., Rogachev S.O.: Kinetics and mechanisms of static failure of cast solebar fragments during bending. Deformatsiya i razrushenie materialov. **3**, 31–35 (2016). [in Russia]
3. Schüttenberg, S., Hunkel, M., Fritsching, U., Zoch, H.-W.: Proceedings of 5th international and European conference on quenching and control of distortion, 25–27 April 2007, pp. 257–264. Berlin
4. Stark, P., Schuettenberg, S., Fritsching, U.: Spray quenching of specimen for ring heat treatment. WIT Trans. Eng. Sci. **70**, 201–212 (2011)
5. Li, Z., Ferguson, B.L., Nemkov, V., Goldstein, R., Jackowski, J., Fett, G.: Effect of quenching rate on distortion and residual stresses during induction hardening of a full-float truck axle shaft. J. Mater. Eng. Perform. **23**(12), 4170–4180 (2014)

Thermotropic Gel-Forming and Sol-Forming Systems for Enhanced Oil Recovery and Technologies of Their Joint Application with Thermal Methods for Oil Production

L.K. Altunina and V.A. Kuvshinov

Abstract The study of the kinetic, physicochemical, and rheological characteristics of solutions, gels and sols for enhanced oil recovery, water shutoff, and stimulation of oil production resulted in the creation of thermotropic systems, based on inorganic and polymer solutions, which are capable to generate a gel or sol in situ, and sol-forming oil-displacing surfactant-based systems with controlled viscosity and alkalinity. The thermal reservoir energy or that of the injected heat carrier is a factor causing solation and gelation. The technologies using the created systems are proposed for complicated operating conditions, including high-viscosity oilfields being developed by thermal-steam stimulation, and a complex of injection options: gradient and component-wise injection, reagent cycling. The technologies were successfully tested in the Permian-Carboniferous reservoir of high-viscosity oil in the Usinsk oilfield, including joint thermal-steam stimulation of the reservoir. The results correspond to the world level.

Keywords Enhanced oil recovery · Physicochemical methods · High-viscosity oil
Thermal methods · Surfactants · Sols · Thermotropic systems · Inorganic and polymer gels · Technologies · Gradient and component-wise injections
Water shutoff

Currently in the world and in Russia, most of major oilfields are at later stages of their development with a high water cut, especially oilfields developed by water flooding and thermal-steam stimulation. The current oil recovery factor is often less than 20%. Therefore application of physicochemical methods for enhanced oil

L.K. Altunina (✉) · V.A. Kuvshinov
IPC SB RAS—Institute of Petroleum Chemistry SB RAS, Tomsk, Russia
e-mail: alk@ipc.tsc.ru

V.A. Kuvshinov
e-mail: vak2@ipc.tsc.ru

© The Author(s) 2018
K.V. Anisimov et al. (eds.), *Proceedings of the Scientific-Practical Conference "Research and Development - 2016"*, https://doi.org/10.1007/978-3-319-62870-7_32

recovery, especially for water shutoff using gel technologies, is becoming actual [3, p. 2; 7, pp. 8–9; 8, p. 51].

The relevance of the research on the project is determined by the necessity to: improve the efficiency of the basic method of oil production in Russia—flooding and thermal methods for production of high-viscosity oils; decrease (stabilize) water cut of the producing oil; expand the scope of application of the technologies for enhanced oil recovery (EOR technologies) and water shutoff and provide the oil industry with effective domestic agents.

The project is aimed to improve the efficiency of oil production by water flooding and thermal methods due to combining them with physicochemical EOR methods and provide the oil industry with effective EOR technologies and domestic agents for their implementation.

Objectives of the project: to develop and study new compositions of the gel- and sol-forming high-viscosity systems intended to enhance oil recovery, decrease water cuttings of well production and stimulate oil production in abnormal operating conditions, including high-viscosity oilfields being developed by thermal-steam stimulation; to create EOR technologies applying new gel- and sol-forming systems and procedures of their injection combined with thermal methods of high-viscosity oil production, such as the injection of hot water and steam, and to test EOR technologies in field conditions.

Research objectives:

- to develop thermotropic inorganic and polymer systems in situ generating gels or sols, which create deflecting screens; surfactant-based gelled system with controlled viscosity and alkalinity to increase an oil displacement factor and conformance at thermal-steam stimulation and to carry out experimental laboratory research;
- to develop technologies for enhanced oil recovery and selective water shutoff, procedures of the systems injection; technology for component-wise system injection for in situ generation of gels due to hydrodynamic dispersion; computer calculation method for injection planning; recommendations concerning the application of EOR technologies.

The fundamental scientific novelty of the research consists: in the use of in situ generation of the thermotropic gels and sols with auto-controlled viscosity based on inorganic and polymer systems to improve the efficiency of water flooding, combined thermal-steam and physicochemical stimulations, as well as surfactant systems enabling to improve the oil recovery due to both increased oil displacement factors and conformance; in technological solutions of the project—joint application of the systems generating in situ mobile sol and immobile gel screens to increase the reservoir coverage and enhance oil recovery at water flooding and thermal-steam stimulation and complex of technological injection options—gradient and component-wise injections, reagent cycling.

Analytical review of scientific information sources: papers in leading foreign and Russian scientific journals, monographs and patents for 2009–2013, the patent research to a depth of 20 years and theoretical studies have shown that the proposed trends are novel and promising.

The study on the kinetics of hydrolysis and gelation, the colloid-chemical and rheological characteristics of the solutions, gels and sols resulted in the creation of thermotropic systems, based on inorganic and polymeric solutions, in situ generating a gel or sol and sol-forming oil-displacing systems based on surfactants with controlled viscosity and alkalinity to enhance oil recovery, to stimulate oil production and for water shutoff.

To improve the efficiency of water flooding and thermal-steam stimulation, we have proposed the thermotropic systems, based on inorganic "Al salt–carbamide–water–surfactant" system, generating gel or sol in situ with auto-controlled viscosity by a mechanism of cooperative process. Using the methods of rotational and vibration viscometry in the temperature range of 70–250 °C to study rheological properties of the gels we have found that aluminum hydroxide gel is a pseudoplastic thixotropic solid-like body of coagulation structure. When surfactants were added to the gel-forming solutions gels viscosity increased considerably.

The studies on rheological properties of sols and gels in "Al salt–carbamide–water" system at different component concentrations using "Reotest-2.1.M" and Haake RheoStress 600 rotational viscometers at different shear stresses have shown that the sols and gels are thixotropic. Thixotropy manifests itself in liquefaction at sufficiently intensive flow or stirring of the gels or sols and their thickening (solidification) after the termination of the mechanical impact. Figure 1 clearly shows the thixotropy manifestation both for the gels and sols—as hysteresis of rheological flow curves: decrease in viscosity is observed with increasing shear rate (forward motion) and increase in viscosity is observed at a subsequent decrease in shear rate (reverse motion). Thixotropy of the sols and gels generated in situ under the influence of the temperature of the injected heat carrier (steam, hot water) in the

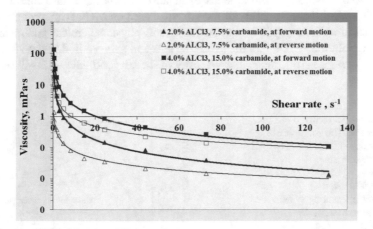

Fig. 1 The viscosity dependence of the sols and gels, prepared from solutions in "Al salt–carbamide–water" system, on the shear rate at the forward and reverse motions

development of high-viscosity oil deposits, gives additional possibilities to form the deflecting screens in the reservoir to equalize the profile of the heat carrier front and to increase the reservoir coverage with thermal-steam stimulation.

At shear rates ranging from 0.01 to 3 s^{-1} at high pressures in a uniform compression, the shear stress and viscosity of the gels, prepared from the solutions of the system based on "Al salt–carbamide–water" system, have values several orders of magnitude higher than at the atmospheric pressure. In the oil reservoir, the high pressure and conditions of uniform compression are realized, therefore the thermotropic gel-forming systems can be used to regulate the injectivity profile in injection wells and for water shutoff in production wells, including the thermal-steam and cyclic-steam reservoir stimulations to increase oil recovery.

The studies of physicochemical properties, rheological characteristics and oil-displacing ability of the sol- and gel-forming systems, based on "Al salt–carbamide–water" system added with substances regulating hydrophilic-lipophilic balance and gelation temperature, showed the possibility of their application in field conditions at a reservoir temperature 20–320 °C, with a high water salinity, up to 300 g/dm^3, to regulate filtration flows [6, p. 1], to enhance oil recovery and reservoir coverage at water flooding and thermal-steam stimulation.

To enhance oil recovery and decrease water cut we have created the thermotropic gel- and sol-forming systems based on a polymer with a lower critical solution temperature, the cellulose ether—methylcellulose (MC), which are capable of forming gels (sols) in situ due to a phase transition "solution—gel (sol)". The temperature and gelation time are regulated by inorganic and organic additives, Figs. 2 and 3, their effect is additive. Upon reaching the gelation temperature the viscosity sharply increases—5–200 times. At the shear rates ranging from 0.5 to 5 s^{-1} the gel is a solid-like body having viscoelastic properties. The gels retain their rheological characteristics at high temperatures—up to 150–220 °C. In the course of time the gels harden.

We have investigated the kinetics of gelation, rheological, and filtration characteristics, oil-displacing ability of the systems based on MC of different grades, we also determined their optimal compositions and physicochemical parameters: the pH of the solutions—6.4–7.0, the pH of the sols and gels—6.8–8.5; the viscosity of the solutions—25–100 mPa·s; the viscosity of the sols and gels—150–3900 mPa·s

Fig. 2 Change in the viscosity of 1% methyl cellulose MC-100 solution of with various carbamide concentrations after thermostatting at 90 °C, using Haake Viscotester iQ rheometer (measuring system of coaxial cylinders, the CR-mode) equipped with a temperature module on the Peltier elements

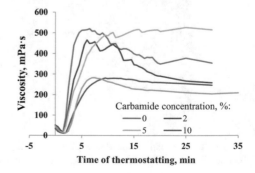

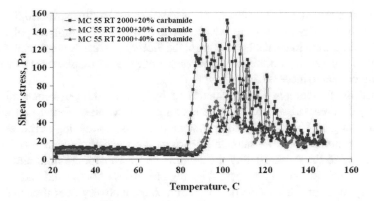

Fig. 3 The temperature dependence of the shear stress for 1.5% solutions of methyl cellulose MC 2000 55 RT added with different amounts of carbamide at a shear rate of 3 s^{-1} using Haake RheoStress 600 rotational viscometer

at low shear rates (0.5–5 s^{-1}), under pressure—up to 90,000–180,000 mPa·s; the density of the solutions—1000–1080 kg/m^3; gelation time—from several minutes to some days depending on the temperature and composition of the solution; freezing temperature—0–minus 20 °C. The increase in the oil displacement factor in heterogeneous reservoir models with a permeability of 0.167–9.456 μm^2 ranged from 3 to 21%, thus high absolute values of the oil displacement factor were achieved [4, p. 3].

To enhance oil recovery and stimulate oil production we have created the thermotropic sol-forming oil-displacing systems based on "surfactants—carbamide—ammonium salt—aluminum salt—water" system with controlled viscosity and alkalinity, which enable to increase not only the oil displacement factor, but also the factor of reservoir coverage by thermal-steam or cyclic-steam stimulations at simultaneous intensification of the development. To create the systems we used a promising concept of using the reservoir energy or that of the injected heat carrier to generate in situ the oil-displacing fluid and sols, which were thermotropic chemical "intelligent" systems, preserving and self-regulating in the reservoir for a long time a complex of colloid-chemical properties, optimal for oil displacement purposes.

In surface conditions the thermotropic sol-forming oil-displacing systems based on "surfactants–carbamide–ammonium salt–Al salt–water" system are low-viscosity liquids, and in reservoir conditions under the influence of the reservoir temperature or that of the injected heat carrier the systems are converted into sols with controlled viscosity and density, high oil-displacing capacity. Thus these systems acquire the rheological and colloidal-chemical properties enabling to use for the formation of mobile blocking screens redistributing filtration flows and increasing the reservoir coverage by the base stimulation and the oil displacement factor. The high oil-displacing properties of the systems are provided by in situ formation of CO_2 and ammonia buffer solution with a maximum buffering capacity in the pH range of 9.0–10.5, which cause a multiple increase in surfactants detergency, decrease oil viscosity and swelling of clay minerals in reservoir-rock. In situ thermotropic solation due to carbamide hydrolysis and the conjugated process of hydrolytic polycondensation of aluminum ions enables to control the viscosity of the systems adjusting it to the specific reservoir conditions. As a result both the oil displacement factor and reservoir coverage increase.

Based on the studies of the kinetics of hydrolysis, the physicochemical and rheological characteristics of the solutions and sols we have found that after the thermostatting at 70–200 °C of the system solutions, containing surfactants, carbamide, ammonium and Al salts, the pH rises, the viscosity increases 6–78 times, Fig. 4. Prior to the thermostating the solution was a Newtonian fluid and after the solation it became a viscoplastic fluid having properties of both solid and liquid and the ability to exhibit the properties of elastic shape recovery after the stress relief. Oil displacement factor was increased by 5–39%. The system solutions have demulsifying effect and the amount of water in the oil was decreased 10–220 times.

The created systems have the following physicochemical parameters: the pH of the solutions—3.4–4.1 pH units.; the pH of the sols and gels—7.7–10.1 pH units; the viscosity of the solutions—1.6–3.5 mPa·s; the viscosity of the sols—9.7–260 mPa·s; the density of the solutions—1161–1178 kg/m³; the gelation time—from several minutes to some days depending on the temperature and composition of the solution; the freezing temperature—minus 20.4–minus 21.2 °C. The systems can be used in a wide temperature range, from 70 to 220 °C. They have high technological and economic efficiency and are environmentally safe.

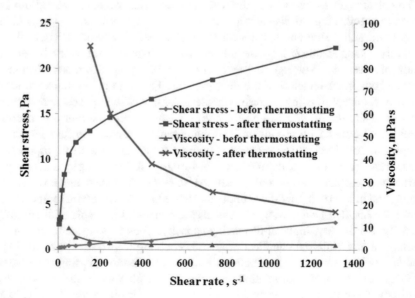

Fig. 4 Total flow curves and viscosity dependence on shear rate for the sol-forming oil-displacing system containing surfactants, carbamide, ammonium and aluminum salts (2.5%), before and after 5 h thermostatting at 150 °C

To enhance oil recovery, decrease water cuttings of well production and stimulate oil production under complicated operating conditions, including deposits of high-viscosity oils developed by thermal-steam stimulation, we have created thermotropic systems based on inorganic and polymer systems generating in situ a gel or sol and the sol-forming oil-displacing surfactant-based systems with controlled viscosity and alkalinity. The thermal reservoir energy or that of the injected heat carrier is a factor causing solation and gelation. Laboratory technological regulations for preparing the systems have been developed. The products of large-tonnage chemical production, with a preference for low-cost domestic products are used as the system components. It enables to create effective competition to the existing chemical EOR technologies such as polymer flooding and others.

The thermotropic sol- and gel-forming systems have the following characteristics: can be used in a wide range of reservoir temperatures (20–320 °C); have a low freezing point, homogeneity; retain their viscosity characteristics; exhibit high thermal-oxidative stability in highly saline reservoir waters, up to 300 g/dm^3; high penetrability, stimulate reservoir microflora and can be used in the formations with polymictic, carbonate and other reservoirs of different structure and permeability, from 1mD to 20 D, including low-permeable, highly heterogeneous and fractured reservoirs developed by water flooding or thermal-steam stimulation.

Under reservoir simulating conditions, the filtration characteristics and the oil displacement factors have been determined for typical oilfields of light and high-viscosity oils, in particular for Permian-Carboniferous reservoir in the Usinsk oilfield developed by thermal-steam and cyclic-steam stimulations. A computer model of the deflecting screens formed during the injection of the gel-forming and sol-forming systems has been developed. A calculation method has been created to schedule the systems injections to enhance oil recovery and for water shutoff: calculation of the required amounts of chemicals, scheme of injections and efficiency forecast. Extensible Markup Language XML of MathSoft development environment MathCad v13 is a source language. The text of the program is presented as a sheet of MathCad package.

We have developed the technologies and technological instructions for application of the gel-forming and sol-forming systems at water flooding and thermal-steam stimulation to enhance oil recovery and for water shutoff. Proposed were different variants of the system injections—reagent cycling, gradient, and component-wise. The technology of the component-wise injection of the systems is based on the in situ gelation in the predetermined reservoir place due to hydrodynamic dispersion. The technologies are environmentally safe and can be implemented using standard field equipment.

To study the influence of the developed EOR technologies on the physicochemical properties of oils and formation waters, reservoir microflora, we regularly analyzed oils sampled from the wells in the test areas. In 2014–2016 we analyzed 237 samples of oil and the formation waters from wells of the Permian-Carboniferous reservoir in the Usinsk oilfield.

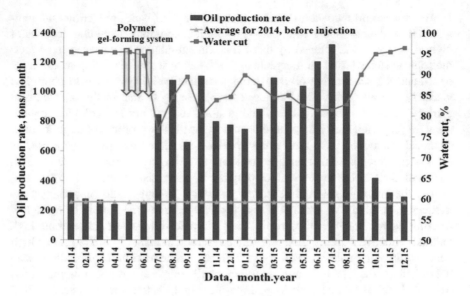

Fig. 5 The results of pilot tests for the water shutoff using the thermotropic polymer gel-forming system total in 5 production wells in the Permian-Carboniferous reservoir of the Usinsk oilfield: increased oil production rate and reduced water cut

In 2014–2016, the EOR technologies were successful tested in the Permian-Carboniferous reservoir of the Usinsk oilfield [1, p. 3; 2, p. 10; 5, pp. 280–282]: using the inorganic gel-forming system in three production wells developed by the natural mode, one steam-injection and two cyclic-steam wells; using the thermotropic polymer gel-forming system—in 10 production wells in the test area developed by thermal-steam stimulation and the sol-forming oil-displacing system with the controlled viscosity and alkalinity—in eight steam-injection wells (the effect was calculated for 75 production wells). The testing proved the efficiency of the systems and EOR technologies in the natural development mode and the thermal-steam stimulation. After the injection of the systems one observed a steady decrease in water cut and increase in oil production, Figs. 5 and 6. Thus, in 2014–2015 due to the injection of the sol-forming oil-displacing system in eight steam-injection wells incremental oil production amounted to more than 76,000 tons. The results obtained correspond to the world scientific and technological level.

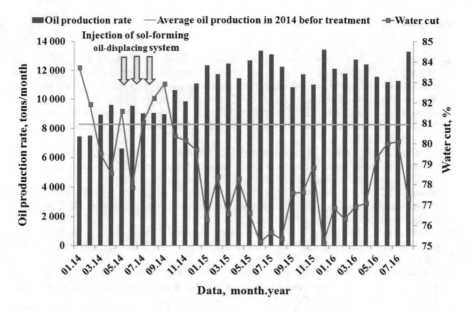

Fig. 6 Oil production rate and water cut before and after the injection of the sol-forming oil-displacing system at the thermal-steam stimulation in the test area of the Permian-Carboniferous reservoir of the Usinsk oilfield in 2014–2015

Conclusions

The use of new environmentally safe technologies intended to enhance oil recovery and stimulate oil production will enable to increase the final oil recovery factor by 2–7%; to decrease water cut of well production by 5–30%; to stimulate oil production 1.2–1.3 times and to carry out technological operations in the temperature range of 20–320 °C and formation water salinity up to 300 g/dm^3. Large-scale commercial application of the newly developed EOR technologies will enable to extend the profitable exploitation of deposits at the later stage of their development and engage in the development deposits with difficult-to-recover hydrocarbon reserves including high-viscosity oil pools and Arctic deposits. It will contribute to the development of the oil industry in Russia and to the expansion of its fuel and energy base.

Acknowlegements Research is carried out with the financial support of the State represented by the Ministry of Education and Science of the Russian Federation. Agreement no. 14.607.21.0022 05. June 2014. Unique project Identifier: RFMEFI60714X0022.

References

1. Altunina L.K., Kuvshinov V.A., Kuvshinov I.V.: «Cold» technologies for enhanced oil recovery from high-viscosity oil pools in carbonate reservoirs. In: 7th EAGE saint petersburg international conference and exhibition: understanding the harmony of the earth's resources through integration of geosciences, pp. 16–20. Saint Petersburg (2016)
2. Altunina L.K., Kuvshinov V.A., Chertenkov M.V., Ursegov S.O.: Integrated IOR technologies for heavy oil pools. In: Abstract Book of the 21st World Petroleum Congress, pp. 10–11. Moscow, Russia (2014)
3. Altunina L.K., Kuvshinov V.A., Kuvshinov I.V., Chertenkov M.V., Ursegov S.O.: Pilot tests of new EOR technologies for heavy oil reservoirs. In: Proceedings of SPE Russian petroleum conference, Moscow. Russia (2015). Paper 176703-MS. Flash-memory
4. Altunina, L.K., Kuvshinov, V.A.: Gel-forming METKA® system for selective water shutoff and enhanced oil recovery from Permocarbonic deposit in Usinskoye oilfield. AIP Conf. Proc. **63**(1), 37–48 (2008)
5. Altunina, L.K., Kuvshinov, V.A.: Improved oil recovery of high-viscosity oil pools with physicochemical methods and thermal-steam treatments. Georesources. **18**(4/1), 277–287 (2008)
6. Korsakova, N.K., Penkovsky, V.I., Altunina, L.K., Kuvshinov, V.A.: Redistribution of filtration flows by thermogel at boundary water flooding of oil reservoirs. AIP Conf. Proc. **1783**, 020106 (2016)
7. Romero-Zeron, L.: Chemical enhanced oil recovery (cEOR)—a practical overview. ISBN 978-953-51-2701-7. Print ISBN 978-953-51-2700-0. InTech, 2016. 200 p
8. Ruzin L.M., Morozyuk O.A., Durkin S.M.: Features and innovative ways of highly viscous oil field development. Neftyanoe khozyaystvo. **8**, pp. 51–53 (2013)

The Mixture of Fatty Acids Conversion into Hydrocarbons Over Original Pt-Sn/Al$_2$O$_3$ Catalyst

A.E. Gekhman, A.V. Chistyakov, M.V. Tsodikov, P.A. Zharova, S.S. Shapovalov and A.A. Pasynskii

Abstract Development of alternative approaches to fuel components and basic organic synthesis precursors producing based on biomass treatment products is important objective for ecology and chemistry. In this work, a number of original Pt–Sn containing catalysts were used for the mixture of fatty acids conversion. The peculiarities of used catalysts are usage of heterometallic precursors that possesses metal–metal bonds. Such kind of catalysts precursors allow obtaining more active and selective catalysts then ones based on a mixture of monometallic Pt and Sn precursors. Structural peculiarities of catalysts were characterized with TEM&EDS and XAS technique. Relations between Pt clusters structure and its catalytic properties were determined.

Keywords Heterogeneous catalysis · Green diesel

A.E. Gekhman (✉) · A.V. Chistyakov · M.V. Tsodikov · P.A. Zharova
MIPT, Moskovskaya, Russia
e-mail: gekhman@mail.ru

A.V. Chistyakov
e-mail: chistyakov@ips.ac.ru

M.V. Tsodikov
e-mail: tsodikov@ips.ac.ru

P.A. Zharova
e-mail: zharova@ips.ac.ru

S.S. Shapovalov · A.A. Pasynskii
IGIC RAS, Moscow, Russia
e-mail: sshap@yandex.ru

A.A. Pasynskii
e-mail: pasynskii@igic.ras.ru

© The Author(s) 2018
K.V. Anisimov et al. (eds.), *Proceedings of the Scientific-Practical Conference
"Research and Development - 2016"*, https://doi.org/10.1007/978-3-319-62870-7_33

Introduction

Nowadays, a significant interest has been concentrated on effective approaches related to renewable biomass conversion into fuels. Development of motor fuel production processes on the basis of promising types of plant oils is being a subject of intensive research nowadays [2]. Well known is the process of oils conversion towards aromatics over zeolite containing catalysts modified with Pd and Zn [1], Ni–Mo or Ni–W [4] and Co [3]. But due to aromatics containing reduction in modern gasoline fuels this route loses its perspective. Fatty acids obtained from different oil crops and waste edible fats are prospective raw materials for biofuel production [2]. A special attention is given to the production of fatty acids from crops capable of growing on non-croplands such as microalgae. The purpose of this study is to develop the chemical basis of a breakthrough technology that will make the move to large-scale production of fuel components and basic petrochemical products from microalgae biomass. The present work proposes catalysts and process for the high-selectivity conversion of fatty acids into hydrocarbons.

Experimental

Hydrothermal hydrolysis of triglycerides of fatty acids was carried out in a flow reaction vessel (internal diameter 4 mm, length 400 mm of the heated part) equipped with the dosing pump Gilson 305 (10 SC head), the back pressure controller at the reactor outlet, the condenser and the products receiver. Suspension of 10% rapeseed oil and 5% microcrystalline cellulose in distilled water fed into the flow reactor. Fatty acids extracted from condensate with diethyl ether, fatty acid content was determined by chromatography according to [5].

In the range of 250–350 °C and 200–300 atm triglycerides of rapeseed oil are converted to fatty acid mixture. The ratio between oleic acid and stearic is equal to the ratio between appropriate fragments in the rapeseed oil. This indicates that unsaturated fragments of fattyacids are stable under reaction condition. At temperatures below 300 °C precipitation of microcrystalline cellulose is observed in the reaction products.

The active components in Sn: Pt molar ratios of 1 and 5 were deposited on γ-Al2O3 by the impregnation method. The platinum loading on the catalysts was 0.4 wt%. Heterometallic Pt–Sn bond-containing complex precursor (PPh4)3[Pt(SnCl3)5] obtained according to a unique procedure, were used as precursors for catalyst preparation of the [6].

Both gaseous and liquid organic products in aqueous and organic phases were identified by GC-MS. Catalyst testing was performed in a PID Eng and Tech microcatalytic fixed-bed flow reactor unit, equipped with relevant instrumentation and control devices, under pressure 50 atm of H_2, temperature 400–480 °C, and substrates space velocity in the range of 1.2 h^{-1}.

High-resolution transmission electron microscopy (HRTEM) (a JEOL JEM_2010 microscope with a grid resolution of 0.14 nm and an accelerating voltage of 200 kV), and energy-dispersive X-ray spectroscopy (EDS) (EDAX attachment to an EDAX microscope, X-ray microanalyzer with a semiconductor Si(Li) detector with a resolution of ~ 130 eV) were used to study the morphology, structure, composition, and particle size distribution of the samples. To construct particle size distribution histograms, the data for 192–243 particles were statistically processed.

The local structure and charge state of platinum were studied by XAFS spectroscopy. The XANES and EXAFS X-ray absorption spectra were measured at the Structural Materials Science station of the Kurchatov Center for Synchrotron Radiation and Nanotechnology. The spectra were measured in the transmission mode using two ionization chambers filled with argon. A monoblock monochromator with a Si(111) cut was used for the monochromatization of a synchrotron radiation beam. To prepare a sample, catalyst powder was pressed into a pellet 1.5 mm thick in an atmosphere of H_2 or Ar; the pellet was covered with a thin polymer film and transferred to the spectrometer under anaerobic conditions. For each particular sample, three to four independent measurements were performed to check the reproducibility of the results.

Results and Discussion

The objectives of the study included

- Development co-hydrolysis process of triglycerides and fatty acids polysaccharide shell microalgal biomass watered to recover a lipid component;
- Development of new catalysts based on polynuclear metal complexes;
- Development of heterogeneous catalytic processes of fatty acids conversion.

In during of hydrothermal hydrolysis of triglycerides of fatty acids dependence of the oil conversion on the contact time is S-shaped (Fig. 1). At contact times of more than 6 min the conversion of oil tends to a certain limit. This can be explained by a decrease, of the oil concentration in the reaction mixture. In the range of 0.5–5 min the conversion of oil nonlinear increases, the second time derivative of the curve is positive. S-shaped form of the dependence cannot be explained by a formation during the reaction of strong acids (bases) capable to catalyze the hydrolysis, the pH of the initial suspension as product and do not differ. An unusual form of dependence might be related to the accumulation in the system of free fatty acids, which changes the properties of the fluid, for example, can increase the solubility of the oil or oil droplets fragmentation occur, and, consequently, accelerate mass transfer processes.

Over Sn/Al_2O_3 catalyst under 400 °C, VHSV 1, 2 h^{-1} the mixture of fatty acids converts into alkanes and olefins C_3–C_{17} fraction and aromatics. As shown in Fig. 2

Fig. 1 The dependence of
the conversion of rapeseed oil
on contact time. 300 atm,
300 °C

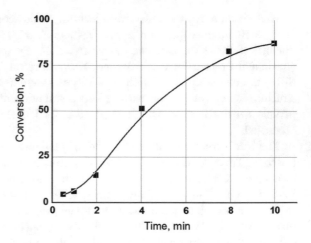

Fig. 2 Composition of
hydrocarbons obtained during
the mixture of fatty acids
conversion over Sn/Al_2O_3
catalyst

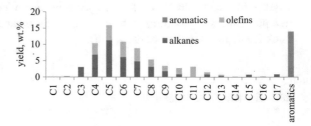

Fig. 3 Composition of
hydrocarbons obtained during
the mixture of fatty acids
conversion over 1Pt–
$5Sn/Al_2O_3$ catalyst

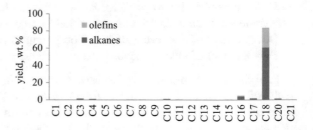

among aliphatic products hydrocarbons C_4–C_8 dominates that testify about intensive cracking of C–C bond of fatty acids. Also considerable amount of aromatics was obtained. Its yield was 14 wt%.

Comparison of products composition of the mixture of fatty acids conversion over Pt–Sn/Al_2O_3 catalysts showed that Sn content increasing led to intensification of C–O bonds hydrogenation and reduction of cracking and decarboxylation processes. Over $1Pt$–$5Sn/Al_2O_3$ the highest yield of C_{3+} hydrocarbons was reached equal 99.5 wt% from which 84 wt% was C_{18} fraction and 4.5 wt% C_3–C_4 fraction (Fig. 3). Among C_{18} fraction was found 23 wt% of olefins of which 7–10 wt% were linear alpha olefins. Obtained products may be used as C_3–C_4 fuel components

extraction followed by disengagement of linear alpha olefins and the rest alkanes may be incorporated into high quality diesel fuel by hydroizomerization process. Should be noted that over 1Pt–5Sn/Al$_2$O$_3$ catalyst take place mainly reduction of fatty acids into alkanes or olefins. Products of cracking and decarboxylation processes did not exceed 8 wt%. Moreover, total yield of C$_1$ products (methane, carbon oxides) observed lower then 0.1 wt%.

Obtained XAS data (Figs. 4 and 5) showed that fine dispersion of Pt in initial and after experiment 1Pt–5Sn/Al$_2$O$_3$ catalyst. In-depth quantitative analysis became more complicated due to absence of order in Pt envelope. In initial catalyst Pt has a wide range of near atoms (O, Cl, Pt) with considerable differences in interatomic distances and coordination numbers. After catalytic experiments Pt reduction into Pt^0 was observed but fine dispersion keep stable.

For the sample of 1Pt–5Sn/Al$_2$O$_3$ catalyst obtained from the heterometallic complex the EDX data show that the quantity of Sn is close to ~ 1 at.% compared to Al. Particles of Pt are present in the form of clusters and small particles (marked with red circles) (Fig. 6). The sample of 1Pt–5Sn/Al$_2$O$_3$ catalyst contains only two types of nanoparticles: Pt nanoparticles with a size of 1–2 nm, and ones with dimensions of ~ 3–5 nm and of PtSn$_{3\pm\delta}$ composition, according to the EDS data. Note that the individual nanoparticles of "pure" tin were not found.

The results allow us to conclude that in the activated state the active ingredients consisting of superfine Pt, Sn2+, Sn4 + particles and particles of PtSn$_{3\pm\,\delta}$ alloy are present on the surface of the catalyst system. Such a high selectivity of the catalyst in the conversion of the mixture of acids resulting in the quantitative yield of hydrocarbon fragments and water is probably caused by the chemisorption of the oxygen atoms of the carbonyl and ether groups on the ions of tin (2$^+$–4$^+$).

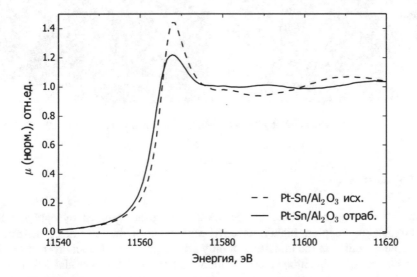

Fig. 4 XANES spectra for 1Pt–5Sn/Al$_2$O$_3$ catalyst

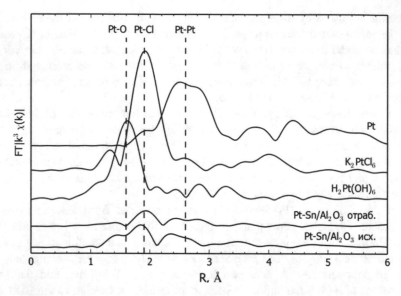

Fig. 5 EXAFS spectra for 1Pt–5Sn/Al₂O₃ catalyst and a number of standards

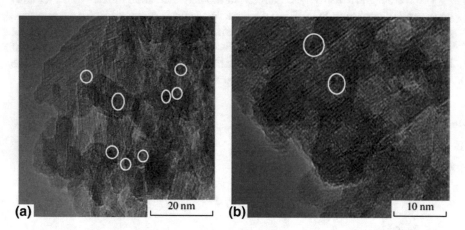

Fig. 6 TEM photomicrographs at different magnifications of a sample of the 1Pt–5Sn/Al₂O₃ catalyst

Conclusions

Thus, in the presence of the 1Pt–5Sn/Al2O3 catalyst prepared by applying the heterometallic complex, highly selective conversion of the fatty acids mixture is achieved, resulting in the formation of only alkane-alkene hydrocarbons that are the same number of hydrocarbon atoms as in initial fatty acid; C1 byproducts formed as a result of cracking reactions and the removal of carbonyl and carboxyl groups of esters are nearly absent. That results in methane and carbon oxides formation suppression. Over 1Pt–5Sn/Al2O3 catalysts during conversion of the fatty acids mixture the C–O hydrogenating selectivity significantly increases. During conversion of the fatty acids mixture under 400 °C a aliphatic hydrocarbons C_3–C_{18} were obtained with total yield about 99% calculated on passed carbon. Noted that only two fractions of C_3 and C_{18} hydrocarbons selectively formed. Obtained results allow minimizing loss of initial carbon weight due to cancelation of carboxyl fragment of fatty acids and glycerol conversion into methane and carbon oxides. Olefins C_{18} were found in products composition. Its yield was 23 wt% of which 7–10 wt% were linear alpha olefins. A high selectivity of the catalyst in the reaction of conversion of the fatty acids mixture is provided by two important factors: particle size factor and the structure of the precursor of the active ingredients. A small size of tin oxide and intermetallic alloy clusters probably set conditions for their interaction only with acid oxygen atoms as the most active centers of the substrate, but the clusters are spatially hindered for the reaction with unsaturated bonds in the carbon chain. The heterometallic complex used as a precursor of the active ingredients comprises a direct bond between the platinum and the tin atoms, which probably favors the formation on the surface of adjoining tin-containing and intermetallic centers having the ability for chemisorption of fatty acid by oxygen atoms, and for its reduction with hydrogen, and a weakened ability for cracking hydrocarbon fragments. This result made conversion of the fatty acids mixture over 1Pt–5Sn/Al$_2$O$_3$ catalyst prospective for fuel components and monomers production.

Acknowlegments Researches are carried out with the financial support of the state represented by the Ministry of Education and Science of the Russian Federation. Agreement (contract) no. 14.575.21.0052. June 2014. Unique project Identifier: RFMEFI57514X0052.

References

1. Chistyakov A.V., Gubanov M.A., Tsodikov M.V.: The direct conversion of rapeseed oil towards hydrocarbons over industrial catalysts. Chem. Eng. Trans. **32**, pp. 1093–1098 (2013)
2. Demirbas A.: Biodiesel: a realistic fuel alternative for diesel engines, p. 208. Springer-Verlad London Ltd., London (2008)
3. Kovács S., Boda L., Leveles L., Thernesz A., Hancsók J.: Catalytic hydrogenating of triglycerides for the production of bioparaffin mixture. Chem. Eng. Trans. **21**, pp. 1321–1326 (2010)
4. Mikulec J., Cvengros J., Joríková L., Banic M., Kleinová A.: Diesel production technology from renewable sources—second generation biofuels. Chem. Eng. Trans. **18**, pp. 475–480 (2009)
5. Nishiyama-Naruke A., Souza J.A., Carnelos M., Curi R.: HPLC determination of underivatized fatty acids saponified at 37C. Analysis of fatty acids in oils and tissues. Anal. Lett. **14**, pp. 2565–2576 (1998)
6. Shapovalov S.S., Pasynskii A.A., Torubaev Yu.V., Skabitskii I.V., Scheer M., Bodensteiner M.: Stannylene complexes of manganese, iron, and platinum. Russ. J. Coord. Chem. **3**, pp. 131–137 (2014)

Beneficiation of Heat-Treated Crushed Brown Coal for Energy Production and Utilities

V.A. Moiseev, V.G. Andrienko, V.G. Piletskii, V.A. Donchenko and A.I. Urvantsev

Abstract The article deals with the problems of increasing the efficiency of electric separators through developing new approached to their design for upgrading of heat-treated crushed brown coal. The rational use of natural resources and efficient energy consumption are two basic requirements of EU Directive 2008/1/EC. It is expected that the techniques for brown coal pre-drying will result in increased energy efficiency of enterprises of up to 5% and upgrading of heat-treated brown coal will increase energy efficiency by the same value. To upgrade heat-treated brown coal it is necessary to design new equipment including high-efficiency electric separators. It was earlier found that the maximum value of coal particle charge was provided in a corona-electrostatic separator, a lower level of coal particle charge was observed in a triboelectrostatic separator and the lowest level of coal particle charge was established in a plate-type electrostatic separator. The basic efficiency constraint for drum-type corona-electrostatic separators is the diameter of a collecting electrode and one operating area. To select the directions for increasing the productivity and efficiency in separating heat-treated crushed brown coal in drum-type corona-electrostatic separators the results of studies of changes in force vectors affecting its charged particles, nature of their motion in the electric field resulting in separated products—organic and mineral components of brown coal, taking this into consideration the factors of the heating temperature, voltage across a corona-producing electrode (drum), diameter of a corona-producing electrode and its rotation speed, have been analyzed. The increased efficiency of a drum-type CES with a slight increase in its weight is associated with removing a constraint from design parameters—the diameter of a collecting electrode (drum). It is realized through the change of orientation of a collecting electrode from horizontal to vertical.

Keywords Heat-treated brown coal · Upgrading · Particle charge
A corona-electrostatic separator · Efficiency · Vertical corona-producing electrode (drum)

V.A. Moiseev · V.G. Andrienko · V.G. Piletskii (✉) · V.A. Donchenko · A.I. Urvantsev
"COMPOMASH-TEK" Company, Moscow, Russia
e-mail: pvg2000@yandex.ru

© The Author(s) 2018
K.V. Anisimov et al. (eds.), *Proceedings of the Scientific-Practical Conference "Research and Development - 2016"*, https://doi.org/10.1007/978-3-319-62870-7_34

305

Introduction

The rational use of natural resources and efficient energy consumption are two basic requirements of EU Directive 2008/1/EC. It is expected that the techniques for brown coal pre-drying will result in increased energy efficiency of enterprises of up to 5% and upgrading of heat-treated brown coal will increase energy efficiency by the same value. To upgrade heat-treated brown coal it is necessary to design new equipment including high-efficiency electric separators. After performing theoretical and laboratory researches of processes of effecting upgrading of heat-treated crushed brown coal HTCBC it has been proven that the basis for changing electrical conductivity in particles of crushed brown coal, which improves the efficiency of upgrading, is its heat treatment. Thus, electrical conductivity for semiconductors and dielectrics increases when the temperature rises, separation of charged particles of HTCBC is improved due to their drying, classification and dedusting [1–3]. Figure 1 shows the dependence of the charge of coal particles of different sizes on the heating temperature. The researchers have identified the reasonability of heating up to 120–160 °C with electric separation of coal particles for effective separation of a mineral component, for example, fractions "−0.5 + 0.0", 14% yield, 9.6% ash content.

Dependence of HTCBC charges and particles of non-heat-treated brown coal on the temperature is shown in Fig. 2. When particle sizes decrease the charge increases. Heating of particles up to 105 °C is accompanied by condensation removal and reduced resistivity of a substance (from 10^6 to 10^3 M Ohm) [4–6]. This results in increase of contact potential difference.

Before separating on drum-type separators sizing is recommended otherwise centrifugal forces, proportional to the cube of diameter of particles, can neutralize the effect of electric forces proportional to the square of the diameter of particles [1, 3].

Fig. 1 Dependence of the charge of coal particles of different sizes on their preheating. *1* to *5*—sizes of particles 0.5–0.25; 0.25–0.175: 0.175–0.15; 0.15–0.1 and 0.1–0.074 mm

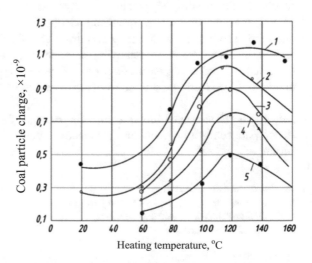

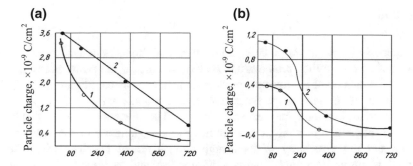

Fig. 2 Dependence of particle charge on particle sizes and the heating temperature. *Particle size 1* particles have been air-dried; *Particle size 2* particles have been heated up to 105 °C

It is known that in case of dedusting and classification of bulk materials, including coal, the disadvantage of sieves with a mesh smaller than 0.15 mm is hole clogging, wearability, expensiveness and low efficiency of sifting, excluding their commercial application. Therefore, corona separators used for dedusting and sizing of different materials are of practical interest [7, 8]. Working with the materials with fractions from 5 to 40%, sized less than 0.07 mm, it was found that extraction of this fraction on drum-type corona separators reaches up to 99%. When the content of fine fractions increases the number of treatment operations increases as well— when their content is from 5 to 20% one or two operations are required, when it is 20% and higher—three operations [7, 8].

The surface electrical conductivity of minerals (especially that of dielectrics and semiconductors) depends on the amount of adsorbed moisture which dramatically increases electrical conductivity. Thus, when the ambient relative humidity is from 15 to 55% the charge value of particles remains almost the same, when the humidity changes by more than 55% the value of the coal particle charge will decrease dramatically. When developing design solutions for efficient separation it is necessary to take into account peculiarities of the particle charge in semiconductors and dielectrics. Mineral conductors are well separated from conductors and nonconductors. It is more difficult to separate semiconductors from nonconductors (mineral components of brown coal) which can complicate HTCBC upgrading and requires the intensified process of formation of electric charges with their increased values. It has been determined that charged particles are separated in the inhomogeneous electric field as a result of interaction of electric and mechanical forces which requires reasoning for selecting HTCBC separation techniques [1, 9]. In the course of theoretical and laboratory researches of triboelectrostatic upgrading of brown coal the following disadvantages were identified:

- reduced efficiency as compared to the corona-electrostatic process of upgrading;
- in operation of the separator and free settling of clean coal the surface of plate electrodes must be free of dust which deteriorates separation;
- the concentrate contains increased mineral components and the emerging tailings—increased organic components.

Method

Triboadhesive separators process conductive and nonconductive, organic, and nonorganic finely milled minerals and materials. Thus, if the air humidity is more than 70% adhesion of microscopic particles is increased by capillary forces [4]. The upper limit of the size of particles, which can retain on the surface, differs in different conditions and may exceed 100 µm. The diversity of factors affecting adhesive interaction of coal particles indicates the complexity of the process concerned and the possibility of its adaptive management [7]. The triboadhesive method is limited by high energy consumption and necessity to operate the equipment $U = 20$–60 kV. Therefore, this method of upgrading of heat-treated crushed brown coal will be unprofitable. When assessing the possibility of reliable separation of brown coal particles in terms of semiconductors from dielectrics when using a triboelectrostatic, triboadhesive and corona-electrostatic methods of HTCBC separation drum-type corona-electrostatic method was selected as the most productive one. The studies of corona-electrostatic separators (CES) with horizontal and vertical drums showed that under equal conditions separators with a vertical drum are 2–3 times more productive. Therefore, to upgrade HTCBC it is recommended to design a CES with a vertical drum—collecting electrode. When developing design solutions of a CES with a vertical drum it is necessary to take into account the following:

- The polarity of a corona-producing electrode affects the operation of separators with a corona discharge. The breakdown voltage is higher with a negative corona than that with a positive corona which is to be grounded.
- When the linear speed of a drum in constant electric field intensity is increased, the efficiency of a separator can be decreased.
- The corona discharge emerges only in the inhomogeneous electric field in a small area near a thin conductor. This discharge does not extend to the opposite electrode and can be regarded as a partial gas breakdown [7].
- One of the factors influencing the charge of HTCBC particles, sufficient for separation, is the corona discharge current. The corona current depends on the shape of a corona-producing electrode, voltage applied and structural features of the area of the separator corona discharge [10].

These dependences (Fig. 3a) for the corona current are applied to the simplest cases when the current is determined only by the voltage across the electrodes, their dimensions and ion mobility, thus, without taking into account such factors as temperature, pressure, humidity, gas velocity and type and the presence of suspended particles. Field intensity increases near a corona-producing electrode and remains almost the same in the rest interval between electrodes (Fig. 3b). To avoid sparking between cylindrical electrodes certain ratio between the wire radius r and the cylinder radius R must be ensured. For gas ionization without short circuit $R : r \geq 2.7$.

Based on the analysis of works of Russian and foreign scientists devoted to the charge and dynamics of separation of mineral particles in the course of

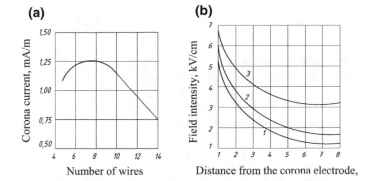

Fig. 3 **a** Dependence of the corona current on the number of corona wires [23]. **b** The curves of field intensity distribution at different current values with the corona discharge between cylinders (cylinder ⌀ 148 mm, wire ⌀ 2 mm) [7]. *1* $U_0 = 26$ kV, $I = 0.67 \times 10^{-6}$ A/cm; *2* $U_0 = 30$ kV, $I = 1.51 \times 10^{-6}$ A/cm; *3* $U_0 = 38$ kV, $I = 10^{-6}$ A/cm

corona-electrostatic separation [1–3, 5, 9, 11–14], one can conclude that mineral particles in the electric field are separated by means of retaining of charged particles on the surface of the grounded rotating electrode (retaining mode) or as a result of their turning towards the electrode with the potential opposed to that of particles (extraction mode).

The above equations characterize the qualitative behavior on a CES horizontal collecting electrode of spherical particles and allow to assess the degree of influence of certain physical factors on the separation process [15–17].

Separation of HTCBC particles in a CES with a vertical drum is similar to that in a separator with a horizontal drum [1, 3, 7, 18–20]. Unlike the diagram of forces in a separator with a horizontal drum, projection of the gravity force affecting the particle in a separator with a vertical drum, the axis of interaction of electrical pressing forces and a centrifugal separating force is equal to zero [18].

The effect of forces on particles retaining on the surface of a vertical collecting electrode

$$m \frac{d\bar{V}}{dt} = \sum \bar{F} = \bar{F}_C + \bar{F}_{mir.} + \bar{F}_{pond} + \bar{F}_{ad.} + \bar{F}_{c.f.} \tag{1}$$

where $\bar{F}_k$ is the Coulomb force of the effect of the electric field on the charged particle

$$F_C = Q_e E_c \tag{2}$$

where Q_e is the equilibrium particle charge in the operating area of a corona-producing electrode; E_c is zero tension of the corona discharge near a collecting electrode.

$$F_C = 4\pi\varepsilon_0 \cdot \left(1 + 2\frac{\varepsilon_1 - 1}{\varepsilon_1 + 2}\right) \cdot r^2 \cdot E^2 \cdot f(R') \tag{3}$$

$\bar{F}_{c.f.}$ is the centrifugal force affecting the particle retaining on the surface of a collecting electrode (drum) and is caused by rotation of the latter.

$$F_{c.f.} = m \cdot v^2 / R \tag{4}$$

where v is a linear speed of drum rotation,

$$v = 2\pi \cdot R_1 \cdot n/60 \tag{5}$$

where R_1 is the radius of a collecting electrode; n is the number of drum revolutions per minute

R is the distance from the drum center to the particle gravity center,

$$R = R_1 + r_1, \tag{6}$$

where r_1 is the distance from the drum surface to the particle gravity center. Since r_1 is very little as compared to R_1, we can assume that $R_1 = R$,

For spherical particles the weight is

$$m = 4/3\pi \cdot r^3 \gamma_p \tag{7}$$

Equation (12) with consideration of (5) and (6) is as follows

$$F_{c.f.} = \frac{\pi^3 \cdot r^3 \cdot \gamma_p \cdot R_1 \cdot n^2}{675}, \tag{8}$$

$\bar{F}_{mir.}$ is the mirroring force resulting from interaction of a particle resulting charge and induced to a collecting electrode of the electric charge equal in value to the resulting charge but opposed to it in its sign [10]. It is known that the particle discharge to the grounding surface of a drum is expressed in the equation [21]

$$Q_{RES} = Q_e \cdot e^{-\frac{t}{R' \cdot c}}, \tag{9}$$

where Q_{RES} is the residual charge after t sec after a particle leaves the operating area of a corona-producing electrode; Q_e is the equilibrium particle charge which it receives in the corona field; R' is the contact resistance between a particle and a collecting electrode; C is the space between a particle and an electrode.

R and C values determine the time constant of charging and discharging of a particle τ and depend only on the properties of a particle itself and do not depend on the external field intensity [22]

$$\tau = \varepsilon_0 \cdot \frac{\varepsilon_1 d_a + \varepsilon_2 \cdot (1 - d_a)}{\rho_{v_1} \cdot d_a + \rho_{v_2} \cdot (1 - d_a)}, \tag{10}$$

for the air at $pv_2 = 0$ and $\varepsilon_2 = 1$ this formula changes to [22]

$$\tau = \varepsilon_0 \frac{\varepsilon_1 + 1 - d_a}{d_a \cdot \rho_{v_1}}, \tag{11}$$

where ε_1 is dielectric conductivity of the particle material; d_a is the depolarization coefficient determined by the ratio of main ellipsoid axises a, b, c (particle of a rotational ellipsoid shape); for a spherical particles $d_a = 0.5$; pv_1 is the specific bulk electrical conductivity of a particle which can be derived using the formula $pv_1 = 1/pv$, where p_v is the bulk resistance of the particle material.

Taking into account particle discharge in the BC area based on Eq. (9) the formula for the mirroring force when the particle leaves the ionization area is as follows:

$$F_{MIR.} = \frac{\left(Qp \cdot e^{-\frac{t}{R^l \cdot c}}\right)^2}{16 \cdot \pi \cdot \varepsilon_0 \cdot r^2} = \pi \cdot \varepsilon_0 \cdot \left(1 + 2\frac{\varepsilon_1 - 1}{\varepsilon_1 + 2}\right)^2 \cdot r^2 \cdot E_{\kappa}^2 \cdot f^2(R) \cdot e^{-\frac{2t}{R^l \cdot c}} \tag{12}$$

In practice, components $-\frac{2t}{Rc}$ of Eq. (20) are neglected since the value is close to 1 [14]. In separation of particles larger than 0.05 mm in air by the medium resistance force, adhesive force, Archimedes (buoyant) force, the ponderomotive force can be neglected, the equation of balance of forces for a separator with a vertical drum is as follows [21]:

$$\sum \bar{F} = \bar{F}_c + \bar{F}_{mir.} + \bar{F}_{c.f.} \tag{13}$$

Equation for the resulting force [1, 9, 17, 23]

$$\sum F_{res} = \pi \varepsilon_0 \cdot r^2 \cdot E_C^2 \left(1 + 2\frac{\varepsilon_1 - 1}{\varepsilon_1 + 2}\right)\left(5 + 2\frac{\varepsilon_1 - 1}{\varepsilon_1 + 2}\right) - \frac{\pi^3 \cdot r^3 \cdot \gamma_p \cdot R_1 \cdot n^2}{675} \tag{14}$$

With $\sum F_{res} \geq 0$ particles will be retained on a vertical collecting electrode, and with $\sum F_{res} < 0$ they will be removed from the drum surface by the centrifugal force.

Therefore, the dependence of the size of nonconductor particles retained on a vertical collecting electrode in the area of the corona discharge:

$$r = \frac{675\varepsilon_0 \cdot r^2 \cdot E_{\kappa}^2 \left(1 + 2\frac{\varepsilon_1 - 1}{\varepsilon_1 + 2}\right)\left(5 + 2\frac{\varepsilon_1 - 1}{\varepsilon_1 + 2}\right)}{\pi^3 \cdot r^3 \cdot \gamma_q \cdot R_1 \cdot n^2}, \tag{15}$$

For example, for quartz particles with the density of $\gamma_p = 2.65 \times 10^3$ kg/m³, dielectric conductivity $\varepsilon_1 = 4.5$, with field intensity of the corona discharge $E_c = 5 \times 10^5$ V/m:

$$r = \frac{0.721}{R_1 \cdot n^2} \tag{16}$$

The content of the mineral component of HTCBC corresponds to the group of inertinite microcomponents—clay, sulfate sand, pyrite, and carbonates, i.e., analogue of quartz sand.

A number of HTCBC properties are similar to those of quartz sand: dielectric conductivity 4.5, density of up to 1.500 kg/m^3, bulk density of up to 1.4 t/m^3, surface area of up to 5 m^2/kg, and the diameter of a quartz sand particles is from 0.05 mm to 1 mm—with fraction particles "−1 mm" of HTCBC with the highest content of non-combustible components. The purpose of this work is to study the principles of separation of non-combustible components of HTCBC, the results of investigation of the quartz sand model can be used for the analysis of the quality of upgrading of HTCBC on a drum-type CES.

In the mode of retaining nonconductor particles when a particle leaves the corona area, only the mirroring force ($\bar{F}_{mir.}$) and the centrifugal force ($\bar{F}_{c.f.}$) affect the latter [21]:

$$\sum \bar{F}_{res} = \bar{F}_{mir.} + \bar{F}_{c.f.} \tag{17}$$

$$\sum F_{res} = 675 \cdot \varepsilon_0 \cdot r^2 \cdot E_c^2 \left(1 + 2\frac{\varepsilon_1 - 1}{\varepsilon_1 + 2}\right)^2 = \pi^3 \cdot r^3 \cdot \gamma_p \cdot R_1 \cdot n^2, \tag{18}$$

For quartz particles with the density of 2.65×10^3 kg/m^3 and dielectric conductivity $\varepsilon_1 = 4.5$, the field intensity of the corona discharge is $E_c = 5 \times 10^3$ V/m

$$r = \frac{0.247}{R_1 \cdot n^2} \tag{19}$$

By comparison, the size of quartz particles outside the corona area in separators with a horizontal drum is as follows:

$$r = \frac{0.247}{R_1 \cdot n^2 + 895.44} \tag{20}$$

Results

Calculation of the size of quartz sand particles retained on a vertical collecting electrode and on a horizontal electrode is shown in Figs. 4 and 5 for different values of the angular velocity of a collecting electrode and the linear speed of a drum [4]. Increase in size of nonconductor particles retained on a vertical drum, as compared to a horizontal drum at the same linear speed, is due to change of the centrifugal force affecting the particles on a vertical drum based on Eq. (12).

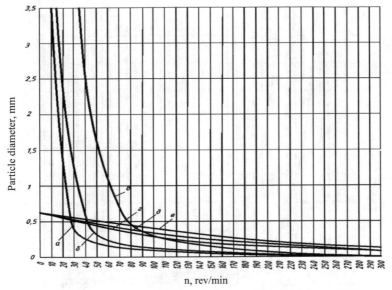

a – for a horizontal drum Ø150 mm, b – for a horizontal drum Ø240 mm;

c – for a horizontal drum Ø356 mm, d – for a vertical drum Ø500 mm;

e – for a vertical drum Ø1000 mm, f – for a vertical drum Ø2000 mm.

Fig. 4 Dependence of the diameter of retained particles outside the corona on the number of revolutions of a collecting electrode

The efficiency of a drum-type CES with a vertical collecting electrode can be determined based on the value of the efficiency of an electric separator with a horizontal electrode according to the formula [24]:

$$Q_{\text{V.S.}} = \sqrt{\frac{R_V}{R_H} \cdot \frac{L_V}{L_H}} \cdot N \cdot C_v \cdot Q_{\text{H.S.}}, \tag{21}$$

where C_v is the coefficient of occupation of the surface of a vertical collecting electrode which is determined by the ratio of the length of an outlet slot of the feeder ($Lo.f$) to the length of the generator of a collecting electrode (Lv) in unit fractions ($Cv = Lo.f./ Lv$); $Q_{\text{H.S}}$ is the efficiency of an electric separator with a horizontal collecting electrode calculated by the formula $Q_{\text{SEP}} = N \cdot L \cdot b \cdot q \cdot 3.6 \times 10^3$ m/hour) or experimentally, t/h; R_V and R_H are the radiuses of a collecting electrode, vertical and horizontal, m; L_V and L_H are the lengths of the generator of a collecting electrode, vertical and horizontal, m; N is the number of individual sections of an electric separator with a vertical collecting electrode (Fig. 6).

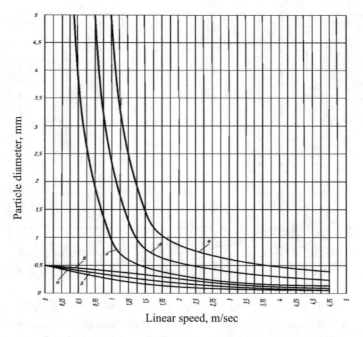

Linear speed, m/sec

a – for a horizontal drum Ø150 mm, b – for a horizontal drum Ø240 mm;

c – for a horizontal drum Ø356 mm, d – for a vertical drum Ø500 mm, e – for a vertical drum
Ø1000 mm;f – for a vertical drum Ø2000 mm.

Fig. 5 Dependence of the diameter of particles retained outside the corona on the linear speed of a collecting electrode

When the linear speed of drums at constant intensity increases, the efficiency of the electric field of separators decreases. This is due to the increase in the centrifugal force—the size of retained particles is decreased, the size of the layer on the drum is decreased, separator efficiency is reduced.

Researches on classification and upgrading were performed using an experimental facility. To retain a collecting electrode of dielectric particles on the surface when the linear speed of a drum is increased the charging process for a material being separated must be [24]. To successfully separate mineral mixtures in a CES the particles must be charged and must contact the surface of a collecting electrode. The flow of ions generated in the "corona hood", affected by the electric field, moves to the collecting (grounded) electrode, where it charges the particles. Thus, the corona discharge current and its value are the factors affecting generation of charges on particles, which are sufficient for their separation. Separation of minerals in drum-type corona separators is determined by the discharge current intensity. However, the current levels of the corona discharge in different points of a collecting electrode differed and in the course of the experiment on a chamber corona separator the following trend was observed: the content of magnetite in the center of

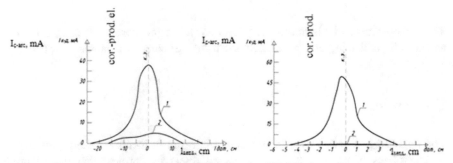

Linear speed of a horizontal drum, m/sec

1 – voltage across a corona-producing electrode of 32 kV; 2 – voltage across a corona-
producing electrode of 16 kV

Fig. 6 a Level and distribution of the current of the corona discharge on the surface of a vertical collecting electrode of ∅ 1 m for the "wire-drum" open system. **b** Level and distribution of the current of the corona discharge on the surface of a vertical collecting electrode of ∅ 1 m for the "wire-drum" open system bounded by glass-cloth plates in two sides

a section of the grounded electrode is usually minimal, it increases near the edges and then it decreases again [25]. Electrical resistivity of brown coal in its nature is ionic and is widely used depending on humidity. Heat-treated crushed brown coal has low electrical resistivity of $10^{-2}-10^{-4}$ Ohm/m, similar to magnetite values, which allows to use the results of experiments for a comparative analysis (Fig. 7).

Fig. 7 Level and distribution of the current of the corona discharge on the surface of a vertical collecting electrode with the diameter of 1 m for the "wire-drum" system bounded by a glass-cloth plate in one side

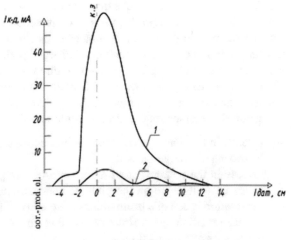

1 – voltage across a corona-producing
electrode of 32 kV; 2 – voltage across a corona
-producing electrode of 16 kV

Discussion

Thus, the value and distribution of the current of the corona discharge on the surface of a collecting electrode affect the results of electric separation [8] (Fig. 7).

Conclusion

The upgrading of heat-treated brown coal will result in increased energy efficiency of enterprises of up to 5%. The conducted researches proved that the effective techniques for separation of semiconductor (organic component) and dielectric materials (mineral component) for upgrading of heat-treated crushed brown coal are electric separation techniques including drum-type corona-electrostatic, plate-type electrostatic, drum-type triboelectrostatic and drum-type triboadhesive ones. The maximum level of coal particle charge was provided in a corona-electrostatic separator, a lower charge level was observed in a triboelectrostatic separator and the lowest ones were established in a plate-type electrostatic separator. Laboratory tests performed to establish the optimal size of HTCBC fed into separators and to select the most efficient upgrading technique showed the following:

– Techniques using a plate-type electrostatic separator and a triboadhesive separator work poorly for separation of heat-treated crushed brown coal.
– A drum-type CES is the most efficient one.

The basic efficiency constraint for drum-type corona-electrostatic separators is the diameter of a collecting electrode and one operating area. To select the directions for increasing the productivity and efficiency in separating heat-treated crushed brown coal in drum-type corona-electrostatic separators the results of studies of changes in force vectors affecting its charged particles, nature of their motion in the electric field resulting in separated products—organic and mineral components of brown coal, taking this into consideration the factors of the heating temperature, voltage across a corona-producing electrode (drum), diameter of a corona-producing electrode and its rotation speed, have been analyzed.

– Increase in the diameter of a corona-producing electrode results in reduction in the corona discharge current.
– Increase in the voltage across a corona-producing electrode of up to 32 kV has the greatest influence on the yield of nonconductors.
– The rotation speed of a drum affects the quality of fractions separated, halving of the drum rotation speed (from 60 to 30 rev/min) results in increase in the yield of a nonconductor fraction.

The increased efficiency of a drum-type CES is associated with removing a constraint from design parameters—the diameter of a collecting electrode (drum). It is realized through the change of orientation of a collecting electrode from horizontal to vertical.

Acknowledgments Researches are carried out with the financial support of the state represented by the Ministry of Education and Science of the Russian Federation. Agreement No. 14.579.21.0036, 05. Jun 14. Unique project Identifier: RFMEF157914X0036.

References

1. Olofinsky, N.F., Novikov, V.A.: Tribo-Adhesive Separation. Nauka, Moscow (1974)
2. Fraus, F. (1962). Electrostatic Separation of Granular Materials. U.S. Dept. of the Interior Bureau of Mines
3. Mesenyashin, A.I.: Electric Separation in Strong Fields. Nedra, Moscow (1978)
4. Simon, A.D.: Dust and Powder Adhesion, 2nd edn. Revised and enlarged. Khimiya, Moscow (1979)
5. Volkova, E.V., Zhus, G.V., Kuzmin, D.V.: Dielectric Separation of Various Polyconcentrates and Materials. Nedra, Moscow (1975)
6. Lower-James, E.: Fundamentals of electrical concentration of minerals. Mines Mag. 1(50) (1966)
7. Plaksin, I.N., Olofinsky, N.F.: Electric corona separation and its application for upgrading, classification and dedusting. In: Energy and Magnetic Separation Methods. Moscow (1965)
8. Kovalev, A.P.: Electrical Upgrading of Coal Dust. Works of the MPEI, 6th edn. (1950)
9. Vereshchagin, I.P., Levitov, G.Z.: Fundamentals of Electro-Gas Dynamics of Disperse Systems. Energia, Moscow (1974)
10. Olefinsky, N.F.: Electric separation of black sands of foundries. J. Mech. Eng. 12 (1947)
11. Egorov, V.L.: Magnetic, Electric and Special Methods of Upgrading of Ores. Nedra, Moscow (1977)
12. Karnaukhov, N.M.: Techniques for Upgrading of Bulk Concentrates using Electric Separation. Nedra, Moscow (1966)
13. Mamedov, A.I.: Physical Methods of Upgrading of Ore Stones. STI of NIITEKHIM, Moscow (1979)
14. Angelov, A.I., Ershov, V.S., Losaberidze, S.I., Pashni, M.M.: Movement of Charged Particles in the Electrostatic Field of a Drum-Type Corona-Electrostatic Separator. Electronic Materials Processing, 4 (1978)
15. Bachkovsky, M.V., Besov, B.V., Novikov, V.A., Olofinsky, N.F.: Electrical conductivity of materials and establishment of electrical resistivity. In: Theory and Practice of Techniques for Electric and Magnetic Separation of Minerals. Moscow (1968)
16. Urvantsev, A.I., Shikhov, N.V., Mushketov, A.A.: On the increase in Capacity of Drum-Type Corona—Discharge Electrostatic Separators. Upgrading of Ores. Special Issue for the XXI IMPC, pp. 62–65 (2000)
17. Cherchintsev, V.D.: Investigation of the triboelectric effect of minerals of dielectrics in their Corona separation. Works Magnitogorsk Inst. Min. 112, 119–122 (1975)
18. Levitov, V.I.: Alternating current Corona. Issues of the theory of research methods and theoretical characteristics. Synopsis of a thesis. MPEI, Moscow (1966)
19. Degtyarenko, A.V., Kashkarov, I.F.: Role and capabilities of electric separation in upgrading of complex titanic ores. In: Improvement of Electric Separation Processes and Concentration of Electric Separators. Interdepartmental Collection of Scientific Papers. Mechanobr (1987)
20. Khopunov, E.A., Revnivtsev, V.I.: Application of modern electrophysical methods in investigation of upgrading process. Works Uralmekhanobr Inst. 18, 54–62 (1972)

21. Urvantsev, A.I., Shikhov, N.V., et al.: Development of a high-performance drum-type corona-electrostatic vertical separator. Tsvetnyie Metally **11**, 71–74 (1995)
22. Kakovsky, I.A., Revnivtsev, V.I.: On the influence of the surface state in the process of electric separation of minerals with low electrical conductivity. In: 5th International Congress on Mineral Processing, Moscow (1962)
23. Komlev, A.M., Urvantsev, A.I., Shikhov, N.V., Zhuravsky, I.V.: Technological Tests of an Electric Separator SE-70/100: Report of the Uralmekhanobr Institute, p. 69. Sverdlovsk (1986)
24. Shikhov, N.V., Urvantsev, A.I. Electric separation as one of promising methods of complex processing of minerals. Abstracts of the Scientific and Technical Conference "Problems of Complex Processing of Titanomagnetite Ores of the South Urals", Magnitogorsk (2001, March 28–29)
25. Karnaukhov, N.M., Tarasova, T.B.: On increasing the efficiency of electric separators. Tsvetnyie Metally **4**, 41–47 (1963)
26. Belov, P.D.: Electrostatic Fluidization Separation. Nedra, Moscow (1977)
27. Komlev, A.M., Urvantsev, A.I., Druzhinin, V.M.: Investigation of Physical Models of Separation in Electric Separators: Report of the Uralmekhanobr Institute, p. 66. Sverdlovsk (1984)
28. Urvantsev, A.I., Shikhov, N.V., Zaitsev, G.V.: Results of researches and practice of upgrading of minerals by means of electric separation. News Higher Educ. Inst. Min. J. **5**, 37–51 (2005)
29. Revnivtsev, V.I., Olofinsky, N.F.: Current state and development trends of electrical separation of minerals. In: Works of the World Electrotechnical Congress. VELK Organizing Committee, Moscow (1977)
30. Meseniashin, A.I.: On the effect of electric forces on particles at the electrode. Electron. Mater. Process. **5**, 65–69 (1982)
31. Plaksin, I.N., Olofinsky, N.F.: On application of corona dedusters and classifiers for upgrading of fine sizes of minerals. In: Application of the Electric Field Forces in Industry and Agriculture, pp. 239–250. Moscow (1964)

NiMo/USY-Alumina Catalysts with Different Zeolite Content for Vacuum Gas Oil Hydrocracking Over Stacked Beds

P.P. Dik, V.P. Doronin, E.Yu. Gerasimov, M.O. Kazakov,
O.V. Klimov, G.I. Koryakina, K.A. Nadeina, A.S. Noskov
and T.P. Sorokina

Abstract The stacked beds comprising hydrotreating catalyst as the top layer, hydrocracking catalyst based on amorphous silica-alumina as the interlayer and hydrocracking catalyst based on USY zeolite as the bottom layer were tested in hydrocracking of mixed feed containing straight-run VGO, heavy coker gas oil, aromatic extract and petrolatum. It is shown that stacked beds with developed catalysts can be successfully used both in the once-through hydrocracking to provide VGO conversion of 70–80% with middle distillates yields up to 50 wt% and in the first stage operation of two stages hydrocracker to provide 35–65% VGO

P.P. Dik · E.Yu. Gerasimov · M.O. Kazakov (✉) · O.V. Klimov ·
G.I. Koryakina · K.A. Nadeina · A.S. Noskov
Boreskov Institute of Catalysis SB RAS, Novosibirsk, Russia
e-mail: kazakov@catalysis.ru

P.P. Dik
e-mail: dik@catalysis.ru

E.Yu. Gerasimov
e-mail: gerasimov@catalysis.ru

O.V. Klimov
e-mail: klm@catalysis.ru

G.I. Koryakina
e-mail: koryakina@catalysis.ru

K.A. Nadeina
e-mail: lakmallow@catalysis.ru

A.S. Noskov
e-mail: noskov@catalysis.ru

V.P. Doronin · T.P. Sorokina
Institute of Hydrocarbons Processing SB RAS, Omsk, Russia
e-mail: doronin@ihcp.ru

T.P. Sorokina
e-mail: sorokina@ihcp.ru

© The Author(s) 2018   319
K.V. Anisimov et al. (eds.), *Proceedings of the Scientific-Practical Conference
"Research and Development - 2016"*, https://doi.org/10.1007/978-3-319-62870-7_35

conversion and produce high-quality middle distillates and feed for the second stage. The commercial partner of this work is Gazprom Neft PJSC (Gazprom Neft Omsk Refinery).

Keywords Hydrocracking · Vacuum gas oil · Stacked bed · Middle distillates Catalyst · Zeolite

Introduction

Hydrocracking of heavy petroleum distillates is one of the key processes in oil refinery. Hydrocracking increases oil refining efficiency and provides the production of high-quality low-sulfur middle distillates and fractions, which are used as a feed for other processes such as reforming and catalytic cracking [15]. Hydrocracking catalysts are bifunctional systems with hydrogenating and acid sites. Supported Ni–Mo(W)-S component performs hydrogenation and hydrodesulfurisation reactions, while acidic support, usually based on amorphous silica-alumina (ASA) or zeolites, performs cracking and isomerization reactions [1]. For the second stage of hydrocracking, when the feed with low-sulfur and nitrogen content (as a rule, less than 10 ppm) is used, Pt and Pd bifunctional catalysts may be effective [5].

Zeolite catalysts are widely used for vacuum gas oil (VGO) hydrocracking, since they are more active and allow the process to be carried out at lower temperatures. Catalysts based on ASA are less active but provide higher selectivity to middle distillates [3, 17]. Zeolite catalysts cannot be used separately for hydrocracking of non-pretreated VGO, since heteroatomic compounds, containing in non-pretreated feed, rapidly deactivate the acid function of zeolites. They should be loaded either as the bottom catalyst bed, with hydrotreating catalyst bed being on the top of the reactor in the case of layer-by-layer loading, or in the second reactor, if a hydrotreating catalyst is loaded in the first reactor [2, 11]. Hydrogenation and partial cracking of polyaromatic compounds along with transformation of nitrogen-containing compounds, which poison zeolite catalysts, occur in the first reactor or in the first catalyst bed. When a catalyst based on ASA is used separately for VGO hydrocracking, higher yield of middle distillates is achieved. However, stacked beds become more popular due to higher activity and stability and improved quality of desired products—middle distillates, particularly higher cetane number for the diesel fuels and higher smoke point for the kerosene [6, 17]. In this work, we study the influence of USY zeolite content on the properties of hydrocracking catalyst and the activity and selectivity of stacked bed in vacuum gas oil hydrocracking to achieve the maximal yield of middle distillates. The stacked bed contains hydrotreating catalyst as the top layer, hydrocracking catalyst based on amorphous aluminosilicate as the interlayer and hydrocracking catalyst based on zeolite as the bottom layer.

Experimental

AlOOH pseudoboehmite produced by ISCZC (Russia), H-USY zeolite, obtained from microcrystalline zeolite NaY, and amorphous aluminosilicate ASA were used for the preparation of supports. ASA was prepared by the method of sequential precipitation according to the technique described in [4]. H-USY preparation procedure included nearly complete removal of sodium ions by ammonium exchange with subsequent dealumination by thermal treatment with water vapor.

Four supports with USY content from 10 to 40 wt% were prepared by mixing of USY zeolite and AlOOH with extrusion of obtained paste using plunger extruder. The obtained trilobe extrudates were dried at 120 °C and calcined at 550 °C. The Al_2O_3 and $ASA-Al_2O_3$ supports were prepared by the same procedure.

NiMo/USY(x)–Al_2O_3 (where x—USY zeolite content in the support) were prepared by impregnation with aqueous solution prepared from nickel carbonate, ammonium heptamolybdate and citric acid. Impregnated catalysts were dried at 120 °C and calcined at 550 °C. NiMo/Al_2O_3 catalyst, which was used as a first layer in a stacked bed, was prepared using similar impregnation solution but without calcination after drying at 120 °C [8]. Ni and Mo content in NiMo/Al_2O_3 catalyst was 3.7 and 12.5 wt% respectively. BET surface area was 148 m^2/g and pore volume was 0.36 cm^3/g with average pore diameter of 104 Å. NiW/ASA–Al_2O_3 catalyst, which was used as a second layer in a stacked bed, was prepared by impregnation with aqueous solution prepared from nickel carbonate, ammonium paratungstate and citric acid with subsequent drying at 120 °C and calcination at 550 °C [13]. Ni and W content in NiW/ASA–Al_2O_3 catalyst was 3.1 and 17.4 wt% respectively. BET surface area was 207 m^2/g and pore volume was 0.51 cm^3/g.

Elemental analysis of catalysts was carried out using atomic emission spectroscopy with inductively coupled plasma on Optima 4300 DV. Textural properties of the catalysts and supports were determined by nitrogen physisorption using an ASAP 2400 Micrometrics instrument. HRTEM images were obtained on a JEM-2010 electron microscope, JEOL. Bulk crushing strength (BCS) according to Shell SMS 1471 or analogous standard ASTM method 7084-4 was measured using Bulk Crushing Strength instrument, VINCI Technologies.

Testing in hydrocracking of vacuum gasoil was carried out in a high-pressure trickle-bed unit. Catalyst trilobe granules with a length of 3–6 mm and without defects were used for reactor loading. The total volume of the catalysts in reactor was 60 cm^3: first layer—NiMo/Al_2O_3 20 cm^3; second layer—NiW/ASA-Al_2O_3 20 cm^3; third layer—NiMo/USY(x)–Al_2O_3 20 cm^3. To minimize a breakthrough of the feed through a catalyst bed, catalyst granules were mixed with SiC. A mixed feed, comprising straight-run VGO (69 wt%), heavy coker gas oil (22 wt%) and fractions from solvent extraction unit (aromatic extract—7 wt%) and solvent dewaxing unit (petrolatum—2 wt%), was used as a feed. The content of sulfur and nitrogen in the feed was 0.97 and 0.11 wt% respectively. Catalysts were sulfided in situ at a pressure of 3.5 MPa using the mixture of straight-run gas oil, dimethyl disulfide and aniline. The catalyst loading and sulfidation procedures are thoroughly described in [6]. Hydrocracking tests were carried out at a pressure of 16.0 MPa, a liquid hourly space velocity (LHSV) of 0.71 h^{-1} and H_2 to oil ratio of 1500 (v/v). The temperature in the reactor was 360 °C during the first 12 h, and then the temperature was increased up to 390 °C and higher. Each experimental temperature (390 and 410 °C) was maintained for 192 h to ensure that steady state conditions were reached. Gas effluent from the separator was analyzed by gas chromatography using FID and capillary column. Liquid products were analyzed by SIMDIS-GC in accordance with the ASTM D7213 standard test method. Product yields were defined by summing up an amount of fractions, determined by SIMDIS-GC, and an amount of fractions from the gas effluent from the separator. According to ASTM D86, liquid samples were fractioned by distillation at atmospheric pressure into naphta (<130 °C), middle distillates (130–360 °C) and unconverted oil (>360 °C). Sulfur content in middle distillates was measured by ultraviolet fluorescence on Xplorer-NS, TE Instruments. The conversion and selectivity to middle distillates were calculated as in [6]. The results of catalyst testing are reported from the average of several gas and liquid product samples taken from 144 to 192 h on stream.

Results and Discussion

The stacked bed, which comprises a layer of zeolite catalyst, can be successfully used in the once-through VGO hydrocracking only in the case, when previous catalyst layers provide the feed quality that does not result in rapid deactivation of zeolite-containing catalyst. Accordingly, layers for preliminary hydrotreating/mild hydrocracking occupy most part of a catalyst bed, while the part of zeolite catalyst does not exceed 40% of the bed volume. To achieve the maximum yields of desired middle distillates, zeolites with low acidity should be used. Typical example of such material is dealuminated zeolite Y with high silica modulus [9, 17]. Zeolites with a silica modulus higher than 20 provide the optimal acidity for hydrocracking. According to the NMR data, framework SiO_2/Al_2O_3 ratio for USY used in this work is 27.4. Thus, the USY sample is characterized by the high dealumination degree and appropriate for the preparation of hydrocracking catalysts. The internal and external surface area of USY sample is 551 and 64 m^2/g respectively.

The data on particle morphology of USY sample and USY(40)-Al_2O_3 support were obtained by HRTEM (Fig. 1). The initial pseudoboehmite (HRTEM images not shown), which is used as a binder, has needle-shaped particles [7]. This morphology provides high bulk crushing strength of the obtained supports. According to HRTEM data, the USY zeolite has high crystallized particles with prismatic shape. The average crystal size of for USY zeolite less than 500 nm is obtained by

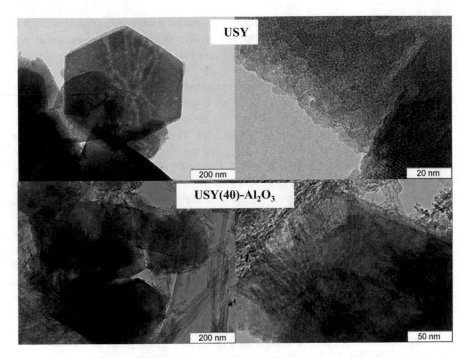

Fig. 1 HRTEM images of USY zeolite and USY(40)–Al$_2$O$_3$ support

the preparation procedure and provides low steric hindrance for bulky molecules transformation during VGO hydrocracking. Mesopores formed by framework dealumination can be observed on HRTEM images. Besides, an amorphous layer at the exterior surface of the zeolite particle is noticeable. On HRTEM images of USY (40)–Al$_2$O$_3$ support (Fig. 1) the areas related to USY zeolite and Al$_2$O$_3$ are observed. Alumina in USY(40)–Al$_2$O$_3$ support has needle-shaped particles, similar to initial pseudoboehmite [7]. The surface of zeolite crystals in the support is decorated with alumina. However, from HRTEM data it can be assumed that the zeolite morphology remains unchanged in comparison with initial zeolite powder.

The composition and textural characteristics of USY-containing supports are given in Table 1. USY–Al$_2$O$_3$ supports have high surface area—higher than

Table 1 Composition and textural characteristics of USY–Al$_2$O$_3$ supports

Characteristic	USY(10)–Al$_2$O$_3$	USY(20)–Al$_2$O$_3$	USY(30)–Al$_2$O$_3$	USY(40)–Al$_2$O$_3$
USY content, wt%	10	20	30	40
BET surface area (m^2/g)	253	275	301	351
Pore volume (cm^3/g)	0.54	0.50	0.50	0.52

250 m^2/g. The increase of USY content provides proportional increase of support surface area. It can be supposed that there is no formation of new phases and Al$_2$O$_3$ and USY present in support as individual phases. This result in good agreement with HRTEM data. The pore volume of all supports is similar.

The elemental analysis data, textural characteristics and bulk crushing strength of the catalysts are given in Table 2. In all the cases the catalysts with the bulk crushing strength higher than 1.2 MPa were obtained. This value is sufficient for industrial application. Incorporation of Ni and Mo in the supports leads to the decrease of the surface area by 46–71 m^2/g comparing with the initial support. The pore volume of all catalysts is similar and from 0.08 to 0.16 cm^3/g lower than the pore volume of corresponding supports.

Nitrogen adsorption-desorption isotherms for the supports are attributed to pseudo-type II isotherms with hysteresis loops of type H3 [16]. The adsorption-desorption isotherms have narrow hysteresis loop, which indicates the wide pore size distribution in the supports. The increase of zeolite content in the support virtually have no influence on the form of adsorption-desorption isotherms and hysteresis loops. Nitrogen adsorption-desorption isotherms for the catalysts have similar form as for the supports. The increase of the width of hysteresis loops at P/P^0 = 0.4–0.6 indicates the presence of micropores in the obtained catalysts.

Pore size distributions of the USY–Al$_2$O$_3$ supports in comparison with corresponding NiMo catalysts are shown in Fig. 2. The assessment of the pore size distribution was made by adsorption branch of isotherm, since no plateau is apparent at high P/P^0. The obtained supports have similar pore size distribution. The supports are characterized by bimodal pore size distribution with pores of less than 4 nm diameter and mesopores with 4–25 nm diameter.

Preparation of catalysts using citric acid prevents undesirable penetration of metals in narrow pores of a support, thus active component is localized in pores with diameter higher than 5 nm [12]. According to obtained results, supported metals are preferentially localized in the pores with diameters higher than 6 nm (Fig. 2). It is confirmed by the decrease of the volume of these pores after supporting of Ni and Mo. As it was observed previously [6], the catalyst preparation

Table 2 Composition and properties of NiMo/USY–Al$_2$O$_3$ catalysts

Characteristic	NiMo/USY (10)–Al$_2$O$_3$	NiMo/USY (20)–Al$_2$O$_3$	NiMo/USY (30)–Al$_2$O$_3$	NiMo/USY (40)–Al$_2$O$_3$
Ni (wt%)	2.1	2.9	2.8	2.6
Mo (wt%)	7.2	9.7	9.0	8.7
BET surface area (m^2/g)	190	204	255	288
Pore volume (cm^3/g)	0.41	0.40	0.42	0.36
BCS (MPa)	1.3	1.4	1.2	1.2

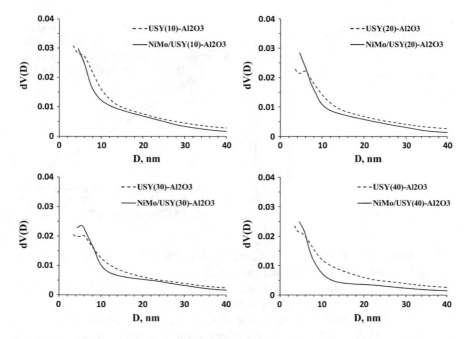

Fig. 2 Pore size distribution in the USY–Al₂O₃ supports and NiMo/USY–Al₂O₃ catalysts

Table 3 The results of VGO hydrocracking—product yield and sulfur content in middle distillates

Sample	NiMo/USY (10)–Al$_2$O$_3$		NiMo/USY (20)–Al$_2$O$_3$		NiMo/USY (30)–Al$_2$O$_3$		NiMo/USY (40)–Al$_2$O$_3$	
Temperature (°C)	390	410	390	410	390	410	390	410
Gas (wt%)	0.1	2.8	0.4	4.0	0.5	5.4	1.0	8.4
Naphta < 130 °C (wt%)	0.7	8.9	1.6	16.8	1.4	23.4	4.7	32.6
Middle distillates 130–360 °C (wt%)	28.0	43.9	29.4	47.3	25.8	44.9	31.6	46.5
Unconverted oil > 360 °C (wt%)	71.2	44.3	68.6	31.9	72.3	26.3	62.8	12.6
S in middle distillates (wt%)	11	9	9	6	10	8	12	7

method, used in this work, provides preferential localization of Ni and Mo on Al₂O₃, with most part of the surface and volume of zeolite pores to be free and available for catalysis.

Stacked beds containing three catalysts: NiMo/Al₂O₃, NiW/ASA–Al₂O₃ and NiMo/USY(x)–Al₂O₃ were tested in VGO hydrocracking. Weight yields of each fraction and sulfur content in the middle distillates are given in Table 3. The dependence of selectivity to middle distillates on conversion of VGO is shown in Fig. 3. Selectivity to middle distillates was referred to the fraction 130–360 °C in the product mixture.

Fig. 3 Selectivity to middle distillates as a function from VGO conversion

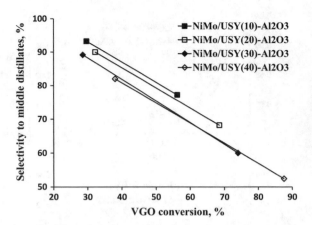

The conversion of VGO at 390 °C was between 29 and 38 wt% for all catalysts and middle distillates yield did not exceed 32 wt%. Increase of the process temperature to 410 °C leads to the significant increase of VGO conversion and middle distillates yield. However, toughening of hydrocracking process conditions inevitably results in undesirable gas formation. This effect was more pronounced for the catalysts with higher USY content, which have lower selectivity. In our case NiMo/USY(10)-Al$_2$O$_3$ catalyst, containing 10 wt% of USY in the support, showed the highest selectivity to middle distillates (Fig. 3) at minimal gas formation (Table 3). However, VGO conversion was the lowest: at 410 °C VGO conversion was less than 60%. Comparing properties of hydrocracking catalysts at higher temperatures is not reasonable because of significant increase of thermal (non-catalytic) cracking rate [14] and deactivation rate of catalysts [10]. Besides, when the zeolite content of the hydrocracking catalyst is relatively low (<15 wt%), the non-zeolitic component of the catalyst can have a substantial impact on activity and selectivity [17].

The highest conversion of 88% was achieved on the catalyst with maximum USY content. At the same time the selectivity to middle distillates was the lowest and gas yield was more than 8 wt%. Although, VGO conversion on NiMo/USY(20)–Al$_2$O$_3$ catalyst was lower than on the catalysts with higher zeolite content at the same hydrocracking temperature, the middle distillates yield was slightly higher and exceeded 47 wt% at temperature of 410 °C along with reasonable gas formation (Table 3). It should be noted that the middle distillate fraction yield did not exceed 50 wt% for all stacked beds with zeolite-containing catalysts because increase of VGO conversion is accompanied by the increase of gas and naphta yield and drastic decrease of selectivity of NiMo/USY–Al$_2$O$_3$ catalysts. Sulfur content in the obtained middle distillate fractions did not exceed 10 ppm at 410 °C process temperature and was between 9 and 12 ppm for 390 °C (Table 3).

Conclusions

NiMo/USY–Al_2O_3 catalysts with different zeolite content, NiMo/Al_2O_3 and NiW/ASA–Al_2O_3 catalysts were prepared by impregnation of shaped supports with solutions containing precursors of active metals and citric acid. The catalysts were tested in hydrocracking in the form of stacked beds, where layers containing hydrotreating catalyst (NiMo/Al_2O_3) and hydrocracking catalyst based on amorphous aluminosilicate (NiW/ASA–Al_2O_3) were loaded before the layer of zeolite catalyst. NiMo/USY(10)–Al_2O_3 catalyst, containing 10 wt% of USY in the support, was established to have the highest selectivity to middle distillates, however the activity in hydrocracking is insufficient. The highest yield of middle distillates was obtained at temperature of 410 °C on NiMo/USY(20)–Al_2O_3 catalyst. Sulfur content in the obtained middle distillate fractions was from 9 to 12 ppm at 390 °C process temperature and from 6 to 9 ppm at 410 °C. Stacked beds with developed catalysts can be successfully used in the once-through hydrocracking to provide VGO conversion of 70–80% and middle distillates yields up to 50 wt%. Besides, such stacked beds can be used in the first stage operation of two stages hydrocracker to provide 35–65% VGO conversion and produce high-quality middle distillates and feed for the second stage.

Acknowledgements Research are carried out with the financial support of the state represented by the Ministry of Education and Science of the Russian Federation. Agreement no. 14.610.21.0008 11 Aug 2015. Unique project Identifier: RFMEFI61015X0008.

References

1. Alsobaai, A.M., Zakaria, R., Hameed, B.H.: Gas oil hydrocracking on NiW/USY catalyst: Effect of tungsten and nickel loading. Chem. Eng. J. **132**, 77–83 (2007)
2. Alvarez, A., Ancheyta, J.: Simulation and analysis of different quenching alternatives for an industrial vacuum gasoil hydrotreater. Chem. Eng. Sci. **63**, 662–673 (2008)
3. Corma, A., Martínez, A., Martínez-Soria, V.: Catalytic performance of the new delaminated ITQ-2 zeolite for mild hydrocracking and aromatic hydrogenation. J. Catal. **200**, 259–269 (2001)
4. Dik, P.P., Klimov, O.V., Budukva, S.V., Leonova, K.A., Pereyma, V.Yu., Gerasimov, E.Yu., Danilova, I.G., Noskov, A.S.: Silica-alumina based nickel-molybdenum catalysts for vacuum gas oil hydrocracking aimed at a higher diesel fraction yield. Catal. Ind. **6**, 231–238 (2014)
5. Dik, P.P., Klimov, O.V., Danilova, I.G., Leonova, K.A., Pereyma, V.Yu., Budukva, S.V., Uvarkina, D.D., Kazakov, M.O., Noskov, A.S.: Hydroprocessing of hydrocracker bottom on Pd containing bifunctional catalysts. Catal. Today **271**, 154–162 (2016)
6. Dik, P.P., Klimov, O.V., Koryakina, G.I., Leonova, K.A., Pereyma, V.Yu., Budukva, S.V., Gerasimov, E.Yu., Noskov, A.S.: Composition of stacked bed for VGO hydrocracking with maximum diesel yield. Catal. Today **220–222**, 124–132 (2014)
7. Klimov, O.V., Leonova, K.A., Koryakina, G.I., Gerasimov, E.Yu., Prosvirin, I.P., Cherepanova, S.V., Budukva, S.V., Pereyma, V.Yu., Dik, P.P., Parakhin, O.A., Noskov, A.S.: Supported on alumina Co-Mo hydrotreating catalysts: dependence of catalytic and

strength characteristics on the initial AlOOH particle morphology. Catal. Today **220–222**, 66–77 (2014)

8. Klimov, O.V., Nadeina, K.A., Dik, P.P., Koryakina, G.I., Pereyma, V.Yu., Kazakov, M.O., Budukva, S.V., Gerasimov, E.Yu., Prosvirin, I.P., Kochubey, D.I., Noskov, A.S.: CoNiMo/Al$_2$O$_3$ catalysts for deep hydrotreatment of vacuum gasoil. Catal. Today **271**, 56–63 (2016)

9. Martinez, C., Corma, A.: Inorganic molecular sieves: preparation, modification and industrial application in catalytic processes. Coord. Chem. Rev. **255**, 1558–1580 (2011)

10. Martínez, J., Ancheyta, J.: Kinetic model for hydrocracking of heavy oil in a CSTR involving short term catalyst deactivation. Fuel **100**, 193–199 (2012)

11. Nishijima, A., Sato, T., Yoshimura, Y., Shimada, H., Matsubayashi, N., Imamura, M., Sugimoto, Y., Kameoka, T., Nishimura, Y.: Two stage upgrading of middle and heavy distillates over newly prepared catalysts. Catal. Today **27**, 129–135 (1996)

12. Pashigreva, A.V., Klimov, O.V., Bukhtiyarova, G.A., Fedotov, M.A., Kochubey, D.I., Chesalov, Y.A., Zaikovskii, V.I., Prosvirin, I.P., Noskov, A.S.: The superior activity of the CoMo hydrotreating catalysts, prepared using citric acid: what's the reason? Stud. Surf. Sci. Catal. **175**, 109–116 (2010)

13. Pereyma, V.Yu., Dik, P.P., Klimov, O.V., Budukva, S.V., Leonova, K.A., Noskov, A.S.: Hydrocracking of vacuum gas oil in the presence of catalysts NiMo/Al$_2$O$_3$-amorphous aluminosilicates and NiW/Al$_2$O$_3$-amorphous aluminosilicates. Russ. J. Appl. Chem. **88**, 1969–1975 (2015)

14. Ramírez, S., Martínez, J., Ancheyta, J.: Kinetics of thermal hydrocracking of heavy oils under moderate hydroprocessing reaction conditions. Fuel **110**, 83–88 (2013)

15. Stanislaus, A., Marafi, A.: Recent advances in the science and technology of ultra low sulfur diesel (ULSD) production. Catal. Today **153**, 1–68 (2010)

16. Thommes, M., Kaneko, K., Neimark, A.V., Olivier, J.P., Rodriguez-Reinoso, F., Rouquerol, J., Sing, K.S.W.: Physisorption of gases, with special reference to the evaluation of surface area and pore size distribution (IUPAC Technical Report). Pure Appl. Chem. **87**, 1052–1069 (2015)

17. Ward, J.W.: Hydrocracking processes and catalysts. Fuel Process. Technol. **35**, 55–85 (1993)

Comparative Mechanical Tests of Samples Obtained by the Domestic Experimental Unit Meltmaster^{3D}-550

A.V. Dub, V.V. Beregovsky, E.V. Tretyakov, S.A. Schurenkova and A.V. Yudin

Abstract The current development of the domestic engineering industry is closely connected with the development of new production technologies and metal processing methods, which corresponds to the transition of the industry to the sixth technological order. One of the priorities and strategically important directions of this development is the introduction of additive technologies into existing production chains. Currently, in the developing Russian market of 3D printing, there is a shortage of qualitative domestic industrial equipment and consumables materials (metallic powders). The present development of additive technologies in Russian companies is carried out with costly imported equipment based on the use of expensive imported consumables. To reduce the share of imported equipment in the Russian market effectively, the development of the technology and prototype model of the experimental unit MeltMaster^{3D}-550 for precision manufacturing of responsible, bulky products with a complex profile based on advanced technologies of additive manufacturing by selective laser melting method (SLM).

Keywords 3D printing · Equipment · Selective laser melting · Metal powder

A.V. Dub (✉)
JSC «Science and Innovations», Moscow, Russia
e-mail: alvdub@rosatom.ru

V.V. Beregovsky · E.V. Tretyakov · S.A. Schurenkova · A.V. Yudin
SSC RF JSC «RPA «CNIITMASH», Moscow, Russia
e-mail: vvberegovsky@cniitmash.com

E.V. Tretyakov
e-mail: evtretyakov@cniitmash.com

S.A. Schurenkova
e-mail: saschurenkova@cniitmash.com

A.V. Yudin
e-mail: avudin@cniitmash.com

© The Author(s) 2018
K.V. Anisimov et al. (eds.), *Proceedings of the Scientific-Practical Conference "Research and Development - 2016"*, https://doi.org/10.1007/978-3-319-62870-7_36

Introduction

Selective Laser Melting (SLM) is a new technology for producing complex-profile products, which allows obtaining a high density, high accuracy of geometric dimensions and mechanical properties of the corresponding molded materials [1]. The preformed layer of the metal powder is melted by laser energy followed by crystallization through extracting heat into massive building platform. Such layers are fused with each other, thereby forming complex three-dimensional products.

Over the past decade of the development of additive technologies a large number of scientific papers have been published, which describe manufacturing of samples and products made of various stainless steels by the selective laser melting. As a rule, authors give the following main process parameters that affect the quality of products and samples [2, 3]:

- original metal powder chemical: chemical composition, shape and size distribution of particles, the thickness of each melted layer for each production cycle;
- laser source: type, power, space-energy beam parameters (light intensity distribution, the size of the laser beam spot in the melting zone as a result of its divergence);
- characteristics of the production process: scanning speed, type of protective gas environment, choice of the type and parameters of the scanning strategy in each melted layer.

The authors of the article [4] state that the area of process parameters (laser power and scanning speed) should be determined for each material experimentally. Figure 1 shows the dependence of the laser power on the scanning speed indicating the area of technological modes of powder processing with austenitic stainless steel in which steady single vectors (tracks) must form.

Fig. 1 Field of technological modes [4]

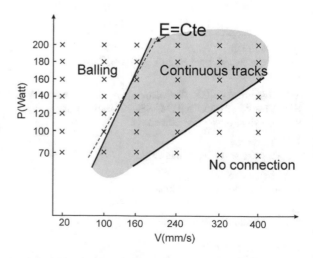

The authors of the article [5] carried out a series of experiments designed to produce bulky samples by combining the methods of «two zones» and «cross-hatch» with the following parameters: input laser power is 50 W, thickness of the layer is 40 microns, scanning speed is120 mm/s, and scanning step is 120 microns. The cylindrical samples were produced in accordance with ISO-7500/1 and their mechanical tests were done. The results are shown in Table 1.

The analysis of the mechanical properties of the samples produced with SLM from stainless steel powder (SS) 316L, showed excellent correspondence to mechanical properties of cast materials.

The study of the influence of the powder layer thickness and the direction of the location of the samples on the build platform during building on the mechanical characteristics of the samples from austenitic stainless steel 316L (see Table 2) was carried out and described in the paper [6]. For this purpose, a series of samples under the same process SLM conditions was produced. Mechanical properties of the samples made at horizontal and vertical positions are shown in Table 2.

It should be noted that both the elongation limit of samples UTS, $\sigma_{0.2}$ and samples with the layer thickness of 60, 80 and 100 microns is higher than that of the samples produced with the thickness of 120 microns and 150 microns [7, 8]. However, the value of elongation for all samples does not reach the standard value of 40%.

The authors [6] explain the increase of elongation and decrease of tensile when the layer thickness grows by the decrease in the number of layers, and as a result reduction in the number of tracks borders, which can be the place of the crack formation.

Table 1 Mechanical properties of SLM samples [5]

Properties of the sample	Material and method of sample production preparation	
	SS 136L	
	SLM	Casting
Tensile strength (MPa)	436 ± 60	480–560
Yield strength (0.2%) (MPa)	366 ± 50	170–290
Elongation (%)	9 ± 2	40

Table 2 Mechanical properties of samples [6]

Layer thickness (μm)	Horizontal position of sample			Vertical position of sample		
	UTS (MPa)	$\sigma_{0.2}$ (MPa)	Elongation (%)	UTS (MPa)	$\sigma_{0.2}$ (MPa)	Elongation (%)
60	720	525	22	690	526	19
80	735	532	32	695	529	16
100	713	528	32	670	518	23
120	700	473	32	660	513	20
150	700	467	35	666	492	25

Table 2 shows that the mechanical properties UTS, $\sigma_{0.2}$ and elongation of the samples grown at the horizontal position are higher than that of similar samples grown in vertical directions [6]. This indicates that the microstructure anisotropy SLM samples occurring during their building plays an important role in determining tensile properties. The similar results were obtained by the authors of the papers [7–10] on such SLM equipment as AM250 (Renishaw) and SLM250 (SLM-Solutions). The samples made at a horizontal position showed higher strength properties due to the tensile strain direction parallel to the cut of layers but perpendicular to dendrites. On the other hand, for the samples produced at the vertical position, the tensile load direction is normal to the layer formed. In this case, the width of the contact area (melting) between two adjacent layers increases with increasing layer thickness, which reduces mechanical characteristics.

The above mentioned works show the dependence of mechanical characteristics of the material on such parameters as layer thickness and direction of product made of powders of austenitic stainless steels by SLM. This article describes the dependence of the material density and its mechanical characteristics on the basic technological parameters of layer building on the example of the samples made of stainless steel 316L, produced on the pilot layer build-up unit by SLM method called MeltMaster^{3D}-550.

Experiment

Equipment Description

The layer build-up process of samples by SLM method took place on the prototype model of the unit MeltMaster^{3D}-550 [11] developed and manufactured by the JSC «RPA «CNIITMASH», with the financial support of the Ministry of Education and Science of the Russian Federation (Unique Identifier PNIER RFMEFI58214H0004). Table 3 shows the main technical characteristics of MeltMaster^{3D}-550 in comparison with the industrial units of the world leaders in the field of producing equipment for 3D printing from metallic powders.

Figure 2 shows a general view of the prototype model of MeltMaster3D-550.

For the preparation and maintenance of layer-by-layer manufacturing of a product with a complex profile using MeltMaster^{3D}-550 in the JSC «RPA «CNIITMASH», the original specialized software, which includes two software modules «SLM-Simulation» and «SLM-Production» was developed.

The functionality of the software includes the automation of all stages of layered manufacturing of products based on the original 3D-model. The following operations are performed in the program «SLM-Simulation»: selection/input of the configuration parameters; check/adjustment of 3D-model; location of 3D-model on the technological platform; cutting 3D-model into layers; hatching in accordance with the chosen strategy; building a technology support; generation of the control code.

Table 3 Technical characteristics of MeltMaster³ᴰ-550

Parameters/name	JSC «NPO «TSNIITMASH»	EOS	SLM-Solutions
	MeltMaster³ᴰ-550	EOS M 400 (EOS M 400-4)	SLM500
Chamber dimensions, mm	550 × 450 × 450	400 × 400 × 400	500 × 280 × 365
Number and power of lasers, pc./kW	1/1 2/0.5 + 1	1/1 4/0.4	2 × 0.4 (или 2 × 0.7); 4 × 0.4 (или 4 × 0.7)
Optical system	3-axis optics/2 × F-Theta	F-Theta/4 × F-Theta, high-speed scanner	3-axis optics
Thickness of a layer (µm)	20–250	20–150	20–75
Productivity (cm³/h)	15–100	100	105
Dimensions W × D × H (mm)	2938 × 2040 (2488) × 3000	4181 × 1613 × 2355	5200 × 2800 × 2700
Applied materials	stainless and tool steel, titanium, and titanium alloys, aluminum alloys, superalloys		
Delivery dates (months)	5–6	6–7	6–7

Fig. 2 General view of MeltMaster3D-550

The control code that describes the parameters necessary for parts production and coordinate ways of scanning on each layer is the result of using the program «SLM-Simulation».

The program «SLM-Production» provides automatic smooth functioning of the layered build-up unit MeltMaster^{3D}-550.

Experimental Technique

The samples were prepared in the layered build-up experimental unit of selective laser melting MeltMaster^{3D}-550 with austenitic stainless steel powder 316L.

Microstructural studies were conducted on the SLM samples prepared in the form of rectangular parallelepipeds with the dimensions $3 \times 3 \times 15$ mm. The hatching was performed with a simple one-pass strategy.

To evaluate the critical scanning speed the following approach was used (1) [12]:

$$\Delta t_{\text{HOM}} \approx \frac{\langle r_{\text{powders}} \rangle^2}{4 \cdot \alpha} \tag{1}$$

where Δt_{HOM}—time alignment in the particle temperature, $\langle r_{\text{powders}} \rangle$—average radius of the powder particles, α—temperature conductivity of bulky powder material (mm^2/s).

The laser exposure time must be greater than the time of homogenization. Then the critical scanning speed can be estimated with the following equation:

$$V_{\text{scan}} < \frac{8 \cdot \alpha \cdot \left(\frac{1}{e^2} d_l + \langle r_{\text{powders}} \rangle \right)}{\langle r_{\text{powders}} \rangle^2} \tag{2}$$

where α—thermal conductivity of the bulk material (mm^2/s), d_l—the size of the laser spot on the surface of the powder layer, $\langle r_{\text{powders}} \rangle$—the average radius of the powder particles.

Thus, for the above conditions, the scanning speed must be less than 1200 mm/s. On this basis, the following technological parameters for producing experimental samples were proposed: the scanning speed from 100 to 1000 mm/s and the laser power from 154 to 491 watts.

The samples were made (in accordance with GOST (State Standards) 1497-84) in two process modes for mechanical testing. Mode 1: the laser power is 264 W, the scanning speed is 600 mm/s; Mode 2: the laser power is 264 W; the scanning speed is 540 mm/s. The thickness of the layers during the production of all the samples was 50 microns. For the first group of samples, the hatching was carried out with a simple one-pass strategy with 160 mm scanning pitch and the rotation angle of 90° when moving to the next layer.

Table 4 Elemental composition of the austenitic steel 316L powder

Powder	Mass content (%)								
	Fe	Cr	Ni	Mo	Mn	Si	P	C	S
316L	66.2	19.4	11.4	–	1.9	1.1	*	*	*

*not determined

The selective laser melting process samples took place in the atmosphere of nitrogen with the residual oxygen concentration of less than 0.07%.

Characteristics of the Original Material

The elemental composition of 316L steel powder was determined by the micro X-ray analysis. The results of the powder steel 316L studies are presented in Table 4.

The powder particles have a rounded shape. The average estimated value of the powder roundness is 1.5 with the standard deviation of 0.5. The dimensions of the powder particles range from 20 to 40 microns.

Methodology of the Study

Density Determination

The sample density was determined by hydrostatic weighing method [GOST 15139-69 Methods for determining density (bulk density)].

Prior to hydrostatic weighing samples were covered with a thin layer of varnish tsapon. The distilled water was used as the immersion liquid. To control the density of air and water accurately approximated dependencies on the temperature were used. A series of weighing was performed for each sample. The true value of the density was chosen as average among three densities. The error of density measurement was calculated by the Least Square Method and was less than 3% for all the samples.

The samples were weighed in air and distilled water with high precision analytical scales 210 G × 0.1 Mg (accuracy ± 0.0001 g), with the special rigging. The sample density was calculated with the following formula (3):

$$\rho_{sample} = \frac{m_{in\ air.}(\rho_{water} + \rho_{air})}{m_{in\ air.} + m_{in\ water}} + \rho_{air}, \tag{3}$$

Mechanical Tests

The mechanical tests were done in accordance with GOST 1497-84 «Testing Stretching Methods».

The universal testing machine Z250 produced by the company Zwick (Germany) in combination with the automatic sensor of longitudinal deformation, high-temperature furnace and heat chamber was used during the experiments.

Results and Discussion

In the manufacturing process of bulk samples by SLM method the following parameters were changed: laser power from 154 to 491 W, scanning speed from 100 to 1000 mm/s (with the step of 100 mm/s). To compare the conditions for obtaining samples with different technological parameters, it is necessary to determine the contribution of specific laser energy. Density measurement results of samples made under the different process SLM conditions are presented in the diagram showing the dependence of sample density on the specific energy (Fig. 3).

The evaluation of the specific laser energy for each tested sample was carried out according to the formula (4) [13]:

$$E_{sp} = P/(V \cdot h), \tag{4}$$

where P—laser power, V—scanning speed, h—diameter of the laser beam.

The analysis of dependency of the sample density on the specific laser energy showed that when the energy changes from 1 to 3 J/mm^2 density increases sharply, but the density is less than 90%. By increasing the energy from 3 to 12 J/mm^2 density also increases, but with a less dependence. Further increases of specific

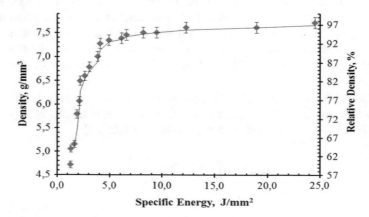

Fig. 3 Dependence of the sample density on the input specific energy

energy practically do not lead to increase in sample density. We can conclude that the optimum value of the laser energy is in the range of 8–12 J/mm^2. These values are coordinated with the data, which were obtained on equipment EOSINT M270 and contained in the published articles, and have a similar character of the sample density dependence on the specific energy [14].

Microstructure Study

The results of the microstructure study of samples produced with the input specific energy to 3.0 J/mm^2 show the presence of open macropores. The samples made with the input energy density from 3.0 to 8.0 J/mm^2 are characterized by a 100-micron pore size. The layered structure also takes place. The samples made with the input specific energy from 8 to 12 J/mm^2 are characterized by the presence of micropores. The samples produced with the specific energy from 12 to 25 J/mm^2 have micropores and remelting tracks. The pore distribution, in this case, is random. Figure 4 shows a typical microstructure of samples with micropores and remelting on the example of the sample produced with the specific energy 11 J/mm^2.

In all the samples observed directed dendritic segregation and fine-grained areas took place. This structure is the result of the fast directionally crystallization during selective laser melting.

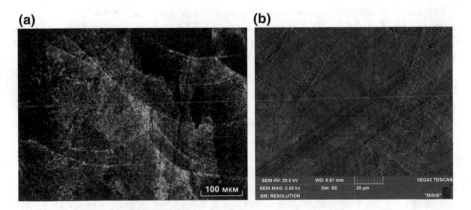

Fig. 4 Typical microstructure of samples: optical microscopy (**a**), SEM (**b**)

Table 5 Mechanical properties of SLM samples

Mode	$\sigma_{0.2}$ (MPa)	Error (%)	UTS (MPa)	Error (%)	δ (%)	Error (%)
1	466	1.38	604	2.02	18.93	2.31
2	461	1.30	601	1.07	16.87	2.42

Mechanical Tests

The relative density of the samples produced in the first process mode is 97.4% ($\pm$0.6%), in the second mode is 98.0% ($\pm$0.1%) (relative density was calculated on the basis of pycnometric density of the stainless steel powder (SS) 316L 7.93 g/cm^3).

The mechanical tests show that the samples produced with the simple one-pass strategy, have a strength comparable to molded material, but at the same time insufficient elongation (see Table 5).

Table 5 shows that the elongation is lower in the second process mode. This is due to higher residual thermal stresses resulting from the application of more specific energy and a possible double remelting during laser layer scanning.

Conclusion

The parameters the prototype model of the experimental equipment MeltMaster^{3D}-550 designed by JSC «RPA «CNIITMASH» is on par with the main representatives of this class of such equipment currently presented in the 3D printing market from metal powders.

The modular architecture of MeltMaster^{3D}-550 provides the production of numerous units with a high potential for adaptation and integration into existing production lines of machine-building enterprises, as well as without additional complex modernization will allow advancing to robotic digital productions on the basis of one or more identical units.

The experimental studies of the technological modes of metal powder selective laser melting based on the austenitic stainless steel 316L in MeltMaster^{3D}-550 have shown the possibility to produce high-density samples. Moreover, the required scanning step has been calculated: when the powder layer thickness is 50 microns it is equal to 160 microns.

The dependence of the sample density on the laser power (input specific energy) and the scanning speed was identified. The optimum range of the input specific energy from 8 to 12 J/mm^2 was determined. These results are consistent with the data presented in the paper [14].

The study of SLM sample mechanical properties has shown that the tensile strength UTS (600 MPa) and yield strength $\sigma_{0.2}$ (460 MPa) are greater than the corresponding values (480–560 MPa, 170–290 MPa) for cast materials. At the same time, the elongation of the product at the level of 18% within the error meets the requirements for construction materials.

On the basis of these results can also be concluded that the physical and mechanical properties of materials produced on the experimental unit MeltMaster^{3D}-550 are comparable with similar characteristics of materials, produced by using imported industrial equipment of such manufacturers as SLM-Solutions, Renishaw, EOS.

This work was financially supported by the Ministry of Education and Science of the Russian Federation (Unique Identifier PNIER RFMEFI58214H0004).

Acknowledgements The research is carried out with the financial support of the state represented by the Ministry of Education and Science of the Russian Federation. Agreement no. 14.582.21.0004 03 Oct. 2014. Unique project Identifier: RFMEFI58214X0004.

References

1. Liu, Y., Yang, Y., Wang, D.: A study on the residual stress during selective laser melting (SLM) of metallic powder. Int. J. Adv. Manuf. Technol., 1–10 (2016)
2. Volosov, M.A., Okounkov, A.A.: Ways of optimization of selective laser melting process by selecting the laser beam processing strategies. Bull. Samara Sci. Cent. Russ. Acad. Sci. **14** (2), 587–591 (2012)
3. Yadroitsev, I., Bertrand, P., Smurov, I.: Parametric analysis of the selective laser melting process. Appl. Surf. Sci. **253**(19), 8064–8069 (2007)
4. Kruth, J.P.: Binding mechanisms in selective laser sintering and selective laser melting. Rapid Prototyp. J. **11**(1), 26–36 (2005)
5. Smurov, I.J., Movchan, I.A., Yadroytsev, I.A., Okounkov, A.A., Cherkasova, N.Y., Antonenkova, G.V.: Additive laser production. Experimental studies. Vestnik MSTU "STANKIN" (1) 36–38 (2012)
6. Ma, M.: Layer thickness dependence of performance in high-power selective laser melting of 1Cr18Ni9Ti stainless steel. J. Mater. Process. Technol. **215**, 142–150 (2015)
7. Guan, K.: Effects of processing parameters on tensile properties of selective laser melted 304 stainless steel. Mater. Des. **50**, 581–586 (2013)
8. Casati, R., Lemke, J., Vedani, M.: Microstructure and fracture behavior of 316L austenitic stainless steel produced by selective laser melting. J. Mater. Sci. Technol. **32**(8), 738–744 (2016)
9. Hanzl, P., et al.: The influence of processing parameters on the mechanical properties of SLM parts. Procedia Eng. **100**, 1405–1413 (2015)
10. Mertens, A., et al.: Mechanical properties of alloy Ti–6Al–4V and of stainless steel 316L processed by selective laser melting: influence of out-of-equilibrium microstructures. Powder Metall. **57**(3), 184–189 (2014)
11. Beregovsky, V.V., Tretyakov, E.V., Schurenkova, S.A.: Equipment for the layer-production of complex articles by selective laser melting MeltMaster^{3D}-550. Additive technology: present and future. Materials of the II International conference. FSUE «VIAM» SRC RF. p. 22 (2016)
12. Fischer, P., et al.: Pulsed laser sintering of metallic powders. Thin Solid Films **453**, 139–144 (2004)

13. Olakanmi, E.O.: Selective laser sintering/melting (SLS/SLM) of pure Al, Al–Mg, and Al–Si powders: effect of processing conditions and powder properties. J. Mater. Process. Technol. **213**(8), 1387–1405 (2013)
14. Spierings, A.B., Levy, G.: Comparison of density of stainless steel 316L parts produced with selective laser melting using different powder grades. In: Proceedings of the Annual International Solid Freeform Fabrication Symposium. Austin, TX, pp. 342–353 (2009)

Development of Lithium-Ion Battery of the "Doped Lithium Iron Phosphate–Doped Lithium Titanate" System for Power Applications

A.A. Chekannikov, A.A. Kuz'mina, T.L. Kulova, S.A. Novikova, A.M. Skundin, I.A. Stenina and A.B. Yaroslavtsev

Abstract Lithium-ion battery based on a new electrochemical system with a positive electrode based on doped lithium iron phosphate and a negative electrode based on doped lithium titanate has been developed. The battery is intended for use in fixed energy storage units. The battery is characterized by the ability to operate at increased charging/discharging currents (up to 30 C). The specific power of the battery was about 2 kW/kg.

Keywords Lithium-Ion battery · Doped lithium iron phosphate · Doped lithium titanate · Specific power

A.A. Chekannikov · A.A. Kuz'mina · T.L. Kulova (✉) · A.M. Skundin
Frumkin Institute of Physical Chemistry and Electrochemistry,
Russian Academy of Sciences, Moscow, Russia
e-mail: tkulova@mail.ru

A.A. Chekannikov
e-mail: andrey.chekannikov@gmail.com

A.A. Kuz'mina
e-mail: nyurka_92@mail.ru

A.M. Skundin
e-mail: askundin@mail.ru

S.A. Novikova · I.A. Stenina · A.B. Yaroslavtsev
Kurnakov Institute of General and Inorganic Chemistry,
Russian Academy of Sciences, Moscow, Russia
e-mail: svetlana_novi@mail.ru

I.A. Stenina
e-mail: irina_stenina@mail.ru

A.B. Yaroslavtsev
e-mail: yaroslav@igic.ras.ru

© The Author(s) 2018   341
K.V. Anisimov et al. (eds.), *Proceedings of the Scientific-Practical Conference*
"Research and Development - 2016", https://doi.org/10.1007/978-3-319-62870-7_37

Introduction

In terms of a specific power traditional electrochemical system of a lithium-ion battery, manufactured since 1991 (lithium cobaltate–graphite), approaches its theoretical limit [1, p. 100]. One of the new electrochemical systems of a lithium-ion battery, such as lithium iron phosphate–lithium titanate, has ultimately higher power. It is conditioned by specific features of current-producing processes in two-phase systems, as well as the essential necessity to use functional electrode materials in the nanosized form [10, pp. 74, 203]. It is obvious that in terms of specific power the lithium iron phosphate–lithium titanate system will lose out to the lithium cobaltate–graphite system due to reduced voltage [10, p. 590]. Simultaneously, a number of applications, such as fixed energy storage units or load leveling systems, require batteries tolerant to high charging/discharging currents, while specific power becomes unimportant for them.

That is why a new electrochemical system for lithium-ion battery with a positive electrode based on doped lithium iron phosphate and a negative electrode based on doped lithium titanate was developed.

Relatively low electronic and lithium conductivity (10^{-13} S cm^{-1}) [25, p. 589; 26, p. 1237; 2, p. A103; 21, p. 1241], as well as low discharge capacity, which is less than that of graphite, can be qualified as disadvantages of $Li_4Ti_5O_{12}$. However, these disadvantages are largely compensated by high cycling stability, especially at high charge/discharge currents, while silicon and tin quickly lose their initial capacity due to degradation resulting from significant volume change at lithium intercalation. To increase electronic and lithium conductivity, there were attempts of heterovalent doping of $Li_4Ti_5O_{12}$ with divalent (Cu^{2+}), trivalent (Cr^{3+}, Sc^{3+}, Al^{3+}, Tb^{3+}), and quintavalent (Ta^{5+}) cations [25, p. 590; 26, p. 1238; 2, p. A103]. In a number of cases, increased conductivity of obtained materials was observed but electrochemical properties of the doped $Li_4Ti_5O_{12}$ were not studied. A number of authors have reported results of electrochemical performances of the doped lithium titanates, but the range of cycling was from 1 to 3 V [22, p. 13198]; [4, p. 396]; [23, p. 1443]; [8, p. 375]; [11, p. 128]; [12, p. 1036]; [9, p. 2250]; [7, p. 748].

Different ways for further improvement of this active material have been extensively studied for the last few years: development of advanced nanostructured LFP-carbon composites; replacement of carbon by conductive, electrochemically active polymers; doping of LFP by the ions of transition metals; and so on. In particular, LFP doping with vanadium has been suggested as a way for the increase in mobility and diffusion coefficient of Li$^+$ ions due to lattice expansion and Li–O interaction weakening [19, p. 2956]. The authors of [20, p. 207] have studied the structure and properties of $LiFe_{0.9}V_{0.1}PO_4$ and found that the cathode properties of the doped counterpart, including reversible capacity, cyclability, and rate capability are better than those of $LiFePO_4$. Later the lengthening and weakening of Li–O bond and improvement of the electrochemical performance especially under the high C rate were confirmed by the examples of $LiFe_{0.95}V_{0.05}PO_4$ [24, p. A730], $LiFe_{0.97}V_{0.03}PO_4$ [18, p. 842], and $LiFe_{0.99}V_{0.01}PO_4$ [13, p. 1019]. Other ions of transition metals, such as Mn [15, p. 446], V [3, p. 280], Mg [16, p. 340], Ni [17, p. 830], Co [6, p. 145], Mo [5, p. 9963] can act as dopants.

Experimental

Lithium iron phosphate of the $Li_{0.99}Fe_{0.98}Y_{0.01}Ni_{0.01}PO_4$ composition was synthesized using the sol–gel method. At the first stage of synthesis, initial reagents were dissolved in stoichiometric ratios in deionized water. $Fe(NO_3)_3 \cdot 9H_2O$ (Sigma-Aldrich, >98%), Li_2CO_3 (Sigma-Aldrich, >98%), $(NH_4)_2HPO_4$ (Sigma-Aldrich, >99%), $Ni(NO_3)_2 \cdot 6H_2O$ (Sigma-Aldrich, >98.5%), Y_2O_3 were used as reagents. When heated, the Y_2O_3 oxide was preliminarily dissolved in the concentrated HNO_3. The solutions obtained after mixing of initial reagents were vaporized with post-heat treatment in an inert atmosphere at (100–1000 °C) in order to obtain the crystallized product. Composite materials with carbon were obtained to increase the electrical conductivity of $Li_{0.99}Fe_{0.98}Y_{0.01}Ni_{0.01}PO_4$. The carbonaceous coat is usually applied using pyrolysis of organic compounds at high temperature (600–900 °C) in the inert atmosphere [14, p. 538]. In this paper, glucose was chosen as the source of carbon. Samples of doped lithium iron phosphate were ground with glucose samples with different weight and were annealed at 800 °C in an inert atmosphere. In these conditions, carbonization is observed. The carbon content in the composites was determined thermogravimetrically and was 6–12%.

Gallium-doped lithium titanate was synthesized using the citrate method. Titanium tetrabutylate (99%, Alfa Aesar) and lithium carbonate (99%, Fluka) were dissolved in the ethanol-nitric acid mixture (volume ratio 5:1), and gallium solutions (99.99%, Aldrich) in nitric acid and citric acid (98%, Sigma) were added in the minimum quantity of water. Lithium carbonate was taken with a 5% surplus to prevent possible losses of lithium during subsequent annealing at high temperatures. The obtained mixture was heated sequentially at 95 °C during 24 h and at 250 °C during 5 h. The so formed precursor was ground in an agate mortar to a smooth paste that was subjected to final annealing at 800 °C during 5 h in air.

The X-ray phase analysis (XPA) of synthesized samples was conducted with a Rigaku D/MAX 2200 diffractometer with the CuK_α radiation. The Rigaku Application Data Processing software package was used for spectra processing. X-ray patterns were processed in FullProf Suite program (WinPlotr), lattice parameters were updated using the Checkcell debugging tool.

The microstructure of samples and study of the composition of elements of materials were analyzed with Carl Zeiss NVision 40 scanning electron microscope at accelerating voltage of 1 kV.

Cathode and anode masses for electrodes were prepared using the ratios as follows: 85 mass% of doped lithium iron phosphate (or doped lithium titanate), 10 mass% of carbon black, 5 mass% of PVDF. The latter was dissolved in N-methylpyrrolidone. Active masses based on doped lithium iron phosphate and doped lithium titanate were heated and applied to aluminum foil substrate with MSK-AFA-II-Automatic Thick Film Coater. After drying, electrode plates were placed in roller press. The semi-finished product was pressed at 2t. Rolled electrode sheets were cut into ready electrodes sized 55 x 55 mm^2 with MiniMarker A2 laser marker, which were subsequently used for assembling batteries. To determine the specific electrochemical capacity of cathode and anode material, small electrodes sized 1.5 x 1.5 cm^2 were cut out, and galvanostatic studies were carried out in three-electrode electrochemical cells. The thickness of the positive electrode's active layer was 90 μm. The thickness of the negative electrode's active layer was about 50 μm. The difference in thickness was conditioned by the difference in specific capacity of doped lithium iron phosphate and doped lithium titanate. Batteries and electrochemical cells were assembled in a glove box in a dry argon atmosphere. The 1 M $LiPF_6$ in a mixture of ethylene carbonate-diethyl carbonate-dimethyl carbonate (1:1:1) prepared in the laboratory of the Frumkin Institute of Physical Chemistry and Electrochemistry of the Russian Academy of Sciences from Battery Grade commercial reagents was used as an electrolyte.

Results and Discussion

X-ray diffraction analysis method showed that the X-ray pattern of the synthesized sample of $Li_{0.99}Fe_{0.98}Y_{0.01}Ni_{0.01}PO_4$ contains reflexes of $LiFePO_4$ (triphylite, orthorhombic modification, Pnma space group). Reflexes of other phases are not detected. X-ray patterns of $Li_{0.99}Fe_{0.98}Y_{0.01}Ni_{0.01}PO_4$ were compared with the Card No. 81-1173 of the powder database of diffraction standards PDF2. Based on the obtained data, the inference should be drawn that the obtained materials represent lithium iron phosphate with the structure of olivine.

When using the sol–gel synthesis method, the chemical composition of obtained materials must be determined by the ratio of initial reagents. However, lithium compounds can be volatile at high temperatures of final annealing, which can result in stoichiometric impurity in the end product. That is why the content of cations and phosphate included as compounds of the material was determined using inductively coupled plasma mass-spectrometry (ICP-MS). For this purpose, the samples were preliminarily dissolved and the solution composition was analyzed. It was demonstrated that the composition of obtained samples corresponds to the initial load (Table 1).

Table 1 Elements content according to inductively coupled plasma mass-spectrometry

Sample	Content in the solution, (mg ml^{-1})				
	Li	Fe	P	Y	Ni
$Li_{0.99}Fe_{0.98}Y_{0.01}Ni_{0.01}PO_4$	7.26	57.84	32.39	0.93	0.61

Following the results of scanning electron microscopy, the size of particles of $Li_{0.99}Fe_{0.98}Y_{0.01}Ni_{0.01}PO_4$ is about 50 nm. The particles form agglomerates from 100 nm to 300 nm (Fig. 1).

X-ray patterns of doped lithium titanate samples contain reflexes of $Li_4Ti_5O_{12}$ only (Card No. 72-0426 PDF-2 database), which evidences that the obtained material is single-phase. Radii of cations of Ga^{3+}, Ti^{4+}, and Li^+ are similar in size, that is why gallium ions can get inserted into both, positions of titanium, and positions of lithium. Ga^{3+} ion attracts oxygen ions more intensively than Li^+ ion considering Coulomb interaction, which results in a greater lattice contraction. To verify this assumption, the structure of doped lithium titanate was updated using the Rietveld method. According to the data obtained as a result of updating the structure using the Rietveld method of the sample of $Li_{4+x}Ti_{5-x}Ga_xO_{12}$ composition at x = 0.1 (Tables 2, 3), gallium ions occupy both, positions of lithium (8a), and positions of titanium (16d).

Moreover, there are about 2.5 times less of gallium ions in octahedral sites than in tetrahedral ones. In accordance with these results, the formula of gallium-doped lithium titanate should be as follows: $Li_{3.812}Ti_{4.972}Ga_{0.1}O_{12}$.

Fig. 1 Micrographs of $Li_{0.99}Fe_{0.98}Y_{0.01}Ni_{0.01}PO_4/C$ composite

Table 2 Results of updating the structure of $Li_{3.812}Ti_{4.972}Ga_{0.1}O_{12}$ at 25 °C

Lattice type	Cubic lattice
Space group, Z	Fd-3 m, 8
Scan space, 2Θ°	10–100
Scanning pitch	0.02
Lattice parameters, A	a = 8.3544(1)
Cell volume, A^3	583.102(16)
Number of reflexes	24
Bragg R-factor, Rf-factor Rp, Rwp	0.0489, 0.0533, 0.0895, 0.133

Table 3 Coordinates of atoms and isotropic parameters of thermal bias (B) at 25 °C

Atom	Symmetry of the site	Population	x	y	z	Biso
Li_1	8a	0.976	0.1250	0.1250	0.1250	0.25(2)
Ga_1	8a	0.024	0.1250	0.1250	0.1250	0.25(2)
Li_2	16d	0.167	0.5000	0.5000	0.5000	0.46(3)
Ti	16d	0.828	0.5000	0.5000	0.5000	0.46(3)
Ga_2	16d	0.005	0.5000	0.5000	0.5000	0.46(3)
O	32e	1.000	0.2618(2)	0.2618(2)	0.2618(2)	0.52(2)

According to the data from the scanning electron microscopy, synthesized samples of $Li_{3.812}Ti_{4.972}Ga_{0.1}O_{12}$ represent a rather homogeneous crystalline mass (Fig. 2). Growth steps are clearly seen in the micrographs. The particle size varies in the range of 450–550 nm.

The results of galvanostatic cycling in Fig. 3 revealed that the specific discharge capacity of lithium iron phosphate doped with yttrium and nickel at the current density of 20 mA/g which corresponds to the current C/8 was about 160 mAh/g. The increased current density logically resulted in the decreased discharge capacity. However, even at the current density of 30 C the discharge capacity was about 56 mAh/g. The discharge potential of lithium iron phosphate doped with yttrium and nickel at low current density (C/8) was about 3.4 V. At increased current densities (30 C), the discharge potential of $Li_{0.99}Fe_{0.98}Y_{0.01}Ni_{0.01}PO_4$ lowered insignificantly and was about 3.2 V.

The ability of lithium iron phosphate to withstand high currents is explained by two factors: first, the high ion conductivity of this material, and second, the small size of particles of synthesized material. The results of galvanostatic cycling of negative electrodes from doped lithium titanate are represented in Fig. 4. Traditionally, lithium titanate is discharged to potential 1 V, which corresponds to the insertion of 3 lithium ions per formula unit. Simultaneously, it was shown in a number of papers that doped samples of lithium nanotitanate can be cycled in a

Fig. 2 Mircograph of $Li_{3.812}Ti_{4.972}Ga_{0.1}O_{12}$ composite

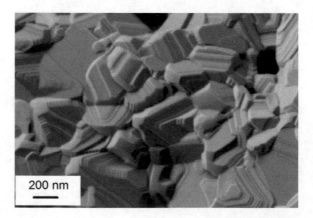

200 nm

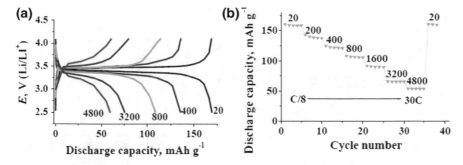

Fig. 3 Charge–discharge curves (a) and dependence of the discharge capacity on the current density (b) $Li_{0.99}Fe_{0.98}Y_{0.01}Ni_{0.01}PO_4$

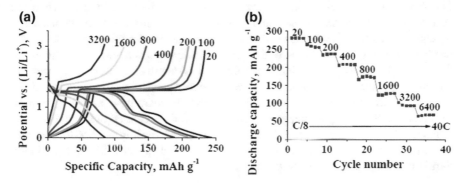

Fig. 4 Charge-discharge curves (a) and dependence of the discharge capacity on the current density (b) $Li_{3.812}Ti_{4.972}Ga_{0.1}O_{12}$

wider potential range (up to 0.01 V), in this case, the discharge capacity is increased up to 275 mAh/g, which is 75% over the discharge capacity registered in a narrower potential range.

Analysis of Fig. 4 shows that extending the cycling range leads to the increase in the discharge capacity, in which case the ability of lithium titanate to operate at high current densities up to 40C (6400 mA/g) is preserved. It is important to emphasize that the extended cycling range does not result in the increased degradation during the cycling. The discharge current equal to 6400 mAh/g corresponds to the charge for 1.5 min (or 40C regime), in which case the discharge capacity remains at the level of about 50 mAh/g. It is important to note that such a high discharge rate corresponds to the current of about 180 mA/cm², which ensures the high power of the battery. Seven (7) negative and seven (7) positive electrodes were used to manufacture a stack battery of 1 Ah rated capacity. The results of cycling are represented in Fig. 5. As Fig. 5a shows, the battery discharge capacity at the current of C/5 equals to the rated capacity, and the average discharge voltage is about

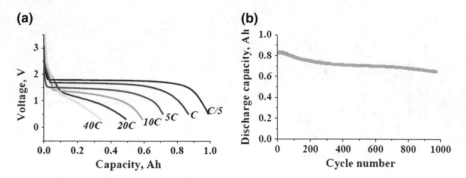

Fig. 5 Charge-discharge curves **a** and change in the discharge capacity **b** of the doped lithium iron phosphate-doped lithium titanate system battery

1.8 V, which evidences of the battery's insignificant ohmic resistance. When the current density is increased, the battery's discharge capacity and average discharge voltage are reduced, however, even at increased current densities up to 30 C the discharge capacity is about 22% of the rated capacity. Degradation and cyclic life of the lithium iron phosphate–lithium nanotitanate system battery were determined at the charging-discharging current 1 A, which corresponded to the so called cycle service 1C. As the Figures show, the increase in charging-discharging current results in the decreased discharge capacity, as well as decreased average discharge voltage. Change in the discharge capacity during the cycling for 950 cycles was 16 mAh on average, which is about of 16% of the rated capacity. Thus, degradation during the cycling was 0.017% per cycle.

Conclusions

In order to develop a battery with increased power specifications, new materials for the lithium-ion battery were synthesized: cathode material based on lithium iron phosphate doped with nickel and yttrium ($Li_{0.99}Fe_{0.98}Y_{0.01}Ni_{0.01}PO_4$) and anode material based on doped lithium titanate ($Li_{3.812}Ti_{4.972}Ga_{0.1}O_{12}$). Both electrode materials were characterized by their ability to operate at increased current densities (up to 30 C). The lithium-ion battery of the doped lithium iron phosphate-doped lithium titanate system was developed based on these materials. The energy density of the battery was about 100 Wh kg^{-1}, while its specific power was about 2 kW kg^{-1}. The battery of this electrochemical system is intended for use in fixed energy storage units.

Acknowledgments Research is carried out with the financial support of the state represented by the Ministry of Education and Science of the Russian Federation. Agreement No. 14.604.21.0126 26.Aug. 2014. Unique project Identifier: RFMEFI60414X0126.

References

1. Bagotsky, V.S., Skundin, A.M., Volfkovich, Yu.M. Electrochemical Power Sources: Batteries, Fuel Cells, and Supercapacitors, 400p. Wiley (2015)

2. Chen, C.H., Vaughey, J.T., Jansen, A.N., Dees, D.W., Kahaian, A.J., Goacher, T., Thackeray, M.M.: Studies of Mg-substituted $Li_{4-x}Mg_xTi_5O_{12}$ spinel electrodes ($0 \leq x \leq 1$) for lithium batteries. J. Electrochem. Soc. **148**, A102–A104 (2001)

3. Chen, M., Shao, L., Yang, H., Ren, T., Du, G., Yuan, Zh.: Vanadium-doping of $LiFePO_4$/carbon composite cathode materials synthesized with organophosphorus source. Electrochim. Acta. **167**, 278–286 (2015)

4. Du, P., Tang, L., Zhao, X., Weng, W., Han, G.: Effect of Tb^{3+} doping on the preferred orientation of lead titanate thin film prepared by sol–gel method on ITO/glass substrates. Surf. Coat. Technol. **198**, 395–399 (2005)

5. Gao, H., Jiao, L., Peng, W., Liu, G., Yang, J., Zhao, Q., Qi, Z., Si, Y., Wang, Y., Yuan, H.: Enhanced electrochemical performance of $LiFePO_4$/C via Mo-doping at Fe site. Electrochim. Acta. **56**, 9961–9967 (2011)

6. Gao, H., Jiao, L., Yang, J., Qi, Z., Wang, Y., Yuan, H.: High rate capability of Co-doped $LiFePO_4$/C. Electrochim. Acta. **97**, 143–149 (2013)

7. Guo, M., Chen, H., Wang, S., Dai, Sh., Ding, L., Wang, H.: TiN-coated micron-sized tantalum-doped $Li_4Ti_5O_{12}$ with enhanced anodic performance for lithium-ion batteries. Alloy. Compd. **687**, 746–753 (2016)

8. Guo, M., Wang, S., Ding, L., Huang, C., Wang, H.: Tantalum-doped lithium titanate with enhanced performance for lithium-ion batteries. J. Power Sources **283**, 372–380 (2015)

9. Hu, G., Zhang, X., Peng, Z.: Preparation and electrochemical performance of tantalum-doped lithium titanate as anode material for lithium-ion battery. Trans. Nonferrous Met. Soc. China **21**, 2248–2253 (2011)

10. Julien, C., Mauger, A., Vijh, A., Zaghib, K.: Lithium Batteries Science and Technology, 619p. Springer International Publishing, Switzerland (2016)

11. Liu, H., Li, C., Cao, Q., Wu, Y.P., Holze, R.: Effects of heteroatoms on doped $LiFePO_4$/C composites. J. Solid State Electrochem. **12**, 1017–1020 (2008)

12. Lin, J., Hsu, C., Ho, H. Sol–gel.: synthesis of aluminum doped lithium titanate anode material for lithium ion batteries. Electrochim. Acta. **87**, 126–132 (2013)

13. Lin, C., Ding, B., Xin, Y., Cheng, F., Lai, M., Lu, L.: Advanced electrochemical performance of $Li_4Ti_5O_{12}$-based materials for lithium-ion battery: Synergistic effect of doping and compositing. J. Power Sources **248**, 1034–1041 (2014)

14. Moldoveanu, S.: Pyrolysis of organic molecules: Applications to health and environmental issue. Hardbound. (2009). 744p

15. Novikova, S., Yaroslavtsev, S., Rusakov, V., Chekannikov, A., Kulova, T., Skundin, A., Yaroslavtsev, A.: Behavior of $LiFe_{1-y}Mn_yPO_4$/C cathode materials upon electrochemical lithium intercalation/deintercalation. J. Power Sources **300**, 444–452 (2015)

16. Örnek, A., Efe, O.: Doping qualifications of $LiFe_{1-x}Mg_xPO_4$-C nano-scale composite cathode materials. Electrochim. Acta **166**, 338–349 (2015)

17. Qing, R., Yang, M., Meng, Y., Sigmund, W.: Synthesis of $LiNi_xFe_{1-x}PO_4$ solid solution as cathode materials for lithium ion batteries. Electrochim. Acta. **108**, 827–832 (2013)

18. Sun, C.S., Zhou, Z., Xu, Z.G., Wang, D.G., Wei, J.P.: Improved high-rate charge-discharge performances of $LiFePO_4$/C via V-doping. J. Power Sources **193**, 841–845 (2009)

19. Wang, D., Li, H., Shi, S., Huang, X., Chen, L.: Improving the rate performance of $LiFePO_4$ by Fe-site doping. Electrochim. Acta **50**, 2955–2958 (2005)

20. Wen, Y., Zeng, L., Tong, Z., Nong, L., Wei, W. J.: Structure and properties of $LiFe_{0.9}V_{0.1}PO_4$. Alloys Compd. **416**, 206–208 (2006)

21. Wilkening, M., Amade, R., Iwaniak, W., Heitjans, P.: Ultraslow Li diffusion in spinel-type structured $Li_4Ti_5O_{12}$—A comparison of results from solid state NMR and impedance spectroscopy. Phys. Chem. Chem. Phys. **9**, 1239–1246 (2007)

22. Wu, F., Li, X., Wang, Z., Guo, H.: Synthesis of chromium-doped lithium titanate microspheres as high-performance anode material for lithium ion batteries. Ceram. Int. **40**, Part B, 13195–13204 (2014)
23. Wu, X., Wen, Zh, Wang, X., Xu, X., Lin, J., Song, Sh: Effect of Ta-doping on the ionic conductivity of lithium titanate. Fusion Eng. Design **85**, 1442–1445 (2010)
24. Yang, M., Ke, W.: The doping effect on the electrochemical properties of $LiFe_{0.95}M_{0.05}PO_4$ ($M = Mg^{2+}$, Ni^{2+}, Al^{3+}, or V^{3+}) as cathode materials for lithium-ion. J. Electrochem. Soc. **155**, A729–A732 (2008)
25. Yang, Z.G., Choi, D., Kerisit, S., Rosso, K.M., Wang, D.H., Zhang, J., Graff, G., Liu, J.: Nanostructures and lithium electrochemical reactivity of lithium titanites and titanium oxides: A review. J. Power Sources **192**, 588–598 (2009)
26. Yi, T.F., Jiang, L.J., Shu, J., Yue, C.B., Zhu, R.S., Qiao, H.B.: Recent development and application of $Li_4Ti_5O_{12}$ as anode material of lithium ion battery. J. Phys. Chem. Solids **71**, 1236–1242 (2010)

Advanced Heat-Resistant TiAl (Nb,Cr,Zr)-Based Intermetallics with the Stabilized β(Ti)-Phase

A.V. Kartavykh, M.V. Gorshenkov and A.V. Korotitskiy

Abstract The paper represents a brief review of authors' research results and publications in the area of materials science and engineering of innovated light-weight heat-resistant TiAl-based intermetallic alloys. The system TiAl(Nb,Cr,Zr) under development is being considered as the advanced basis for the creation of TiAl-intermetallics of 3rd generation (TNM) TiAl(Nb,Mo)-like alloys, those being the most promising nowadays for an application in aviation jet engines design. This research is implemented within the frame of Federal Targeted Program for R&D in Priority Areas of Development of the Russian Scientific and Technological Complex for 2014–2020 (Russian FTP for R&D 2014–2020).

Keywords Structural intermetallics · Gamma-Titanium aluminides Multicomponent alloying · Heat resistance · Microstructure · High-Gradient float-zone processing · Aerojet turbine design

Introduction

In the review [1, 2] in 2013, we analyzed the state-of-the-art of materials science and technology of alloyed TiAl-intermetallics intended for extreme performances. In this review, we have predicted the technology transfer of these materials within the nearest future from the research stage to industrial application in aviation jet turbine-building corporations of the most developed countries. Indeed, the prospective at that time TiAl-intermetallics of the second generation have firstly been used in the low-pressure turbine (LPT) of serial engine GEnx-1B by General

A.V. Kartavykh · M.V. Gorshenkov (✉) · A.V. Korotitskiy
National University of Science and Technology "MISIS", Moscow, Russia
e-mail: mvg@misis.ru

A.V. Kartavykh
e-mail: karta@korolev-net.ru

A.V. Korotitskiy
e-mail: akorotitskiy@gmail.com

© The Author(s) 2018   351
K.V. Anisimov et al. (eds.), *Proceedings of the Scientific-Practical Conference*
"Research and Development - 2016", https://doi.org/10.1007/978-3-319-62870-7_38

Electric for the equipping of long-range passenger aircraft Boeing 787 Dreamliner [3]. This breakthrough event, being almost synchronous with the review publication, has marked the new stage of TiAl-alloys application—their successful worldwide commercialization beginning. The strategy of application of the new class material in GEnx-1B turbine is rather cautious: from the lightweight alloy GE48-2-2 (of Ti-48Al-2Nb-2Cr at.% composition) only 2 final LPT stages (turbine blade discs) are manufactured out of 5 in total, with the "softest" temperature service mode up to 650 °C. Even such limited γ-TiAl application allows saving 180 kg per engine when compared to its precursor CF6-80 that best illustrates the cutting edge of γ-TiAl-alloys potential. The GEnx offers up to 15% better fuel consumption, which translates to 15% less CO_2 emission [3]. However, the heat-resistant structural potential of the GE48-2-2 alloy is seemingly near to be exhausted and limited by the final LPT stages operating at the "softest" thermal and pressure modes. The following progress of γ-TiAl application in aircraft turbines requires the improvement of heat resistance and heat strength within the temperature range expanded towards 800 °C and higher. That led to the recent development of novel γ-TiAl-based alloys family, so-called TNM, i.e., TiAl(Nb,Mo)-like alloys which define the upper strength limit of titanium aluminides. Present paper represents the brief review of authors' research and publications in the area of properties engineering of TiAl(Nb,Cr,Zr) system. The intermetallics of this elemental system could offer the improved basis for the γ-TiAl materials creation of 3rd generation.

The Concept of Development and Engineering the Composition of β-Stabilized Alloys

The development of new generation of TiAl-intermetallics is being addressed to solve the problem of insufficient mechanical ductility, strength and stability of items at elevated temperatures, as well as adaptation of thermo-mechanical regimes of alloys processing (forging, rolling, etc.) to the technical specifications of industrial metalworking equipment.

Innovated gamma-titanium aluminides (TNM-like alloys) contain 42–46 at.% of aluminum, and up to 10–12 at.% functional additives of transient metals, those stabilize the primary β-Ti phase (known also as B2 phase being in low-temperature ordered state). Apart from the obligatory Nb, such β-stabilizers as Mo, Ta, Zr, Cr, W, and V could be used. Such alloying leads to the retaining in solidified alloy of a relatively low volumetric fraction of residual B2-phase on the base of *bcc* lattice, which is ductile at high temperatures. The development concept and achieved properties of TNM-alloys are represented rather comprehensively in recent monograph [4] and in the review [5].

Historically for the first time the molybdenum was used as a β-stabilizing additive, which possesses utmost stabilizing activity (and the acronym TNM = TiAl-Nb-Mo is originated from hence). However, it was revealed that Mo is worsening the corrosion resistance of γ-TiAl [6]. Nevertheless, the effect of β-stabilizers on the phase diagram transformation of an alloyed material was first studied using the Ti-43.5Al-4Nb-1Mo (at.%) alloy as the typical example [7].

Authors in collaboration with the E.O. Paton Electric Welding Institute have elaborated the alternative alloy Ti-44Al-5Nb-3Cr-1.5Zr (at.%) as the basic composition possessing enhanced corrosion stability due to a homogeneous distribution of anti-corrosion additives Cr and Zr through the constituting intermetallic phases. Particularly, this feature results also in high tribological and tribochemical resistance of the material at the high-temperature friction against heavily alloyed chromium steels [8].

Strong β-stabilizing doping results in a specific change of phase diagram and schematic of solid-state temperature-phase transformations of cast alloy with the participation of stabilized β(Ti)/B2 phase in all technologically important domains (see Fig. 1a, b in comparison), that is the general trait of TNM-like alloys [7].

Key features of TiAl(Nb,Cr,Zr) diagram are the following: strong expansion of primary β(Ti) phase area toward Al and lower temperatures; narrowing and shift of high-temperature α, (α + γ) domains toward Al, that leads to their excluding from the transformation pathway of the Ti-44Al-5Nb-3Cr-1.5Zr composition (this pathway is marked by the arrow in Fig. 1b). The latter feature is defined substantially the specifics of microstructure forming of Ti-44Al-5Nb-3Cr-1.5Zr in comparison with conventional compositions, including the Ti-46Al-8Nb—precursor of our elaborating alloy. The mechanism of solid-state phase transformations of TiAl(Nb,Cr,Zr) could be simplified by the following sequence of stages:

$$\beta(Ti) \xrightarrow{(a)} \beta(Ti)/B2 + \alpha \xrightarrow{(b)} B2 + \alpha_2 - Ti_3Al + \gamma - TiAl \qquad (1)$$

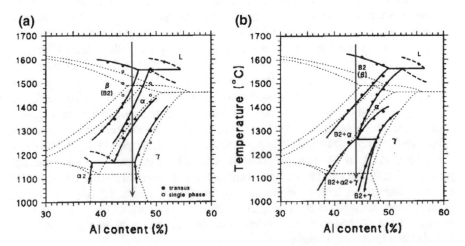

Fig. 1 The quasi-binary isopleths of systems TiAl-8at.%Nb (**a**) and TiAl-10at.%(Nb + Cr + Zr) (**b**), where phase transformation pathways are marked by the arrows for Ti-46Al-8Nb and Ti-44Al-5Nb-3Cr-1.5Zr (at.%) alloys, respectively. The reference binary TiAl diagram is drawn with *dotted lines*

It will be shown in Section 2.2 that stage (b) is characterized by two kinetic mechanisms, proceeding at cooling of the solid alloy generally in a nonequilibrium manner. That leads to the formation of γ-granular and (α_2-Ti$_3$Al + γ-TiAl)-lamellar substructures, being in coexistence with a residual B2-interlayer phase that is located along the boundaries of former transformed α-grains:

$$\alpha \rightarrow (\alpha_2 - Ti_3Al + \gamma - TiAl) - lamellar. \tag{2}$$

$$B2 \rightarrow \gamma - TiAl - granular. \tag{3}$$

Since the volumetric fractions of these substructures are being formed as a result of nonequilibrium reactions, one can control them to some extent, applying different cooling rates and targeted thermal treatments [7, 9].

Engineering the Microstructure and Mechanical Properties of TiAl(Nb,Cr,Zr)

Microalloying with LaB$_6$

It is well known that ductility of a multiphase intermetallic alloy could be improved by the grain refinement of isotropic equiaxed-granular microstructure [10]. For this purpose, one can apply the doping of Nb-alloying titanium gamma-aluminides with boron-containing additives (usually with TiB$_2$). That leads to the precipitation of micro/nanoscaled particles of monoboride-based solid solution (Ti,Nb)B within high-temperature domains of the phase diagram. These micro-crystals act as the local (point) seeds at the appearance of major intermetallic phases, leading to the numerous nucleation events and competitive statistic growth of fine equiaxed structural grains. This approach was earlier successfully applied, and the structure-forming mechanism was studied in detail by us in the different technological routes of the TiAl(Nb)B system solidification [11–13]. Additionally, the ductility can be enhanced by the lowering of interstitial embrittle impurity content of oxygen. Working with the complicated TNM-like system TiAl(Nb,Cr,Zr), we firstly applied joint doping with boron and rare-earth element (lanthanum) in the form of LaB$_6$ compound [14]. That results in the ultra-fine microstructure formation of derivative alloys Ti-44Al-5Nb-2Cr-1.5Zr-0.4B-0.07La and Ti-44Al-5Nb-1Cr-1.5Zr-1B-0.17La (at.%) with the smallest achieved mean diameter of a structural grain of 30 μm, and low (300 wt.ppm) content of dissolved oxygen due to the joint effect of boron and lanthanum internal getter. The enhanced efficiency of LaB$_6$ ligature application when compared to the microalloying with TiB$_2$ is clearly seen in Fig. 2 from the comparison of microstructures refinement degree of TiAl(Nb)B vs. TiAl(Nb,Cr,Zr)B,La alloys.

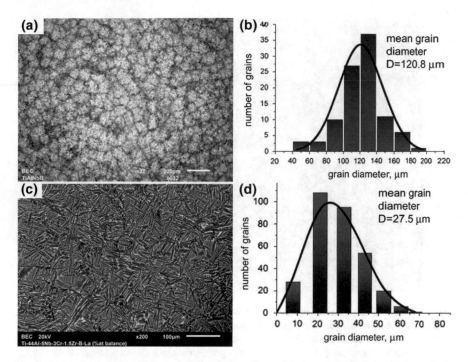

Fig. 2 Isotropic refined microstructures of the alloys (electron microscopy data), and the functions of their structural grains distribution by the diameter: **a, b**—Ti-44Al-7Nb-2B (at.%), micro-alloyed with TiB$_2$ [11, 12]; **c, d**—Ti-44Al-5Nb-1Cr-1.5Zr-1B-0.17La (at.%), with LaB$_6$ [15]

The difference in microstructures of Ti-44Al-7Nb-2B and Ti-44Al-5Nb-1Cr-1.5Zr-1B-0.17La alloys, as displayed in Fig. 2, is related to the different phase-forming mechanisms at the microalloying of TiAl(Nb) and TiAl(Nb, Cr,Zr) systems with boron. In the TiAl(Nb) system the borides (Ti,Nb)B act as the seeds of high-temperature α-phase at the transformation β(Ti) $\rightarrow$ α (Fig. 1a). When avoiding the domains α and (α + γ) in the system TiAl(Nb,Cr,Zr) (Fig. 1b), the ribbon-like borides display the activity in the more low-temperature field (B2 + α_2 + γ). Here they act as the centers of solid-state germination (seeding) of α_2-Ti$_3$Al phase, which is growing afterward through the matrix of basic γ-TiAl phase as lamellae (Fig. 3). This particular boride-induced mechanism of phase- and structure-forming in a TNM-like alloy have been studied and published by us for the first time in papers [15, 16].

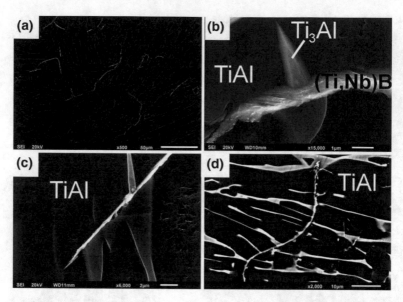

Fig. 3 Boride ribbons resolved in Ti-44Al-5Nb-2Cr-1.5Zr-0.4B-0.07La (**a**) and three consecutive stages of structure refining mechanism development: event of solid-state germination (seeding) of α_2-Ti$_3$Al phase on boride facet (**b**); morphological development of α_2-laths and their growth through γ-TiAl matrix (**c**); formation of new grain boundary by elongated curved boride surrounded by fine $\gamma + \alpha_2$ phase structure inside the delimited grains (**d**). The electron microscopy micrographs [15, 16]

High-Gradient Float-Zone Processing

Another fruitful approach in structure and properties engineering is using of controlling thermodynamic impact to the intermetallic system, i.e., solidification and cooling (annealing) of alloys in the strong directional fields—in magnetic or/and thermal ones. The directional solidification within thermal gradient is the most studied and therefore is most applicable.

At the directional solidification of TiAl-alloys by Bridgman method or by power-down technique [11, 12], the maximally achieved value of axial thermal gradient amounts to 50–70 °C/cm. In the Ti-46Al-8Nb (at.%) alloy it results in the partial ordering of microstructure by the formation of elongated columnar primary β-grains aligned to the crystallization direction. Finally, the microstructure gets obtained consisting of elongated lamellar colonies, inside of which the separation of (α_2-Ti$_3$Al + γ-TiAl) lamellae in the low-temperature-phase domain (Fig. 1a) proceeds chaotically, in disagreement with the neighboring colony.

The application of float-zone technique (FZ) with a narrow zone allows increasing considerably the axial thermal gradient near the solid/melt interface. Induction FZ in an argon stream with the gradient of 300 °C/cm was applied by authors firstly to the TNM-like alloy Ti-44Al-5Nb-3Cr-1.5Zr (at.%) in papers [17, 18], and led to the promising results. Let us consider them in more details.

Figure 4a displays the irregular microstructure of initial virgin ingot of Ti-44Al-5Nb-3Cr-1.5Zr, which was manufactured by semi-continuous electron-beam synthesis/casting technique from the pure metals [14]. Basic phase γ-TiAl is imaged here by gray, α_2-Ti$_3$Al—by black and B2—as bright phase. For comparison in Fig. 4b the oriented duplex (lamellar-granular) microstructure is given of the same alloy after FZ-processing. It consists of the axially aligned lamellar α_2-Ti$_3$Al + γ-TiAl matrix (80% volumetric), granular γ-TiAl microstructure (15%), and 5% of bright inter-granular layers of stabilized β-Ti (B2) phase. Thus, the sub-structural volumetric formula of the FZ-processed Ti-44Al-5Nb-3Cr-1.5Zr alloy can be expressed as $(γ + \alpha_2)/γ/B2 = 85:15:5$.

The lamellar matrix is magnified in Fig. 4d, being consisted of alternating lamellae of γ-TiAl and α_2-Ti$_3$Al phases of sub-micron thicknesses, aligned with the direction of the high-temperature gradient. Figure 4c represents the magnified

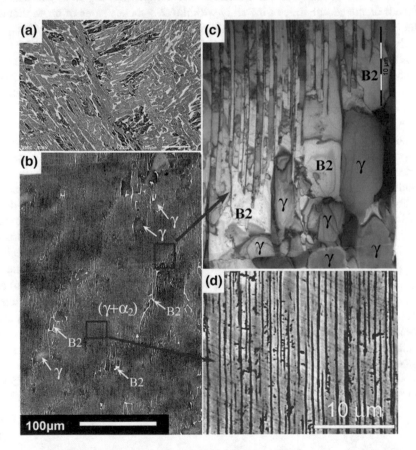

Fig. 4 **a**—Irregular microstructure of cast Ti-44Al-5Nb-3Cr-1.5Zr; **b**—ordered microstructure of the alloy after FZ-processing; **c**—magnified transition between the lamellar area and (γ + B2) interlayer; **d**—magnified axially aligned (γ + α_2) lamellae. The electron microscopy micrographs after [17, 18]

transient area between the lamellar and γ-granular fractions, where the details of structural transition and "relaxing" grains of B2-interlayer are seen; the latter are ductile at elevated temperature. Evidently, as a result of FZ-processing, the microstructure of the alloy is cardinally modified, refined and ordered at the conservation of an unchanged set of constituting phases. Therefore, the microstructural engineering is carefully being performed.

The duplex structure, represented in general view in Fig. 4b was formed within the low-temperature domain (B2 + α_2 + γ) of phase diagram (Fig. 1b) during the nonequilibrium cooling under high thermal gradient impact. The gradient promotes the ordering, "pooling" of growing α_2 + γ lamellae along the ingot axis at their formation from α-phase according to the reaction (2). The γ-granular fraction is being formed at the transformation of B2 phase according to reaction (3) and is situated along the former boundaries of columnar textured α-grains, those elongated as well in the thermal gradient direction. The completion degree of this reaction depends on the cooling rate of the alloy within (B2 + α_2 + γ) area of phase diagram after FZ passing. Therefore it depends on the zone movement rate. Thus, the relative volumetric ratio of sub-structural fractions (γ + α_2)/γ/B2 could be optimized rather easily by the kinetic way, i.e., by the zone movement rate variation.

The ordered microstructure with optimized (γ + α_2)/γ/B2 volumetric quotas (Fig. 4b) possesses more balanced properties when compared to the cast material. Fine lamellar matrix (Fig. 4d) is responsible for the improved strength and creep especially under axial loading. Meanwhile, the heat resistance and plasticity get enhanced thanks to the incorporated interlayers composed of γ-grains and ductile B2-phase (Fig. 4c). At the expense of limited elastic mobility of γ-grains within the medium of B2-phase, such interlayers promote the relaxation of stresses in basic lamellar structure, thus raising the high-temperature threshold of its destruction.

The specimens of virgin and structurally modified alloy Ti-44Al-5Nb-3Cr-1.5Zr were comparatively examined by the uniaxial compression along the ingot's axis at the temperatures from 750 to 1050 °C. The results of high-temperature tests are given in Figs. 5a–d, illustrating the comparative degradation of physical–mechanical properties of the alloys vs. temperature. FZ-processing led to substantial improvement of deformability (Fig. 5a), an increase of yield strength (Fig. 5b) and Young's modulus (Fig. 5c) at the same temperatures. At the same loads, the creep resistance rises. Exemplarily, under the load of 200 MPa the first signs of creep of FZ-alloy appear only at 950 °C (Fig. 5d). In other words, FZ-alloy possesses the identical level of deformability parameters at the temperatures by 100–150 °C higher when compared to cast material. Thus the upper-temperature limit of Ti-44Al-5Nb-3Cr-1.5Zr structural applicability could be expanded from 750 to 800 °C towards 900–950 °C.

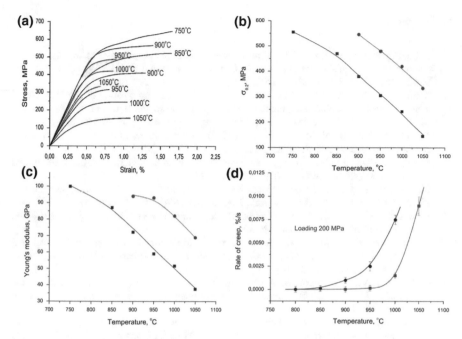

Fig. 5 Plots of physical–mechanical properties of cast (in *black*) and FZ-processed (in *red*) alloys vs. temperature [18]: **a**—axial deformation curves; **b**—yield strength; **c**—Young's modulus; **d**—creep rate

Conclusions

We have demonstrated in laboratory scale that the system TiAl(Nb,Cr,Zr) under development could represent the advanced basis for the creation of TiAl-intermetallics of third generation, those being the most promising nowadays for an application in aviation jet engines design. Considering the Ti-44Al-5Nb-3Cr-1.5Zr (at.%) composition, the features of phase diagrams and structure formation have been discussed of a new class of γ-TiAl-intermetallics with stabilized β phase, those allow applying the new effective principles of the necessary structural properties creation of the material. The brief review of authors' innovative research and publications is presented in the field of engineering the new types of heat-resistant microstructures in gamma-titanium aluminides by joint microalloying with boron and lanthanum, and by the high-gradient float-zone processing application. The experimental specimens of microstructured alloy possess both increased high-temperature strength, and creep resistance at the uniaxial loading, thus exhibiting the substantial extension of thermal service range when applying γ-TiAl in turbine blades and other crucial components of aircraft jet engines design.

References

1. Kartavykh, A.V., Kaloshkin, S.D., Cherdyntsev, V.V. et al.: Application of microstructured intermetallides in turbine manufacture. Part 1: present state and prospects (a review). Inorg. Mater.: Appl. Res. **4** № 1, 12–20 (2013)
2. Kartavykh, A.V., Kaloshkin, S.D., Cherdyntsev, V.V. et al.: Application of microstructured intermetallides in turbine manufacture. Part 2: problems in development of heat-resistant alloys based on TiAl (a review). Inorg. Mater.: Appl. Res. **4** №1, 36–45 (2013)
3. The GEnx Commercial Aircraft Engine (2015). [Electronic resource] URL. http://www.geaviation.com/commercial/engines/genx/. Accessed 25 Oct 2016
4. Appel, F., Paul, J.D.H., Oering, M.: Gamma titanium aluminide alloys: science and technology, p. 745. WILEY-VCH, Weinheim (2011)
5. Clemens, H., Mayer, S.: Design, processing, microstructure, properties, and applications of advanced intermetallic TiAl alloys. Adv. Eng. Mater. **15**(4), 191–215 (2013)
6. Brotzu, A., Felli, F., Pilone, D.: Effect of alloying elements on the behaviour of TiAl-based alloys. Intermetallics **54**, 176–180 (2014)
7. Schwaighofer, E., Clemens, H., Mayer, S., et al.: Microstructural design and mechanical properties of a cast and heat treated intermetallic multi-phase γ-TiAl based alloy. Intermetallics **44**, 128–140 (2014)
8. Kartavykh, A.V., Gorshenkov, M.V., Danilov, V.D. et al.: Tribochemistry of dry-sliding wear of structural TiAl(Nb,Cr,Zr)B,La intermetallics family against the chromium steel. Tribol. Int. **90**, 270–277 (2015)
9. Erdely, P., Werner, R., Schwaighofer, E. et al.: In-situ study of the time-temperature-transformation behaviour of a multi-phase intermetallic β-stabilised TiAl alloy. Intermetallics. **57**, 17–20 (2015)
10. Morris, M.A., Leboeuf, M.: Grain-size refinement of γ-Ti-Al alloys: effect on mechanical properties. Mater. Sci. Eng. **A224**, 1–11 (1997)
11. Kartavykh, A.V., Tcherdyntsev, V.V., Gorshenkov, M.V. et al.: Tailored microstructure creation of TiAl-based refractory alloys within VGF solidification. Mater. Chem. Phys. **141** (2–3), 643–650 (2013)
12. Kartavykh, A.V., Tcherdyntsev, V.V., Gorshenkov, M.V. et al.: Microstructure engineering of TiAl-based refractory intermetallics within power-down directional solidification process. J. Alloys Compd. **586**, S180–S183 (2014)
13. Kartavykh, A.V., Gorshenkov, M.V., Tcherdyntsev, V.V. et al.: On the state of boride precipitates in grain refined TiAl-based alloys with high Nb content. J. Alloys Compd. **586**, S153–S158 (2014)
14. Kartavykh, A.V., Asnis, E.A., Piskun, N.V. et al.: Lanthanum hexaboride as advanced structural refiner/getter in TiAl-based refractory intermetallics. J. Alloys Compd. **588**, 122–126 (2014)
15. Kartavykh, A.V., Gorshenkov, M.V., Podgorny, D.A.: Grain refinement mechanism in advanced γ-TiAl boron-alloyed structural intermetallics: the direct observation. Mater. Lett. **142**, 294–298 (2015)
16. Kartavykh, A.V., Gorshenkov, M.V., Podgorny, D.A.: The direct observation of grain refinement mechanism in advanced multicomponent γ-TiAl based structural intermetallics doped with boron. In: Polychroniadis, E.K. et al. (eds.) 2nd International Multidisciplinary Microscopy and Microanalysis Congress. Springer Proc. in Physics. **164**, 175–181 (2015)
17. Kartavykh, A.V., Asnis, E.A., Piskun, N.V. et al.: Microstructure and mechanical properties control of γ-TiAl(Nb,Cr,Zr) intermetallic alloy by induction float zone processing. J. Alloys Compd. **643**, S182–S186 (2015)
18. Kartavykh, A.V., Asnis, E.A., Piskun, N.V. et al.: A promising microstructure/deformability adjustment of β-stabilized γ-TiAl intermetallics. Mater. Lett. **162**, 180–184 (2016)

Structural and Magnetic Properties of As-Cast Fe–Nd Alloys

V.P. Menushenkov, I.V. Shchetinin, M.V. Gorshenkov and
A.G. Savchenko

Abstract The effect of composition on the magnetic properties and microstructure of as-cast Nd–Fe alloys was investigated. The temperature dependence of the hysteresis loops was studied. The magnetic phases with ordering temperatures in the range from 7 to 50 K and from 420 to 580 K are detected from zero-field cooled (ZFC) and field cooled (FC) dependencies of magnetization. At temperatures below 100 K, the increase of the magnetizing field leads to a sharp increase of the magnetization, which does not saturate in the magnetization field 90 kOe. The correlations between the microstructure and coercivity of the as-cast Nd–Fe alloys are discussed.

Keywords Nd–Fe alloys · As-Cast · Microstructure · Hysteresis loop
Magnetization · Coercivity

Introduction

The Nd–Fe alloys have great interest because of their important role in the sintering process and formation of Nd–Fe–B permanent magnets with large coercivity. The as-cast $Fe_{16}Nd_{84}$ alloy first reported by Drozzina and Janus [1] [P. 36] in 1935 had coercive force $H_{ci} \approx 4,5$ kOe at room temperature (RT). High coercivity ($H_{ci} \approx 8,6$ kOe at RT and $H_{ci} \approx 59$ kOe at 20 K) in rapidly quenched $Nd_{40}Fe_{60}$ alloys was reported by Croat [2] [P. 125, 3 P 3161]. The coercive force of the as-cast Nd-rich Nd–Fe alloys is associated with the metastable anisotropic A_1 phase which is formed during crystallization and cooling below the eutectic temperature by decomposition of the Fe-rich regions within the metastable eutectic [4] [P. 215],

V.P. Menushenkov · I.V. Shchetinin (✉) · M.V. Gorshenkov · A.G. Savchenko
National University of Science and Technology "MISIS", Moscow, Russia
e-mail: ingvar@misis.ru

V.P. Menushenkov
e-mail: menushenkov@gmail.com

363

K.V. Anisimov et al. (eds.), *Proceedings of the Scientific-Practical Conference
"Research and Development - 2016"*, https://doi.org/10.1007/978-3-319-62870-7_39

[5] [P. 5971], [6] [P. L1], [7] [P. 169], [8] [P. 209], [9] [P. 245], [10] [P. L5], [11] [P. 273], [12] [P. 97], [13] [P. 149], [14] [P. MM3.2.1]. Kumar [14] [P. MM3.2.1] proposed that the A_1 regions behave as an assembly of randomly oriented uniaxial Nd–Fe clusters or nanocrystallites with the nature of single-domain particles. According to [3] [P 3161], [15] [P. 2088], [16] [P. 2302], [17] [P. 2483] high H_{ci} of the melt-spun Fe–Nd alloys results from the different nonequilibrium hard magnetic phases. The size of the clusters or nanocrystallites in rapidly quenched Fe–Nd alloys is extremely small and its composition analysis was beyond the limits of TEM capacity at the time. Despite the numerous investigations, the elucidation of the physical origin of the high coercivity in Nd–Fe alloys is still remaining as a problem. A comparison of the microstructure and magnetic properties of the melt-spun Nd–Fe alloys should help to clarify the origin and mechanism of the high coercivity at room and cryogenic temperatures.

In the present paper, we examine the effect of the composition of as-cast Fe_xNd_{100-x} ($x = 14$–50) alloys on their microstructure and magnetic properties. The relationship between the coercivity and microstructure is also discussed.

Experimental

The Fe_xNd_{100-x} alloys with $x = 14, 28, 38, 50$ were arc melted under an Ar atmosphere. The purity of the raw materials was Nd 99.9% and Fe 99.95%. The ingots were cast into a copper mold to prepare bulk samples. Magnetic measurements in the temperature range from 5 to 650 K were measured using a PPMS EverCool-II magnetometer in the maximum magnetizing field up to 90 kOe. The structure of the melt-spun ribbons was characterized by X-ray diffraction (XRD) using Co-Kα radiation. The study of the microstructure was detected by a JEOL JSM-6610LV scanning electron microscope (SEM) and a JEM-1400 transmission electron microscope (TEM) operated at 120 kV. The thin foils for TEM were prepared by ion milling with Ar ions and a glancing angle of $3°$.

Results and Discussion

The SEM images of the as-cast alloys with $x = 14$ and 50 are represented in Fig. 1. The microstructures of the both the samples consist of different regions. The microstructure of $x = 14$ sample consists of the Nd phase crystals and intergranular eutectic. The SEM image of the $x = 50$ sample show the crystals of the Nd_2Fe_{17} phase surrounded Nd-rich shell and intergranular eutectic phase. The composition of the phases in the alloys were measured by EDX analysis attached to the SEM.

The TEM images of the eutectic regions in the as-cast $x = 14$ and 50 samples are shown in Fig. 2. The interplanar spacing (d_{HKL}) calculated from SAED patterns of the as-cast samples $x = 14$ (0.322, 0.280, 0.22, 0.197, 0.167, 0.161, 0.127, 0.122,

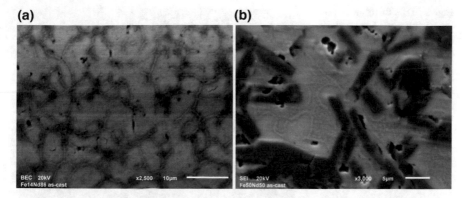

Fig. 1 SEM micrographs of the as-cast alloys $x = 14$ (**a**) and 50 (**b**)

0.112 nm) and $x = 50$ (0.286, 0.190, 0.166, 0.126, 0.155, 0.118 nm) correspond to the fcc phase with $a = 5.5$ nm. The fcc phase was observed in melt spun Nd-Fe alloy [7, 18] and in as-cast Nd-Fe alloys [19, 20]. But the presence in the SAED patterns of the broadened rings with $d_{HKL} = 0.3$ and 0.21 nm could be corresponded to the unknown A_{un} phase. According to Fig. 2, the structure of the as-cast samples is comprised of the grains of the crystalline Nd along with amorphous-like regions. The amorphous-like regions in turn consists of nanocrystals of the A_{un} phase embedded in amorphous matrix.

The hysteresis loops for the $x = 14$ and 50 as-cast samples are shown in Fig. 3. The hysteresis loop of the as-cast alloys $x = 14$, 28 (Fig. 3a, b) and $x = 32$, 50 (Fig. 3c, d) differ significantly. First, the as-cast alloys $x = 14$, 28 have at RT a high coercive force $H_{ci} = 4,7$ and 4,3 kOe, respectively. The alloys $x = 32$, 50 have low coercive force $H_{ci} = 2,4$ and 0,1 kOe, respectively. Second, in the as-cast alloys $x = 14$, 28 during the temperature decrease from 300 to 100 K the value of M_r is maintained, but the width of the hysteresis loops continuously increases along the X axis, that resulting in the growth of coercive force up to 28 kOe ($x = 14$) and 37 kOe ($x = 28$) after magnetization in a field of 50 kOe. However, at temperatures below 100 K, a sharp increase of the magnetization (3–4 times) in strong external fields is observed, which is followed by a decrease in coercive force.

In the as-cast alloys $x = 38$, 50 during the temperature decrease from 300 to 100 K the coercive force practically does not change and remains low ($H_{ci} = 2,4$ and 0,1 kOe, respectively), but the M_s continuously increases and the hysteresis loops have a constricted shape. Low remanence and coercive force of the $x = 38$ and 50 alloys are the result of the presence in their structure of the soft magnetic Nd_2Fe_{17} phase, the amount of which exceeds 50% (Fig. 1). The amount of Nd_2Fe_{17} phase was calculated from the SEM images. It can also be assumed that the reason for the expansion of the hysteresis loop at higher fields is the resistance of the spin moments rotation of Nd-based phases, which are ferromagnetic below 50 K and provide a sharp increase of the magnetic moment in a field $H \geq 50$ kOe. At a temperature below 100 K, the magnetization of the $x = 14$ and 28 ribbons do not reach saturation in the field as high as 90 kOe.

Fig. 2 TEM micrographs of the as-cast $x = 14$ (**a**) and 50 (**b**) samples. Darkfield image from the intergranular region 1, taken in first ring reflection (**c**) and SAED patterns of the intergranular region 1 (**d**) and crystalline region 2 (**e**). The scale bar in Fig. 2a is 1 μm

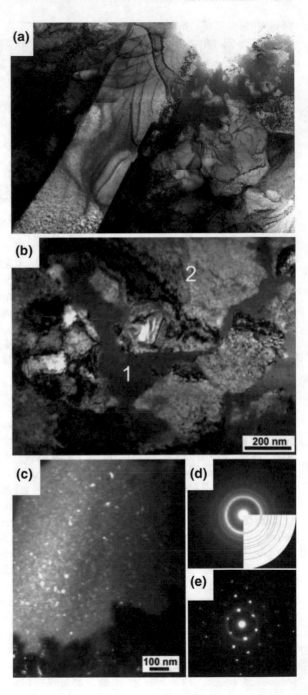

Fig. 3 The temperature dependence of M for the as-cast Fe_xNd_{100-x} samples with $x = 14$–50

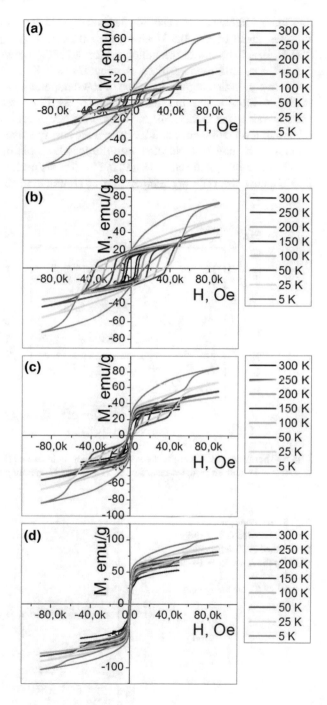

Figure 4 shows a ZFC and FC thermomagnetic curves for the as-cast $x = 14–28$ samples, taken in $H = 0.5$ kOe operating field in the temperature range from 5 to 350 K. Insert (b) shows the initial part of the M(T) curves on a large scale. Three magnetic transitions take place at about $T_1 \approx 7$ K, $T_2 \approx 17$ K and $T_3 \approx 33$ K, respectively, indicating the presence of low-temperature magnetic phases. The data in Fig. 4 allow to assume that the bend of M(T) at around $T_1 \approx 7$ K and 17 K corresponds to the antiferromagnetic ordering temperature of the dhcp Nd (the onset of ordering on cubic sites and hexagonal sites, respectively). Curie temperatures at around 33 K may be connected with the fcc allotropic of Nd [21] [P. 1260].

The thermomagnetic curves M(T) in Fig. 5 measured at a temperature higher than 300 K show the ordering temperature at around 550 K. These high-temperature

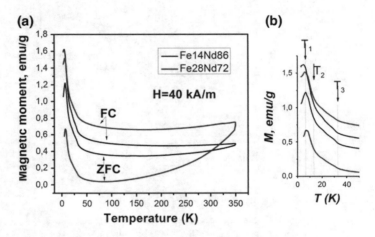

Fig. 4 Thermomagnetic zero-field cooled (ZFC) and field cooled (FC) dependencies of *M* for the as-cast $x = 14$ and 28 samples measured in $H = 0.5$ kOe operating field

Fig. 5 Termomagnetic zero-field cooled (ZFC) and field cooled (FC) dependencies of *M* for the as-cast $x = 14–50$ samples measured in $H = 40$ kA/m operating field

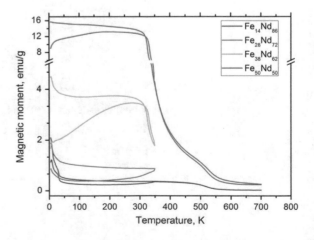

ferromagnetic–paramagnetic transitions should be attributed to the well-known metastable highly anisotropic A_1 phase which is associated with the high coercivity of the as-cast Nd–Fe alloys at RT. Due to the presence of these magnetic phases the magnetization of the Nd–Fe ribbons and as-cast alloys does not saturate in highest magnetization field 90 kOe at temperatures below 100 K.

The temperature dependence of coercive force for the $x = 14$ and 28 ribbons from 50 to 300 K is shown in Fig. 6a. As in the case of the melt-spun alloys, the temperature dependence shows a linear relationship between $H_{ci}^{1/2}$ and $T^{2/3}$ in the temperature range 50–300 K for the samples $x = 14$ and 28 (Fig. 6b).

These data also demonstrate a good agreement with strong pinning model of domain walls, proposed by Gaunt [22] [P. 261]. The coercivity of the as-cast samples both at room temperature and at lower temperatures may be associated with a domain wall pinning on the nanocrystals inside amorphous-like regions. However, the cause of the high coercivity at temperatures below 50 K is still not clear, since at these temperatures the nanocrystals of Nd-rich phase are ferromagnetic.

Fig. 6 The temperature dependence of H_{ci} (**a**) and $(H_{ci})^{1/2}$ as a function of $(T)^{2/3}$ (**b**), for as-cast Fe_xNd_{100-x} ($x = 14$–28) samples over the temperature range 50–300 K

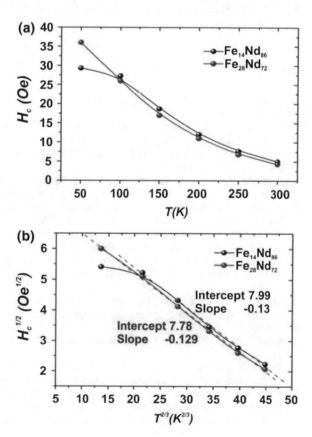

Conclusion

The comparison of the Fe_xNd_{100-x} alloys, prepared by crystallization in the mold, allowed revealing the influence of the composition on the microstructure and properties of the synthesized alloys.

The coexistence of some magnetic phases with ordering temperatures below 50 K takes place in as-cast Nd–Fe alloys. The curves of M(T) at around 7 K and 17 K corresponds to the antiferromagnetic ordering temperature of the dhcp Nd (the onset of ordering on cubic sites and hexagonal sites, respectively). Curie temperature at around 33 K may be connected with the fcc allotropic of Nd, and $T_C \approx 47$ K may be related to the magnetic transition of the Nd-rich nanocrystals.

The strong pinning model of domain walls can describe the temperature dependence of coercive force of the as-cast alloys in the temperature range 50–250 K, but the cause of the high coercivity at temperatures below 50 K is still not clear.

Acknowledgements Research are carried out with the financial support of the state represented by the Ministry of Education and Science of the Russian Federation. Agreement (contract) no. 14.575.21.0043 27 Jun 2014. Unique project Identifier: RFMEFI57514X0043.

References

1. Drozzina, V., Janus, R.: A new magnetic alloy with very large coercive force. Nature **135**, 36–37 (1935)
2. Croat, J.J.: Crystallization and magnetic properties of melt-spun neodymium-iron alloys. J. Magn. Magn. Mater. **24**, 125–131 (1981)
3. Croat, J.J.: Magnetic hardening of Pr-Fe and Nd-Fe alloys by melt-spinning. J. Appl. Phys. **53** (4), 3161–3169 (1982)
4. Schneider, G., Martinek, G., Stadelmaier, H.H., Petzow, G.: High magnetic coercivity due to a new phase in cast eutectic Fe-Nd alloys. Mater. Lett. **7**, 215–218 (1988)
5. Tsoukatos, A., Strzeszewski, J., Hadipanayis, G.: High coercivities in as-cast Nd-Fe and Nd-Fe-Ti alloys. J. Appl. Phys. **64**, 5971–5973 (1988)
6. Hadipanayis, G., Tsoukatos, A., Strzeszewski, J., Long, G.J., Pringle, O.A.: A new hard magnetic phase in binary Nd-Fe and Pr-Fe alloys. J. Magn. Magn. Mater. **78**, L1–L5 (1989)
7. Schneider, G., Landgraf, F.J.G., Missell, F.P.: Additional ferromagnetic phases in the Fe-Nd-B system and the effect of 600°C annealing. J. Less. Common Met. **153**, 169–180 (1989)
8. Landgraf, F.J.G., Schneider, G., Villas-Boas, V., Missell, F.P.: Solidification and solid state transformations in Fe-Nd: revised phase diagram. J. Less-Comm. Met. **163**, 209–218 (1990)
9. Moreau, J.M., Paccard, L., Nozieres, J.P., Missell, F.P., Schneider, G., Villas-boas, V.: A new phase in the Nd-Fe system: Crystalline structure of Nd_5Fe_{17}. J. Less. Comm. Met. **163**, 245–251 (1990)
10. Givord, D., Nozieres, J.P., Rossignol, M.F., Tailor, D.W., Harris, I.R., Fruchart, D., Miraglia, S.: Structural analysis of the hard ferromagnetic phase observed in quenched Nd-Fe alloys of hyper-eutectic composition. J. Alloys Comp. **176**, L5–L11 (1991)
11. Delamare, J., Lemarch, D., Vigier, P.: Structural investigation of the metastable compound A1 in an as-cast Fe-Nd eutectic alloy. J. Alloys Comp. **216**, 273–280 (1994)

12. Menushenkov, V.P., Anderson, S.J., Hoier, R.: Electron-microscopy investigations of microstructure in Fe-Nd alloys, Proc. of the 20-th Int. Symposium on Magnetic Anisotropy and Coercivity in Rare-Earth Transition Metal Alloys, edited by L. Schultz and K.-H. Müller, Werkstoff-Infromationsgesellschaft, Frankfurt, pp. 97–106 (1998)
13. Menushenkov, V.P., Lileev, A.S., Oreshkin, M.A., Zhuravlev, S.A.: Metastable nanocrystalline A1 phase and coercivity in Fe-Nd alloys. J. Magn. Magn. Mater. **203**, 149–152 (1999)
14. Kumar, G., Kerschl, P., Rößler, U.K., Nenkov, K., Müller, K.-H., Schultz, L.: TEM and XAS characterization of hard magnetic phase in Fe-Nd alloys, Mat. Res. Soc Simp. Proc. **806**, MM3.2.1–MM.3.2.6 (2004)
15. Sellmyer, D.J., Ahmed, A., Muench, G., Hadjipanayis, G.: Magnetic hardening in rapidly quenched Fe-Pr and Fe-Nd alloys. J. Appl. Phys. **55**(6), 2088–2090 (1984)
16. Siratori, K., Nagayama, K., Ino, H., Saitoii, N., Nakagawa, Y.: Appearance of high-coercivity in Fe-Nd amorphous alloys, IEEE Trans. Magn., Vol. MAG-**23**, 2302–2304 (1987)
17. Nagayama, K., Ino, H., Saito, N., Nakagawa, Y., Kita, E., Siratori, K.: Magnetic properties of amorphous Fe-Nd alloys. J. Phys. Soc. Japan **39**(7), 2483–2495 (1990)
18. Menushenkov, V.P., Shchetinin, I.V., Gorshenkov, M.V., Savchenko, A.G., Ketov, S.V.: Microstructure and magnetic properties of melt-spun Nd-rich Nd-Fe alloys, IEEE Magn. Let. **7** (2016). doi:10.1109/LMAG.2015.2512980
19. Liao, L.X., Altaunian, Z., Ryan, D.H.: Formation of high presure phases in rapidly quenched Fe-Nd alloys. J. Appl. Phys. **67**, 4821–4823 (1990)
20. Menushenkov, V.P., Gabay, A.M., Lileev, A.S., Obrucheva, E.V., Jalnin, B.V.: Magnetic properties and structure of hydrogenated Nd-Fe alloys. J. Alloys Comp. **209**, 299–304 (1994)
21. Bucher, E., Chu, C.W., Maita, J.P., Andres, K., Cooper, A.S., Buehler, E., Nassau, K.: Electronic properties of two new elemental ferromagnets: fcc Pr and Nd. Phys. Rev. Lett. **22**, 1260–1263 (1969)
22. Gaunt, P.: Ferromagnetic domain wall pinning by a random array of inhomogeneties. Philos. Mag. B **48**, 261–276 (1983)

Laser Technology of Designing Nanocomposite Implants of the Knee Ligaments

A.Yu. Gerasimenko, U.E. Kurilova, M.V. Mezentseva,
S.A. Oshkukov, V.M. Podgaetskii, I.A. Suetina, V.V. Zar
and N.N. Zhurbina

Abstract We describe a laser method for constructing a biocompatible implant of the knee ligaments based on synthetic-braided fiber structure of polyethylene terephthalate (PET) coated with a nanocomposite coating. A coating based on albumin aqueous dispersion of carbon nanotubes (CNTs) was applied to the synthetic fibers using ultrasound and then formed by laser evaporation of the aqueous dispersion component. The structure of the nanocomposite implants was studied by optical and atomic force microscopy. Composite implant based on single-walled CNTs (SWCNTs) contains pores with a diameter of 10–20 nm, and based on multi-walled CNTs (MWCNTs)—40–60 nm. We conducted in vitro studies of proliferative activity of human fibroblast cells (HFb) during their colonization on

A.Yu. Gerasimenko (✉) · U.E. Kurilova · V.M. Podgaetskii · N.N. Zhurbina
Biomedical Systems Department, National Research University
of Electronic Technology, Moscow, Russia
e-mail: gerasimenko@bms.zone

U.E. Kurilova
e-mail: kurilova_10@mail.ru

V.M. Podgaetskii
e-mail: podgaetsky@ya.ru

N.N. Zhurbina
e-mail: Natalia93Zhurbina@gmail.com

M.V. Mezentseva · I.A. Suetina
Tissue Culture Laboratory, Ivanovsky Institute of Virology,
Moscow, Russia
e-mail: marmez@mail.ru

I.A. Suetina
e-mail: ikas@inbox.ru

S.A. Oshkukov · V.V. Zar
Trauma Department, M. Vladimirsky Moscow Regional Research
Clinical Institute, Moscow, Russia
e-mail: sergey0687@mail.ru

V.V. Zar
e-mail: vzar66@gmail.com

K.V. Anisimov et al. (eds.), *Proceedings of the Scientific-Practical Conference*
"Research and Development - 2016", https://doi.org/10.1007/978-3-319-62870-7_40

the surface of the implant and into the space between synthetic fibers. The highest value of the HFb proliferation was observed on the implant based on MWCNTs with a large pore size and amounted to 55.435 pcs., in contrast to the implant based on SWCNTs (54.931 pcs.) and control one (54.715 pcs.), as shown by fluorescence microscopy and MTT test. A histological study of the interaction of the nanocomposite implant implanted into rabbit knee joint with bone canal was carried out. The bone germination in the implantation area at 2, 4 and 8 weeks after surgery was shown.

Keywords Knee ligament implant · Nanocomposite material · Bioengineering Carbon nanotubes · Bovine serum albumin · Laser structuring · Scaffold Cells · Microscopy

Introduction

Injuries of the intraarticular knee joint ligaments (KJL) constitute one of the best-studied problems in contemporary orthopedics. This relates to the high incidence of these injuries and their social importance. In the USA, isolated injuries of the anterior cruciate ligament in the knee occur in 43 cases per 100,000 of the population per year and have increased significantly in the under 20s and middle-aged women [6]. In Germany, every traumatologist performs an average of 35 knee joint ligament reconstructions per year [11]. Data from Russian medical institutions show that the incidence of ligament injuries is one case per 3000 of the population, most patients being active working people aged 16–45 years.

One important cause of unsatisfactory surgical outcomes for these injuries is the poor biological integration of bone channel wall tissues into ligament implants [5]. Knee joint ligament implants are usually developed using synthetic-braided materials as flat ribbons or seamless tubes made of polyethylene terephthalate (PET) fibers. Advantages of using PET include its resistance to the acidic and alkaline environments of the body and to mechanical damage. This material has a density of 1360–1400 kg/m^3 and a breaking point on stretching of 172 MPa. However, the strength characteristics of braided PET fibers is much higher than the strength of the canal's bone tissue, in which the implant is fixed. Therefore there is the problem of destruction of the bone canal surface, particularly at the first time after the operation.

Contemporary regenerative methods are being used to overcome this problem, specifically tissue engineering, which is the method of partial or complete restoration of the physical and biological functions of injured or lost organs by stimulating regeneration of the three-dimensional tissue structure [8]. This method is conceptually based on creating artificial tissue-engineered scaffolds to support cell proliferation and maintain the growth of biological tissues [9]. In order to fulfill their function, scaffolds must be biocompatible and biodegradable, must have a porous structure, and must have mechanical characteristics corresponding to specific biological tissues [2].

Carbon nanotubes (CNT) distributed in a biological matrix are one of the perspective materials for creating of tissue-engineered scaffolds [1, 4, 7]. These scaffolds are potential materials for solving the problem of KJL restoration by below described method of constructing biocompatible knee ligament implants when nanocomposite coatings are formed on the PET fibers surface. The coating based on tissue-engineered scaffolds of CNTs in protein matrix provides the best adhesion and cell growth of connective and bone tissues, which can lead to reduction of a postoperative recovery period of patients.

Materials and Methods

The method of the nanocomposite coating forming on PET fibers involves three steps. The first step consists of preparing a homogeneous dispersion of CNTs in the protein matrix. We used two types of nanotubes: single-walled carbon nanotubes (SWCNTs) and multi-walled carbon nanotubes (MWCNTs) with purity of 98%. SWCNTs were synthesized by an electric arc method on a Ni/Y catalyst and are presented as a black paste in deionized water (concentration is $\sim 2.5\%$ by weight). The mean diameter of SWCNTs was ~ 1.5 nm and length was 0.5–1.5 nm. Nanotubes were aggregate as filaments with mean length 1–10 μm and cross-sectional diameters of 67 nm [10]. MWCNTs were synthesized by a low temperature thermo-catalytic method on a Ni catalyst and consisted of a black powder. The external diameter of MWCNTs was 40–120 nm, length was 1–10 μm, and internal diameter was 10–30 nm [3]. As the technology for coatings making requires thermal heating of the dispersion, bovine serum albumin (BSA) was selected as a protein with maximal resistance to temperatures. The purity of the BSA was 98.1% by weight and the heavy metals content was no greater than 0.001% by weight.

Apparatus for forming nanocomposite coating on synthetic construction is illustrated in Fig. 1. The mass proportions of the components of the dispersion were weighed out on an analytical balance (1) and were 0.02–0.1% CNTs, 25% for BSA, and 74.98–74.9% for water. CNTs were initially mixed in water with a magnetic

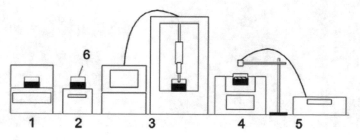

Fig. 1 Apparatus for forming composite coatings on synthetic PETF fiber constructs (see text for explanation)

stirrer (2) and ultrasound homogenizer (3) at a power level of 40 W/cm^2 to prepare a homogeneous dispersion of CNTs in water, as the presence of nanotube agglomerates would lead to an inhomogeneous coating. The CNTs dispersion was then supplemented with powder of BSA, and the mixture was processed gently with a magnetic stirrer (2) and an ultrasound bath (4) at a power of 2 W/cm^2.

The second step in forming the composite nanocoating consisted of ultrasound application of the dispersion, prepared in the first step, to the surface of PET fibers braided as a flat ribbon or tubular constructs. The synthetic construct was placed in a vessel with the dispersion (6), placed in an ultrasound bath with liquid, such that the levels of the dispersion and the liquid were the same. Ultrasound treatment was at a power of 1–2 W/cm^2 for 10 min. At the third step, the dispersion was removed from the vessel containing the synthetic construct and the dispersion on PET surface was treated by uniform laser radiation. Irradiation was with a continuous infrared diode laser (5) at a wavelength of 810 nm and a power level of 35 W. The spatial profile of the laser beam had a Gaussian distribution and its diameter was comparable to the width of the synthetic construct.

Figure 2 shows images of the structure of a synthetic construct bearing an MWCNTs-based dispersion (Fig. 2a, b) or an SWCNTs-based dispersion (Fig. 2c, d) produced by optical microscopy at magnifications of ×40 (Fig. 2a, c) and ×5 (Fig. 2b, d).

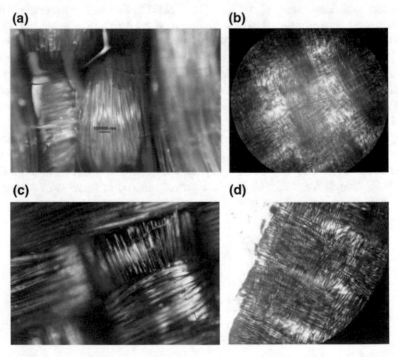

Fig. 2 Optical microscope images of nanocomposite coating with composite nanocoatings based on MWCNT (**a, b**) and SWCNT (**c, d**)

The ultrasound method of applying the nanocarbon dispersion generated a uniform coating on the surface of the synthetic construct. The aqueous component of the dispersion was evaporated during the laser treatment. Under the influence of electric field of the targeted laser irradiation organization of the nanotubes, scaffold was carried out in synthetic construct coating.

Figure 2 shows that the SWCNTs-based composite nanocoating on PET fibers had a more uniform light gray coloration than the MWCNTs-based coating. This may be due to the presence of hydrophilic functional groups on the nanotube surfaces, resulting in better homogenization in the aqueous matrix.

Because of the evaporation of water, large numbers of pores were seen throughout the volume of the scaffold, these being needed for adhesion of cellular material (Fig. 3). Form and quantity of pores depend on manufacturing process parameters. Basically, CNTs adding to BSA matrix resulted in decreasing of pores quantity and size. This effect can be due to reinforcing the action of CNTs when the scaffold is formed by laser irradiation. An average size of scaffold nanoelements was less than 500 nm, a volume density was in the range 10^6–10^8 cm^{-3}.

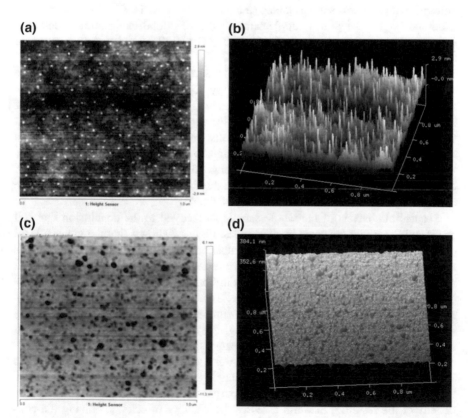

Fig. 3 2-D (**a, c**) and 3-D (**b, d**) AFM topograms of the composite coating, based on SWCNT (**a, b**) and MWCNT (**c, d**) formed by the laser irradiation

Detailed studies of the porosity of the samples by capillary adsorption–desorption of nitrogen yielded the following results. Mean pore diameter for SWCNTs-based samples was 10–20 nm, while the value for MWCNTs-based samples was almost three times greater, at 40–60 nm. Specific pore volume in the SWCNTs-based sample was greater (0.005 ml/g) than in MWCNT-based samples (0.0033 ml/g).

Results

The biological characteristics of the synthetic constructs with nanocomposite coatings were assessed by studying the proliferative activity of connective tissue cells—human fibroblasts (HFb)—in the presence of samples. Samples were fragments of synthetic constructs of size 1 × 1 cm with and without nanocomposite coatings.

Prepared samples were sterilized by UV irradiation for 1 h. HFb cells were cultured with experimental samples and then incubated in DMEM (Dulbecco's Modified Eagle's Medium) supplemented with 10% fetal calf serum. Incubation was in culture plates in an incubator at 37 °C for 72 h in a moisture-saturated atmosphere containing 5% CO_2.

Colonization of HFb cells occurred on the surfaces and between the PET fibers and was assessed by fluorescence microscopy. Cells on samples of synthetic constructs were treated with fluorescein diacetate (FDA) to a final concentration of 25 μg/ml and then with ethidium bromide (EtBr) to a final concentration of 1 μg/ml with incubation during 5 min at 37 °C. Cells were then washed three times with medium without calf serum and were examined by an Olympus BX43 fluorescence microscope with a TRITC (CY3) red-orange filter. The resulting images were converted to grayscale images, which showed staining nuclei of HFb on the surface of the sample and between the PET fibers (Fig. 4).

Sequential analysis of the whole sample surface led to the conclusion that cell proliferative activity occurred in all samples. Figure 3 shows, that the greatest level of cell activity was seen in 24 h after incubation. The largest number of cells was colonized on samples of the synthetic construct with the SWCNTs-based nanocomposite coating, though the distributions of cells over all samples were inhomogeneous and did not differ greatly. There was a decrease in HFb proliferative activity at 96 h of incubation, larger numbers of cells were presented on the sample with the MWCNTs-based coating. Smaller numbers of cells for the two incubation periods were seen on the uncoated sample.

The proliferative activity of HFb cells was assessed using the photocolorimetric MTT test [12]. The MTT test consists of staining living cells with crystals of yellow tetrazole (MTT), which they reduce to purple formazan. Analysis of the optical density of the resulting medium assesses the viability of cells colonizing the constructs and, thus, its cytotoxicity.

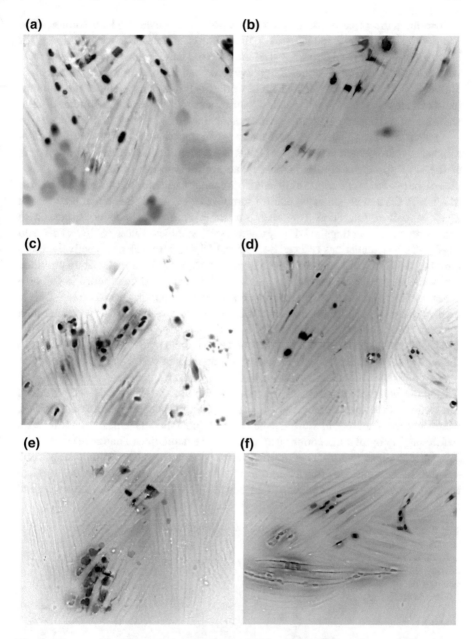

Fig. 4 Fluorescence microscope images of stained HFb colonizing the surfaces and between the PETF fibers of synthetic constructs (**a**, **b**) and constructs with nanocomposite coatings based on SWCNTs (**c**, **d**) and MCWNTs (**e**, **f**) at 24 (**a**, **c**, **e**) and 96 h (**b**, **d**, **f**) of cell incubation

The numbers of cells proliferated on samples were evaluated by adding MTT solution to the culture medium to a final concentration of 0.25 mg/ml. 3 h of incubation was followed by removal of the culture medium and washing of the cells with phosphate-buffered saline. Samples were then supplemented with 100 μl of dimethyl sulfoxide and incubated for a further 30 min. This method allows complete extraction of dye from the cells. Determination of the absolute number of HFb was run on a Scepter 2.0 counter with sensor tips.

Optical densities were measured using an Immunochem-2100 microplate photometer. Excitation was at 545 nm, which is corresponding to the first absorption peak of the MTT spectrum. Figure 5 shows mean optical densities for experimental samples of the synthetic constructs with nanocomposite coatings based on SWCNTs (2) or MWCNTs (3) and an uncoated control sample (1). Figure 5 shows that the greatest level of cell proliferation was seen on samples of synthetic constructs with nanocomposite coatings. HFb growth on samples with SWCNTs- and MWCNTs-based coatings reached 54,931 and 55,435 HFb cells, respectively, compared with 54,715 cells on the control sample. The MWCNTs-based coating provided the best cell growth. These data are clearly linked with a tendency for the cell processes of fibroblasts to attach to the most porous surface coating based on MWCNTs. Cell viability on samples bearing the SWCNTs-based coating was 2.7% greater than that on the uncoated sample, while cell viability on constructs with the MWCNTs-based coating exceeded the uncoated sample by 8%. This result points to the ability of nanocomposite coatings to provide the best cell adhesion and growth.

To study the interaction between nanocomposite implant and bone canal, we carried out a surgery to insert the implant into rabbit knee joint. Then we estimated interaction of the implant material with surrounding tissues by histological studies. Four Russian Giant rabbits were selected as biological models. Anesthesia was made before surgery by an intravenous injection. Then the depilation of hair in the area of the left and right knee joints was performed. Through bone canal 2–5 mm in length and 4.5 mm in diameter was formed in the right knee joint of each animal by surgical method. Nanocomposite implant was introduced in the bone canal, and the bone canal

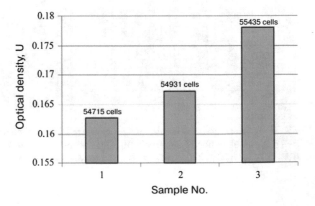

Fig. 5 Mean optical densities of samples of synthetic constructs (*1*) and constructs bearing nanocomposite coatings based on SWCNT (*2*) and MWCNT (*3*)

was filled by water-protein CNTs dispersion and treated by laser radiation with the temperature not higher than 55 °C to evaporation of the water component.

At the end of the operation, the incision was closed with sutures. The left knee joint of each animal was used as a control, and we inserted and secured there the commercial version of the traditional tape for ligament replacement with identical geometrical dimensions to experimental joint. Removing animals from the experiment was carried out at 2, 4, and 8 weeks, followed by preparation for histological research. Histological sections were made using microtome. Slice thickness was 5–10 microns. Sections of the samples were placed on a slide and stained with hematoxylin-eosin dye.

We observed bone tissue with the adjoined implanted artificial fiber material, on all histological sections. Two weeks after the implantation, there was no bone response in experimental and control samples (Fig. 6a, b). After 4 weeks, the test specimen had a moderate lobular proliferation of immature cellular fibrous tissue (Fig. 6c). In the control sample (Fig. 6d) obvious proliferation of immature cellular fibrous tissue was detected. After 8 weeks, the control sample had an obvious proliferation of immature cellular fibrous tissue with small areas of bone formation and small focal calcification (Fig. 6d). The test specimen had a proliferation of immature cellular fibrous tissue with areas of mature fibrous tissue and obvious bone formation and focal calcification (Fig. 6e).

Conclusions

A method for making knee joint ligament implants based on synthetic constructs made of PET fibers bearing nanocomposite coatings was developed. The coatings consisted of water-albumin dispersions of CNTs and were applied to the synthetic fibers using ultrasound. Coatings were fixed by laser evaporation of the liquid component of dispersion. The electric field of the irradiation promoted self-organization of the nanotubes on the scaffold. Evaporation of the liquid led to the formation of pores in the coating, and mean pore diameters for SWCNTs-based samples were 10–20 nm, compared with 40–60 nm for MWCNTs-based samples. In vitro studies of the activity of HFb cells, colonized on the surface and between the synthetic fibers, showed that a higher level of HFb proliferation was seen on samples of synthetic constructs with composite coatings based on MWCNTs. MTT test demonstrated, that cells amount in this sample was about 55,435 cells, which is a lot more than on SWCNTs-based coatings and control samples. Thus, these results lead to the conclusion that fibroblast adhesion and growth are better on the MWCNTs-based composite coatings, which has a greater pore size than the other samples.

After nanocomposite implants implantation in rabbit knee joints, there was a pronounced proliferation of immature cellular fibrous tissue with lots of mature fibrous tissue and obvious osteogenesis unlike control sample in 8 weeks. As a

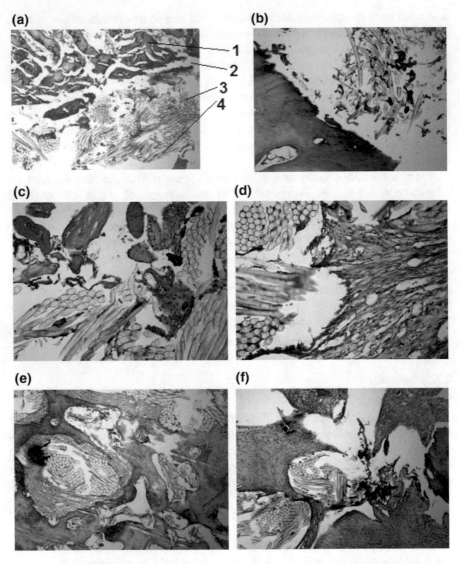

1- bone tissue; 2- empty space; 3- cross section of a synthetic fiber structure;
4- longitudinal section of a synthetic fiber structure

Fig. 6 Histological sections of nanocomposite implant in the bone canal. Magnification of ×100 (**a**, **e**, **f**), and ×200 (**b–d**). Animal No 1 (**a**), control in 2 weeks (**b**). Animal No 2 (**c**), control in 4 weeks (**d**). Animal No 3 (**e**), control in 8 weeks (**f**)

result, the studied nanocomposite implants provide high osteoconductivity and can be used for the reconstruction of the knee ligaments.

Acknowledgements Research is carried out with the financial support of the state represented by the Ministry of Education and Science of the Russian Federation. Agreement no. 14.575.21.0089 21.Oct 2014. Unique project Identifier: RFMEFI57514X0089.

References

1. Abattategi, A., Gutierrez, M.C., Moreno-Vicente, C., Hortiguela, M.J., Ramos, V., Lopes-Lacomba, J.L., Ferrer, M.L., Monte, F.: Multiwall carbon nanotube scaffolds for tissue engineering purposes. Biomaterials **29**, 94–102 (2008)
2. Fomin, A., Dorozhkin, S., Fomina, M., Koshuro, V., Rodionov, I., Zakharevich, A., Petrova, N., Skaptsov, A.: Composition, structure and mechanical properties of the titanium surface after induction heat treatment followed by modification with hydroxyapatite nanoparticles. Ceram. Int. **42**(9), 10838–10846 (2016)
3. Gerasimenko, A.Yu., Ichkitidze, L.P., Podgaetsky, V.M., Savelyev, M.S., Selishchev, S.V.: Laser nanostructuring 3-D bioconstruction based on carbon nanotubes in a water matrix of albumin. Proc. SPIE **9887**, 988725-1–988725-10 (2016)
4. Hirata, E., et al.: Development of a 3D collagen scaffold coated with multiwalled carbon nanotubes. J. Biomed. Mater. Res. B: Appl. Biomater. **90**(2), 629–634 (2009)
5. Jiang, J., Wan, F., Yang, J., Hao, W., Wang, Y., Yao, J., Shao, Z., Zhang, P., Chen, J., Zhou, L., Chen, S.: Enhancement of oseointegration of polyethylene terephthalate artificial ligament by coating of silk fibroin and depositing of hydroxyapatite. Int. J. Nanomed. **9**, 80–4569 (2014)
6. Mall, N.A., Chalmers, P.N., Moric, M., Tanaka, M.J., Cole, B.J., Bach, B.R., Paletta, G.A.: Incidence and trends of anterior cruciate ligament reconstruction in the United States. Am. J. Sports Med. **42**(10), 70–2363 (2014)
7. Martino, A., Sittinger, M., Risbud, M.V.: Chitosan: a versatile biopolymer for orthopaedic tissue-engineering. Biomaterials **26**(30), 5983–5990 (2005)
8. Mooney, D.J., Mikos, A.G.: Growing new organs. Sci. Am. **280**, 38–43 (1999)
9. Peters, M.C., Mooney, D.J.: Synthetic extracellular matrices for cell transplantation. Mater. Sci. Forum. **250**, 43–52 (1997)
10. Podgaetsky, V.M., Selishchev, S.V., Bobrinetskii, I.I., Nevolin, V.K.: Volumetric nanodesign by new laser method. Application for medical purposes. Opt. Mem. Neural Netw. (Information Optics) **17**(2), 147–151 (2008)
11. Shafizadeh, S., Jaecker, V., Otchwemah, R., Banerjee, M., Naendrup, J.H.: Current status of ACL reconstruction in Germany. Arch. Orthop. Trauma. Surg. **136**(5), 593–603 (2016)
12. Wilson A.P.: Chapter 7: Cytotoxicity and viability. In: Masters, J.R.W. (ed.) Animal Cell Culture: A Practical Approach, vol. 1, 3rd. edn. Oxford: Oxford University Press (2000)

Properties of Structural Steels with Nanoscale Substructure

T.V. Lomaeva, L.L. Lukin, L.N. Maslov, O.I. Shavrin and
A.N. Skvortsov

Abstract To increase the reliability of products, the structural integrity of structural steels are relevant scientific challenges for materials specialists all over the
world. A new direction of dealing with these challenges, i.e., making steels with
superdispersed, including nanosized, structures, was formed during the past decade.
Methods for obtaining such materials define their structural features (grain sizes,
grain boundary interface development) and strength characteristics under different
types of loading

Keywords Nanosizes · Structural steels · Nanopatterning · Processing methods
Heating · Strain · Polygonization · Strength characteristics · Thermal strain
processing

Introduction

Increasing the structural integrity of structural steel and engineering products due to
nanotechnology in the world material science become a new scientific direction
[5, p. 71; 9, p. 888; 3, p. 914].

T.V. Lomaeva · L.L. Lukin · L.N. Maslov · O.I. Shavrin · A.N. Skvortsov (✉)
FSBEI HPE Kalashnikov Izhevsk State Technical University, Izhevsk, Russia
e-mail: scv@istu.ru

T.V. Lomaeva
e-mail: lomaeva@yandex.ru

L.L. Lukin
e-mail: pmm@istu.ru

L.N. Maslov
e-mail: pmm@istu.ru

O.I. Shavrin
e-mail: shavrin@istu.ru

© The Author(s) 2018
K.V. Anisımov et al. (eds.), *Proceedings of the Scientific-Practical Conference
"Research and Development - 2016"*, https://doi.org/10.1007/978-3-319-62870-7_41

Nanopatterning strengthening effect of structural bulk metallic materials can increase the reliability and durability of engineering, energy-saving and source-saving products.

Researches carried out in this area [5, p. 71; 9, p. 888; 6, p. 546] allowed to develop methods of nanoscale structural materials production by dividing them into four groups:

- amorphous condition crystallization;
- powder metallurgy (nanopowder compacting);
- intensive plastic strain;
- various methods of nanocoating processes.

The authors think that grain size, morphology, and texture may change with respect to corresponding technological parameters of nanoscale material production process. Boundary interface volume fraction (grain boundaries and triple points) increases considerably with decreasing grain size, it has an essential influence on nanoscale material properties.

The criteria for present division into groups is only grain boundaries of materials and the effect they have on strength on structural materials. The works of Bernshtein M.L. school of thought [1, p. 211; 3, p.140] strongly indicate that the grain fine structure, formed by polygonal and cellular sub-boundaries, is the strength affecting factor even without significant grain size decrease.

Thermal Strain Processing

The size of fine structure elements, i.e., polygons and cells is determined by processing methods used for fine structure patterning. Types of thermomechanical processing from polygonal substructure patterning viewpoint can be united by a notion of thermal strain processing which becomes the fifth method of nanoscale structure patterning in structural materials.

The structure and operations used for this process, in general, may be similar to thermomechanical processing, especially for high-temperature thermomechanical processing (HTTMP) only in the set of operations, however, their parameters and some features are different.

As one can see from the literature, if the process of thermomechanical processing is performed at a conventional metallurgical production including steel making, crystallization, and multi-stage working of the finished section with quenching, then parameters of these operations being set by the desire to their optimization due to general production strategy, can provide only physical and mechanical characteristics improvement of metal.

Thermal strain process performed as a special strengthening processing the object of strengthening is a standard metal treated in accordance with the corresponding Standard GOST, i.e., a wire, hot-rolled, and calibrated steel, used as a workpiece for parts (e.g. spring wire), or some semi-finished product subjected to surface forming operations of a part, being subjected to strengthening thermal strain processing (TSP) provides improvement of its physical and mechanical characteristics due to nanoscale substructure patterning in the material of the part.

In case of thermomechanical processing, [4, p.76] performed during metallurgical production, the final operation in the processing chart is preceded by a number of technological conversions that are usually far from their optimal modes from the final result viewpoint—reaching the high-level strength characteristics.

In case of TSP, when the processing is performed as a special (separate) strengthening operation, one has to deal with metal subjected to all operations of metallurgical processing method. The TSP processing model, in this case, depends on part design, its end-use requirements and type of the workpiece used.

For a TSP object, like wire or calibrated steel, the processing model of thermal strain nanoscale substructure patterning includes basically the same list of operations. A workpiece for TSP operation to strengthen wire and calibrated steel can be:

1. hot-rolled wire or hot-rolled product;
2. cold-worked wire or calibrated steel.

Both workpiece types should be subjected to surface processing to remove the defect surface layer (decarburization, scaling removal, etc.). The quality of surface preparation is also determined by metal application area after strengthening processing.

The processing model of strengthening for wire and calibrated steel due to thermal strain nanoscale structure patterning includes the following operations:

1. fast heating of metal up to a temperature of homogeneous austenite formation (by its chemical composition), with full carbide solution.

Heating parameters (speed, temperature, holding) should provide austenite homogeneity without grain growth. The optimum is required to ensure homogeneity at the minimal size of initial austenite grain being further subjected to straining.

2. plastic metal strain. One should take into consideration the following parameters when choosing the value of strain:
   a. feasibility;
   b. loading straining model (SM) during operation of the produced part, which workpiece is being strained.
   c. the time of plastic straining.

The strain value may vary from 10–20% when using the plastic strain model taking into account SM of part operation.

Minimization of the time taken to plastic straining excluding development of annealing processes according to dynamic recrystallization model keeps the substructure nanoscale dimensionality.

The combination of three process parameters, i.e., heating temperature, strain value, and time and the time of strain holding are to provide steel polygonized substructure patterning prior to recrystallization. Under optimal combination, the substructure dispersity may correspond to nanoscale-dimensional range.

4. final operation is quenching.

The requirement to the final stage of nanoscale structure process is cooling, i.e., implementation of complete martensitic transformation.

Requirements to Parameters of Process Stages

I heating

1. temperature range 900–1000 °C;
2. speed V_{phase}°/sec 200–300.

The task is to provide disperse, homogeneous structure of chemical content.

Hence, the temperature should be minimal, while heating rate should be maximum. Primary chemical content homogenization should not contain unsolved secondary phase.

II straining

– Strain degree 10–25% by components of shear.

The task is to pattern a dislocation structure corresponding to hot–cold work dislocation structure [2, p. 20, 31] excluding recrystallization structure elements. To minimize strain value coincidence of straining models under strengthening high-temperature straining and operational loading.

III strain holding

– duration within 5–10 s.

The task is polygonal dislocation rearrangement to pattern a polygonized substructure of nanoscale dispersion.

IV cooling to quench the strained metal.

The cooling rate exceeds the critical quenching rate for the given steel.

Wire and Calibrated Steel TSP

In accordance with TSP concept, nanoscale structure patterning for structural steels can be performed on specialized equipment.

To perform TSP methods, a typical processing method variable for wire and calibrated steel strengthening. The processing method includes the following operations: initial material surface preparation, induction heating for straining, straining, strain holding, cooling (quenching) in continuous and sequential mode, and tempering.

Hot straining is performed as follows: for wire by drawing, for calibrated steel by spring setting in rotating head [8, p. 58; 7, p. 35] (Figs. 1 and 2).

The design scheme for calibrated steel with TSP production is shown in Fig. 1, it consists of a workpiece feeding device 2 to feed the workpiece 1, high-frequency heating device 3 and spring setting (SS) straining device 4, cooling sprayer 5 and a pulling mechanism 6. The deforming head consists of 3 rollers located at an angle of 120°, the head rotates at 500–700 rpm. The workpiece heating temperature before straining (900–1000 °C) and the diameter setting degree (15–25%). The workpiece being stretched after straining is cooled with water in the sprayer within controlled time to guarantee its straightness ($\Delta \leq 0.2$ mm/m) and exclude further restriking.

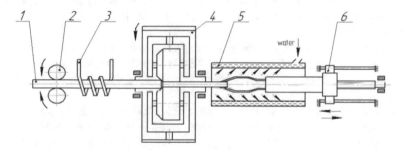

Fig. 1 Design scheme of TSP calibrated steel *1* workpiece, *2* feeding rollers, *3* induction block, *4* deforming head, *5* sprayer, *6* axial motion drive

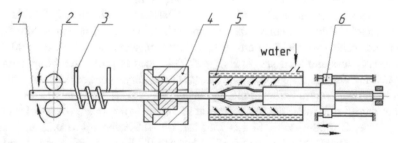

Fig. 2 Design scheme of TSP wire *1* workpiece, *2* feeding rollers, *3* induction block, *4* deforming drawing block, *5* sprayer, *6* axial motion drive

The design scheme for making TSP wire (Fig. 2) consists of the same functional assemblies as shown in Fig. 1., however, wire straining with strain degree up to 20% is performed in the hard-alloy die.

Wires made of 51ХФА (C 0.47–0.55%, Si 0.15–0.3%, Mn 0.3–0.6%, Ni up to 0.25%, S up to 0.025%, P up to 0.025%, Cr 0.75–1.1%, V 0.15–0.25%, Cu up to 0.2%) and 60C2A (Fe ~ 96%, C 0.58–0.63%, Si 1.6–2%, Mn 0.6–0.9%, Ni up to 0.25%, S up to 0.025%, P up to 0.025%, Cr up to 0.3%, Cu up to 0.2%) steel grades with finishing diameter 5 mm, and calibrated steel 9X (C 0.9%, Cr 1%) with diameter 16 mm were subjected to TSP.

Strength and plasticity characteristics were determined in accordance with Standard GOST 10446-80. Strength characteristics of steel 9X (C 0.9%, Cr 1%) for low-temperature tempering state were determined under concentrated load bending test due to low plasticity.

Fine structure of steels was studied on foils by means of electronic microscope EM-125 M under accelerating voltage of 100 kV.

Substructure images were computer processed, the crosswise size was measured— the shortest distance between elongated subgrain boundaries—elements of α-phase and carbides [8, p. 120]. The obtained results were statistically processed with the determination of the mean crosswise size of substructural element and root mean square deviation.

Electron microscopical studies showed that tested steels exhibited generally polygonal substructure patterning under TSP of the developed model—induction heating up to 900–1000 °C, strain degree 10–20% and rigidly controlled cooling, as shown in Fig. 3.

The study of wire heating temperature (920–1000 °C) made of steel 51ХФА (C 0.47–0.55%, Si 0.15–0.3%, Mn 0.3–0.6%, Ni up to 0.25%, S up to 0.025%, P up to 0.025%, Cr 0.75–1.1%, V 0.15–0.25%, Cu up to 0.2%) at TSP (Fig. 3a–c, e) showed that wire structure is of principally same type, i.e., polygonal structure of parallel elements fragmented by equiaxial cells.

The average polygon crosswise size for wire heated before straining up to 920 °C is 51.9 nm, and wire heated up to 1000 °C it is 80.5 nm.

The principal structural difference of wires heated up to 920 and 1000 °C is in the occurrence of carbides and their sizes. Carbide sizes in wires heated up to 920 °C are bigger and reach 150 nm, the number of carbides in wires heated up to 1000 °C is considerably less and their size is also less. This goes to prove that temperature of 920 °C is not enough for carbide solution. It is necessary to increase the time of wire under high temperature before straining during TSP.

The study of strain degree effect at TSP of wire made of steel 60C2A (Fe ~ 96%, C 0.58–0.63%, Si 1.6–2%, Mn 0.6–0.9%, Ni up to 0.25%, S up to 0.025%, P up to 0.025%, Cr up to 0.3%, Cu up to 0.2%) (Fig. 3e) it was found that nanoscale substructure being patterned even under strain degree 10% do not principally change under strain degree of 20%.

The structure is both polygonal with parallel polygons and cellular. Substructure element sizes correspond to nanoscale dimensionally criterion, i.e., average size is 70 nm.

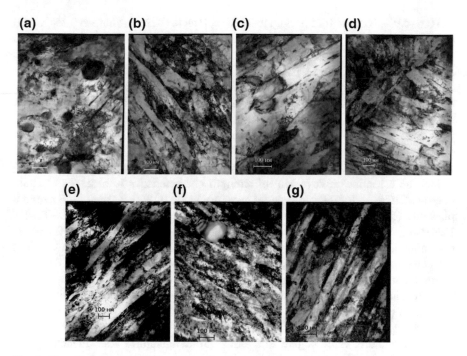

Fig. 3 Fine structure of steels after TSP. **a, b** wire made of steel 51ХФА (C 0.47–0.55%, Si 0.15–0.3%, Mn 0.3–0.6%, Ni up to 0.25%, S up to 0.025%, P up to 0.025%, Cr 0.75–1.1%, V 0.15–0.25%, Cu up to 0.2%) T_{strain} − 920 °C; **c, d** wire made of steel 51ХФА (C 0.47–0.55%, Si 0.15–0.3%, Mn 0.3–0.6%, Ni up to 0.25%, S up to 0.025%, P up to 0.025%, Cr 0.75–1.1%, V 0.15–0.25%, Cu up to 0.2%) T_{strain} = 1000 °C; **e** wire made of steel 60С2А (Fe ∼ 96%, C 0.58–0.63%, Si 1.6–2%, Mn 0.6–0.9%, Ni up to 0.25%, S up to 0.025%, P up to 0.025%, Cr up to 0.3%, Cu up to 0.2%) $\lambda = 10\%$; **f** steel 9Х (C 0.9%, Cr 1%); **g** steel 38ХС (Fe ∼ 95%, C 0.34–0.42%, Si 1.0–1.4%, Mn 0.3–0.6%, Ni up to 0.3%, S up to 0.035%, P up to 0.035%, Cr 1.3–1.6%, Cu up to 0.3%)

Ring electron diffraction pattern defines elementary α-phase cells with dominating orientation 110 with zone axis [001]. Azimuthal component of subgrain disorientation in this direction is 0.077 …. 0.096 rad.

The study of structural steels structure under TSP [38ХС (Fe ∼ 95%, C 0.34–0.42%, Si 1.0–1.4%, Mn 0.3–0.6%, Ni up to 0.3%, S up to 0.035%, P up to 0.035%, Cr 1.3–1.6%, Cu up to 0.3%) and 9Х (C 0.9%, Cr 1%)] shown in Fig. 3f, g exhibited nanoscale substructure patterning. Steel grade affects the type of substructure elements either polygonal or cellular. Steel grade 9Х (C 0.9%, Cr 1%) after TSP showed the occurrence of unsolved carbides under induction heating.

Test results of specimens made of steel 60С2А (Fe ∼ 96%, C 0.58–0.63%, Si 1.6–2%, Mn 0.6–0.9%, Ni up to 0.25%, S up to 0.025%, P up to 0.025%, Cr up to 0.3%, Cu up to 0.2%) 5 mm in diameter are given in Table 1. The effect of strain degree and tempering temperatures was studied. The check specimens were treated at conventional modes for steel 60С2А (Fe ∼ 96%, C 0.58–0.63%, Si 1.6–2%,

Mn 0.6–0.9%, Ni up to 0.25%, S up to 0.025%, P up to 0.025%, Cr up to 0.3%, Cu up to 0.2%), i.e., furnace heating temperature constitutes 860 °C and oil cooling. Both check specimens and TSP specimens were tempered at temperature values of 200, 300, 400, and 460 °C.

The check specimens after low-temperature tempering at 200 °C show brittle fracture without plasticity features (absence of necking on fractured specimens). Plasticity features appear only after tempering at temperature 300 °C, i.e., $\delta = 3\%$, $\psi = 23\%$. Values of both δ and ψ correspond to the specified requirements and are observed after tempering at 460 °C.

The study of strain degree effect showed that for the applied TSP processing model the essential improvement of strength characteristics is observed at strain degree of 10%. After tempering at temperature value of 300 °C under the required plasticity ($\delta = 6.5\%$, $\psi = 36\%$) values of s_u, $s_{0.2}$, s_{pr} are 2480, 2070, 1990 MPa. Increasing strain degree up to 20% at slightly low values of δ (4.7%) allow obtaining s_u, $s_{0.2}$, s_{pr} values of 2500, 2290, and 1980 MPa correspondingly. The increased results were obtained for other tempering temperatures. This is the result of induction heating that allows obtaining relative combinations of structures and strength characteristics during industrial technology elaboration by varying the number of parameters.

Studying the TSP methods for calibrated steel 9X (C 0.9%, Cr 1%) the effect of strain temperature, strain degree, and tempering temperatures after quenching was examined.

The check specimens were taken after conventional heat treatment used for 9X (C 0.9%, Cr 1%) steel grade. The tempering temperatures were assigned by the necessity to study steel properties at high hardness stipulated by operational conditions.

Testing metals of high hardness under tension has two obstacles: difficulty in specimen manufacturing and noninformative results due to brittle fracture.

Taking all these facts into account strength characteristics were determined at bending test loaded by a concentrated load. Bending moment M, ultimate bending strength $s_{u\ bending}$ and deflection f mm were measured.

The results of specimen bending tests are given in Table 2.

As it follows from the results given above, strength characteristics are essentially improved after TSP, both strength—M, $s_{u\ bending}$ and plasticity—f. Maximum strength increase is obtained under 10% straining at straining temperature value of 900 °C after tempering at 150 °C, i.e., characteristics of M, $s_{u\ bending}$ and f are 3270 Nm, 3930 MPa and 18.2 mm correspondingly. Check specimens after the same tempering conditions showed the values of 1860 Nm, 2150 MPa and 4.4 mm correspondingly.

When raising straining temperature up to 1000 °C strength characteristic values decrease up to M, $s_{u\ bending}$ and f to 2910 Nm, 3430 MPa and 10.0 mm correspondingly.

The effect of tempering temperatures within the studied interval showed that increasing tempering temperatures leads to increasing of all values—M, $s_{u\ bending}$ and f—after tempering at 180 °C they reach values of 4190 Nm, 5230 MPa and

Table 1 Properties of wire specimens made of steel 60C2A (Fe ~ 96%, C 0.58–0.63%, Si 1.6–2%, Mn 0.6–0.9%, Ni up to 0.25%, S up to 0.025%, P up to 0.025%, Cr up to 0.3%, Cu up to 0.2%), d = 5 mm

Processing mode	$T_{tempering}$, °C	s_u (MPa)	$s_{0.2}$ (MPa)	s_{pr} (MPa)	d (%)	y (%)	$(s_{u(TSP)} - s_{u(CHT)})/s_{u(CHT)}$ (%)	$(s_{0.2(TSP)} - s_{0.2(CHT)})/s_{0.2(CHT)}$ (%)	$(s_{pr(TSP)} - s_{pr(CHT)})/s_{pr(CHT)}$ (%)
Conventional heat treatment (CHT), $T_{heating}$ = 860 °C furnace, oil cooling	220	brittle fracture							
	300	2100	1970	1910	3	23			
	400	2000	1900	1800	4.5	40			
	460	1620	1520	1400	7.6	41			
TSP, 10%, T_{strain} = 1000 °C	220	2520	2120	2000	4.8	32			
	300	2480	2070	1990	6.5	36	18	5	4
	400	2280	2130	2010	9.2	42	14	12	12
	460	1650	1550	1420	10.1	40.1	2	2	1
TSP, 20%, T_{strain} = 1000 °C	220	2610	2330	2210	4.9	32			
	300	2500	2290	1980	4.7	37.5	19	16	4
	400	2280	2040	1960	5.7	45.1	14	7	9
	460	1640	1580	1440	10.7	42.6	1	4	3

Table 2 Properties of specimens made of calibrated steel 9X (C 0.9%, Cr 1%) (bending test)

Processing mode	$T_{tempering}$ (°C)	M (Nm)	$s_{u\ bending}$ (MPa)	F (mm)	HRC	$(s_{u\ bending\ (TSP)} - s_{u\ bending\ (CHT)})/s_{u\ bending\ (CHT)}$(%)	$(M_{(TSP)} - M_{(CHT)})/M_{(CHT)}$(%)	$(F_{(TSP)} - F_{(CHT)})/F_{(CHT)}$(%)
CHT, $T_{heating}$ = 800 °C oil cooling	100	1650	1970	3.0	64.9			
	150	1800	2150	4.4	64.7			
	180	2110	2510	7.6	62.2			
TSP, 10%, T_{strain} = 900 °C	150	3270	3930	18.2	68.1	83	81	313
TSP, 10%, T_{strain} = 950 °C	150	3040	3710	13.7	66.0	72	69	211
TSP, 10%, T_{strain} = 1000 °C	150	2910	3430	10.0	66.0	59	62	127

18.5 mm. Straining was performed at a temperature of 950 °C, the strain degree constituted 20%. However, increasing tempering temperatures reduced the HRC hardness by 4 units.

Conclusions

1. In the result of TSP according to the elaborated processing model, it was found that nanoscale substructure patterning in spring wire made of steel 51ХФА (C 0.47–0.55%, Si 0.15–0.3%, Mn 0.3–0.6%, Ni up to 0.25%, S up to 0.025%, P up to 0.025%, Cr 0.75–1.1%, V 0.15–0.25%, Cu up to 0.2%) and steel 60С2А (Fe ∼ 96%, C 0.58–0.63%, Si 1.6–2%, Mn 0.6–0.9%, Ni up to 0.25%, S up to 0.025%, P up to 0.025%, Cr up to 0.3%, Cu up to 0.2%) and for structural steel 9Х (C 0.9%, Cr 1%).
2. The TSP method used for nanoscale substructure patterning provides an improvement of strength and plastic characteristics of spring and structural steels.
3. Maximum strengthening effect is observed when comparing the strength of steels subjected to TSP and to conventional heat treatment under the same plasticity level.

Acknowledgements Research are carried out (conducted) with the financial support of the state represented by the Ministry of Education and Science of the Russian Federation. Agreement (contract) no. 14.577.21.0011 06. June 2014. Unique project Identifier: RFMEFI57714X0011.

References

1. Bernshtein, M.L.: Strain structure of metals, p. 432. Metallurgiya, Moscow (1977)
2. Bernstein, M.L., Dobatkin, S.V., Kaputkina, L.M., Prokoshkin, S.D.: Hot strain diagrams, structure and properties of steels. Metallurgia, Moscow. 544 p (1989)
3. Bernshtein, M.L., Zaymovsky, V.A., Kaputkina, L.M.: Thermomechanical treatment of steel, p. 480. Metallurgiya, Moscow (1983)
4. Kodzhaspirov, G.E., Khlusova, E.I., Orlov, V.V.: Physical simulation of thermomechanical processing and structure control of structural steels. J. "Voprosy Materialovedeniya" 3, 59 (2009)
5. Lyakishev, N.P., Alymov, M.I.: Nanoscale structural materials. Nanotechnol. Russ. 1(1–2), 71–81 (2006)
6. Machine building. Encyclopedia. Frolov, K.V. et al. (eds.) Mashinostroenie, Moscow. Manufacturing technology of composite, plastic, glass products, vol. III-6. Endorsed by Bogolyubov, V.S., p. 576/Section 6. Nanotechnologies in machine building, pp. 544–555 (2006)
7. Shavrin, O.I., Maslov, L.N., Lukin, L.L.: Influence of thermal strain strengthening on spring steel substructure and strength characteristics. Nanotechnol. Russ. 3–4, 56–61 (2016)
8. Shavrin, O.I., Skvortsov, A.N., Maslov, L.M.: Manufacturing schemes of nanoscalestructure formation in machine parts. Mach. Technol. Mater. 6, 34–37 (2015)

9. Valiev, R.Z.: Nanomaterial advantage. Nature. **419**, 887–889 (2002)
10. Wang, Y., Chen, M., Zhou, F., Ma E.: High tensile ductility in a nanostructured metal. Nature. **419**, 912 (2002)

Near-Net Shapes Al$_2$O$_3$–SiC$_w$ Ceramic Nanocomposites Produced by Hybrid Spark Plasma Sintering

E. Kuznetsova, P. Peretyagin, A. Smirnov, W. Solis and R. Torrecillas

Abstract This article describes the process and demonstrates the possibility to obtain a complex square-shaped nanostructured ceramic cutting composite by spark plasma sintering. Microstructure, mechanical, and wear properties of complex shape inserts were studied and compared with the properties of inserts which were cut from the SPS-sintered cylinder by diamond disk. Both types of inserts exhibited similar properties, meanwhile, fabrication of complex-shaped sample is less expensive and time-consuming process due to the absence of diamond disk cutting operation.

Keywords Spark plasma sintering · Complex shape · Cutting tools
Ceramic composites · Al$_2$O$_3$–SiC$_w$

E. Kuznetsova · P. Peretyagin (✉) · A. Smirnov · W. Solis · R. Torrecillas
Laboratory of Electric Currents and Sintering Technologies (LECAST),
Moscow State University of Technology "STANKIN", Moscow, Russian Federation
e-mail: p.peretyagin@stankin.ru

E. Kuznetsova
e-mail: ev.kuznetsova@stankin.ru

A. Smirnov
e-mail: a.smirnov@stankin.ru

W. Solis
e-mail: washsolis@gmail.com

R. Torrecillas
e-mail: r.torrecillas@stankin.ru; r.torrecillas@cinn.es

R. Torrecillas
Nanomaterials and Nanotechnology Research Centre (CINN),
CSIC-Universidad de Oviedo, Madrid, Spain

© The Author(s) 2018
K.V. Anisimov et al. (eds.), *Proceedings of the Scientific-Practical Conference
"Research and Development - 2016"*, https://doi.org/10.1007/978-3-319-62870-7_42

Introduction

Spark plasma sintering is a high-speed powder consolidation/sintering technology capable of processing a wide variety of conductive and nonconductive materials [1]. The main advantage of the spark plasma sintering technique is very high heating and cooling rates that allow to produce highly dense traditionally difficult-to-sinter materials. A wide range of material types such as nanostructured materials [2], functional-graded materials [3], hard alloys [4], titanium alloys [5], bioceramics [6], porous materials [7] for various applications was fabricated by spark plasma sintering.

Specimens produced by spark plasma sintering have shown improvements in microstructure (including decreased grain growth) [8] corrosion resistance [9], and mechanical properties [10], compared to conventional methods. On the other hand, wide application of the spark plasma sintering in industrial field is limited due to the extreme difficulty inobtaining near-net-shape ceramic samples. Ceramic cutting inserts are one of the examples of complex shape material. Previous research showed enhancement of endurance limit and reduction of the probability of catastrophic failure in ceramic cutting inserts produced by spark plasma sintering compared to conventional ones. This behavior was explained by the lower grain size and homogeneous microstructure of spark plasma sintering sintered ceramic cutting inserts. The main problem of sintering samples of complex shape is to design special molds and technological processes of consolidation to obtain the desired homogeneity of the microstructure and, consequently, physical and mechanical properties of the sintered sample. At present, typical spark plasma-sintered samples have a cylindrical shape and should be machined after sintering. Nevertheless, the high hardness and brittleness of ceramics make machining very difficult or even impossible.

Therefore, the solution proposed in this paper, to create a special graphite mold and sinter ceramic cutting inserts with square shape. In addition, in order to compensate thermal gradients of spark plasma sintering additional heating system was used. This hybrid system leads to enhanced sintering behavior with optimized homogeneity.

The objective of the present work was to investigate the sintering behavior of ceramic cutting inserts with square shape and study microstructure and mechanical properties of sintered samples.

Materials, Methods, and Characterization

Modeling

Numerical modeling was performed by finite element method of the SPS Al_2O_3–SiC_w samples in the form of square plates SNGN standard geometry. The temperature distribution and mechanical stresses in the sample and in the mold were obtained. Numerical modeling was performed using the software COMSOL Multiphysics.

The properties of materials—graphite, alumina, and silicon carbide were obtained from standard materials library of COMSOL Multiphysics software.

Numerical simulations were performed by solving the dual problem of thermoelectric and static loading in the COMSOL Multiphysics software. The thermoelectric task of heating sample was solved using a special module Joule heating. The task of modeling the stress–strain state of the system at the applications of static loads to the upper punch size of 100 MPa for 120 s was solved in the Solid Mechanics module of the COMSOL Multiphysics software.

Raw Materials

Al$_2$O$_3$–SiC$_w$ was used of Ceramtuff (grade HA9S) *"ready-to-press powder"*, a commercial blend of alumina (Al$_2$O$_3$) powder and 17 vol.% of silicon carbide whiskers (SiC$_w$), fabricated by the company Advanced Composite Materials, LLC (Greer, SC, USA), was chosen for the production of ceramic-graphene composites. The typical properties of HA9S after sintering by hot press at 1850 °C are presented in Table 1.

Spark Plasma Sintering

Powder densification was performed by SPS (FCT Systeme GmbH, KCE FCT-H HP D-25 SD, Rauenstein, Germany) at a maximum temperature of 1780 °C, reached under vacuum at a heating rate of 100 °C/min, and an applied pressure of 80 MPa. The final temperature and pressure were maintained for 3 min. Sintering temperature was chosen based on a previous study [11].

Table 1 Properties of the Ceramtuff blend after densification

Density (% φ_{th})	Flexural strength (MPa)	Young modulus (GPa)	Hv (GPa)	K$_{IC}$ (MPa √m)	Thermal conductivity (W/m K)	Thermal shock resistance ΔT (°C)	Coefficient of thermal expansion (10^{-6}/ °C)
99	550–700	400	20.7	7–9	35	1000	6.8

Microstructural Characterization

Scanning electron microscopy (SEM) characterization was carried out on polished down to 1 μm and thermally etched surfaces (1250 °C for 3 min) by VEGA 3 LMH (SEM Tescan, Brno, Czech Republic). The density of the sintered samples (ρ) was measured in distilled water using Archimedes' principle and was compared with the theoretical value, calculated according to the rule of mixtures.

Mechanical Properties

Vickers hardness, Hv, was measured on polished surfaces using a Vickers diamond indenter (QNess A10 Microhardness Tester, Salzburg, Austria), applying a load of 98 N and an indentation time of 10 s. The magnitude of the Vickers hardness was determined according to:

$$Hv = 0.1891 P/d^2, \tag{1}$$

where P is the applied load (in N) and d is the average length of the two diagonals (in mm). The sizes of the corresponding indentations were determined via SEM. The hardness results were averaged over 10 indentations per specimen.

Fracture toughness (K_{1c}) was measured using single edge notched beams (SENB, dimension $3.0 \times 4.0 \times 45$ mm^3). Tests were performed at room temperature, using the same testing machine applied for flexural strength determination, at a crosshead speed of 0.5 mm/min with a span of 40 mm. Specimens were notched with a diamond blade saw. The method and formulas for calculating K_{1c} have been reported elsewhere [12].

Machining Testing Conditions

The efficiency of cutting tools is determined by measuring wear occurring on the contact surface between the tool and the machined material. It is important to note that the wear observed in the tool depends on the properties of the tool material as well as the machined material, but also on wear testing conditions.

As a criterion of quality of the cutting inserts was chosen the wear of its rear surface, which occurs the turning of cylindrical samples of X8NiCrAlTi32-21 ISO 4955 heat-resistant steel.

The longitudinal turning of steel specimens was performed with the following testing conditions: cutting speed $V = 300$ m/min and traverse $S = 0.15$ mm/rev, depth of cut $t = 0.5$ mm. The maximum limit in wear on the backside of the cutting edge of ceramic materials was taken to be $h = 0.5$ mm, as higher values are considered as catastrophic wear, Outbreaks of wear plates were observed and measured with an optical microscope Zeiss discovery v12.

For comparison purpose, inserts from standard disk Al$_2$O$_3$–SiC$_w$ SPS samples were also used.

Results and Discussion

Figure 1a, b shows the FEM modeling results. Figure 1c exhibits a square cross sectional near-net shape spark plasma sintering graphite die for ceramic inserts fabrication. Figure 2 shows the SEM micrographs of a fracture surface of (a) standart disk-shaped and (b) near-net shape samples of the Al$_2$O$_3$-SiC$_w$ ceramic composites sintered by SPS.

The theoretical density for the composite of Al$_2$O$_3$ matrix with 17 vol.% of SiC whisker reinforcements was calculated, and its value is $\rho_{teor} = 3.86$ g/cm^3. The measured density of the sintered samples was 3.83 ± 0.1 g/cm^3, and this value is

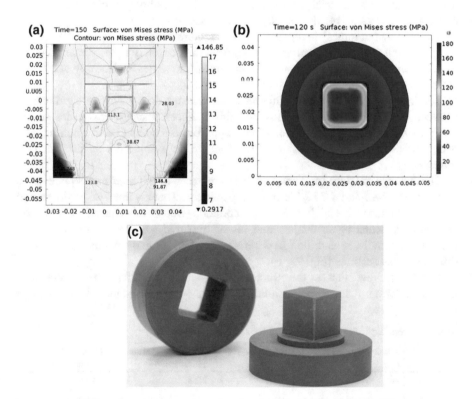

Fig. 1 The FEM modeling results and (**a**), (**b**), and near-net shape graphite mold (**c**)

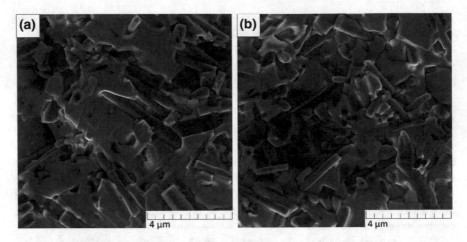

Fig. 2 SEM micrograph of the fracture surface of the standard disk Al_2O_3–SiC_w composites (**a**) and near-net shape Al_2O_3–SiC_w composites (**b**)

the 99.2% of the theoretical. This value indicate the possibility of obtaining density material by the using molds with square cross section.

The sintered samples' hardness and the fracture toughness were measured in three different regions of the cross section, from the periphery to the center, in order to determine the presence of possible mechanical properties' anisotropy caused by temperature gradients during SPS process (Table 2).

The hardness values of two different cross sections do not change significantly between the three different regions and this indicates that the properties in the volume of the material are practically the same for both cases (Table 2); moreover, it confirms a uniform distribution of the heat on the sample during the sintering process.

The average fracture toughness of the samples with circular and square cross sections are 7.82 and 7.74 MPa m$^{1/2}$ respectively. These values of hardness and fracture toughness are higher than the common values of samples obtained by conventional sintering methods. Analysis of fractured surfaces shows a uniform distribution of silicon carbide fibers throughout the volume for the samples with different cross sections (Fig. 2a, b).

Figure 2b shows that, in the structure, the grains have different sizes, but there are no abnormally large dimensions. Thereby, with a selection of the optimal

Table 2 Properties of the composites with different molds design

	Density (% φ_{th})	Hardness, Hv	Toughness, K_{IC} (MPa $\sqrt{m}$)	Life time of cutting inserts, τ, s	Wear major flank, h (mm)
Standard disk	99	2165	7.80	2.8	0.51
Near-net shape	99	2198	7.83	3	0.48

parameters for the hybrid SPS process we were able to obtain a uniform microstructure in the sintered ceramic sample with a complex shape.

The machining test shows (Table 2) that the tool life of the ceramic cutting inserts with a complex shape (square cross section) was 2.8 min, and the tool life for the inserts obtained from the sample with circular section was 3 min. These results indicate that the tool life of the sample with the square cross section is not much different from the second sample.

It should be noted that properties of the ceramic cutting inserts with complex shape were achieved by pre-selecting the optimum geometrical parameters of the graphite die for the SPS equipment, which provide a uniform temperature gradient in the volume of the sintered sample.

Conclusion

Fully dense and homogeneous Al$_2$O$_3$–SiC$_w$ square-shaped ceramic composites have been successfully fabricated by Hybrid Spark Plasma Sintering. No significant differences in hardness (2209 HV) and fracture toughness values (7.82 MPa m$^{1/2}$) were found in comparison with the composites produced from traditional SPS sintered cylindrical samples by diamond cutting. Therefore, SPS graphite mold with complex shape offers multiple advantages over traditional cylindrical forms and enables advanced ceramic materials to be intricately shaped with required accuracy and machining cost reduction.

Acknowledgements Research are carried out with the financial support of the state represented by the Ministry of Education and Science of the Russian Federation. Agreement no. 14.577.21.0089 22.June 2014. Unique project Identifier: RFMEFI57714X0089.

References

1. Risbud, S.H., Han, Y.-H.: Preface and historical perspective on spark plasma sintering. Scr. Mater. **69**, 105–106 (2013)
2. Nygren, M., Shen, Z.: Novel assemblies via spark plasma sintering. Silic. Indus. **69**, 211–218 (2004)
3. Watari, F., Yokoyama, A., Omori, M., Hirai, T., Kondo, H., Uo, M., Kawasaki, T.: Biocompatibility of materials and development to functionally graded implant for bio-medical application. Compos. Sci. Technol. **64**, 893–908 (2004)
4. Zhang, F., Shen, J., Sun, J.: The effect of phosphorus additions on densification, grain growth and properties of nanocrystalline WC–Co composites. J. Alloys Compd. **385**, 96–103 (2004)
5. Zhang, F., Weidmann, A., Nebe, J.B., Beck, U., Burkel, E.: Preparation, microstructures, mechanical properties, and cytocompatibility of TiMn alloys for biomedical applications. J. Biomed. Mater. Res. B. **94B**, 406–413 (2010)
6. Gu, Y.W., Khor, K.A., Cheang, P.: Bone-like apatite layer formation on hydroxyapatite prepared by Spark Plasma Sintering (SPS). Biomaterials **25**, 4127–4134 (2004)

7. Zhang, F., Otterstein, E., Burkel, E.: Spark plasma sintering, microstructures, and mechanical properties of macroporous titanium foams. Adv. Eng. Mater. **12**, 863–872 (2010)

8. Han, Y.H., Nagata, M., Uekawa, N., Kakegawa, K.: Eutectic Al_2O_3–$GdAlO_3$ composite consolidated by combined rapid quenching and spark plasma sintering technique. Brit. Ceram. Trans. **103**, 219–222 (2004)

9. Yue, M., Zhang, J.X., Liu, W.Q., Wang, G.P.: Chemical stability and microstructure of Nd-Fe-B magnet prepared by spark plasma sintering. J. Magnetism Magnet. Mater. **271**, 364–368 (2004)

10. Nygren, M., Shen, Z.: On the preparation of bio-, nano- and structural ceramics and composites by spark plasma sintering. Solid State Sci. **5**, 125–131 (2003)

11. Gutiérrez-González, C.F., Suarez, M., Pozhidaev, S., Rivera, S., Peretyagin, P., Solís, W., Díaz, L.A., Fernandez, A., Torrecillas, R.: Effect of TiC addition on the mechanical behaviour of Al_2O_3–SiC whiskers composites obtained by SPS. J. Eur. Ceram. Soc. **36**, 2149–2152 (2016)

12. Smirnov, A., Bartolomé, J.F.: Mechanical properties and fatigue life of ZrO_2-Ta composites prepared by hot pressing. J. Eur. Ceram. Soc. **32**(15), 3899–3904 (2002)

Development of Technical and Technological Solutions in the Field of Multilayer Graphene for Creating Electrode Nanomaterial Energy Storage Devices

N.R. Memetov, A.V. Schegolkov, G.V. Solomakho and A.G. Tkachev

Abstract The technologies of production of graphene nanoplatelets and nanocomposite materials (nano-, meso-porous carbon)/(graphene nanoplatelets, carbon nanotubes) were developed. The nanocomposite materials obtained possess specific surface area as high as 2000–3000 m^2/g and more, and exceed the existing carbon materials by parameters of surface area, pore volume, and pore size. Supercapacitors based on the nanocomposite materials developed were made and tested.

Keywords Graphene nanoplatelets · Carbon–carbon composites
Electrode materials · Supercapacitors

Introduction

Power engineering and energy technologies in the past years came to the forefront of technical and social progress. The efficiency of technologies of energy conversion, transfer, and storage become more and more important.

Modern energy-intensive electrical and electronic systems put forward strict requirements to the power supplies. A variety of equipment—from digital cameras

N.R. Memetov · A.V. Schegolkov · G.V. Solomakho · A.G. Tkachev (✉)
Federal State Budgetary Educational Institution of Higher Professional Education, Tambov State Technical University, Tambov, Russia
e-mail: nanotam@yandex.ru

N.R. Memetov
e-mail: mnr979@gmail.com

A.V. Schegolkov
e-mail: energynano@ya.ru

G.V. Solomakho
e-mail: nanotam@yandex.ru

© The Author(s) 2018
K.V. Anisimov et al. (eds.), *Proceedings of the Scientific-Practical Conference "Research and Development - 2016"*, https://doi.org/10.1007/978-3-319-62870-7_43

and portable electronic devices to electric power trains, hybrid buses, trucks, and cars—needs to collect and supply the required energy. One solution to these challenges is the use of a relatively new class of devices called supercapacitors.

The capacity of modern supercapacitors reaches thousands of Farads. The accumulation of energy in supercapacitor occurs in the process of charging due to the polarization of electric double layer at the anode–electrolyte and cathode–electrolyte interfaces. In the liquid electrolyte, electric double layer has a thickness of up to nearly 1 nm that can be attributed to the principle of operation of the supercapacitors in the area of nano-electrochemistry. Supercapacitors should have a large area distributed in the volume of the device of the porous electrodes. Porous substances such as activated carbon or foam, with a specific surface area up to $1000–3000 \ m^2/g$ are used as electrode materials in supercapacitors.

Modern supercapacitors generally have relatively low values of specific energy, so their use is limited and cannot fully meet the requirements of the market. To meet wide market needs, it is necessary to increase the specific energy up to 20–30 W-h/l, which is 2–4 times higher than existing values (5–10 W-h/l). One of the main approaches to solving this problem is to increase the specific capacitance of the electrodes and increase the operating voltage.

A significant increase in operational characteristics of supercapacitors and hybrid power sources is possible through the using of new nanostructured materials, primarily of graphene materials (particularly, graphene nanoplatelets—GNPs), carbon nanotubes, and combinations thereof, and nanocomposite materials in which conductive carbon nanomaterial is a carrier for nano-sized particles or layers of electrochemically active components possessing highly developed surface and porosity, the electrically conductive polymers, compounds of transition metals, metalloids.

The most promising materials for the creation of conductive nanocomposites can be graphene nanoplatelets (GNPs), since the special methods of their processing allow to develop their specific surface area to $2500–3000 \ m^2/g$ with retaining high electrical conductivity. Porous structure and surface area of these materials can be adjusted through the methods of processing (activation) of the GNPs themselves and through the creation of nanocomposites with other nano-sized particles or layers.

In the present work, the development of production technology of graphene nanoplatelets and their modified forms for use as electrodes of supercapacitors is described.

Method of Obtaining GNPs

A prototype of the method of obtaining the GNPs is described in [1–3]. According to this method, graphene nanoplatelets were received by exfoliation of expanded graphite materials, which were obtained by intercalation of the crystalline (natural) graphite GSM-1 with peroxo-sulfate compounds. Water or aqueous solutions of surfactants were used as a liquid medium for exfoliation. Ultrasonic treatment was used as a physical treatment method for exfoliation.

Using graphene nanoplatelets in their original form (obtained after exfoliation) as electrode materials of supercapacitors is problematic due to the fact that they do not provide the required characteristics of porosity and specific surface area. For using electrode materials as energy storage devices, the materials should possess developed surface, porosity, and high electrical conductivity. The resolving of these problems was found by creating nanocomposite materials in which graphene nanoplatelets are electrically conductive and structure-forming component, while the layer of nanoporous (preferably mesoporous) carbon on the surface of the GNPs provides the necessary characteristics of specific surface area and porosity. To obtain such nanocomposites, organic precursors of the nanoporous (mesoporous) carbon were combined with GNPs and/or carbon nanotubes and then the composition was activated to achieve the desired nanocomposite structure.

The Synthesis of the Nanocomposite GNPs/ (Nano-, Meso-Porous Carbon)

As a precursor of nanoporous or mesoporous carbon water-soluble phenol-formaldehyde resin (PFR) was used in combination with components regulating the porous structure. Water-soluble carbohydrates or their derivatives (sugar, dextrin, carboxymethylcellulose were used as regulators of the porous structure. The aqueous solution of a carbohydrate was mixed with the PFR, and then mixed with the aqueous dispersion of graphene nanoplatelets, carbon nanotubes, or combination of them. Then the PFR was cured and preliminary carbonization of the mixture at a temperature of up to 300 °C was conducted.

The next stage was the chemical activation of the pre-carbonized material in molten potassium hydroxide at temperature up to 750 °C. Then, the activated material was washed with water and treated with hydrochloric acid to remove iron impurities which arise from the walls of the steel reactor. Then the product was washed with water again, dried first in air at 110 °C, and then in flow of argon at 350 °C to remove adsorbed water. Finally material with high surface area and porosity, with good electrochemical characteristics, containing 15–20 wt% GNPs was obtained.

Taking into account the known literature data about the synergy of graphene nanoplatelets and carbon nanotubes in various applications, we have also synthesized hybrid materials containing graphene nanoplatelets, carbon nanotubes, and mesoporous carbon. The total content of nanocarbon components (GNPs + CNTs) in these materials amounted to 15–30%. The introduction of nanocarbon components improved the porous structure of the material in terms of the transport pores, accelerating the penetration of ions in solution to the surface of the mesopores which mainly creates the electric double layer.

The parameters of the surface and porosity of the synthesized materials were determined with using the device Autosorb-IQ-MP (Quantachrom) by nitrogen adsorption at 77 K. Specific surface area was determined by using multipoint BET method and the DFT method (calculation programs were integrated into the software of the device). Specific pore volumes and distribution of pores by width were determined using DFT method, assuming that the pores are slit like.

In Table 1 the parameters of the surface and porosity for several synthesized nanocomposite materials are compared. For comparison, the same data for the commercially available carbon materials for supercapacitors are presented (determined with using the same apparatus and calculation methods).

It should be noted that theoretically physical surface area for carbon materials cannot exceed the value of 2630 m^2/g (monolayer of graphene). However, in literature BET surface areas of activated carbon materials up to 3000–4000 m^2/g are often reported. This contradiction arises from the fact that the BET model is not adequate for such materials. However, it is used because it is the standard model for calculating surface area.

In Figs. 1 and 2, SEM images of the nanocomposite materials mesopores carbon/carbon nanotubes and mesopores carbon/graphene nanoplatelets are presented.

We can see from the images that the texture of the materials greatly depends on the nature of the nano-carbon component (nanotubes or graphene). One can assume that the lamellar structure of the material formed by graphene nanoplatelets will provide better diffusion of electrolyte ions in the volume of the electrode and thus provide the best power characteristics.

Table 1 Characteristics of porosity and specific surface area of carbon materials

Carbon material	Specific surface area (Multipoint BET) (m^2/g)	The specific pore volume (DFT) (cm^3/g)	The average pore width (DFT) (nm)
Materials available on the market			
Energ2 P2	1884	1.544	1.178
Energ2 V2	1217	0.601	0.899
XHAC	1865	0.916	1.178
Kuraray YP-50F	1356	0.700	0.899
Kuraray YP-80F	2058	1.043	1.126
Norit DLC Supra30	1856	0.943	0.899
Materials synthesized by our technology (laboratory labels of samples are given)			
G_157—mesoporous carbon/carbon nanotubes	3616	2.611	3.627
S_048—mesoporous carbon/graphene	3202	2.496	3.627
S_014—mesoporous carbon	2479	2.486	4.543
S_020—mesoporous carbon	2517	3.069	4.152

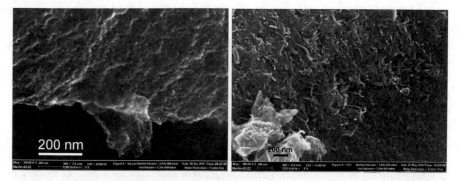

Fig. 1 SEM image of the nanocomposite material mesopores carbon/carbon nanotubes

Fig. 2 SEM image of the nanocomposite material mesopores carbon/graphene nanoplatelets

Production and Testing of Electrode Materials

The electrochemical characteristics of nanocomposite materials obtained were investigated together with "RICON" Ltd. (Voronezh, Russia). For manufacturing supercapacitor electrodes, the investigated carbon materials weighing 0.5 g were mixed with 0.05 g of the conductive filler (carbon SN-210), 1 ml PTFE suspension F-4D (40 ml of isopropyl alcohol, 60 ml of distilled water, 4.6 g of F-4D), 10 ml of isopropyl alcohol was added to the mixture, and the mixture was treated 5 min by ultrasonic apparatus USG-15 0.1/22.

Then, the mixture was dried to remove the solvents from the obtained carbon mass containing the tested nanocomposite material. The electrodes were formed by pressing the carbon mass obtained with a force of 3 tons on a metal mesh made of stainless steel and having a cell size of 0.5 mm. Before the experiment, the metal mesh was etched in the mixture of acids HF, HNO_3, H_2SO_4 (1:1:1), repeatedly washed with distilled water and isopropyl alcohol. Before the electrochemical measurements, the electrode was dried at 120 °C for 24 h to remove traces of adsorbed water.

3 M aqueous sulfuric acid and 1 M solution of tetraethylammonium tetrafluoroborate in acetonitrile (water content <10 ppm) were used as electrolytes. For effectively filling the pores of the electrode material by the electrolyte solution vacuum impregnation was used. The electrochemical testing was performed in a standard three-electrode cell. To test the materials in aqueous electrolyte platinum wire was used as an auxiliary electrode and silver chloride electrode was used as a reference electrode. For testing the samples in nonaqueous electrolyte a platinum wire was used as an auxiliary electrode and Ag/Ag+ electrode RE-7 (ALS Co., Ltd, Japan) as a reference electrode. Cyclic voltammetric plots were obtained with using Elins P-30 J potentiostat. The researching electrochemical impedance is impedancemetry Elins Z-500P. Various current measurements were performed in the frequency range from 105 Hz to 1.4×10^{-2} Hz with an amplitude of 10 mV.

Figure 3 shows the cyclic voltammetry curves with different scan speed potentials (5–100 mV/s) obtained in the sulfuric acid solution, and Fig. 4—cyclic voltammperometry in 1 M TEABF4 solution in acetonitrile (AN).

On the basis of results of CVA curves, we found specific capacitance by the formula:

$$C_{y\partial} = \frac{1}{mv(E_{\mathcal{K}} - E_{\mathcal{H}})} \int_{E_{\mathcal{H}}}^{E_{\mathcal{K}}} I(E)\mathrm{d}E$$

by integration method in Origin 8 Pro program.

The calculated data are shown in Table 2.

The obtained results prove that these nanocomposite materials are promising materials for use as electrode materials in supercapacitors. It should be noted also that with increasing the potential sweep speed, there is a significant fall in capacity,

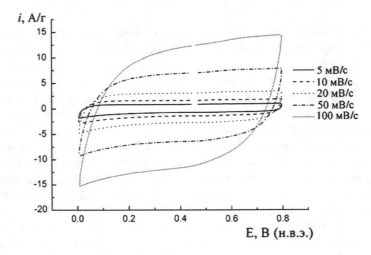

Fig. 3 Cyclic voltammperometry in 3M H_2SO_4

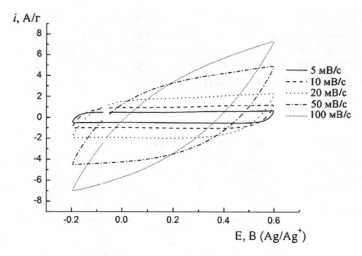

Fig. 4 Cyclic voltammperometry in 1M TEABF4 AN solution

Table 2 Specific capacitance of the investigated material in different electrolytes

Electrolyte	Specific capacity C, F/g at the potential sweep rate v, mV/c				
	5	10	20	50	100
3M H_2SO_4/water	161 ± 8	149 ± 7	137 ± 7	116 ± 8	97 ± 8
1M $TEABF_4$/AN	96 ± 3	89 ± 4	78 ± 6	50 ± 9	25 ± 7

because of lack of transport pores in the material. For full usage of highly developed surface of synthesized materials at high sweep speed (high frequency) it is necessary to modify the texture of these materials so as to create large transport pores, which can be achieved by choosing the optimal concentration of additives of nanocarbon components.

Manufacturing and Testing Supercapacitors Based on Developed Materials

For testing the supercapacitors of various of sizes with a diameter of 11 and 15 mm, height 23 and 34 mm were manufactured. We added the soot Vulcan to the developed carbon material for manufacturing electrodes of supercapacitors. The addition of the soot in the composite mixture reduces its specific capacity, but improves its mechanical and technological properties. Manufacturing of electrode material on the basis of only the materials developed and further work with him is almost impossible due to the lack of mechanical strength.

For fabricated experimental samples of supercapacitors it was measured as the internal resistance ESRAC alternating current frequency of 1 kHz, and recorded the number of included electrodes (anodes, cathodes) of the carbon material. The results for the calculation of specific power and energy density and their values are given in Tables 3 and 4, respectively.

Table 3 Experimental data and calculated specific power

Characteristics of experimental samples			Specific power P
Limit voltage $U_\text{н}$, v	ESR_{AC}, Oм	Weight M (kg)	(KW/kg)
2.7	0.015	0.00387	31.4
		0.00397	30.6
	0.010	0.00600	30.2
		0.00600	30.1
	0.0060	0.00990	30.7
		0.00980	31.0
	0.014	0.00372	35.0
		0.00379	34.3
	0.010	0.00610	29.9
		0.00600	30.4
	0.0065	0.00868	32.3
		0.00884	31.7
	0.0055	0.00884	37.5

Table 4 Experimental data and calculated specific energy

Characteristics of experimental samples			Calculated values of specific energy, E (wt h/kg)
Limit voltage U_lim (v)	Capacity C (F)	The mass of carbon in the electrode of the sample m (kg)	
2.7	7.3	0.000271	26.9
	7.6	0.000285	26.7
	14.6	0.000533	27.4
	13.0	0.000520	25.0
	25.5	0.000934	27.3
	25.9	0.000952	27.2
	7.9	0.000317	24.9
	8.1	0.000324	25.0
	16.3	0.000644	25.3
	15.0	0.000595	25.2
	24.5	0.000953	25.7
	25.4	0.000966	26.3

The specific energy of the test subject, referred to the weight of carbon in sample was calculated by the formula

$$E_{уд.} = \frac{C \cdot U^2}{2 \cdot 3600 \cdot m} \quad \text{W h/kg,}$$

C   Capacitance of the sample f
U   Limit voltage v
M   Weight of carbon in the sample kg

The specific capacity of the tested objects was calculated by the formula

$$P_{уд.} = \frac{U^2}{4 \cdot ESR_{AC} \cdot M} \quad \text{KW/kg,}$$

U Limit voltage w;
ESR_{AC} the internal resistance of the alternating current, Ом;
M the mass of sample, kg.

Conclusions

The technology of producing graphene nanoplatelets based on the method of intercalation of crystalline graphite and subsequent ultrasonic exfoliation in aqueous medium was developed.

The technology of production of nanocomposites GNPs/(nano-, meso-porous carbon) for use as electrodes in supercapacitors was developed. It is established that the introduction of GNPs in the composition of the nanoporous (mesoporous) carbon changes its structure with formation of large-pore plate-like texture, which apparently provides better diffusion of electrolyte ions in the volume of the electrode.

Electrochemical studies of the developed composite materials in aqueous and nonaqueous electrolytes were conducted and cyclic voltammetric curves were obtained, which allow calculating the specific capacitance of the carbon nanocomposites.

It was made samples of supercapacitors based on the developed materials, identified by their operating characteristics: power density of supercapacitors experimental samples was more than 30 kW/kg (in terms of carbon material), the specific energy more than 25 W h/kg.

Acknowledgements The research was conducted with the financial support represented by the Ministry of Education and Science of the Russian Federation. Agreement no. 14.577.21.0091 22 Jul. 2014. Unique project Identifier: RFMEFI57714X0091.

References

1. Melezhyk, A.V., Tkachev, A.G.: Synthesis of graphene nanoplatelets from peroxosulfate graphite intercalation compounds. Nanosyst.: Phys. Chem. Math. **5**(2), 294–306 (2014)
2. Melezhyk, A.V., Kotov, V.A., Tkachev, A.G.: Optical properties and aggregation of graphene nanoplatelets. J. Nanosci. Nanotechnol. **16**(1), 1067–1075 (2016)
3. Melezhik, A.V., Pershin, V.F., Memetov, N.R., Tkachev, A.G.: Mechanochemical synthesis of graphene nanoplatelets from expanded graphite compound. Nanotechnol. Russ. **11**(7–8), 421–429 (2016). © Pleiades Publishing, Ltd., 2016. – Original Russian Text © Melezhik, A.V., Pershin, V.F., Memetov, N.R., Tkachev, A.G., 2016, published in Rossiiskie Nanotekhnologii **11**, 7–8 (2016)

Carbon Fiber-Reinforced Polyurethane Composites with Modified Carbon–Polymer Interface

A.R. Karaeva, N.V. Kazennov, V.Z. Mordkovich, S.A. Urvanov and E.A. Zhukova

Abstract Carbon fiber-reinforced polyurethane composites were received by means of technique, which includes modification of polyurethane–carbon fiber interface. The modification was done by carbon nanotube grafting onto a surface of the fiber. A sophisticated grafting technique allowed to avoid almost inevitable grafting-induced deterioration of the fiber properties. The technique implies the introduction of an intermediate protective aluminum oxide layer. The measurement of interfacial shear strength (IFSS) was used for estimation of polymer–fiber interface properties. It was shown that IFSS doubled due to nanotube grafting. The enhancement of both thermal conductivity and mechanical properties including delamination resistance was registered for composites with the modified interface, which allows to state that the resulting materials can be considered as novel flexible composites.

Keywords Composite · Polyurethane · Carbon fiber · Carbon nanotube Interface · Mechanical properties · Thermal conductivity

A.R. Karaeva · N.V. Kazennov · V.Z. Mordkovich (✉) · S.A. Urvanov · E.A. Zhukova
Technological Institute for Superhard and Novel Carbon Materials, Moscow, Russia
e-mail: mordkovich@tisnum.ru

A.R. Karaeva
e-mail: karaevaar@tisnum.ru

N.V. Kazennov
e-mail: kazennov@tisnum.ru

S.A. Urvanov
e-mail: urvanov@tisnum.ru

E.A. Zhukova
e-mail: katyazhu@tisnum.ru

© The Author(s) 2018
K.V. Anisimov et al. (eds.), *Proceedings of the Scientific-Practical Conference "Research and Development - 2016"*, https://doi.org/10.1007/978-3-319-62870-7_44

Introduction

The carbon fiber-based (CF-based) composites have recently become a choice for the most responsible applications. Despite growing demand for such composites, many application areas are still closed for the CF-based composites due to delamination, which makes all the composite properties poorer because the load in the composites is transferred from the matrix to the fiber through the shear stress. Delamination, in turn, is determined by weak CF-to-matrix binding. This weak CF-to-matrix binding maybe improved and a better interfacial shear strength (IFSS) maybe reached through the control of the fiber–matrix interface [1, 2].

Earlier attempts to improve the situation by carbon nanotube (CNT) grafting onto the CF surface were reported in literature [1, 3, 4]. It is obvious that grafting would increase the IFSS due to CNT-assisted increase of active surface area, local stiffening and mechanical interlocking at the fiber-matrix border. However, a drastic drop in the CF tensile strength was reported in [3, 5, 6] although some authors such as [7] showed more optimistic results. The negative results are most probably induced by the CF surface erosion due to interaction with catalyst particles. Such erosion can be prevented by the introduction of an interlayer at the surface of CF as suggested in [5].

The purpose of this work was to show that a flexible CF-reinforced composite can be produced if a proper technique is elaborated for CF–polymer interface modification. Polyurethane was suggested as a matrix for the composites due to its resistance to moisture, corrosion, and wear.

Experimental

Polyurethane with value 90 of Shore A hardness was used as a matrix. That polyurethane was based on di(methylthio)toluene diamine and 2,6-toluene diisocyanate (by Smooth-on, Inc.). The so-called mid-grade polyacrylonitrile-derived carbon fibers UKN-M-12k (provided by Argon Ltd., Russia) were used in this work. The fibers were arranged in bundles of 12,000 monofilaments each (specific weight 1.75 g/cm^3, average diameter of a filament 7 μm). These fibers can be characterized by Young modulus of 220 GPa, tensile strength of 3.0 GPa, and strain-to-failure of about 1.2%. Both fibers and composites were characterized by scanning electron microscopy (SEM). A tensile testing machine "Instron 5980" was used for the measurement of mechanical properties of both CF (with 100 mN sensitive sensor) and composites (with 1 kN sensitive sensor, ASTM D3039 measurement). Every single value of stiffness or strength was received by averaging at least 20 different measurements.

A modified "microdroplet" test [3] was used for interfacial shear strength (IFSS). It implied single filament pull-out tests, i.e., a monofilament was fixed in grips, then the bottom part was immersed into polyurethane precursor. After polymer curing, the bottom part was carefully cut for obtaining samples where the length of immersed fiber was measured. Usually, this length was about 1–2 mm. The values of IFSS were calculated from a maximum load, a monofilament diameter, and the gauge length of monofilament. 20 specimens were tested for each material. These measurements were also carried out at Instron 5980.

Thermal conductivity of composites was determined in 45–50 °C temperature range through heat capacity [DSC8000 (Perkin Elmer)] and temperature conductivity (NETZSCH LFA 457/2/G MicroFlash analyser).

The fiber tows were precoated by soaking into 5% aluminum hydroxide sol. The formed hydroxide layer was dried at 300 °C and then annealed at 800 °C in Ar atmosphere. The after-annealing thickness of the formed Al_2O_3 was 70 nm (estimated by SEM). The Al_2O_3 content was 3.3% weight.

The Al_2O_3-precoated CF was impregnated with the catalyst precursor [aqueous Fe (II) acetate] solution and then went through CVD process (carbon nanotubes growth). The precursor was decomposed in H_2 at 550 °C, which provides formation of Fe nanoparticles at the surface of Al_2O_3. The atomic ratio of Fe/Al varied 1/20–1/10 along the fiber as determined by EDS spectroscopy.

The CVD process employed the He/H_2 mixture (33 mol% of H_2) as a carrier gas. Ethanol vapor was introduced into the carrier gas by means of bubbling the gas through absolute ethanol at the temperature of 25 °C. The duration of CVD process was 20 min at the temperature of 700 °C. This temperature value was chosen as an optimal value, which provides high enough CNT yield without any significant deterioration of CF mechanical properties. More details of the grafting can be found elsewhere [7].

The composites were fabricated from polyurethane and modified CF in the following way: putting a CF tow into a plastic tooling; then filling the tooling with polymer precursor; and then rolling and tolling down to the required shape. The CF weight content in the composite was 60%. As a result, a composite sample looked like a flat strip.

Results and Discussion

Modification of CF surface with carbon nanotubes resulted in significant changes in the CF surface morphology and structure. Figure 1 manifests that smooth and featureless surface of pristine fibers (Fig. 1a) becomes grafted with multiple nanotubes (Fig. 1b). The nanotubes have length from 0.1 to 5 µm, and the diameter of CNT varies 5–30 nm.

The grafted fibers did not undergo serious deterioration although some decrease (less than 10%) in tensile strength σ was still observed as shown by single filament tensile tests. Neither elastic modulus E nor failure strain ε showed any strong variations—see Table 1.

The interface shear strength results are shown in Table 2. One can see a slow increase of pulling load due to CF monofilament tightening while pulling from the

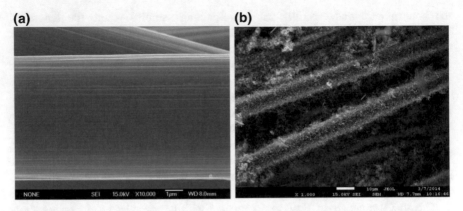

Fig. 1 Electron microscopy of carbon fibers: **a** Pristine. **b** CNT-grafted

Table 1 Single filament tensile test results

Sample	Carbon fiber		
	σ (MPa)	E (GPa)	ε (%)
Pristine	3040	220	1.2
CNT-grafted	2820	220	1.0

Table 2 Composite properties versus interfacial shear strength

Property	CF polymer	
	With pristine CF	With CNT-grafted CF
Interface properties		
IFSS (MPa)	12.0	29.3
ε (%)	3.1	5.5
Composite properties		
σ (MPa)	313	303
E (GPa)	2.44	2.86
λ/Wm^{-1} K^{-1}	0.289	0.532

polymer, then a steady increase of the pulling load because of elongation of the CF monofilament, and, finally, a failure. This is, actually, a classical behavior. Modification of the fiber–polymer interface with carbon nanotubes leads to drastic increase of IFSS by 144%.

The composites were fabricated as elongated strips with carbon fibers aligned inside. SEM of such materials is especially informative if a fracture surface is pictured just after destruction. As a result, one can make conclusions not only on the structure, but also on the destruction mechanism, which becomes obvious after analyzing SEM of fractured surfaces in Fig. 2. It can be seen that pristine fibers slip out of the matrix due to weak adhesion and very easy delamination between monofilaments and polyurethane. The surface of the pristine fibers is smooth and clean after slipping away.

(a)   **(b)**

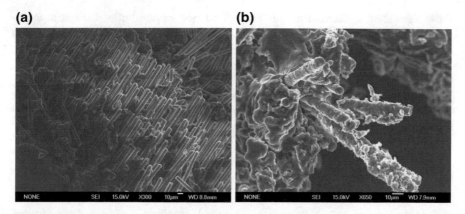

Fig. 2 SEM of tensile fracture: **a** Composite with pristine fiber. **b** Composite with nanotube-grafted fiber

The composites with nanotube-grafted fibers, in contrast, undergo fracture in a very different way. Even after fracture the modified fibers are covered with polymer even after fracture as manifested in Fig. 2b. It means that the load transferability is indeed important. CNT remains on CF surface and penetrate through residual polymer, which improves the resistance to delamination.

The powerful interlocking at the polymer–carbon interface and penetration of CNT into the polymer led to thermal conductivity jump (see Table 2). This jump happens due to better fiber–polymer heat transfer through the improved CNT-enhanced interface. It is also possible that percolation takes place due to long enough CNT, which can span through the sample.

The composite with CNT-grafted fibers manifested lower tensile strength than the pristine CF-based composite; the difference was as small as 14%. This value is very similar to the value of the monofilament strength decrease, so there is a reasonable assumption that better interlocking at the fiber–matrix interface was compensated by lower strength of the fibers proper. It is remarkable that the stiffness increased for composites with modified CF up to 23% as shown in Table 2. The increase of stiffness suggests significant load transfer after fiber modification.

Conclusions

Grafting carbon nanotubes onto carbon fiber surface can be accomplished with preservation of the pristine fiber properties due to the application of an improved grafting technique. Although some loss in tensile strength of modified composites is still observed, it can be improved in the future. It means that composites with promising properties become possible due to modification of the fiber–polymer interface and doubling the IFSS. Both composites stiffness and thermal conductivity show significant enhancement and thus afford new opportunities for future applications. The development of flexible composites with outstanding delamination resistance due to improvement of the interfacial properties of carbon fibers in elastomeric matrices is revealed by this study.

Acknowledgements Research is carried out (conducted) with the financial support of the state represented by the Ministry of Education and Science of the Russian Federation. Agreement (contract) no. 14.577.21.0094, October 25, 2014. Unique project Identifier: RFMEFI57714X0094.

References

1. Qian H., Bismarck A.-J., Greenhalgh E.S., Kalinka G., Shaffer M.S.: Hierarchical composites reinforced with carbon nanotube grafted fibers: the potential assessed at the single fiber level. Chem. Mater. **20**, 1862–1869 (2008)
2. Zhao, J., Liu, L., Guo, Q., Shi, J., Zhai, G., Song, J., Liu, Z.: Growth of carbon nanotubes on the surface of carbon fiber. Carbon **46**, 365–389 (2008)
3. An, F., Lu, Ch., Li, Y., Guo, J., Lu, X., Lu, H., He, S., Yang, Y.: Preparation and characterization of carbon nanotube-hybridized carbon fiber to reinforce epoxy composite. Mater. Des. **33**, 197–203 (2012)
4. Thostenson E.T., Li W.Z., Wang D.Z., Ren Z.F., Chou T.W.: Carbon nanotube/carbon fiber hybrid multiscale composites. J. Appl. Phys. **91**, 6034–6037 (2002)
5. Urvanov S.A., Alshevskiy Yu.L., Karaeva A.R., Mordkovich V.Z, Chernenko D.N., Beyilina N.Yu.: Carbon fiber modified with carbon nanotubes and fullerenes for fibrous composite application. J. Mater. Sci. Eng. **3**, 725–731 (2013)
6. Zhang, Q., Liu, J., Sager, R., Dai, L., Baur, J.: Hierarchical composites of carbon nanotubes on carbon fiber: influence of growth condition on fiber tensile properties. Compos. Sci. Technol. **69**, 594–601 (2009)
7. De Greef N., Magrez A., L. Couteau E., Locquet J.-P., Forro L., Seo J.W.: Growth of carbon nanotubes on carbon fibers without strength degradation. Phys. Status Solidi B. **249**, 2420–2423 (2012)

Development and Research
of Multifrequency X-ray Tube
with a Field Nanocathode

**T.A. Gryazneva, G.D. Demin, M.A. Makhiboroda, N.A. Djuzhev
and V.E. Skvorcov**

Abstract The conceptual model of X-ray source, consisting of the field-emission cathode and transmission-type thin film target, which is combined with X-ray transparent window, has been proposed. By means of numerical simulation methods, it was shown that the proposed design makes it possible to generate X-rays under the influence of an electron beam of the field-emission cathode. It is possible to get a small focal spot on the target and, therefore, a high resolution. The experimental sample of X-ray source was made and its measurement tests were conducted. The following results of the experimental studies of the sample of X-ray source were obtained: the power supply voltage is 37 kV, the power consumption is 2.77 W, the cathode current is 74.80 mA, the sample dimensions are 65 × 22 mm, the focal spot size is 439 mm, and the cathode current is about 75.2 μA after exposure to high and low temperatures.

Keywords X-ray source · X-ray · Field nanocathode · Autoemission
Emission current · MEMS · Computer modeling

T.A. Gryazneva · G.D. Demin · M.A. Makhiboroda (✉)
Science–Technology Center «Nano- and Microsystem Technique», National Research
University of Electronic Technology «MIEE», Moscow, Russia
e-mail: m.makhiboroda@gmail.com

T.A. Gryazneva
e-mail: gryazneva@ntc-nmst.ru

G.D. Demin
e-mail: demin@ntc-nmst.ru

N.A. Djuzhev
Science–Technology Center «Nano- and Microsystem Technique», Multi-Access Center
«Microsystem Technology and Electronic Component Base», Moscow, Russia
e-mail: djuzhev@unicm.ru

V.E. Skvorcov
LLC «MELZ», Moscow, Russia
e-mail: melz-zap@mail.ru

© The Author(s) 2018   421
K.V. Anisimov et al. (eds.), *Proceedings of the Scientific-Practical Conference
"Research and Development - 2016"*, https://doi.org/10.1007/978-3-319-62870-7_45

Development of Mathematical Models and Numerical Experiment

In this paper, the concept of the X-ray source is presented where the X-ray source consists of an array of field-electron emitters separated by a vacuum gap from the transmission-type thin-film metal target which is formed on the silicon membrane and operates as an X-ray window. The mathematical models of nano-sized cathode, transmission-type thin film target and silicon X-ray window, were developed. The simulation was conducted using software package COMSOL Multiphysics [1], based on which the important results were obtained needed for the optimization and further development of multifrequency X-ray tube.

In the proposed concept of X-ray source, the array of needle-type field-electron emitters with the tip radius of a few nanometers acts as a field-emission cathode [2, p. 37]. The control grid electrode general for the field-emitter array is made of a thin metal film with self-alignment holes opposite to each emitter, and it is isolated from the field-electron emitters by dielectric layer. Such an array of field emission units can be technologically realized based on silicon MEMS technology. X-window represents a square silicon membrane formed on the wafer by anisotropic etching, with the membrane's side length of 2 mm. Thin metal film, which is deposited on the inner surface of membrane, is used as a transmission-type target. Schematically, the proposed model is shown in Fig. 1.

Fig. 1 The investigated conceptual model of the X-ray source

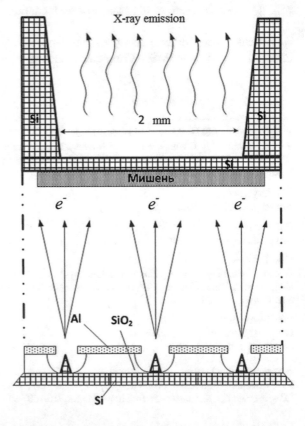

The proposed form of nanocathode is shown in Fig. 2. The main elements are silicon substrate with the array of tips formed thereon, the insulating dielectric layer, and metal film of the control grid electrode.

The geometrical dimensions of the cathode are shown in Table 1.

During numerical experiments, voltages applied to the system were varied as follows: for the grid electrode the potential varies from 100 to 150 V, and for the transmission-type target—from 0 to 40 kV correspondingly. It was found that a detectable current occurs when the voltage on the control grid electrode is 100 V, and it reaches value of about 90 nA at 150 V. Therefore, to ensure the total emission current of 75 mA, an array of ~850 nanocathodes is required. Since the linear dimensions of the single nanocathode is less than 10 μm, the allocation of that amount of cathodes within a small area is technically achievable task.

Tungsten W (Z = 74) and molybdenum Mo (Z = 42) were chosen as target materials. Trajectories of electrons that penetrate into the target material at a beam energy of 40 keV were calculated based on the Monte Carlo simulation using Win X-ray software [3, p. 1498].

It was found that the effective generation of radiation is achieved in the tungsten film with the thickness of 0.25 μm and in molybdenum film with the thickness of 0.13 μm, respectively. Experimental measurements have shown that the 1-μm thickness silicon membrane with dimensions of 1 × 1 mm can withstand pressure more than 2 atmospheres. Thus, the silicon membrane of 1-μm thickness provides not only sufficient transparency for X-rays generated by the target of almost all types of materials but also has a required mechanical strength.

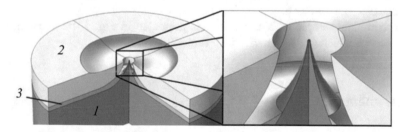

Fig. 2 Structure of nanocathode. **1** silicon (substrate); **2** metal grid electrode; **3** dielectric insulating layer

Table 1 The main dimensions of the elements of nanocathode

Characteristic	Value
Spherical radius of tip	10 nm
Radius of the hole in the grid electrode	4.5 μm
Height of tip	1 μm
The thickness of dielectric layer	0.2 μm
The thickness of the electrode film	0.9 μm
The distance between the cathode tip and target	5 mm

Fig. 3 Experimental sample of X-ray source

Table 2 Specifications of X-ray source	Operating voltage	From 30 to 40 kV
	Power consumption	Less than 3 W
	Cathode current	75 ± 5 uA
	Size of focal spot	Less than 500 μm
	Operating temperature range	From −40 to +85 °C
	Overall dimensions	70 ± 2 × 20 ± 2 mm

Preparation of Experimental Sample of X-ray Source

With the assistance of industrial partner of the project, Limited Liability Company "MELZ", an experimental sample of X-ray source was made using solutions developed in our research.

Experimental sample of X-ray source is presented in Fig. 3 the working characteristics of which are summarized in Table 2.

Experimental Results

For experimental research of test samples of field nanocathode and the sample of X-ray source special stands were fabricated (Fig. 4).

During the experimental research of the samples of nanocathode, the threshold voltage required for the field emission (the level of the detectable current ~ 10 nA) was 20–25 V at the initial state, and it was decreased after a short measurement test to less than 20 V. Thus, the measured values of current from a single cathode are more than 1.5 mA, which corresponds to a current density of about $5–10^5$ A/cm^2. Such current density of field-electron emission is considered to be very high for the operation of the emitter in a static mode and lies on the boundary of the current range, in which significant heating and degradation of the emitters occur [4, p. 22; 5, p. 387].

The following results from the experimental research of the fabricated sample of X-ray source were obtained: the power supply voltage is 37 kV, the power

Fig. 4 Stands for the experimental research: *Left* part of the photo is for X-ray source, *Right* part of the photo is for Cathode

consumption is 2.77 W, the cathode current is 74.80 μA, the sample dimensions is of 65 mm × 22 mm, the focal spot size is 439 μm, and the cathode current is about 75.2 μA after exposure to high and low temperatures.

The Potential Practical Application of Research Results

At the present time in various fields of science and technology there is an urgent need for the miniature X-ray source with low power consumption and the possibility of adjustment of radiation frequency. Such X-ray sources are in demand in a number of practical applications, such as medical technology, scientific analytical equipment, security systems and counter-terrorism systems. X-ray sources, providing a focal spot and resolution in the nanometer range, will find applications in advanced analytical equipment and technologies for the fabrication of nanostructures and new materials.

Electron and scanning probe microscopy allows mainly obtaining an image of the surface of objects, but in many respects the properties of nanomaterials are related to their internal structure.

Information about the internal structure can be obtained only with special preparation of samples such as cleavage of sample or preparation of microscopic section. Furthermore, with layer-by-layer ion beam etching of the surface, it is possible to obtain this kind of information about the objects under study. But these are destructive and expensive methods do not give full and online information about the nanomaterial. There are a number of problems with the preparation of facilities to conduct such studies, as well as with and interpretation of the results. By means of the X-ray radiation it is possible to examine the internal structure of the objects and to obtain three-dimensional images. This radiated emission practically does not interact with the objects and in many cases does not require special preparation of samples. Research can be done in the air, including a liquid phase, and in a vacuum.

In addition, at the present time research is being conducted by the combination of local exposure of the substrate to the radiation of the X-ray source and processes of ALD for the realization of additive technologies (3D printing) for the formation of topological elements of the functional layers depending on the type of reactive chemical, surface, and the wavelength of the emission.

Conclusion

As a part of the research and development of multifrequency X-ray tube with a field nanocathode, there was developed a mathematical model of nanocathode, transmission-type thin film target, and silicon X-ray window. Simulation was performed and results were obtained, which show the fundamental performance of the proposed concept. The combination of field emission cathode as a source for narrow electron beam, electron optics, and transmission-type thin film target, combined with a silicon X-ray window, opens the possibility of creating a new class of devices—scanning and multifrequency microfocus X-ray sources.

Acknowledgements Research is conducted with the financial support of the Ministry of Education and Science of the Russian Federation. Agreement No 14.578.21.0001 (RFMEFI57814X0001).

References

1. Comsol MultiPhysics [Electronic resource] URL:http://www.comsol.ru
2. Dyuzhev, N.A., Makhiboroda, M.A., Gusyм, E.E., Gryazneva, T.A., Demin, G.D.: The process flow simulation of the cathode-grid system and its emission properties. Problems of development of perspective micro and nanoelectronic systems—2016. Collected works under the General editorship of academician RAS A.L.Stempovskogo, M., Part IV. pp. 37–42. (in Russia) (2016)
3. Demers, H., Horny, P., Gauvin, R., Lifshin, E.: Microsc. Microanal. **8**(S02), 1498 (2002)
4. Dyuzhev, N.A., Makhiboroda, M.A.: Mathematical modeling of heat processes involved in field emission of nano-sized tip. In: Proceedings of universities. Electronics, No 2(88), pp. 22–26. Russia (2011)
5. Dyuzhev, N.A., Makhiboroda, M.A., Kretov, V.I., Churilin, M.N., Rudnev, V.Y.: Investigation of the thermal degradation of the silicon field-emission cathode as a two-phase system. Russian Microelectron. **41**(7), 387–392 (in Russia) (2012)

Quasicrystalline Powders as the Fillers for Polymer-Based Composites: Production, Introduction to Polymer Matrix, Properties

A.A. Stepashkin, D.I. Chukov, L.K. Olifirov, A.I. Salimon and V.V. Tcherdyntsev

Abstract Powders of icosahedral $Al_{65}Cu_{23}Fe_{12}$ and decagonal $Al_{73}Cu_{11}Cr_{16}$ quasicrystalline intermetallics were synthesized by the mechanical alloying in combination with subsequent annealing. The conditions of mechanical alloying were purposely chosen to obtain the composite materials filled by dispersed (<3 µm) quasicrystalline particles. A number of silanes were tested for the surface treatment of quasicrystalline particles in order to provide the uniform distribution of quasicrystals over the polymer melt and chemical binding with the polymer matrix and the most efficient silane type was found. The composites based on ethylene-vinyl acetate EVA, polysulphone PSU, and polyphenylene sulfide PPS were produced by the filling with quasicrystalline powders. The study of rheological characteristics has shown that high fluidity of the melt is retained, while uniform distribution of quasicrystalline particles over the polymer is provided. The data of mechanical and physical properties are reported.

Keywords Quasicrystals · Composite materials · Mechanical alloying
Silane · Extrusion · Wettability

A.A. Stepashkin · D.I. Chukov (✉) · L.K. Olifirov · A.I. Salimon · V.V. Tcherdyntsev
National University of Science and Technology "MISIS", Bld 4, Leninsky prospect,
Moscow, Russia
e-mail: dil_chukov@mail.ru

A.A. Stepashkin
e-mail: a.stepashkin@misis.ru

A.I. Salimon
e-mail: a.stepashkin@misis.ru

V.V. Tcherdyntsev
e-mail: vvch@misis.ru

© The Author(s) 2018
K.V. Anisimov et al. (eds.), *Proceedings of the Scientific-Practical Conference
"Research and Development - 2016"*, https://doi.org/10.1007/978-3-319-62870-7_46

429

Introduction

In spite of expectations based mainly on scientific intuition and more than 30 years of fundamental studies worldwide quasicrystals (QC) have not found the applications at industrial scale. In general, the quasicrystals possess the combination of some attractive properties for general uses: high hardness and wear resistance, low surface energy, low friction coefficient, significant radiation and corrosion resistance, low electrical and thermal conductivity, and unusual optical properties [1–7]. Possible areas of application of an unusual combination of thermal, electrical, and optical properties of quasicrystals were considered in [4, 7–12]. The QC phases can be used as wear-resistant, heat-shielding coatings operating at the temperatures above 450–600 °C. Recent discovery of superplasticity in the nanoscaled single quasicrystals [13] may provoke a fresh start in the search of commercially promising structural applications, since their intrinsic brittleness is considered as the main obstacle for processing, shaping and final use as the bulk product.

The application of quasicrystals in the form of powders and namely as the fillers for composites rests, perhaps, the only option for further development of structural materials which utilize the quasicrystalline form of matter. Al matrix composites reinforced with Al based quasicrystals [14–22] have demonstrated merely good combination of mechanical and tribological properties, although still far from the best conventional precipitation hardened Al alloys or Al-ceramics composites. The polymer-based composites filled with powder quasicrystalline fillers seem to be much more promising since they simultaneously improve physical, mechanical, tribological and thermal properties [23–27]. The development of these composites, besides the proper characterization, presumes the solution of a number of engineering problems: mass production of quasicrystalline powders, search for the suitable binding agents and the optimization of the filler content in order to remain the processability with the highly productive technologies. This article reports and discusses the recent progress reached along this methodology pass.

Experimental

Pure Al, Cu, Fe, and Cr (98.5–99.5 at.%) powders (d50 about 498 μm for Al and d50 about 35 μm for others) were used for the mechanical alloying (MA) of $Al_{65}Cu_{23}Fe_{12}$ and $Al_{73}Cu_{11}Cr_{16}$ compositions (in at.%) by means of water-cooled APF-3 planetary ball mill. The time of MA varied from 20 to 180 min. Loading and unloading of the powders into the vials was provided in an inert atmosphere (argon gas). Pure ethanol was used as process control agents (PCA) to prevent cold welding of powder particles to each other and to the walls of the vials and balls. The annealing of the powders obtained by mechanical alloying (precursors of the quasicrystals) was performed in a pure argon atmosphere, at the temperatures of about 600–800 °C.

Surface treatment of the obtained quasicrystalline powders was carried out using following silanes: triethoxyvinylsilane Geniosil GF 56, gamma-methacryloxypropyltrimethoxysilane Silquest A-174, and polydimethylsiloxane PDMS 200. Silanization of the quasicrystalline fillers was performed in 20% ethanol solution at a temperature of 40–45 °C for 6 h under intensive stirring.

Ethylene-vinyl acetate Evatane 28-05 and Evatane 28-40 (Arkema), PPS (DIC DSP B-100-C) and PSU (Ultrason S2010) were used as the polymer matrix.

X-ray diffraction analysis was performed using an automated X-ray diffractometer DRON-4-07, with CoKα monochromatic radiation. The spectra were analyzed using reduced Rietweld refinement.

The "thermoplastic polymer—quasicrystal" concentrates were prepared by the extrusion using Thermo Scientific HaakeMiniLab and Scientific LTE-16 laboratory extruders. The degree of granulate filling with the quasicrystals was about 10–40 wt %. The mixing procedure was performed at the temperatures of about 110–130 °C, in a HaakeMiniLab Thermo Scientific extruder. The time of the mixing was varied in the range from 3 to 15 min. The resulting melt flow of concentrates was measured with a CEAST MFT7025 rheometer, in accordance with ISO 1133: 2011 [Plastics—Determination of the melt mass-flow rate (MFR) and the melt volume-flow rate (MVR) of thermoplastics]. The measurements were carried out at a temperature of 190 °C for the following set of loads: 1.2–2.16–3.8–5 kg.

The microstructure of the samples of the powder compositions and polymer concentrates was examined with Hitachi TM-1000 and Tescan Vega3 scanning electron microscopes. The microstructure of the concentrates was studied on the chips made in liquid nitrogen.

The FTIR spectroscopy of the thermoplastic polymer—quasicrystals concentrates was performed using a Nicolet 380 IR—Fourier spectrometer (the spectral range of 4000–450 cm^{-1} with a resolution of about 0.9 cm^{-1}, the accuracy of the wave number of 0.01 cm^{-1}) in attenuated total reflection (ATR) mode.

Results and Discussion

Figure 1 summarizes the data on the phase composition of the mechanically alloyed powder mixtures after the annealing at the temperatures varying over range 25–900 °C. The quasicrystalline intermetallics are formed due to the solid state reactions and their content can reach up 90% after the annealing at 700–750 °C in the Al-Cu-Fe system and at 850–900 °C in the Al-Cu-Cr. This provides the mass production of quasicrystalline powder fillers and potentially mechanical alloying as the technology may recycle the debris, chips, and other rests of machining to reduce the costs for composite materials. The optimal powder fillers' size to reinforce the thermoplastic polymers is achieved by annealing at 700–800 °C. It consists of weakly bound agglomerates of quasicrystalline particles with the thickness of about 20–30 μm and length size of about 300–400 μm, whereas the individual particles are 1–3 μm in size—see Fig. 2. During the extrusion mixing under the action of

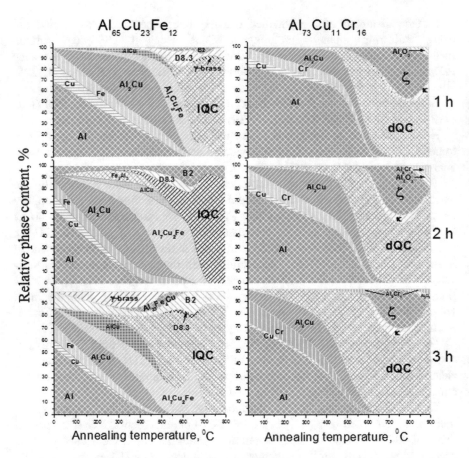

Fig. 1 Evolution of phase compositions of as-milled powders after annealing (*IQC* icosahedral quasicrystalline phase, *dQC* decagonal quasicrystalline phase, *D8.3* Al_4Cu_9, *B2* AlFe, *γ brass* $Al_{17}Cu_{33}$, *ζ* $Al_{72.3}Cu_{10.9}Cr_{16.8}$, *κ* $Al_{67.4}Cu_{14.3}Cr_{18.3}$)

shear, these agglomerates decompose into individual particles and are distributed over the polymer.

Figure 3 shows the FTIR spectra of the $Al_{65}Cu_{23}Fe_{12}$ quasicrystalline powders treated with silanes (observed spectra of icosahedral $Al_{65}Cu_{23}Fe_{12}$ phase and decagonal $Al_{73}Cu_{11}Cr_{16}$ phase treated by silanes are identical). The absorption at 790 cm^{-1} associated with Al-O symmetric stretching, the broad absorption at 1200–900 cm^{-1} range associated with Al-O-Si and Si-O-Si bonds and the absorption at 930–970 cm^{-1} associated Al-OH bond were revealed. This confirms the fact that the bond with the silane is carried out through the interaction with Al mainly and the formation of Al^{3+} ions. The strongest line of asymmetric stretching vibration of Al-O-Si and Si-O-Si is observed in the case of Geniosil GF 56. In the case of silane Silquest A174, the peaks, which are characteristic to silane, remain, but their weaker intensity reflects the weaker interaction between silane and

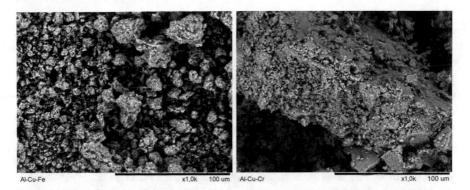

Fig. 2 The appearance of mechanically alloyed and annealed quasicrystalline powders

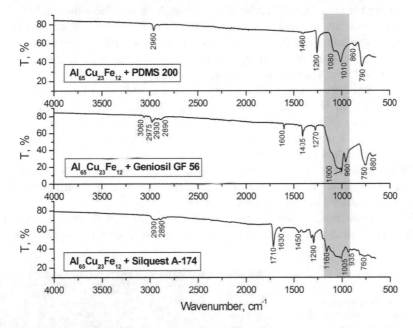

Fig. 3 FTIR spectra of the $Al_{65}Cu_{23}Fe_{12}$ quasicrystalline powders treated with triethoxyvinylsilane Geniosil GF 56, gamma-methacryloxypropyltrimethoxysilane Silquest A-174, and polydimethylsiloxane PDMS 200

quasicrystals. In case of silane PDMS 200, the strong peaks of silane are observed and their intensity remains almost unchanged. Moreover, in contrast to PDMS200 the Geniosil GF 56 and Silquest A174 silanes do form the suspension of the quasicrystalline particles and thin film on the particle surface. Finally, Geniosil GF 56 as the strongest agent was used further in the composite synthesis.

Figure 4 shows that due to the high shear stress in the extrusion machine the aggregated particles are destructed and the homogenization is completed by 12–15 min. Increasing the viscosity of the melt allows one to manage the processes of destruction of the aggregated particles and further homogenization of the filled polymer. The melt flow of the concentrates filled with the quasicrystals is slightly reduced by 12–15% with the increase of filler concentration up to 40 wt%. The differences in the melt flow of the concentrates based on icosahedral $Al_{65}Cu_{23}Fe_{12}$ phase and a decagonal $Al_{73}Cu_{11}Cr_{16}$ phase having a different shape of the particles, is statistically insignificant, when the particles are less than 3 μm in size.

As it seen from the Table 1 the filling with quasicrystals results in the increase of stiffness and almost does not affect tensile strength for PPS and PSU. One can find no influence of quasicrystals' nature—both icosahedral and decagonal intermetallics return the same shift of the properties. Both for PPS and PSU Brinell hardness rises from about 130 N/mm^2 for pure polymers to about 160 N/mm^2 for the composites containing 20 w.% of quasicrystals, that reflects the modest.

Table 1 increase of yield strength. In general, this tendency reflects insufficient binding of the matrix and the filler and does not give promising perspective for the structural applications of the composites. The tribological applications, nevertheless, may still be considered and studied in future.

The measurements reveal the increase of thermal diffusivity from 0.132 mm^2/s for pure PPS to 0.154 mm^2/s for the composite containing 20 w.% of filler regardless the nature of quasicrystals (from 0.115 mm^2/s for pure PSU to 0.137 mm^2/s for the composite with 20 w.% of filler). These data as well as the DSC measured data on heat capacity and density were used for the calculation of thermal conductivity. The increase of thermal conductivity for about 25% for PPS and for 32% for PSU was found which reflects the leading contribution of the density rise rather than heat capacity decrease.

The increment of thermal properties in combination with the expected low friction coefficient and surface energy points on the tribological applications for

Fig. 4 The uniform distribution of quasicrystalline particles in Evatane 28-05 polymer after 12 min of extrusion mixing

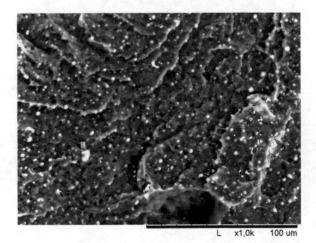

L   x1,0k   100 um

Table 1 Mechanical properties of the PPS and PSU at tension-based composites filled with quasicrystalline powders

QC composite		$Al_{65}Cu_{23}Fe_{12}$			$Al_{73}Cu_{11}Cr_{16}$		
		σ_B (MPa)	E (MPa)	ε (%)	σ_B (MPa)	E (MPa)	ε (%)
PPS		81.0 ± 1.7	2894 ± 33.8	10.3 ± 0.3	81.0 ± 1.7	2894 ± 33.8	10.3 ± 0.3
PSU		78.1 ± 1.1	2688 ± 24.8	4.3 ± 0.2	78.1 ± 1.1	2688 ± 24.8	4.3 ± 0.2
PPS	+1 w.% QC	81.2 ± 1.3	2904 ± 10.1	10.5 ± 0.6	81.4 ± 1.1	2914 ± 25.7	10.2 ± 0.1
PSU		77.5 ± 0.7	2728 ± 30.5	4.2 ± 0.2	77.8 ± 0.9	2732 ± 24.8	4.2 ± 0.3
PPS	+2 w.% QC	81.3 ± 0.8	2944 ± 4.9	10.2 ± 0.4	81.6 ± 1.2	2948 ± 16	10.0 ± 0.2
PSU		77.3 ± 1.7	2758 ± 20.4	4.1 ± 0.3	77.6 ± 1.4	2746 ± 13.5	4.0 ± 0.4
PPS	+5 w.% QC	82.1 ± 0.7	3114 + 15	9.9 ± 0.5	81.9 ± 0.5	3112 ± 17.2	9.6 ± 0.5
PSU		77.6 ± 2.4	2812 ± 31.2	3.5 ± 0.3	76.9 ± 1.6	2830 ± 22.8	3.4 ± 0.1
PPS	+10 w.% QC	80.9 ± 0.6	3144 ± 22.4	6.5 ± 0.3	80.8 ± 1.0	3148 ± 24	6.6 ± 0.2
PSU		74.0 ± 2.7	2998 ± 34.3	3.1 ± 0.3	72.9 ± 2.6	2996 ± 20.6	3.2 ± 0.4
PPS	+20 w.% QC	81.7 ± 0.5	3414 ± 20.6	6.3 ± 0.3	81.3 ± 0.3	3424 ± 22.5	6.0 ± 0.7
PSU		66.5 ± 3.5	3214 ± 35.5	2.5 ± 0.4	67.3 ± 3.3	3242 ± 35.4	2.6 ± 0.3

composite bearings or other high demanding movable mechanical parts. Additionally to higher hardness the facilitated heat release from the tribological contact will reduce the local overheating and by this will protect the matrix against temperature softening and degradation.

Conclusions

The technology for the production of quasicrystalline powder fillers and their introduction to the polymer matrix to finally gain the composite material was developed. The production technology combines the mechanical alloying of powder metals and subsequent heat treatment of the alloyed powders in an inert atmosphere. The technology for the introduction of quasicrystalline to the polymer matrix presumes the treatment with the optimal silane and high energy extrusion mixing at appropriate time in order to reach the homogeneous distribution of the powder filler. The agglomerates obtained at the production stage are crushed and the powder particles of 3 μm size are uniformly distributing in the polymer matrix. The rheological characteristics of the polymers filled with modified icosahedral $Al_{65}Cu_{23}Fe_{12}$ and decagonal $Al_{73}Cu_{11}Cr_{16}$ quasicrystals show that the melt flow of the concentrates depends on the particle size and does not depend on particle' shape.

The resulting concentrates, which are based on highly filled thermoplastic polymers and silane-treated quasicrystalline powders, make possible to synthesize the composite materials based on polyolefin material like PSU and PPS. The mechanical properties of the composites show almost no enhancement as a result of filling. The thermal diffusivity and conductivity, however, rise with the quasicrystals' content what in combination with the hardness increase opens the field of tribological applications like bearings or guiding parts.

It is believed that possible areas for application of these materials are also the composite coatings of the pipes with the reduced rate of deposition of organic deposits due to a low surface energy of the reinforcing quasicrystalline phase, which is comparable to that of pure polymers.

Acknowledgements Research are carried out with the financial support of the state represented by the Ministry of Education and Science of the Russian Federation. Agreement (contract) no. 14.578.21.0003 05.07.2014. Unique project Identifier: RFMEFI157814X0003.

References

1. Shaitura, S., Enaleeva, A.A.: Fabrication of quasicrystalline coatings: a review. Cryst Rep **52**, 945–952 (2007)
2. Samavat, F., Tavakoli, M.H., Habibi, S., Jaleh, B., Ahmad, P.T.: Quasicrystals. Open J. Phys. Chem. **2**, 7–14 (2012)
3. Dubois, J.M., Kang, S.S., Perrot, A.: Towards applications of quasicrystals. Mater. Sci. Eng. A. **179**(180), 122–126 (1994)
4. Vekilov, Y.K., Chernikov, M.A.: Quasicrystals. Phys. Usp. **53**, 537–560 (2010)
5. Koester, U., Liu, W., Hertzberg, H., Michel, M.: Mechanical properties of quasicrystalline and crystalline phases in Al-Cu-Fe alloys. J. Non Cryst. Solids **153**(154), 446–452 (1993)
6. Jenks, C.J., Thiel, P.A.: Surface properties of quasicrystals. MRS Bull. **22**(11), 55–58 (1997)
7. Huttunen-Saarivirta, E.: Microstructure, fabrication and properties of quasicrystalline Al–Cu–Fe alloys: a review. J Alloys Comp. **363**, 150–174 (2004)
8. Thiel, A., Dubois, J.M.: Quasicrystals. Reaching maturity for technological applications. Mater. Today **2**(3), 3–7 (1999)
9. Roy, M.: Formation and magnetic properties of mechanically alloyed $Al_{65}Cu_{20}Fe_{15}$ quasicrystal. J. Magn. Magn. Mater. **302**:52–55 (2006)
10. Maciá, E.: Optimizing the thermoelectric efficiency of icosahedral quasicrystals and related complex alloys. Phys. Rev. B **80**, 205103 (2009)
11. Takagiwa, Y., Kamimura, T., Hosoi, S., Okada, J.T., Kimura, K.: Thermoelectric properties of polygrained icosahedral $Al_{71-x}Ga_xPd_{20}Mn_9$ ($x = 0,2,3,4$) quasicrystals. J. Appl. Phys. **104**, 073721 (2008)
12. Dubois, J.M.: New prospects from potential applications of quasicrystalline materials. Mater. Sci. Eng. A **294–296**, 4–9 (2000)
13. Zou, Y., Kuczera, P., Sologubenko, A., Sumigawa, T., Kitamura, T., Steurer, W., Spolenak, R.: Superior room-temperature ductility of typicallybrittle quasicrystals at small sizes. Nat. Commun. **7**, 12261, DOI:10.1038/ncomms12261, www.nature.com/naturecommunications
14. Mordyuk, B.N., Prokopenko, G.I., Milman, Y.V., Iefimov, M.O., Grinkevych, K.E., Sameljuk, A.V., Tkachenko, I.V.: Wear assessment of composite surface layers in Al–6 Mg alloy reinforced with AlCuFe quasicrystalline particles: effects of particle size, microstructure and hardness. Wear **319**, 84–95 (2014)
15. Lityńska-Dobrzyńska, L., Dutkiewicz, J., Stan-Głowińska, K., Wajda, W., Dembinski, L., Langlade, C., Coddet, C.: Characterization of aluminium matrix composites reinforced by Al–Cu–Fe quasicrystalline particles. J. Alloys Comp. **643**, S114–S118 (2015)

16. Ali, F., Scudino, S., Anwar, M.S., Shahid, R.N., Srivastava, V.C., Uhlenwinkel, V., Stoica, M., Vaughan, G., Eckert, J.: Al-based metal matrix composites reinforced with Al–Cu–Fe quasicrystalline particles: strengthening by interfacial reaction. J. Alloys Comp. **607**, 274–279 (2014)
17. Laplanche, G., Joulain, A., Bonneville, J., Schaller, R., El Kabir, T.: Microstructures and mechanical properties of Al-base composite materials reinforced by Al–Cu–Fe particles. J. Alloys Comp. **493**, 453–460 (2010)
18. Qi, Y.H., Zhang, Z.P., Hei, Z.K., Dong, C.: The microstructure analysis of Al–Cu–Cr phases in $Al_{65}Cu_{20}Cr_{15}$ quasicrystalline particles/Al base composites. J. Alloys Comp. **285**, 221–228 (1999)
19. Tsai, A.P., Aoki, K., Inoue, A., Masumoto, T.: Synthesis of stable quasicrystalline particle-dispersed Al base composite alloys. J. Mater. Res. **8**, 5–7 (1993)
20. Kaloshkin, S.D., Tcherdyntsev, V.V., Laptev, A.I., Stepashkin, A.A., Afonina, E.A., Pomadchik, A.L., Bugakov, V.I.: Structure and mechanical properties of mechanically alloyed Al/Al–Cu Fe composites. J. Mater. Sci. **39**, 5399–5402 (2004)
21. Kaloshkin, S.D., Tcherdyntsev, V.V., Stepashkin, A.A., Gulbin, V.N., Jalnin, B.V., Laptev, A.I., Obrucheva, E.V., Danilov, V.D.: Mechanical alloying of metal matrix composites reinforced by quasicrystals. J. Metast. Nanocr. Mater. **24–25**, 113–116 (2005)
22. Tang, F., Anderson, I.E., Biner, S.B.: Microstructures and mechanical properties of pure Al matrix composites reinforced by Al–Cu–Fe alloy particles. Mater. Sci. Eng. A **363**, 20–29 (2003)
23. Bloom, P.D., Baikerikar, K.G., Anderegg, J.W., Sheares, V.V.: Fabrication and wear resistance of Al–Cu–Fe quasicrystal-epoxy composite materials. Mater. Sci. Eng. A **360**, 46–57 (2003)
24. Bloom, P.D., Baikerikar, K.G., Otaigbe, J.U., Sheares, V.V.: Development of novel polymer/quasicrystal composite materials. Mater. Sci. Eng. A **294–296**, 156–159 (2000)
25. Anderson, B.C., Bloom, P.D., Baikerikar, K.G., Sheares, V.V., Mallapragada, S.K.: Al–Cu–Fe quasicrystal/ultra-high molecular weight polyethylene composites as biomaterials for acetabular cup prosthetics. Biomaterials **23**, 1761–1768 (2002)
26. Kenzari, S., Bonina, D., Dubois, J.M., Fournée, V.: Quasicrystal–polymer composites for selective laser sintering technology. Mater. Design **35**, 691–695 (2012)
27. Kaloshkin, S.D., Vandi, L.J., Tcherdyntsev, V.V., Shelekhov, E.V., Danilov, V.D.: Multi-scaled polymer-based composite materials synthesized by mechanical alloying. J. Alloys Comp. **483**, 195–199 (2009)

Selection of Aluminum Matrix for Boron–Aluminum Sheet Alloys

N.A. Belov, K.Yu. Chervyakova and M.E. Samoshina

Abstract The problem of substantiating the aluminum matrix composition for obtaining the hardenable by heat-treatment boron–aluminum alloys in the form of ingots and sheet products. Alumanation materials alloyed by boron are promising radiation-resistant structural materials. Analysis of basic systems of the hardenable by heat-treatment aluminum alloys was carried out. With the use of the calculations (Thermo-Calc software) and experimental methods (including scanning electron microscopy and microprobe analysis), justified has been an unreasonableness of obtaining the boron–aluminum alloys based on magnesium-containing systems because of an active interaction of that element with boron. An experimental study has been focused on the boron–aluminum alloys based on Al–Zr–Sc (with magnesium, manganese and titanium additives) and Al–Cu systems. It was found that titanium introduction into the systems with zirconium and scandium does not assist in preventing their interaction with boron, which hampers the aluminum matrix hardening. The Al–Cu system meets the requirements best of all since copper doesn't interact with boron and does not affect on composition of the boron-containing phases. It was determined that such system allows to obtain ingots and sheet products of aluminum boron-containing alloy possessing high mechanical properties. The maximum achievable hardness on ingots and sheet products amounts to ~ 130 HV, and the tensile strength (sheet) equals to 430 MPa.

Keywords Boron–Aluminum · Sheet products · Microstructure
Phase diagrams · Heat treatment · Mechanical properties · Hardness

N.A. Belov · K.Yu. Chervyakova · M.E. Samoshina (✉)
Department of Casting Process Technology, National University of Science and Technology, Moscow, Russia
e-mail: samoshina@list.ru

© The Author(s) 2018
K.V. Anisimov et al. (eds.), *Proceedings of the Scientific-Practical Conference "Research and Development - 2016"*, https://doi.org/10.1007/978-3-319-62870-7_47

Introduction

Today, the boron-doped materials are considered as promising radiation-resistant structural materials for use in different fields: atomic engineering industry, at aerospace enterprises, in several lines of electrical engineering, instrument-making industry and electronics [1, p. 1109; 2, p. 358]. The field-performance data of this class of materials should meet very high requirements. In addition to the capability for absorbing thermal neutrons, corrosion resistance, thermal conductivity, boron–aluminum materials should possess high mechanical properties [3, p. 52; 4, p. 470].

Abroad, the volume of production with the use of the boron-filled materials measures by many tens thousands ton per year. Lately, the liquid-phase technologies of boron–aluminum production become widely (in particular, in the form of ingots meant for sheet products making thereof), since they often are essentially cheaper, technologically simpler and guarantee high mechanical properties of materials owing to the strong connection on the matrix-filler border. Some companies producing boron–aluminum use a mixing technology of powder particles of the boron-containing compounds (for example, B_4C) into liquid smelt [5]. In Russia, the wide production of boron-containing aluminum alloys is lacking until now, notwithstanding the fact that a need for them is evidently felt. Specifically, during transportation of proceeded radioactive wastes, the transit of which is currently carried out in the old-type containers made of the boron-filled steel.

Using binary Al–B alloys as an example, in the paper [6, p. 24] it is shown that borides do not make unalloyed aluminum manufacturability worse during cold rolling. However, the strength of the binary boron–aluminum alloys (without additional doping) is not high. To obtain the required mechanical properties (the strengthening ones first of all), an additional doping is necessary [7, p. 647].

It is known that one can gain the highest strength in aluminum alloys due to forming of the nano-sized particles, especially in the aging process [8, p. 208; 9, p. 42]. The main problem of obtaining the boron–aluminum with increased strength is caused by the fact that boron actively interacts with many elements, such as magnesium [10, p. 90], titanium [11], zirconium and scandium [8, p. 391; 12, p. 40]. That is why the optimum concentrations of introduced elements may essentially differ from the compositions of grade alloys, which can provide the desired level of properties on their own (that is without boron).

From the above reasoning, the objective of the present paper was a substantiation of an aluminum matrix composition for obtaining the hardenable by heat-treatment boron–aluminum alloys obtained in the form of ingots and sheet products with the strength level above 300 MPa.

Analysis of Basic Systems of the Hardenable by Heat-Treatment Aluminum Alloys

The main basic systems of the hardenable by heat-treatment aluminum alloys are Al–Cu (reinforcing phase Al_2Cu—Θ', Θ''), Al–Cu–Mg (reinforcing phase Al_2CuMg—S'), Al–Mg–Si (reinforcing phase Mg_2Si—$ß'$, $ß''$), Al–Zn–Mg (reinforcing phase $Al_2Mg_3Zn_2$—τ', τ'' and $MgZn_2$ η', η''), Al–Mg–Si–Cu (reinforcing phase $Al_5Cu_2Mg_8Si_6$—Q'), Al–Zn–Mg–Cu (reinforcing phase AlCuMgZn—T and AlCuMgZn—M), Al–Zr–Sc (reinforcing phase $Al_3(Zr,Sc)$—$L1_2$).

Inasmuch as high temperatures of the smelt are required for the boron–aluminum alloys preparing [6, p. 24], the systems containing zinc and magnesium seem to be unwanted because of significant losses during the melting process. Interaction of this element with boron can also be considered as a weakness of the systems with magnesium in composition. In particular, in the paper [10, p. 90] it is shown that achievement of necessary reinforcement in a boron-containing alloy based on Al–Mg–Si–Cu matrix required the doubled magnesium concentration in comparison with 6xxx grade alloys. Nevertheless, alloys based on this system are used in a number of developments (for example, in [13]).

In Fig. 1 isothermal sections of Al–B–Mg and Al–B–Cu systems at 1100 °C are represented. It is known that there is a continuous series of solid solutions between AlB_2 and MgB_2 borides in an Al–B–Mg system [14, p. 92]. Therefore, an $(Al,Mg)B_2$ compound with particles of adverse needle-shaped form arises even at small magnesium concentrations, as shown in Fig. 1a [6, p. 24]. On the other side, copper is completely in the melt, as it appears from Fig. 1b. Hence, only two basic systems remain for obtaining the hardenable by heat-treatment boron–aluminum alloys, namely: Al–Cu and Al–Zr–Sc (+Mg, Mn). The selection of the latter is based on the paper [4, p. 470].

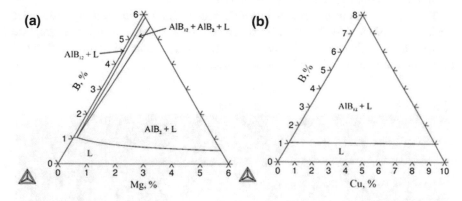

Fig. 1 Isothermal sections of ternary systems at 1100 °C: Al–B–Mg (**a**) Al–B–Cu (**b**) (calculation by Thermo-Calc software)

Experimental Procedures

The main subjects of experimental investigation were the boron–aluminum alloys based on Al–Zr–Sc (+ additives) and Al–Cu matrix systems.

Alloys have been prepared based on high-purity aluminum A99 (99.99%Al) and Al–5%B master alloy. Copper and magnesium have been introduced in pure form (99.9%Mg and 99.9%Cu respectively), titanium has been introduced in the form of T80F20 alloying tablets, the resting elements have been introduced as Al–2%Sc, Al–10%Zr, Al–10%Mn master alloys. Melting has been carried out in a graphite-fireclay crucible at the temperature of 900–950 °C in a RELTEK induction furnace, which provides an intensive melt mixing, required to exclude a possibility of refractory boron-containing particles deposition. The master alloy and melting technology have been selected taking into account the results of the previously fulfilled experiments [6, p. 24]. The melt was poured into a graphite molds to obtain flat ingots of 40 × 120 × 200 mm in size, Fig. 2a. Samples for structural investigations have been cut thereof. Later on, the ingots have been treated by strain processing to a thickness of 0.3 mm, Fig. 2b [6, p. 24].

Experimental alloys have been investigated both in cast condition and after thermal treatment, carried out with the use of a SNOL 8,2/1100 muffle electric furnace and a SNOL 58/350 low-temperature laboratory electric furnace.

Polished sections were prepared by mechanical polishing. Primary microstructure analysis of samples has been fulfilled on an Axio Observer MAT optical microscope, whereas the detailed metallographic research has been conducted on a TESCAN VEGA 3 scanning electron microscope (SEM). The TESCAN microscope, completed by an energy dispersive microanalyser device manufactured by Oxford Instruments and Aztec software, has also be used for the electro-microprobe analysis (EMPA).

The Thermo-Calc program (with a TCAL4 database) has been used for calculating phase composition of the systems [15].

The hardness has been measured on a NEMESIS 9000 multi-purpose hardometer made by INNOVATEST. In order to determine mechanical properties (rupture strength UTS, yield stress 0.2YS and elongation El), an uniaxial tension testing has been conducted on a Zwick Z250 tearing machine.

(a)   **(b)**

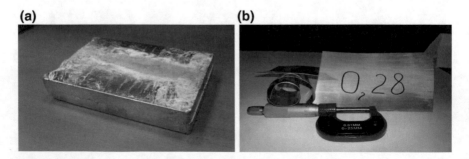

Fig. 2 General form: ingot (**a**), sheet products (**b**)

Results and their Examination

Composition of alloy Al–2%Mn–0.25%Zr–0.1%Sc has been chosen relying on papers [8, p. 369; 16; 17] its heat resistance and high mechanical properties of Al–2%Mn–0,25%Zr–0,1%Sc basic alloys are conditioned by the fact that the doping elements form Al_6Mn and $Al_3(Zr,Sc)$ dispersoids, possessing high thermal stability. However, as it follows from the paper [4, p. 470], zirconium and scandium interacts with boron. To avoid losses of these elements, titanium has been introduced, suggesting its interaction with the boron for a TiB_2 compound formation as a result. An Al–4%Mg–1%Mn–2%B–1.5%Ti–0.3%Sc alloy containing scandium only has been complimentary examined after both introducing magnesium to guarantee an additional reinforcement of alloy due to the rise of the solid solution hardness and the scandium content increasing for providing a dispersion hardening.

In the alloys microstructure primary crystals of zirconium with manganese are distinguished. In addition to the aluminum solid solution, presence of ZrB_2, TiB_2, and $Al_6(Fe,Mn)$ phases has been revealed by X-ray phase analysis. In spite of titanium presence in the composition, zirconium interacts with the boron. Studies of composition of aluminum matrix of the melt have shown that it practically does not contain other elements except Mn. Obviously, high mechanical properties and heat resistance could not be achieved with such a structure by means of zirconium additives.

It is evident from the theoretical calculation of quantitative phase analysis of the Al–2%Mn–0.25%Zr–0.1%Sc–1.5%Ti–2%B alloy that all doping components interact with the boron except scandium. Only manganese and noninteracting with the boron scandium are resting in an aluminum solid solution at the temperature of 655 °C.

Measuring hardness of the samples under consideration has shown that the basic alloy gets an increased hardness after heterogenizing annealing, though alloys with the boron additives do not become reinforced. This is connected with the boron interaction with zirconium and scandium, which leads to the lack of these elements in a quantity required to form the reinforcing phases.

Thus, the introduction of small addition of boron negatively effects on capability of alloys, containing zirconium and scandium, for a precipitation strengthening. In [4, p. 470] there is considered a process of boron–aluminum obtaining when the boron has been introduced in the form of B_4C, with the increased zirconium and scandium content. The obtained result meets the requirements upon high mechanical properties, but the redundant introduction of zirconium and scandium results in a rise in the cost of the produced material.

Hence, the Al–Cu system rests the only one to be considered. It is evident from the calculations of the quantitative phase analysis of the Al–6%Cu–2%B alloy that three phases are observed in the microstructure, but for all that copper does not interact with boron and is contained in a solid aluminum solution (or smelt) and in an Al_2Cu compound.

Study on the Al-6%Cu-2%B alloy microstructure shows a uniform distribution of AlB_{12} boride particles, crystals of which do not exceed 30 μm (Fig. 3a).

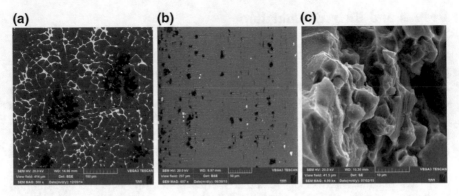

Fig. 3 Images of microstructure of Al-6%Cu–2%B alloy on a scanning electron microscope: ingot (**a**) Sheet products (**b, c**) Polished samples (**a, b**) Fractogram (**c**)

The needle-shaped impurities of AlB_2 are observed in a small proportion (as in a source master alloy [6, p. 24]). In the structure, there are revealed light Al_2Cu streaks of eutectic origin, which are well observable in fractograms (Fig. 3c).

Annealing at 540 °C doesn't affect the morphology and composition of borides. The main part of Al_2Cu from a non-equilibrium eutectic has been dissolved in (Al); the remaining impurities have been shaped into a globular form.

Strengthening level has been estimated on the ingots after applying various heat treatment conditions. As one can see in Fig. 4a, the maximum value of hardness is observed at the aging temperature of 180–210 °C. With the temperature increase, a softening (or overaging) takes place. Studying on the strengthening level on the sheet products (Fig. 4b) has confirmed the results obtained on the ingots.

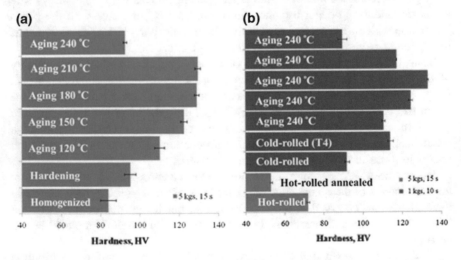

Fig. 4 Ingots (**a**) and sheet products (**b**) of Al-6%Cu-2%B alloy hardness dependence of a heat-treatment mode

Table 1 Mechanical properties of the Al-6%Cu–2%B alloy sheet products

Aging temperature	0.2YS, MPa	UTS, MPa	El, %
T4 (hardening and natural aging)	241±6	368±5	10.2
120 °C, 3 h	255±8	357±9	7.8
150 °C, 3 h	287±6	401±11	8.9
180 °C, 3 h	307±2	430±14	9.5
210 °C, 3 h	270±4	379±8	8.4
240 °C, 3 h	206±3	289±13	6.4

Testing on uniaxial tension has showed high strength properties (Table 1), achievable by forming the nanoscale phases as a result of heat treatment.

Conclusions

Systems of the hardenable by heat-treatment aluminum alloys have been analyzed with reference to obtaining the boron–aluminum alloys on their base. There were indicated the weaknesses of traditional systems containing magnesium and zinc as well as of those containing zirconium and scandium.

It has been found experimentally that introducing zirconium and scandium additives to the boron–aluminum alloys is inexpediently for these elements form primary crystals, which practically completely remove them from the composition of the solid aluminum solution.

It was shown that the most promising matrix which can be used as a base for obtaining the boron-aluminum alloys is an Al–Cu system, since copper does not interact with boron and that permits to achieve the same hardening as that of the grade alloys of a AA2219 type.

Using a model Al–6%Cu–2%B alloy as an example, it was shown that the alloying system under review allows reaching a combination of high manufacturability during rolling and high mechanical properties, including that after the heating at 210 °C.

Acknowledgements This article was written within the implementation of the Agreement on subsidies between the National University of Science and Technology "MISiS" and the Ministry of Education and Science of the Russian Federation within the framework of realization of the Federal Target Program "Research and Development in the Priority Directions of Progress of the Scientific and Technological Complex of Russia for 2014–2020", approved by the Decree of the Government of the Russian Federation of November 28, 2013, No. 1096. Agreement No. 14.578.21.0004. Unique identifier of the project RMEF157814X0004.

References

1. Eichler, J., Lesniak, C.: Boron nitride (BN) and BN composites for high-temperature applications. J. Eur. Ceram. Soc. **28**, 1105–1109 (2008)
2. Peng, Z., Yuli, L., Wenxian, W., Zhanping, G., Baodong, W.: The design, fabrication and properties of B₄C/Al neutron absorbers. J. Nucl. Mat. **437**, 350–358 (2013)
3. Mohantya, R.M., Balasubramaniana, K., Seshadrib, S.K.: Boron carbide-reinforced alumnium 1100 matrix composites: fabrication and properties. Mat. Sci. Eng. A. **498**, 42–52 (2008)
4. Lai, J., Zhang, Z., Chen, X.-G.: The thermal stability of mechanical properties of Al–B4C composites alloyed with Sc and Zr at elevated temperatures. Mat. Sci. Eng. A. **532**, 462–470 (2012)
5. Skibo, M.D., Schuster, D.M., Bruski, R.S.: Apparatus for continuously preparing castable metal matrix composite material. Patent USA, No. 5531425. F27D27/00, B22D11/11, B01F7/16, C22C32/00, F27D3/00, C22C1/10, B22D1/00, C22C1/00. Patent holder: Alcan Aluminum Corporation. Asserted 07.02.1994. Published 02.07.1996
6. Samoshina, M.E., Belov, N.A., Alabin, A.N., Chervyakova, K.Yu.: Struktura i mekhanicheskiye svoystva listovogo prokata iz splava Al-3% B, poluchennogo zhidkofaznym metodom (Structure and mechanical properties of alloy Al—3% B flats, obtained by liquid-phase method). Tsvetn. Met. Non-Ferrous Met. **10**, 19–24 (2015)
7. Ömer, S., Ramazan, K.: Production and wear properties of metal matrix composites reinforced with boride particles. Mat. Des. **51**, 641–647 (2013)
8. Belov, N.A.: Fazovyy sostav promyshlennykh i perspektivnykh alyuminievykh splavov (Phase Composition of Industrial and Prospective Aluminium Alloys). Publishing House of "MISiS", Moscow, (2010). 511 p
9. Mondolfo, L.F.: Struktura i svoystva alyuminievykh splavov (Aluminium Alloys: Structure and Properties) Translated from English. Moscow : Metallurgiya, (1979), 640 p
10. Kurbatkina, E.I., Belov, N.A., Alabin, A.N., Sidun, I.A.: Osobennosti plavki i litiya bor-soderzhashchikh alumomatrichnykh kompositov na osnove splavov 6xxx serii (Peculiarities of melting and casting of boron-containing aluminum-matrix composites based on 6xxx alloys). Tsvetn. Met. Non-Ferrous Met. **1**, 85–90 (2015)
11. Chen, X.-G., Dube, G., Steward, N.: Neutron absorption effectiveness for boron content aluminum materials. Patent US, No. 20080050270. G21F 1/08, C22B 21/00, C22C 21/00, C22C 32/00. Assertes 21.04.2005. Published 12.06.2007
12. Alabin, A.N., Belov, N.A., Tabachkova, N.Yu., Akopyan, T.K.: Heat resistant alloys of Al–Zr–Sc system for electrical applications: analysis and optimization of phase composition. Non-ferrous Met. **2**, 36–40 (2015)
13. Aruga, Y., Kajihara, K., Sugizaki, Y.: Aluminum base alloy containing boron and manufacturing method thereof. Patent US. No. 7125515. B22D 30/00, C22F 1/04. Asserted 15.04.2003. Published 24.10.2006
14. Belov, N.A., Samoshina, M.E., Alabin, A.N., Chervyakova, K.Yu.: Vliyanie medi i magniya na strukturu i fazovyi sostav slitkov boraluminiya (Copper and magnesium effect on structure and phase composition of boron-aluminum ingots). Met. Met. **1**, 86–92 (2016)
15. Information at www.thermocalc.com
16. Alabin, A.N., Belov, V.D., Belov, N.A., Mishurov, S.S.: Termostoykiy liteynyi aluminievyi splav (Heat-resistant casting aluminum alloy). Patent RF, No. 2478131. IPC B82B 3/00, C22C 21/06. Patent holder: National University of Science and Technology "MISiS". Asserted 29.10.2010. Published 27.03.2012
17. Belov, N.A., Alabin, A. N. Termostoykiy splav na osnove aluminiya i sposob polucheniya iz nego deformirovannykh polufabrikatov (Aluminum base heat-resistant alloy and strained half-finished product manufacturing method thereof). Patent RF, No. 2446222. IPC C22C 21/14, C22F 1/057. Patent holder: National University of Science and Technology "MISiS". Asserted 29.10.2010. Published 27.03.2012

Features of Carbide Precipitation During Tempering of 15H2NMFA and 26HN3M2FA Steels

S.V. Belikov, V.A. Dub, P.A. Kozlov, A.A. Popov, A.O. Rodin,
A.Yu. Churyumov and I.A. Shepkin

Abstract Thermodynamic calculation of the equilibrium phase composition and evolution of the size and composition of carbide particles was carried out using Thermo-Calc and TC-Prisma 2.0 in order to identify the optimal modes of the final heat treatment for 15H2NMFA and 26HN3M2FA steels. The formation of structure and mechanical properties complex of the investigated steels was experimentally studied after tempering. The model describing the precipitation and growth of carbide particles was suggested based on the experimental results. This model will be used as part of the developed control complex of thermodynamic and kinetic conditions for the formation of micro grains and nano-sized hardening phases.

Keywords Tempering · Carbides · Thermo-Calc · Sem

S.V. Belikov (✉) · A.A. Popov
Ural Federal University, Ekaterinburg, Russia
e-mail: s.v.belikov@urfu.ru

A.A. Popov
e-mail: a.a.popov@urfu.ru

V.A. Dub · A.O. Rodin · A.Yu. Churyumov
NUST MISiS, Moscow, Russia
e-mail: rodin@misis.ru

A.Yu. Churyumov
e-mail: churyumov@misis.ru

P.A. Kozlov · I.A. Shepkin
RF State Research Centre JSC "RPA "CNIITMASH", Moscow, Russia
e-mail: paul.kozlov@gmail.com

I.A. Shepkin
e-mail: shtepkin@mail.ru

K.V. Anisimov et al. (eds.), *Proceedings of the Scientific-Practical Conference "Research and Development - 2016"*, https://doi.org/10.1007/978-3-319-62870-7_48

449

The aim of the present study is to improve the performance of responsible parts of power engineering made from alloyed structural steels such as 15H2NMFA (increasing of the yield strength at 350 °C is not less than 10% compared to the specifications 0893-013-00212179-2003: not less than 440 MPa; reduction of a brittle-ductile transition temperature to −55 °C); and 20–30CrNiMoV (reduction of differences in the properties of the center and the periphery of the drum forging Ø > 900 mm at least to 20%). Science-based correction of metal products manufacturing technology with a large mass on the basis of thermodynamic and kinetic conditions control for the formation of micro grains and nano-sized hardening phases should provide a reduction of production costs (reduction of the total time of the manufacturing cycle up to 20%, including the heat treatment up to 30%; reduction of charge materials weight characteristic up to 10…15% (relative) as a result of optimization of an alloying elements content). The physical and mathematical model used for technology optimization should allow calculating the phase composition of selected steels, microstructure evolution, and grain size depending on the parameters of technological influences. The work was carried out by a consortium consisting of Ural Federal University, Ekaterinburg, NUST "MISiS" and JSC RPA CNIITMASH, Moscow, an industrial partner of the project—JSC "OMZ-Special Steels", St. Petersburg.

At the first stage of the present work advanced compositions of steels were selected and quantitative estimation of the steels phase composition of Fe–Cr–Ni–Mo–V–Mn–Si–C system in the temperature range of 400–1600 °C was made by method of numerical thermodynamic simulation. Table 1 shows the chemical composition of investigated steels.

The software for thermodynamic simulation Thermo-Calc (TCW 5.0) with a database of steels thermodynamic properties TCFE6.0 was used for the tests mentioned above.

The input data for the calculation in the Thermo-Calc:

– The chemical composition of the simulated steels (11 compositions with average values of alloying elements in the range of grade and with the content of ferrite and austenite forming elements located at the top or bottom of the grade composition);

Table 1 The chemical composition of investigated steels

Steel grade	Weight content of chemical elements (%)						
	C	Si	Mn	Cr	Ni	Mo	
15H2NMFA-A	0.15–0.18	0.15–0.25	0.30–0.60	1.8–2.3	1.30–1.50	0.5–0.7	V 0.10–0.12 S ≤ 0.004 P ≤ 0.004 As ≤ 0.010 Cu ≤ 0.10 Co ≤ 0.03
26HN3M2FA	0.22–0.36	≤0.3	≤0.4	2.4–3.8	3.4–3.8	0.2–0.7	V 0.05–0.15 S ≤ 0.01 P ≤ 0.01

- The test temperature range of 400–1600 °C;
- The pressure of 101325 Pa;
- Amount of substance in a system −1 mol.

Calculations have shown that at tempering range from 600 to 700 °C 15H2NMFA-A steel has a heterophase structure: special carbides such as M_7C_3 and $M_{23}C_6$ as well as a minor amount of carbides of MC-type based on vanadium and molybdenum were observed in addition to the basic ferrite matrix (Fig. 1a). The total content of the carbides at a temperature of 600 °C is 2.97%. Moreover, the complete dissolution of these phases occurs above temperatures of 800–840 °C. Modifying of the alloying elements content within grade composition do not lead to the changing of secondary phases type and their volume fraction ranges in a narrow range from 2.52 to 3.08%. The calculated values of the critical points A_1 and A_3 were found to be 707–741 °C and 794–808 °C respectively.

The 26HN3M2FA steel contains greater number of austenite-promoting elements (nickel, carbon black) than 15H2NMFA steel, which leads to increased austenite stability and expansion of austenite field: the temperature A_3 was found to be in a range from 725 to 772 °C. Moreover, a complete decomposition of austenite was observed in steels with a minimum content of γ-stabilizing elements (nickel, carbon, manganese). In this case, the temperature A_1 ranged from 623 to 664 °C.

The content of carbides (special carbides of M_7C_3 type and $M_{23}C_6$, and a minor amount of MC carbides based on vanadium and molybdenum Fig. 1b) is in a wider range than that of 15H2NMFA steel: in a range of 2.67–3.56%.

Thus, for the composition corresponding to the content of alloying elements on the upper limit (for grade), the presence of austenite is also possible (up to 4% at a tempering temperature −600 °C).

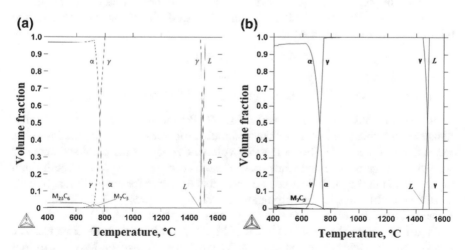

Fig. 1 The effect of temperature on the phase composition of steels **a** 15H2NMFA; **b** 26HN3M2FA

A volume fraction and a size of carbide particles which are formed as a result of the final heat treatment has a major influence on the precipitation hardening of investigated steels. Programs like DICTRA, PrecipiCalc, MatCalc, PanPrecipitation and TC-Prisma received a wide dissemination in the field of kinetics of the phase formation and growth in multicomponent systems (steels and alloys).

A common feature of the mentioned above software is the use of CALPHAD approaches as a basis for determination of the driving forces for the formation and growth of phase particles, the chemical potentials and diffusion mobility of system components [1].

For calculations of the growth kinetics of carbides, the following phases are considered:

- $M_{23}C_6$ and MC carbides in the ferrite matrix of 15H2NMFA steel at a temperature of 650 °C;
- $M_{23}C_6$, M_7C_3 and MC carbides in a ferritic matrix of 26HN3M2FA steel at a temperature of 600 °C.

Simultaneously, during the calculations the following assumptions were used:

- the precipitation and growth of carbides takes place in the body of the grain (i.e., diffusion in the volume of grain is controlling process of the phase precipitation and growth);
- calculations of precipitation and growth kinetics are carried out under isothermal conditions (i.e., for maximum constant tempering temperature), without taking into account the prehistory of heating and cooling for quenching and tempering;
- phase precipitations have an equiaxed shape and close to the stoichiometric composition or correspond it.

Furthermore, the definition of the interface energy is a requirement for the calculation of precipitation and growth kinetics of carbides: based on an analysis of literature data [1–3] for 15H2NMFA and 26HN3M2FA steels the value of the interface energy was assumed to be:

- for carbides M23C6-$\gamma^{M23C6/\alpha} = 0.27-0.29$ J/m^2;
- for carbides M7C3-$\gamma^{M7C3/\alpha} = 0.41$ J/m^2;
- for carbides MC-$\gamma^{VC/\alpha} = 0.56$ J/m^2 and $\gamma^{(Mo,V)C/\alpha} = 0.36$ J/m^2.

All calculations were carried out using TC-Prisma 2.0 program with thermodynamic and kinetic database of the materials properties TCFE7.0 and MOB2.0, respectively. Calculation shows that the volume fraction of the carbides increases with the passage of time and tends to asymptotes corresponding to the equilibrium phase content in the steel. It should be noted that exposure for about 90 h at 650 °C is sufficient for complete precipitation of $M_{23}C_6$ carbides ($\sim$3.3%), while the vanadium carbide precipitation does not occur completely (Fig. 2a).

Figure 2b shows the changes in size of $M_{23}C_6$ and MC carbide precipitates. The kinetics of changes in precipitates size is parabolic. A size of $M_{23}C_6$ carbides

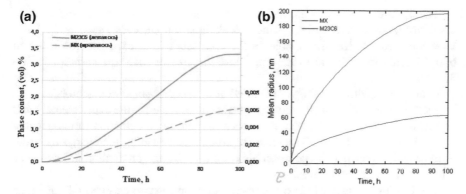

Fig. 2 Changing of carbides content (**a**) and the kinetics of their growth (**b**) for 15H2NMFA steel during isothermal aging for 100 h at a temperature of 650 °C

reaches about 195 nm (after 100 h) while a size of vanadium carbides reaches 62–65 nm.

Calculation made for 26HN3MFA steel shows that processes of carbides precipitation do not proceed completely at a given temperature and exposure time: the curves do not indicate a transition to a process of "saturation" (Fig. 3a).

It should be noted that the values obtained for the phases in the steel does not correspond to the values obtained as a result of thermodynamic equilibrium calculations: according to the calculations of the kinetics of phase precipitation, the main phase is the molybdenum MC-type carbide, whereas the calculations made in the Thermo-Calc program indicate that the main secondary phase in this steel is a carbide such as M_7C_3. This deviation may be related to a necessity to clarify the initial conditions of the settlement—the interphase energy. Due to a lack of information about the value of $\gamma_{600\,°C}^{M23C6/\alpha}$, $\gamma_{600\,°C}^{M7C3/\alpha}$ and $\gamma_{600\,°C}^{MC/\alpha}$ the experimental studies for subsequent calibration of the model is required for the investigated steels at a given temperature.

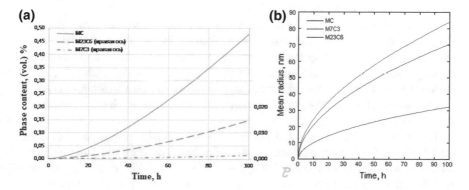

Fig. 3 Changing of carbides content (**a**) and the kinetics of their growth (**b**) in 26HN3M2FA steel during isothermal aging for 100 h at a temperature of 600 °C

The kinetic curves of carbides size are parabolic as well as in the case of 15H2NMFA steel. According to the calculations the carbides M_7C_3 and $M_{23}C_6$ have the greatest size. The carbide sizes after isothermal exposure for 100 h at 600 °C value from 70 up to 85 nm, while the size of MC-type carbides is 30–35 nm.

Laboratory melting steels were studied at the second stage of research.

Melting and casting of laboratory ingots were produced at the experimental basis of JSC RPA CNIITMASH. Melting was carried out in an induction furnace with a basic magnesite lining of 50 kg capacity crucible, by melting the charge material. Pure or highly purified charge materials were used for the production of laboratory ingots of hull and rotor steels.

The feeding of materials into the open induction furnace was carried out in the following order: armco iron, metallic chromium, nickel, synthetic iron. The ferrosilicon for pre-deoxidation was added during melting. Slag mixture from lime, fluorspar and aluminum powder was partly added during melting. Molybdenum was introduced during the process of the liquid bath appearance. Ferrovanadium was introduced in the liquid melt. Ferrosilicon and manganese were introduced into the furnace behind 3–5 min to release. Aluminum was added before the release, then a temperature measurement was carried out. The metal temperature was estimated to be 1610 °C.

The metal was tapped from the furnace to 4 ingots with a mass of 12 kg each. After an ingots casting, the exothermal mixture was introduced to the metal surface in an amount of 100 g per an ingot. Ladle samples were selected to carry out the chemical analysis, the results of which are presented in Tables 2 and 3.

Ingots were cold before forging. Heating of ingots in an electric furnace was performed under the regime: furnace loading at 700 °C, heating up to 1190 °C (speed of a furnace temperature rising −130 °C/h), 60 min exposure at 1190 °C. Heating time was about 5 h.

Interim heating of forgings was performed up to 1180 °C. Heating of forgings from the temperature of forging closing (~ 850–900 °C) was performed in the same furnace, heated to 1180 °C. Final deformation temperature, including cutting operation was not lower than 850 °C.

The investigated steels have shown a sufficient technological plasticity. However, the defects as fine cracks were formed in several bars after the last removal from furnace. Figure 4 shows the martensitic-bainitic microstructure of investigated steels. An insignificant amount of vanadium carbide with a size of 30–50 nm was observed in the structure. Table 4 shows the heat treatment modes.

The effect of heat treatment parameters was considered by an example of 15H2NMFA steel.

Tempering for 2 h leads to the precipitation of special carbides particles on the boundaries of the former martensite laths (Fig. 5a). The obtained properties complex: $\sigma_{0.2} = 820$ MPa; $\sigma_b = 925$ MPa; $\delta = 17.5\%$; $\psi = 64\%$. Increasing of exposure time to 4 h leads to the additional precipitation of carbides dispersed particles of another type (Fig. 5b) and a certain increase in tensile strength ($\sigma_{0.2} = 850$ MPa; $\sigma_b = 955$ MPa) while keeping high ductility ($\delta = 17\%$; $\psi = 65\%$). A further

Table 2 The specified and the actual chemical composition of 15H2NMFA-A ingots

Melting	Weight content of chemical elements (%)												
	C	Si	Mn	Cr	Mo	Ni	P	S	V	As	Cu	Co
68	0.15	0.23	0.53	2.11	0.63	1.42	0.002	0.003	0.11	<0.01	<0.01	<0.01
125	0.16	0.23	0.54	2.28	0.63	1.44	0.002	0.003	0.11	<0.01	0.01	0.01
Standard	0.15–0.18	0.15–0.25	0.45–0.6	2.1–2.3	0.6–0.7	1.35–1.5	<0.004	<0.004	0.1–0.12	<0.01	<0.03	<0.03

Table 3 The specified and the actual chemical composition of 26HN3M2FA ingots

Melting	Weight content of chemical elements (%)										
	C	Si	Mn	Cr	Mo	Ni	P	S	V	Al	Cu
79	0.28	0.03	0.45	1.69	0.61	3.71	0.002	0.003	0.16	0.02	
111	0.29	0.04	0.51	1.70	0.58	3.71	0.003	0.004	0.16	0.05	0.03
Standard	0.25–0.30	<0.04	0.3–0.6	1.3–1.7	0.5–0.7	3.4–3.8	<0.01	<0.012	0.12–0.18		<0.2

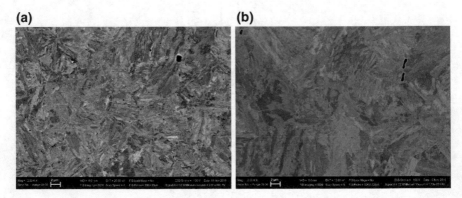

Fig. 4 SEM images showing the initial structure of forged steels **a** 15H2NMFA; **b** 26HN3M2FA

Table 4 Modes of heat treatment of forged bars made from 15H2NMF and 20–30CrNiMoV steels

Steel	Melting	Heat treatment mode		
15H2NMF	68	Preparation of samples for investigation of the grain structure evolution using Gleeble System 3800		
		Fine grain	Average grain	Oversize grain
		900 °C (exposure for 1, 2 and 3 h. respectively)	1050 °C (exposure for 1, 2 and 3 h. respectively)	1250 °C (exposure for 1, 2 and 3 h. respectively)
	125	Preparation of samples for creating of the heat treatment model: Quenching from 910–920 °C (after exposure for 3 h) with rapid cooling. Tempering at 650 °C for 2–4–8–16–32–62 h.		
26HN3M2FA	79	Preparation of samples for investigation of the grain structure evolution using Gleeble System 3800		
		Fine grain	Average grain	Oversize grain
		900 °C (exposure for 1, 2 and 3 h. respectively)	1050 °C (exposure for 1, 2 and 3 h. respectively)	1250 °C (exposure for 1, 2 and 3 h. respectively)
	111	Preparation of samples for creating of the heat treatment model: Quenching from 840–860 °C (after exposure for 3 h) with cooling rate (100–200 °C/h). Tempering at 650 °C for 2–4–8–16–32–62 h		

increase in exposure time leads to a weakening associated with the process of coalescence. After 62 h exposure strength reduces to $\sigma_{0.2} = 600$ MPa; $\sigma_b = 715$ MPa, while ductility increases insignificantly: $\delta = 22\%$; $\psi = 71\%$. This fact is connected with the coarsening of carbides precipitated on the boundaries laths (Fig. 5c). A similar dependence is observed for 26HN3M2FA steel, however a slight increase in strength is observed after exposure for 16 h.

The experimental results are in good agreement with the calculated, however, the secondary hardening is not predicted.

The calculation of the average rate of nucleus formation with a critical size shows that this is not a controlling factor for the process. However, the calculation indicates sensitivity to the value of ΔG (the Gibbs energy) and explains why the primary growth always occurs at the grain boundaries and dislocations.

The calculation of the grain boundaries energy increases the nucleation rate by 4 orders. A consideration of the dislocations presence shows that particles can appear earlier practically at all dislocations.

Particle growth is described by the following equation in case if they are isolated from each other, i.e., due to the diffusion supply of substance (Particles in a volume of grain with the average distance corresponding to a specified density of dislocations):

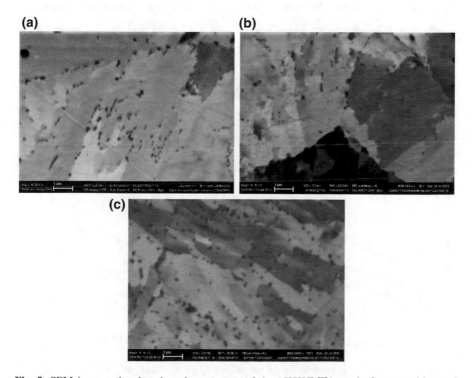

Fig. 5 SEM images showing the microstructure of the 15H2NMFA steel after quenching and tempering at a temperature of 650 °C: **a** 2 h; **b** 4 h; **c** 62 h

$$c(r,t) = c_0 + (c_1 - c_0) \times \frac{\varphi_j\left(r/(Dt)^{1/2}\right)}{\varphi_j\left(R/(Dt)^{1/2}\right)}, \tag{1}$$

where, function

$$\varphi_j(\xi) = \int_x^\infty \xi^{1-j} \exp\left(-\xi^2/4\right) d\xi \tag{2}$$

If temperature process assumed to be constant in the volume, then D (diffusion coefficient) is constant, and $j = 3$.

Calculation of particle size in the grain boundaries, taking into account the "absorption" of some by other particles. According to the developed model the particle size can be calculated as:

$$r = \sqrt[4]{\frac{4\gamma\Omega C\delta D_{g.b.}t}{3ABkT}}, \tag{3}$$

where γ—interfacial tension; C—the equilibrium concentration of the substance in the border; $D_{g.b.}$—Grain boundary diffusion coefficient; k—Boltzmann's constant; δ—the width of the grain boundary; $A = \frac{2}{3} - \frac{\gamma_{g.b.}}{2\gamma} + \frac{1}{3}\left(\frac{\gamma_{g.b.}}{2\gamma}\right)^3$—geometrical constant; $B = \frac{1}{2}\ln\left(\frac{n}{f}\right)$—morphological constant; T—temperature; r—the radius; t—time; Ω—the atomic volume (Fig. 6).

It can be seen that the obtained values are around 200 nm for 100 h, and extremely sensitive to temperature exposure, while exposure time in the latter stages has almost no effect.

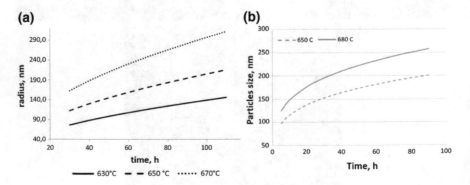

Fig. 6 Calculated size of particles (**a**) in the volume of grain; (**b**) at grain boundaries

Conclusions

The values of the volume fraction and size of carbide particles which were formed as a result of tempering were calculated using Thermo-Calc and TC-Prisma 2.0 programs. These values allow to predict the main tendencies and relatively roughly estimate the temperature and time parameters of the final heat treatment modes. However, the inability to adjust the simulation models incorporated in commercial products, provides an opportunity to improve the superiority of the developed national control complex of thermodynamic and kinetic conditions for the formation of micro-sized and nano-sized hardening phases. The features of carbide precipitation were experimentally studied. The conditions for obtaining the specified properties of 15H2NMFA and 26HN3M2FA steels after tempering were established.

Acknowledgements Research are carried out (conducted) with the financial support of the state represented by the Ministry of Education and Science of the Russian Federation.

Agreement (contract) no 14.578.21.0114 27.10.2015.

Unique project Identifier: RFMEFI57815X0114.

References

1. Kozeschnik, E.: Modeling Solid-state Precipitation. Momentum Press, New York, (2013), 464 p
2. Hou, Z.Y., Hedstrom, P., Xu, Y.B., Wu, D., Odqvist, J.: Microstructure of martensite in Fe–C–Cr and its implications for modelling of carbide precipitation during tempering. ISIJ Int. **54** (11), 2649–2656 (2014)
3. Prat, O., Garcia, J., Rojas, D., Sanhueza, J.P., Camurri, C.: Study of nucleation, growth and coarsening of precipitates in a novel 9%Cr heat resistant steel: experimental and modeling. Mater. Chem. Phys. **143**(2), 754–764 (2014)

Improvement of the Mechanical and Biomedical Properties of Implants via the Production of Nanocomposite Based on Nanostructured Titanium Matrix and Bioactive Nanocoating

E.G. Zemtsova, A.Yu. Arbenin, R.Z. Valiev and V.M. Smirnov

Abstract The work describes the complex approach for production of the biomaterial that is able for the accelerated osteosynthesis (i.e., rate of the implants engraftment into the bone tissue). The combined approach is based on bulk nanostructuring of titanium matrix and surface nanostructuring based on bioactive porous coating. This approach leads to enhancement of both mechanical and biomedical properties of implants. We created bioactive coating on nano-Ti substrates using modeling the oxide layer structure on nano and micro level. The developed bioactive surfaces with two-level surfaces allow to control surface relief both on micro and nano level with high precision (1 nm). These surfaces do not lead to the degradation of the mechanical properties of nanotitanium. Biomedical studies showed that the composite coating demonstrates high surface adhesion properties for the osteoblasts MC3T3-E1 cell line. Along with adhesion, also initial differentiation of osteoblasts is observed. This indicates the ability of the surface to the accelerated osteosynthesis. The developed nano-Ti-based composite nanomaterial with bioactive nanocoating can be used for the production of the new generation implants for dentistry, reconstructive surgery and orthopedics.

E.G. Zemtsova (✉) · A.Yu. Arbenin · R.Z. Valiev · V.M. Smirnov
Saint Petersburg State University, Saint Petersburg, Russia
e-mail: ezimtsova@yandex.ru

A.Yu. Arbenin
e-mail: aua47@yandex.ru

R.Z. Valiev
e-mail: rzvaliev@gmail.com

V.M. Smirnov
e-mail: vms11@yandex.ru

K.V. Anisimov et al. (eds.), *Proceedings of the Scientific-Practical Conference "Research and Development - 2016"*, https://doi.org/10.1007/978-3-319-62870-7_49

461

Keywords Composite biomaterial · Nanostructured titanium · Bioactive nanocoating · Chemical modification of surface · Surface relief Titanooragnic coatings · Calcium–Phosphate structures · Method of molecular layering (ml-ald) · Sol-gel synthesis · Osteointegration · Titanium implants

To the moment, traditionally used metallic biomaterials mostly reached their maximum of tensile strength. For application in medicine, these biomaterials must satisfy the set of requirements. On the one hand, they should be bioactive and biocompatible. On the other hand, such materials should demonstrate high mechanical strength especially under cyclic loads. This is an important property to ensure durability of medical products.

The increase of human life expectancy as well as the progress of modern surgery require the development of the metallic biomaterials (implants) that demonstrate accelerated engraftment (improved bioactive properties). According to the clinical statistics, the problem of rejection is relevant even for the latest models of implants. There is also the problem of the huge healing period, which is not comfortable for the patients.

Most of bioactive coating studies in Russia and worldwide are dedicated either to change of surface relief or to variation of the surface layer composition. However, acceptable coating from the point of view of the osteointegration acceleration is still not found. It was shown earlier that elemental composition of the implant surface and the titanium surface relief play important role in the osteoblasts (young bone cells) formation and in the increase of their biocompatibility [7, p. 2727]. Qualitative and quantitative indicators of osteointegration directly depend on Ti-implant surface topography and chemical composition [8, p. 87; 10, p. 5]. In this regard, an important direction of the medical materials science is to develop surface modification methods, which will increase osteocompatibility between implants and tissues.

The goal of this work is the development of the scientific basics of nanotitanium-based composite material with bioactive nanocoating. The combined approach is based on bulk nanostructuring of titanium matrix by means of severe plastic deformation and surface nanostructuring based on bioactive porous coating. This approach leads to increase of mechanical and biomedical properties of implants. We reached the accelerated osteosynthesis for enhancement of the rate of implants engraftment. Also the life of the product is increased and whole-life implantation is provided. Improved mechanical properties of the material give miniaturized implant construction. This effect decreases injury surface and accelerates osteosynthesis.

The developed bioactive surfaces with two-level surfaces allow to control surface relief both on micro and nano level with high precision (1 nm). Moreover, these surfaces do not lead to the degradation of the mechanical properties of nanotitanium.

For application in medicine, the titanium-based biomaterials must satisfy the set of requirements. On the one hand, they need to be bioactive and biocompatible. On the other hand, such medical materials must possess considerable mechanical strength, especially under action of the cyclic loads. This is important to ensure the durability of medical products.

In this work, we applied severe plastic deformation (SPD) approach that represents powerful mechanical action on the material. By means of ECAP-Conform, the structure is formed with the average grain size of 100 nm, and high level of the mechanical characteristics (tensile strength (σB) −1240 MPa). It is established that high level of the mechanical characteristics without loss of plasticity is provided by the specific properties of the formed grain boundaries and their high densities in the nano-Ti structure. Application of nanostructured Ti allows to miniaturize implants construction keeping unchanged its exploitation properties.

The joint replacement is actively developed within the modern orthopedics. One of the novel important materials providing fast implants engraftment is bioactive coatings. A lot of scientific groups work in this direction: there are researches on the possibility of introducing of various organic compounds into the surface layer [8, p. 87], coating implants with polymers—both biogenic [9, p. 1025], or synthetic [11, p. 4135]. However, as it is mentioned in some works [2, p. 217; 3, p. 889], along with the chemical composition, another key factor that determines bioactivity is the coating relief. Cytological and histological analysis indicates the efficiency of the coatings with two-level surface relief organization on micro- and nanolevel [4, p.].

In this work, the complex approach is suggested for the generation of bioactive coating on nanotitanium matrices. This attained by means of modelling of the oxide layer structure and composition, as well as coating thickness on the nano scale.

As a matrix, nanotitanium samples were used that were produced from Grade 4 titanium according to [12, p. 28]. Surface topography was examined by scanning probe microscopy (Solver P47 Pro in the tapping mode on air) and scanning tunnel microscopy (Zeiss Supra 40VP). The research was performed in the Interdisciplinary Research Center for Nanotechnology of Saint Petersburg State University (SPbSU). According to atomic force microscopy (AFM), the surface of initial Ti substrate after mechanical and chemical treatment is characterized by very low roughness level. The average height difference was ∼3 nm.

We used ALD method in order to produce the nanocoating with controlled structure relief. ALD is based on the surface chemical reactions that take place on the selected substrate [13, p. 590]. The main advantage of this method is the ability to produce nanostructures (thin films) with high precision (Angstroms) with controlled surface geometry and roughness.

It is also worth to note high adhesion of nanocoatings to the titanium substrate. This effect is reached due to consecutive cycling chemisorption of low molecular reagents from the gas phase. Surfaces with predicted roughness were produced by controlling the number of treatment cycles, which led to increasing length of Ti-organic nanostructures (from 5 to 20 cycles) [14, p. 374].

It was shown that after 20 cycles of ALD treatment of nanotitanium, initial surface is fully covered by Ti-organic nanostructures. The distance between nanostructures varies from 75 to 120 nm; the average size of nanostructures is 120 nm. Roughness height lies between 75 and 200 nm (Fig. 1).

In vitro studies of the samples were used to examine structure characteristics on the adhesion, proliferation and differentiation rate of osteoblasts MC3T3-E1. Osteocompatibility of the samples was compared with one for titanium having the natural TiO_2 layer on the surface. The cell state (adhesion and spreading of the cells on the sample surface) was done using SEM.

The sample of nano-Ti with Ti-organic nanostructures demonstrates the pronounced cells monolayer with high adhesion properties of the surface for osteoblasts MC3T3-E1 cell line (Fig. 2). Almost twofold growth of proliferation and viability of cells MC3T3-E1 is observed compared to the control samples without coating. Along with adhesion, also initial differentiation of osteoblasts cells is observed. This effect indicates the ability of the surface to the accelerated osteosynthesis.

So, almost twofold increase of osteogenic differentiation markers production in vitro comparing to the control, allows to predict that time of engraftment will be decreased. This, in turn, leads to reducing duration of hospitalization and

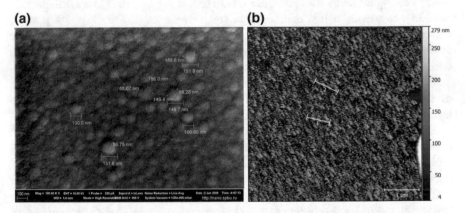

Fig. 1 Analysis of nanotitanium surface with coated Ti-organic nanostructures: **a** microtopography of the sample surface, **b** AFM reconstruction of the sample surface

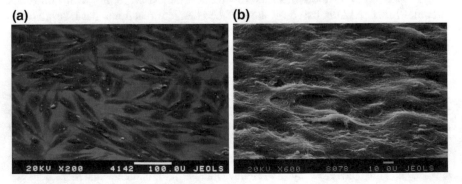

Fig. 2 Electronic microphotograph of the surface after osteoblasts incubation: **a** initial Ti, **b** surface with coated Ti-organic nanostructures

rehabilitation of patients. Our experimental data suggest that cells viability and their growth depend not only on the chemical composition of the coating but also on surface relief (i.e., its roughness).

Taking into account the requirement that the metallic implants (e.g., titanium dental implants) should have strongly rough surface on the nanolevel, we can conclude that ALD method is really prospective for the production of the bioactive nanocoatings.

Also we developed the complex approach of the modelling of oxide layer structure on the nano and micro level. TiO_2-based nanocoating with two-level relief organization are produced by sol-gel method in the dip coating mode. For this nanocoating, roughness is observed both on nano and micro level. Two-level hierarchy is realized by applying shock drying on hot plate at 150–500 °C [1, p. 2453]. Micron-level structure is raised due to erosion of the continuous film. Nano-level structure is caused by granular relief of the film (pore size of 5–20 nm). Oxide layer thickness is varied by means of cyclization of dip coating procedure. As a result, the samples were produced after 1–5 cycles of the nano-Ti substrate treatment. The coatings with thickness between 70 and 200 nm were obtained. Spectral ellipsometry data indicate the linear dependence of the TiO_2 film thickness on the number of layers. The developed approach can lead to large variation of film textures on the nano-Ti substrate: from nearly smooth film until the dense cracks net on the micro level. For the first time, we produced composite coating on the nano-Ti surface (TiO_2/calcium-phosphate composite), that looks like TiO_2 islands with inclusions of calcium-phosphate structures.

The coating with the thickness of 150 nm with the developed net of microcracks and pore size of 5–20 nm (Fig. 3) demonstrates high adhesion properties for the studied cells line MC3T3-E1. Also, along with adhesion, we observed continuous cells layer with osteocyte channels.

The induction period for the osteoblasts differentiation is diminished due to the two-level relief hierarchy (Fig. 4). This indicates the ability of the surface to the accelerated osteosynthesis.

(a) **(b)**

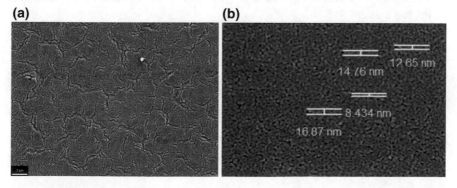

Fig. 3 Microphotographs of the surface of TiO_2 layer produced by sol-gel method in the dip coating mode (thickness 150 nm): **a** microstructure, **b** nanostructure

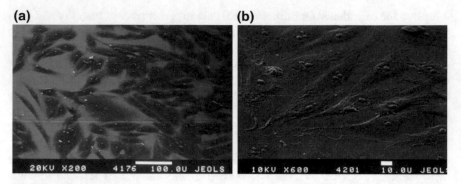

Fig. 4 Electronic microphotograph of the surface after osteoblasts incubation: **a** initial Ti, **b** TiO₂ layer produced by sol-gel method in the dip coating mode (thickness is 150 nm)

According to some references [6, p. 387; 7, p. 2727] the presence of surface roughness (both on micro and nanolevel) is the key factor for osteoblasts adsorption on the implant surface.

Our data suggest the absence of cytotoxicity of the studied coatings. This is consistent with the earlier data [5, p. 397].

The developed technology of the production of nano-Ti based composite nanomaterial with bioactive nanocoating as well as this material itself, can be used for the production of the new generation implants for dentistry, reconstructive surgery and orthopedics. While variating chemical composition of the bioactive coating, the material can be used as the implant that substitutes mineralized tissues of various skeletal parts.

Competitive Advantages of Our Approach

- Possibility to produce nanostructured coating with the high precision (1 nm);
- possibility for directed regulation of geometry and roughness of the surface in the nano scale;
- coating with two-level surface relief that is controlled both on micro and nanolevel;
- absence of cytotoxicity of suggested coatings;
- the coatings do not lead to diminishing mechanical properties of nanostructured titanium, tensile strength is retained (1240 MPa);
- miniaturization of constructions, enhancement of mechanical properties of the implants;
- improvement of functional properties due to bulk and surface nanostructuring allows twofold accelerating of the implants engraftment into human body. The service life of the implant grows manifold.

- high adhesion properties of the implants surface for the studied cell line, ability to the accelerated osteosynthesis (rate of implants engraftment in the human body).

Conclusions

We developed the method of production of nanostructured titanium matrix (nanotitanium) with enhanced strength without plasticity loss. The improved mechanical properties of nanostructured titanium can assist in the future in the miniaturization of implant construction.

Nanotitanium with titan-organic nanostructures on the surface was developed. Biomedical studies showed that the composite coating with surface roughness between 75 and 200 nm with mean pore size of 120 nm demonstrates high surface adhesion properties for the osteoblasts MC3T3-E1 cell line. It is shown that together with adhesion, initial differentiation of osteoblasts is observed. This indicates the ability of the surface to the accelerated osteosynthesis.

Based on the analysis of the mechanical properties, we established that synthesized coatings do not lead to the degradation of mechanical characteristics of nanotitanium. We developed the approach of the synthesis of the nanotitanium samples with porous nanostructured TiO_2 films on the surface with inclusion of calcium-phosphate structures. The modification of the surfaces by porous mesostructured TiO_2 films up to 180 nm thick, with pore size from 5 nm to 20 nm and microrelief is optimal not only for the stimulation of MC3T3-E1 cells proliferation but also for their differentiation in the osteogenic direction. The study of the adhesion activity of MC3T3-E1 cells demonstrates that osteoblasts undergo adhesion on the surface of all the studied samples for all the cultivation times (24 h, 7 or 14 d). They have morphology characteristic for this cell line. The adhesion ability of the osteoblasts MC3T3-E1 is more than 85% for 72 h.

This work is a scientific study that contributes to the development of a new generation of clinically significant metallic biomaterials (bioactive and biocompatible implants). These materials play important role in the human life enhancement and prolongation. The materials are used in the medicine (dentistry, orthopedics, traumatology) in order to keep life and normal functioning of the organism.

Acknowledgements Research are carried out with the financial support of the state represented by the Ministry of Education and Science of the Russian Federation. Contract no. 14.604.21.0084 30. Jun 2014. Unique project Identifier: RFMEFI 60414X0084.

References

1. Arbenin, A.Yu., Zemtsova, E.G., Valiev, R.Z., Smirnov, V.M.: Russ. J. Gen. Chem. **84**(12), 2453–2454 (2014)
2. Bucci-Sabattini, V., Cassinelli, C., Coelho, P.G., Minnici, A., Trani, A., Dohan Ehrenfest, D.M.: Oral Maxillofac. **109**, 217–224 (2010)
3. Buser, D., Schenk, R.K., Steinemann, S., Fiorellini, J.P., Fox, C.H., Stich, H.: J. Biomed. Mater. Res. **25**, 889–902 (1991)
4. Dalby, M.J., McCloy, D., Robertson, M., Wilkinson, C.D.W., Oreffo, R.O.C.: Biomaterials **27**, 1306–1315 (2006)
5. Geetha, M.: Ti based biomaterials, the ultimate choice for orthopaedic implants. A review. Prog. Mater. Sci. **54**, 397–425 (2009)
6. Hench, Larry L., Jones, Julian R. (eds.): Biomaterials, Artificial Organs and Tissue Engineering, p. 304. Woodhead publishing limited, Cambridge, England (2005)
7. Lilja, M., Genvad, A., Astrand, M., Strømme, M., Enqvist, H.: Influence of microstructure and chemical composition of sputter deposited TiO_2 thin films on in vitro bioactivity. J. Mater. Sci. Mater. Med. **22**, 2727–2734 (2011)
8. Chen, Q., Thouas, G.A.: Metallic implant biomaterials. Mater. Sci. Eng. R. **87**, 1–57 (2015)
9. Rammelt, S., Schulze, E., Bernhardt, R., Hanisch, U., Scharnwebe, Dr., Worch, H., Zwipp, H., Biewener, A.: J. Orthop. Res. **22**, 1025–1034 (2004)
10. Singh, P.P.: Drug alluding dental implant Patent № US 0208148-A1. (2012)
11. Singh Harris, L.G., Tosatti, S., Wieland, Textor, M., Richards, R.G.: Biomaterials **25**, 4135–4148 (2004)
12. Valiev Ruslan, Z., Zhilyaev Alexander, P., Langdon Terence, G.: Bulk Nanostructured Materials: Fundamentals and Applications. Wiley, Hoboken, 439 p, (2014)
13. Smirnov, V.M.: Russ. J. Gen. Chem. **72**, 590–607 (2002)
14. Zemtsova, E.G., Morozov, P.E., Smirnov, V.M.: Regulation of surface topography of nanostructured titanium using the method of ML-ALD to create bioactive nanocoatings. Mater. Phys. Mech. **24**, 374–381 (2015)

Nanopowders Synthesis of Oxygen-Free Titanium Compounds—Nitride, Carbonitride, and Carbide in a Plasma Reactor

N.V. Alexeev, D.E. Kirpichev, A.V. Samokhin, M.A. Sinayskiy
and Yu.V. Tsvetkov

Abstract The synthesis of titanium nitride, carbonitride, and carbide nanopowders from titanium tetrachloride vapor in the stream of hydrogen or hydrogen–nitrogen plasma, generated by an electroarc torch, in a confined-jet flow reactor has been experimentally studied. Single-phase nanopowders with a NaCl-type cubic crystal lattice as assemblies of preferably cube-shaped nanoparticles of a 20–150 nm size and aggregates based on them have been obtained in the experiments. By varying the synthesis parameters, it has been possible to prepare titanium nitride nanopowders with a specific surface area in the range of 11–39 m^2/g containing 18.8–22.5 wt% nitrogen, which corresponds to the empirical formula TiN0.79–TiN0.99. The titanium carbonitride nanopowders had a specific surface area of 13–23 m^2/g, carbon and nitrogen contents of 7.5–13.6 and 13.5–5.1 wt%, respectively. The titanium carbide nanopowders had a specific surface area of 14–45 m^2/g and carbon contents of 17–21 wt%. Most reached yield of main products was 94%.

Keywords Titanium · Nitride · Carbide · Carbonitride · Tetrachloride Nanopowder · Synthesis · Thermal plasma · Reactor

N.V. Alexeev (✉) · D.E. Kirpichev · A.V. Samokhin · M.A. Sinayskiy · Yu.V. Tsvetkov
A.A.Baikov Institute of Metallurgy and Material Science RAS, Moscow, Russia
e-mail: nvalexeev@yandex.ru

D.E. Kirpichev
e-mail: dym2004@bk.ru

A.V. Samokhin
e-mail: samokhin@imet.ac.ru

M.A. Sinayskiy
e-mail: ms18@mail.ru

Yu.V. Tsvetkov
e-mail: tsvetkov@imet.ac.ru

© The Author(s) 2018
K.V. Anisimov et al. (eds.), *Proceedings of the Scientific-Practical Conference*
"Research and Development - 2016", https://doi.org/10.1007/978-3-319-62870-7_50

Introduction

Oxygen-free titanium compounds (OFTC)—Nitride, carbonitride, and carbide—have a unique combination of physicochemical and mechanical properties: high melting point, thermal conductivity, chemical stability, and hardness, including those at high temperatures [9, 12]. Owing to these properties, OFTC are of great interest for the creation of materials used in the manufacture of cutting tools, wear-resistant articles and coatings, biocompatible materials and coatings, structural elements operating at high temperatures, etc. For the creation of new nanostructured materials with improved properties compared to the conventional ones, their designers pay a good deal of attention to nanosized powders of inorganic compounds of elements, including OFTC [3, 4]. To date, various methods for the synthesis of titanium nitride, carbonitride, and carbide nanopowders have been proposed, of which processes carried out in a flow of thermal electric-discharge plasma using various titanium-containing feedstocks—titanium metal and its compounds, such a hydride, tetrachloride, and dioxide—are the most widespread place [2, 5–7, 13–17]. Nonetheless, there is no information about commercialization of plasma processes for producing OFTC nanopowders.

For practical implementation of nanopowders synthesis of titanium nitride, carbonitride, and carbide, the most appropriate feedstock is titanium tetrachloride manufactured on an industrial scale to produce titanium metal and titanium dioxide (global production of the titanium sponge alone was about 190,000 tons in 2014 [8]). Titanium tetrachloride has a low boiling point (410 K) and can be easily transferred to the vapor state, making it possible to conduct the plasma-assisted synthesis involving gaseous reactants to produce the desired nanopowders with a uniform particle-size composition. The practical implementation of the production of nanopowders in thermal plasma requires the creation of equipment that ensures the necessary performance of the process and has a long service life. The most effective solution in this line is the use of plasma reactors based on arc plasma torches. To date, electric arc gas heaters (plasma torches) are among the most common devices for generating a low-temperature plasma [18]. This is due to a number of advantages provided by the use of plasma torches: the possibility of heating any gas or mixture to a relatively high-weighed average temperature (1000–5000 K), a high efficiency of heating (90%), a long continuous operating life (to 1000 h), relative simplicity of design of experimental facilities, and sufficient ease in managing operating modes with simultaneous high reliability and robustness.

Plasma-enhanced nanopowder production processes can be conducted in the steady-state continuous mode using a confined-jet flow reactor, in which the nanopowder is deposited on the reactor wall having a temperature that does not permit sintering of the deposited particles and the resulting layer is periodically removed from the wall [1].

The aim of this study was to implement on a laboratory-scale process for the manufacturing of titanium nitride, carbonitride, and carbide nanopowders in an embodiment that is the most suitable for its subsequent commercialization. In view of the foregoing, a process of this kind can be the synthesis of the desired products by reacting titanium tetrachloride vapor (or its mixture with methane) with arc torch-generated hydrogen–nitrogen thermal plasma in a confined-jet flow reactor. The possibility of commercialization of the process implies the availability of source of raw materials, existing technical solutions to create efficient thermal plasma generators operating over a wide power range, and technical solutions for the plasma reactor design with long service life.

Technique of Experimental Researches

Experimental study of the synthesis of titanium nitride, carbonitride, and carbide nanopowders was performed using a plasma unit, based on an arc torch-generating thermal plasma at a rated power of 25 kW, by reacting a mixture of titanium tetrachloride vapor and nitrogen with a hydrogen–nitrogen–argon thermal plasma. In the synthesis of titanium carbonitride and carbide, methane was added to the titanium tetrachloride–nitrogen mixture. To generate plasma, a dc arc plasma torch at a nominal power of 25 kW was used. The experimental setup is shown in Fig. 1. The reactor of 600 mm in length and 200 mm in diameter had water-cooled walls; the reactor design provided for the possibility of placing inside the reactor a quartz tube of a 150 mm diameter with attached thermocouples for measuring the temperature of the powder deposition surface.

Fig. 1 Experimental plasmachemical unit. *1* TiCl$_4$ supply and evaporation system, *2* Plasma generator, *3* Powder removal system from reactor walls, *4* Reactor, *5* Product collectors, *6* Cleaning system of plasma generator nozzle, *7* Filtration device, *8* Off-gases purification system

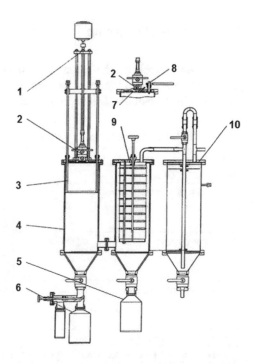

In the reactor volume, the mixing of the reactants and their chemical transformation occurred to form OFTC nanoparticles, which were deposited on the reactor walls and carried over on the bag filter. After the filter, the gaseous products entered into a bubbling absorber, where the chlorine-containing products were absorbed. The range of variation of the parameters of the titanium nitride and carbonitride nanopowder production processes is shown in Table 1.

Complex instrumental analysis of the obtained nanopowders included the following:

X-ray diffraction (XRD) performed using a RIGAKU Ultima-4 diffractometer in filtered Cu Kα radiation with a high-speed detector D/teX, the software package PDXL, and the PDF-2 database;
measurement of the BET specific surface area of the powders with a Micromeritics TriStar 3000 surface area and porosity analyzer;
measurement of the particle-size composition of the powders with a Mastersizer 2000 M laser-diffraction particle-size analyzer;
electron microscopic examination with an FEI Versa 3D, Helios 650 and Scios scanning microscope (SEM) and an FEI Tecnai G2 F20 transmission microscope (TEM);
determination of total carbon and nitrogen using LECO (model CS-400 and TS-600 analyzers, respectively);
determination of total chlorine.

Results and Discussion

The results of earlier calculations for OFTC receiving processes in the approximation of equilibrium thermodynamic model testify to a possibility of synthesis of titanium nitride, carbide, and carbonitride at the interaction of titanium tetrachloride

Table 1 Variation range of process parameters

Parameter		Range	Unit
Flow rate of TiCl$_4$		0.1–1.0	kg/h
Flow rates of plasma gases	H$_2$	0.2–1.4	n. m^3/h (STP)
	N$_2$	0.2–0.8	n. m^3/h (STP)
	Ar	0.0–2.5	n. m^3/h (STP)
Flow rate of gases fed to TiCl$_4$ supply line	N$_2$	0.0–1.1	n. m^3/h (STP)
	Ar	0.1–0.2	n. m^3/h (STP)
	CH$_4$	0.00–0.17	n. m^3/h (STP)
Atomic ratio of elements	H/Cl	2–47	
	N/Ti	10–129	
	C/Ti	0–2.1	
Plasma jet enthalpy, I_{pl}		3.0–6.3	kW*h/m^3
Temperature of nanopowder deposition surface, T_{quar}		400–815	°C

with nitrogen, methane, and their mixes, respectively, in the presence of hydrogen [10]. For the supply of main products yield come nearer to 95%, tenfold hydrogen excess in comparison with required stoichiometrical is necessary; thus, synthesis coproducts are the lowest titanium chlorides. Dependence of all main products yield on temperature has extreme character and the maximum of titanium nitride, carbide, and carbonitride yield are in the temperature range of 1100–1500, 2000–2200, and 1200–1800 K, respectively. During titanium carbide and carbonitride synthesis, there can be a condensed carbon as a part of equilibrium products; however, the condensed carbon is absent on the overhead temperature boundary line of the maximum yield of titanium carbide and carbonitride and at higher temperatures.

Strong inhomogeneity of the temperature and flow-rate fields in the confined-jet flow plasma reactor [11] can be responsible for the presence of titanium chloride impurities in the product because of differences in time–temperature conditions of chemical transformations. Since the titanium chlorides $TiCl_3$ and $TiCl_4$ have boiling points of 1230 and 410 K, respectively, the separation of titanium nitride nanopowder from the gas dispersion stream of the products at a temperature above 1000 K (given the significant dilution of $TiCl_3$ vapor with nitrogen and hydrogen) should exclude the presence of chloride impurities in the final product; however, the following reaction is thermodynamically allowed in this case:

$$TiN(cond.) + 4HCl(gas) = TiCl_4 + 2H_2 + 0.5N_2. \tag{1}$$

The results of preliminary experiments were carried out in the reactor with the water-cooled wall, onto which the produced titanium nitride was directly deposited, showed that the yield of TiN under these conditions was no more than 50% and the product powder contained a significant amount of total chlorine as an impurity. The product yield was determined as the ratio of the mass of the resulting nanopowder to the theoretical mass of titanium nitride $TiN_{1.0}$, that could be obtained from the titanium tetrachloride consumed. When a quartz tube of a 180-mm-diameter was inserted in the reactor and maintained at a temperature of 670–1090 K, the titanium nitride yield increased to 70–94% and the total chlorine content was reduced to percent shares. These results suggest that reaction (1) hardly proceeds under the experimental conditions and all further experiments were carried out in the reactor with the quartz insert. Using the quartz insert not only elevated the temperature of the product deposition surface, but also provided an extension of the high-temperature zone in the reactor, leading to an increased yield of titanium nitride.

It was established experimentally that the interaction of $TiCl_4$ vapor with hydrogen–nitrogen plasma in the presence of nitrogen, methane, and their mixes in the confined-jet flow reactor results in the formation of titanium nitride, carbide, and carbonitride nanopowders, respectively. According to X-ray diffraction data, the product nanopowders are single phase and have a cubic crystal lattice of the NaCl type (Fig. 2a); their total chlorine content is at the level of tenths parts of percent, which was reached at a titanium tetrachloride flow rate of 0.2 kg/h. Increasing the feedstock flow rate to 0.4 kg/h resulted in decrease of main product yield and increase in the total chlorine content of a few percent; X-ray diffraction pattern manifests the presence of hydrolyzed titanium trichloride $TiCl_3 * 6H_2O$ phase (Fig. 2b). The appearance of this phase is due to the sorption of $TiCl_3$ on the surface of nanoparticles and their interaction with water vapor present in the air during the removal of the nanopowder from the reactor wall, an operation that is carried out in contact with air.

The results of electron microscope examinations have shown that all powders are assemblies of nanoparticles of preferably cubic shape with a size of 20–150 nm and aggregates on their basis (Fig. 3). In the nanopowders found to contain $TiCl_3\ 6H_2O$,

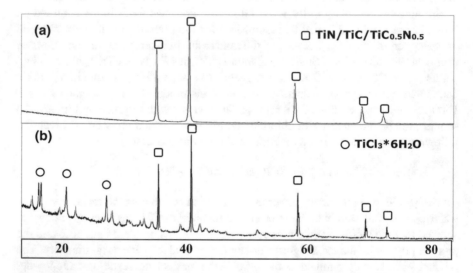

Fig. 2 X-ray diffractograms of titanium nitride, carbonitride, and carbide nanopowders

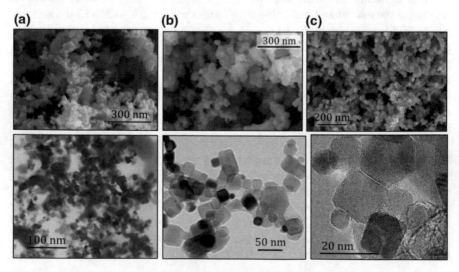

Fig. 3 Typical microphotographs of titanium nitride (**a**), carbonitride (**b**), and titanium carbide (**c**) nanopowders

a change in their morphology was noted, mostly faceted shape of the particles was replaced by a rounded shape, which was caused by the formation of a $TiCl_3 \cdot 6H_2O$ shell on the originally cubic nanoparticles of titanium nitride (Fig. 4).

The nitrogen content of the titanium nitride nanopowders varied in the range of 18.8–22.5 wt%, which corresponds to the empirical formulas $TiN_{0.79}$–$TiN_{0.99}$ and it is within the region of homogeneity of titanium nitride. The specific surface area of titanium nitride, carbonitride, and carbide nanopowders ranged within 11–39, 13–23, and 14–45 m^2/g, which corresponds to the average particle-size range of 100–27 nm.

The energy level of the processes in thermal plasma, which is defined by the plasma flow enthalpy, determines the temperature distribution in the reactor volume and is one of the most important factors on which both the physicochemical properties of the resulting nanopowders and the characteristics of the implemented process (reactants conversion, yield of the end product, and power consumption for its production) depend. As a result of the experiments, it was found that the change in enthalpy of the plasma jet has the strongest influence on the particulate composition of the titanium nitride. As the plasma jet enthalpy increases, the specific surface area of the resulting OFTC nanopowders decreases and, hence, the average nanoparticle size calculated from the measured values of the specific surface area increases (Fig. 5).

Fig. 4 Microphotographs of titanium nitride nanopowder with a $TiCl_3 \cdot 6H_2O$ impurity

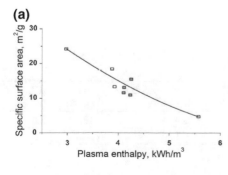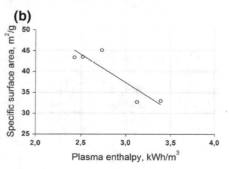

Fig. 5 Dependence of the specific surface area of titanium nitride (**a**) and carbide (**b**) nanopowder on the plasma flow enthalpy. The $TiCl_4$ flow rate is 0.2 kg/h, the element ratios are H/Cl = 11–24, N/Ti = 22–71 (**a**), H/Cl = 38–41, C/Ti = 1.4–1.5 (**b**)

For the titanium tetrachloride flow rate of 0.2 kg/h, increase in the plasma jet enthalpy from 3.0 to 5.6 kW*h/m^3 leads to decrease in the specific surface area from 24.0 to 4.7 m^2/g; this corresponds to change in the average particle size from 46 to 230 nm. Since the formation of titanium nitride nanoparticles in the plasma process occurs via the crystal–vapor macroscopic mechanism (indicated by the presence of faceted particles), the increase in the average particle size with an increase in the plasma jet enthalpy suggests a prevalence of the particle growth rate over to the rate of their formation. Unlike the case of titanium nitride synthesis, an increase of the plasma flow enthalpy in the receiving of titanium carbonitride results in increasing the specific surface area of the nanopowder, although the chemical composition remains definitely unchanged (Table 2). The changes in the specific surface area and, consequently, the average size of nanoparticles could be due to the formation of carbon nuclei during the pyrolysis of methane, which are additional centers of condensation in the formation of titanium carbonitride nanoparticles.

Increase in the specific surface area of the titanium carbonitride nanopowder from 13 to 20 m^2/g was observed with increase in the C/Ti atomic ratio from 0.6 to 2, which could be caused by change in the condensed-phase nucleation rate with increase in the concentration of methane in the high-temperature zone and also the emergence of free carbon as an impurity.

The phase composition of the products remained unchanged with increase in the plasma jet enthalpy and, hence, the average particle size (Fig. 6a), but this led to some reduction in the nitrogen content of the resulting nanopowder: at maximum enthalpy of 5.6 kW*h/m^3, the nitrogen content was 18.8 wt%. The introduction of methane admixed to titanium tetrachloride vapor into the plasma during the synthesis resulted in the substitution of carbon for nitrogen atoms in the nitride lattice; thus with increase in the C/Ti atomic ratio increased the carbon content and decreased the nitrogen content, maintaining their overall concentration at almost constant level (Fig. 6). The carbonitride nanopowders contain 7.5–13.6 wt% of carbon and 13.5–5.1 wt% nitrogen and the carbide nanopowders contain 17–21 wt % of carbon; their chemical composition was determined predominantly by the C/Ti atomic ratio in the reaction mixture (Fig. 6).

Table 2 Properties of titanium carbonitride nanopowders obtained by different plasma flow enthalpies

Enthalpy kW*h/m^3	TiCl$_4$ flow rate (kg/h)	H/Cl ratio	C/Ti ratio	Specific surface area (m^2/g)	[N] (wt%)	[C], (wt%)
3.5	0.2	17.9	0.68	15.9	12.1	8.0
5.2	0.2	18.1	0.64	21.1	13.1	7.5

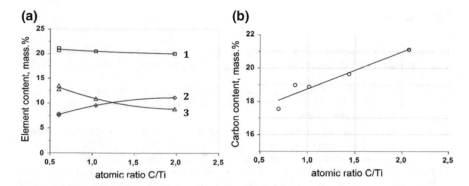

Fig. 6 Dependence of carbon (*2*), nitrogen (*3*), and the total content (*1*) in the titanium carbonitride (**a**) and carbon content in the carbide (**b**) nanopowders depending on the C/Ti ratio. **a** The plasma flow enthalpy is 4.6 kW*h/m³ (**a**), 3.0–3.2 kW*h/m³ (**b**), the TiCl₄ flow rate is 0.2 kg/h; the atomic ratios are H/Cl = 18 (**a**), 38–43 (**b**), N/Ti = 95

Conclusions

In this experimental study, we have shown the feasibility for synthesizing of titanium nitride, carbonitride, and carbide nanopowders from titanium tetrachloride vapor in the hydrogen, hydrogen–nitrogen plasma stream, generated by an arc plasma torch, in the confined-jet reactor. The yield of the main nanopowders has been achieved up to 94% by upgrading the reactor design with providing of wall temperature elevation in the reactor in the range of 670–1090 K, with the amount of total chlorine impurities being reduced to less than 0.1 wt%.

Single-phase nanopowders with a NaCl-type cubic crystal lattice as assemblies of preferably cube-shaped nanoparticles of a 20–150 nm size and aggregates on their basis have been obtained in the experiments. Varying the synthesis parameters has made it possible to prepare titanium nitride nanopowders with a specific surface area in the range of 11–39 m²/g containing 18.8–22.5 wt% nitrogen, which corresponds to the empirical formula TiN0.79–TiN0.99. The titanium carbonitride nanopowders had a specific surface area of 13–23 m²/g, carbon and nitrogen contents of 7.5–13.6 and 13.5–5.1 wt%, respectively. The titanium carbide nanopowders had a specific surface area of 14–45 m²/g and carbon contents of 17–21 wt%. Most reached yield of main products was 94%.

The study has demonstrated the feasibility of the synthesis OFTC nanopowders from titanium tetrachloride vapor in a reactor having a long service life. Further investigation should focus on optimizing the process to ensure the maximum yield of the desired products and minimum power consumption for the manufacturing of the products with a given set of physicochemical properties.

Acknowledgements Research was carried out with the financial support of the state represented by the Ministry of Education and Science of the Russian Federation. Grant agreement no. 14.607.21.0103. Unique project identifier: RFMEFI60714X0103.

References

1. Alekseev, N.V., Samokhin, A.V., Tsvetkov, Yu.V.: Plazmennaya ustanovka dlya polucheniya nanoporoshkov [Plasma installation for nanopowders production]. RU Patent no. 2311225, (2007)
2. Alekseev, N.V., Samokhin, A.V., Tsvetkov, Yu. V: Synthesis of titanium carbonitride nanopowder by titanium tetrachloride treatment in hydrocarbon-air plasma. High Energy. Chem. **33**(3), 194–197 (1999)
3. Andrievskii, R.A.: Nanokompozity na osnove tugoplavkikh soedinenij: sostoyanie razrabotok i perspektivy [Nanocomposites on the basis of refractory compounds: condition of developments and prospect]. Materialovedenie **4**, 20–27 (2006) (In Russ.)
4. Andrievskii, R.A.: Nanomaterialy na osnove tugoplavkikh karbidov, nitridov i boridov [Nanomaterials on the basis of refractory carbides, nitrides and borides]. Uspekhi khimii (Russ. Chem. Rev.) **74**, 1163–1174 (2005). (In Russ.)
5. Batenin, V.M., Klimovskii, I.I., Lysov, G.V., Troitskii, V.N.: Poluchenie ul'tradispersnykh nitridov v plazme SVCH-razryada [Receiving ultradisperse nitrides in microwave plasma]. SVCh-generatory plazmy: Fizika, tekhnika, primenenie (Microwave Plasma Generators: Physics, Equipment, and Applications), Moscow: Energoatomizdat, 178–213 (1988) (In Russ.)
6. Daquan, Z., Suduan, H., Qiongqiong, L., et al.: Preparation of ultrafine powder of titanium carbonitride using an arc hydrogen plasma. ISPC-9, 854–859 (1989)
7. Fil'kov, M.N., et al.: Sintez karbonitrida titana v trekhstrujnom plazmo-khimicheskom reaktore [Synthesis of titanium carbonitride in the three-jet plasmochemical reactor.] Elektorofizika, elektromekhanika prikladnaya elektrotekhnika. Sbornik nauchnykh trudov (Electrophysics, Electromechanics, and Applied Electrical Engineering: Collection of Articles), Alma-Ata: KPI, 17–25 (1982)
8. Mineral Commodity Summaries 2015, Reston, Va.: U.S. Geological Survey, 2015. 196 pp
9. Pierson, H.O.: Handbook of Refractory Carbides and Nitrides: Properties, Characteristics, Processing and Applications. William Andrew, Park Ridge (1997) 363 pp
10. Samokhin, A.V., Alekseev, N.V., Sinaiskiy, M.A., Tsvetkov, Y.V.: Equilibrium energy and technological characteristics of plasma synthesis of titanium nitride, carbide, and carbonitride from titanium tetrachloride. Inorg. Mater.: Appl. Res. **7**(3), 344–349 (2016)
11. Samokhin, A.V., Polyakov, S.N., Alekseev, N.V., Astashov, A.G., Tsvetkov, Yu.V.: Modelirovanie protsessa sinteza nanoporoshkov v plazmennom reaktore strujnogo tipa. I. Postanovka zadachi i proverka modeli [Simulation analysis of process of nanopowders synthesis in the plasma reactor of jet type. I. Problem definition and check of model]. Fizika i khimiya obrabotki materialov, **6**, 40–46 (2013) (In Russ.)
12. Shackelford, J.F., Alexander, W.: CRC Materials Science and Engineering Handbook, p. 1980. CRC, Boca Raton (2000)
13. Shin, D.H., Hong, Y.C., Uhm, H.S.: Production of nanocrystalline titanium nitride powder by atmospheric microwave plasma torch in hydrogen/nitrogen gas. J. Am. Ceram. Soc. **88**, 2736–2739 (2005)
14. Tavares, J., Coulombe, S., Meunier, J.-L.: Synthesis of cubic-structured monocrystalline titanium nitride nanoparticles by means of a dual plasma process. J. Phys. D Appl. Phys. **42**, 102001–102004 (2009)
15. Vollath, D.: Plasma synthesis of nanopowders. J. Nanopart. Res **10**, 39–57 (2008)
16. Zeng, D.-Q., Hu, S.-D., Luo, Q.-Q., et al.: Plasma synthesis of ultrafine titanium carbonitride powders. Acta Chimica Sinica—Chin. Ed. **49**(11), 1103–1106 (1991)
17. Zhu, L., Jin, P., Xiao, X.: The pilot scale experiment on producing ultrafine powder of titanium nitride by RF plasma. ISPC-8, 2087–2092 (1988)
18. Zhukov, M.F., Zasypkin, I.M., Timoshevskii, A.N.: Elektrodugovye generatory termicheskoi plazmy [Thermal Plasma Torches]. Nauka, Novosibirsk (1999). 712 p. (In Russ.)

The Technology and Setup for High-Throughput Synthesis of Endohedral Metal Fullerenes

D.I. Chervyakova, G.N. Churilov, A.I. Dudnik, G.A. Glushenko,
E.A. Kovaleva, A.A. Kuzubov, N.S. Nikolaev, I.V. Osipova
and N.G. Vnukova

Abstract The article presented the technology and setup for high-throughput synthesis of carbon nanostructures. It was shown that the plasma-chemical synthesis of fullerenes and endohedral metallofullerenes (EMF) in high-frequency arc discharge can be controlled by changing helium pressure in the chamber. The methods of extraction using Lewis acids and separation of individual EMF by HPLC were suggested as the most effective.

Keywords Endohedral metallofullerenes · High-frequency arc discharge
Pressure

D.I. Chervyakova · E.A. Kovaleva · A.A. Kuzubov · N.S. Nikolaev
Siberian Federal University, Krasnoyarsk, Russia
e-mail: asteralin@gmail.com

E.A. Kovaleva
e-mail: kovaleva.evgeniya1991@mail.ru

A.A. Kuzubov
e-mail: alexxkuzubov@gmail.com

N.S. Nikolaev
e-mail: churilov@iph.krasn.ru

G.N. Churilov (✉) · A.I. Dudnik · G.A. Glushenko · I.V. Osipova · N.G. Vnukova
Kirensky Institute of Physics Federal Research Center KSC Siberian Branch Russian
Academy of Sciences, Krasnoyarsk, Russia
e-mail: churilov@iph.krasn.ru

A.I. Dudnik
e-mail: churilov@iph.krasn.ru

G.A. Glushenko
e-mail: churilov@iph.krasn.ru

I.V. Osipova
e-mail: churilov@iph.krasn.ru

N.G. Vnukova
e-mail: churilov@iph.krasn.ru

K.V. Anisimov et al. (eds.), *Proceedings of the Scientific-Practical Conference
"Research and Development - 2016"*, https://doi.org/10.1007/978-3-319-62870-7_51

Introduction

Endohedral metallofullerenes (EMF) have stimulated great scientific interest [1]. From the fundamental point of view, this is because they engender new spherical molecules with structure which is similar to the electronic structure of the atom. Indeed, there is a positively charged nucleus—a metal ion and the excess electron density at a quasi-spherical surface, where the carbon atoms have a generalized π-system. EMF offers a broad range of properties that make them appealing for use in different fields, such as material science and nanomedicine [2]. Already EMF-base substances are reagents for medicinal drugs and contrast agents for magnetic imaging. In the future, due to the unique properties, we can expect an even greater requirement of EMF, than is available today. This requires considerably reduce the cost of producing EMF, i.e., develop more efficient technologies of synthesis and isolation.

   This can be done on the basis of the available experimental results and theoretical concepts that define the model representation assembly of molecules and the formation of substances in powder form. We believe that there are the following most important theoretical concepts defining the process of formation of EMF:

1. Fullerenes are formed through a process of clustering from C_2 [3].
2. First, large clusters are produced. That clusters stabilized in shape and size through the loss of C_2. This is described by "shrinking hot giant fullerene" concept [4].

   The most important experimental results:

3. Effective synthesis of fullerene and EMF may be only in a helium atmosphere.
4. Plasma must be carbon arc.
5. EMF, as well as fullerenes, are formed during quenching in the cooling of the plasma layer, wherein the temperature gradient, and for arc plasma and the electron density, have the most important control value [5].
6. EMF extraction must be in a Soxhlet apparatus, and concentration by using Lewis acids.
7. Separation of individual EMF may be by HPLC.

   How effectively we have answered these questions, so will be our efficient method of producing EMF.

Kinetic Study of Metallofullerene Formation

It is well known that fullerenes are made of sp^2 hybridized atoms. We report the results of QM/MD simulations leading to completely closed cage fullerene structures and identify key steps during these dynamics (Fig. 1).

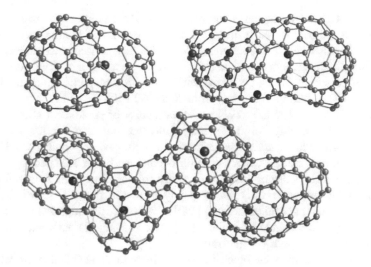

Fig. 1 Giant metallofullerene structures obtained during the simulation

The process of fullerene formation can be considered as a three-step process: C_2 units coalescence leads to the formation of chains, chains form sp^2 structures, and finally, sp^2 structures bend in order to form the cage.

We suggested that the first step follows second-order kinetics while the third step is first-order chemical reaction and studied the dependence of the rate constant for them as a function of temperature and He concentration.

Rate of the chain formation increases rapidly up to 1800 K due to the larger number of effective C_2 collisions. Further increasing of the temperature leads to the break of the bonds and thus lowers the rate constant. The same effect is observed for the cage formation.

Helium atoms act as thermostat taking the extra energy from forming carbon chains and thus helping the formation of larger chains and sp^2 structures. However, as the number of He atoms in 100*100*100 Å box reaches 150, they start to "freeze" sp^2 fragments decreasing the possibility of regrouping barrier overcoming.

Description of the Experiment

Method of EMF Synthesis

Our setup, as well as other setups of this type, is based on the process of cooling the carbonaceous plasma in a helium atmosphere. The plasma is obtained as a result of arc discharge evaporation of graphite rods, or graphite rods are filled with additional substances—dopants [5, 6], i.e., the operation of the setup is based on the W. Kraetchmer's modified method. The distinguished feature of the proposed setup, compared to most of the fullerene generators used nowadays, is the application of medium frequency alternate-current arc. Symmetrical arrangement of the electrodes allows to reach up to a 100% conversion of the electrode material into the fullerenes and EMF containing CC [7, 8]. As a consequence, this allows to avoid the main disadvantage typical for the setups operating on a direct current. The parameters of synthesis can be varied within the wide range of limits: arc current from 50 to 400 A, current frequency from 20 to 160 kHz, the helium pressure in the chamber from 30 to 400 kPa. This broad variability of parameters allows finding optimal values not only for the regular fullerene, but also for the EMF. The chamber, in which the synthesis is performed, is equipped with quartz windows for visual and spectral studies when the setup is combined with the spectrograph.

The profound increase in the setup capacity combined with the enlarged variety of combination options results into a visibly increased productivity of CC along with the high amount of fullerenes and EMF as well to obtain renewed experimental results revealing the degree of influence of various synthesis parameters upon the process of various type EMF generation.

Figure 2 demonstrates the block scheme of the setup. The chamber, in which the synthesis is performed, is cooled by water. Two pairs of graphite electrodes placed on water-cooled rods on the ends are evenly placed into the chamber. The feeding of each rod is made by a special mechanical drive, activated by the stepping motor connected to the control unit. The control unit adjusts the flow rate of the rods so that the arc current remains constant corresponding to a predetermined initial current in the range of 2–3%.

The pressure of the helium, which is fed directly from the cylinder through a pressure regulator and a nitrogen trap into the chamber, is also kept constant, with a

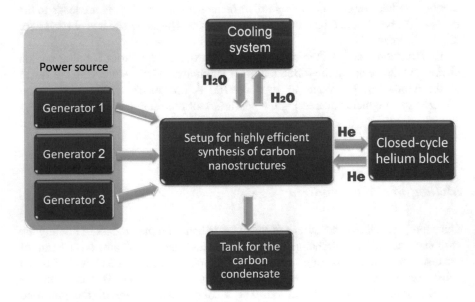

Fig. 2 Block scheme of the setup for the synthesis of CC

special device, made on the basis of a controlled nonreturn valve and matching to given number, in the range of 2–3%. Power of each pair of electrodes is carried out by a separate generator G1 and G2, 25 kW, via the matching blocks. Figure 3 shows a circuit diagram of the installation, in which we can see that a matching unit consists of two reducing transformers T1 and T2 with a transformation ratio of 3:1. The secondary windings of transformers are connected sequentially to two capacitors C1 and C2, an inductance coil L1, a ballast resistor R1, and two graphite electrodes with the arc in a gap between them. Another matching unit also consists of two reducing transformers T3 and T4 with a transformer ratio 3:1 and an analogous set of capacitors (C3 and C4), resistor (R2), inductive coil (L2) and another pair of graphite electrodes. The voltage of current transformer TA1 and TA2 is supplied to the unit controlling the feed rate [9].

Synthesis of nano-dispersed substances based on carbon, performed on setup developed in this work, is possible using graphite rods with a diameter of 4–20 mm. Rods of 6 mm in diameter and 100 mm long were evaporated. The rods were annealed preliminarily in a water-cooled chamber at 10 Pa, at a temperature of 1800–2000 K for 30 min. The same rods, only predrilled along the central axis, were used for the EMF synthesis. A hole with a diameter of 3 mm and a length of 85 mm was filled with a mixture of graphite powder and metal oxide in mass ratio of 1:1. Annealing was performed under the same conditions as those used in the preparation of the rods to obtain conventional fullerenes.

Prior to the synthesis, the chamber is warmed up with the water heated to a temperature of 60–70 °C during 15 min. Then through the top entrance nitrogen trap, the chamber is purged with helium for 1–2 min at the speed of 4–6 l/min. During the synthesis process, the chamber is cooled by the cold water with a

Fig. 3 Electrical scheme of the setup

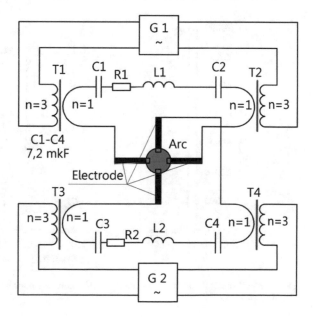

temperature of 10–30 °C, and helium is supplied at a rate of 9 l/min through the flowmeter and pressure regulator, directly from the cylinder container. Cooling of the trap is carried out by liquid nitrogen.

Fullerene content was estimated by atom emission analysis. Atomic emission analysis was performed on the setup, consisting of plasma atomizer—HF discharge in a stream of argon with copper and graphite electrodes, the spectrograph PGS-2 and a computerized spectral registration [10]. The mass spectra of sample were recorded by Bruker Autoflex time of flight mass spectrometer with laser desorption. To optimize the condition of synthesis, we studied the influence of the current range and frequency and helium gas pressure in the chamber on the structure of the formed CC and the content of different fullerenes and EMF. This analysis showed that within the limits of parameters available for our setup, variation of the current parameters and gas pressure lead to the same results. In other words, the structure of the CC, the quantitative content of the fullerenes and EMF in the CC, and their composition depend largely on the pressure. All the changes that can be obtained changing the current and the frequency can be also obtained by changing the pressure in the chamber. Mass spectrum of fullerene, synthesized with the addition of Gd_2O_3 under chamber pressure 64.8, is presented in Fig. 4 [11].

Method of EMF Extraction

Fullerenes were extracted by CS_2 in a Soxhlet apparatus. The fullerene solution was evaporated and the fullerene yield was determined by the weight ratio of fullerite and soot (prior to extraction). In order to enrich of FM with EMF, the complexation

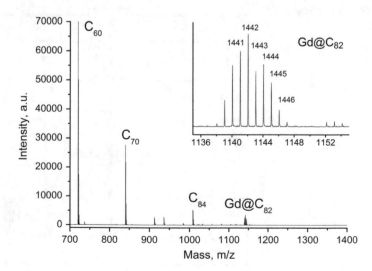

Fig. 4 Mass spectral (positive ion) analysis of fullerenes, detached from CC synthesis under 64.8 chamber pressure

reaction with Lewis acids was used [12, 13]. The resulting filtrate is a mixture of EMF and empty fullerenes, which accounted for most of the EMF. Isolation of individual EMF was carried out by high-performance liquid chromatography (HPLC). It should be noted that $TiCl_4$ Lewis acid is most effective and powerful in purification for the various types of pure metallofullerenes. Other Lewis acids ($CuCl_2$, $AlCl_3$) did not demonstrate such results.

Method of Quantitative Analysis of EMF

The technique for rapid determination of endohedral metallofullerene (EMF) content in a fullerene mixture (FM) was used by [14]. Samples of the CC with Me (for example Gd, Y) obtained at pressures of 353, 225, 98, 64.8, and 32.4 kPa were subject to extraction. The extracts were dried, weighted, and then dissolved in the initial solvent to the concentration of 1 mg/mL. Sample solutions were placed into the holes in the graphite rods and then atomized in the arc discharge in a setup for emission analysis. Standard samples were prepared beforehand using solutions with known Gd concentration in the holes in the graphite rods. Obtained data were used for building the concentration curve for analysis using standard samples [15]. For example, gadolinium content in a given probe was found from the intensity of its line at $\lambda = 335.86$ nm and the concentration curve. Allowing for the mass spectrum showed that gadolinium is contained only in $Gd@C_{82}$, we thus measure mass fraction of this compound in the fullerene extract. Results of investigation of the EMF content in the fullerene mixture extracted from the CC at various pressures are given in Table 1.

Table 1 Dependence of the content of EMF in the FM extracted from the CC obtained under various helium pressures

Helium pressure in chamber, kPa							
360		120		60		30	
Extraction							
C_5H_5N	CS_2	C_5H_5N	CS_2	C_5H_5N	CS_2	C_5H_5N	CS_2
$Gd@C_{82}$, w.%							
2.3	2.0	5.4	4.9	3.0	2.3	1.3	0.9
$Sc@C_{82}$, w.%							
0.5	5.8	9.6	3.1	6.2	5.0	2.5	2.1
$Y@C_{82}$, w.%		$Y@C_{82}$, $Y_2@C_{82}$, $Y_2C_2@C_{82}$, w.%					
0.8	0.33	9.7	7.8	8.5	11.3	4.1	11.3
$Er@C_{82}$, w.%							
–	–	–	5.2	3.3	3.1	–	1.2
$Ti@C_{82}$, $Ti_2C_2@C_{82}$, w.%							
–	–	–	–	–	0.21	–	–

Conclusion

A method for high-throughput synthesis of carbon nanostructures in a high-frequency arc discharge in the flow of helium was presented. It was shown that the plasma-chemical synthesis of fullerenes and EMF in high-frequency arc discharge can be controlled by changing helium pressure in the chamber. The results may also form the basis of the concepts of formation EMF.

Acknowledgements Research is carried out with the financial support of the state represented by the Ministry of Education and Science of the Russian Federation. Agreement no. 14.613.21.0010, 27 Aug 2014. Unique project Identifier: RFMEFI61314X0010.

References

1. Popov, A.A., Yang, S., Dunsch, L.: Endohedral fullerenes. Chem. Rev. **113**, 5989–6113 (2013)
2. Yang, S., Lui, F., Chen, C., Jiao, M., Wei, T.: Fullerenes encaging metal clusters—Clusterfullerenes. Chem. Commun. **47**, 11822–11839 (2011)
3. Afanas'ev, D.V., Bogdanov, A.A., Dyuzhev, G.A., Kruglikov, A.A.: Formation of fullerenes in an arc discharge. Tech. Phys. **42**, 234–241 (1997)
4. Irle, S., Zheng, G., Wang, Z., Morokuma, K.: The C_{60} formation puzzle "Solved": QM/MD simulations reveal the shrinking hot giant road of the dynamic fullerene self-assembly mechanism. J. Chem. Phys. **110**, 14531–14576 (2006)
5. Churilov, G.N., Vnukova, N.G.: The method of synthesis of endohedral fullerenes. Patent. **2582697**
6. Churilov, G.N.: Synthesis of fullerenes and other nanomaterials in arc discharge. Fullerenes, Nanotubes Carbon Nanostruct. **16**, 395–403 (2008)
7. Bezmelnitsyn, V., Davis, S., Zhou, Z. Efficient synthesis of endohedral metallofullerenes in 3-phase arc discharge. Fullerenes, Nanotubes Carbon Nanostruct. **23**. 612–617 (2014)
8. Churilov, G.N., Kratschmer, W., Osipova, I.V., Glushenko, G.A., Vnukova, N.G., Kolonenko, A.L., Dudnik, A.I. Synthesis of fullerenes in a high-frequency arc plasma under elevated helium pressure. Carbon. **62**, 389–392 (2013)
9. Churilov, G.N., Popov, A.A., Vnukova, N.G., Dudnik, A.I., Samoylova, N.A., Glushenko, G. A.: Controlled synthesis of fullerenes and endohedral metallofullerenes in high frequency arc discharge. Fullerenes, Nanotubes Carbon Nanostruct. **24**(11), 675–678 (2016)
10. Sichenko, D.P., Vnukova, N.G., Lopatin, V.A., Glushchenko, G.A., Marachevskiy, A.V., Churilov, G.N.: A facility for atomic emission spectral analysis and methods for spectrum processing. Instrum. Exp. Tech. **47**(4), 489–492 (2004)
11. Churilov, G.N., Popov, A.A., Vnukova, N.G. Dudnik, A.I., Glushchenko, G.A., Samoilova, N.A., Dubinina, I.A., Gulyaeva, U.E.: Method and setup for high-controlled synthesis of fullerenes and endohedral Metallofullerenes. JETP Lett. **9**, 64–70 (2016)
12. Akiyama, K., Hamano, T., Nakanishi, Y., Takeuchi, E., Noda, S., Wang, Z., Kubuki, S., Shinohara, H.: Non-HPLC rapid separation of metallofullerenes and empty cages with $TiCl_4$ Lewis Acid. J. Am. Chem. Soc. **134**, 9762–9767 (2012)
13. Stevenson, S., Rottinger, K.A., Fahim, M., Field, J.S., Martin, B.R., Arvola, K.D.: Inorg. Chem. **53**, 12939–12946 (2014)

14. Churilov, G.N., Popov, A.A., Guliaeva, U.E., Samoylova, N.A., Vnukova, N.G., Kolonenko, A.L., Isakova, V.G., Dudnik, A.I., Koravanets, V.S.: Express analysis of endohedral fullerenes amount contained at fullerene mixture. Nanosystems: physics, chemistry, mathematics. **7**, 140–145 (2016)
15. Torok, T., Mika, J., Gegus, E.: Emission Spectrochemical Analysis. Akademia Kiado, Budapest (1978); Mir, Moscow (1982)

On Some Features of Nanostructural Modification of Polymer-Inorganic Composite Materials for Light Industry and for Building Industry

M.V. Akulova, S.A. Koksharov, O.V. Meteleva and S.V. Fedosov

Abstract The specifics and general patterns of change in physical and physico-chemical properties of polymer-inorganic systems and effective methods of composite materials nanoengineering for light industry and construction industry have been revealed. A quantitative assessment is given of the impact of mechanoacoustic processing of calcium chloride hydrosol on the change in nanoparticles size at the dispersed phase, in parameters of nanoporous structure of cement stone formed. A method has been proposed for obtaining composite gasket materials based on formation of specific comblike structure of interphase layer of reinforcing block copolymer with the introduction of side branches in the pore spaces of the fibrous component. The method allows creating on the basis of a small set of textile carriers a wide range of gasket materials with technologically necessary level of elastic-deformation properties of the duplicated materials and a high degree of discreteness of their regulation. The substantiation of multilayer composite materials structure for the design of special garments, that allow creating new functional energy reflecting and sealing materials, is given.

M.V. Akulova (✉)
Department of Building Materials Production, Special Technologies
and Technological Complexes, Ivanovo State Polytechnical University,
Ivanovo, Russia
e-mail: m_akulova@mail.ru

S.A. Koksharov
Innovation Department, G.A. Krestov Institute of Solution Chemistry
of the Russian Academy of Science, Ivanovo, Russia
e-mail: ksa@isc-ras.ru

O.V. Meteleva
Department of Sewing Garments Technology,
Ivanovo State Polytechnical University, Ivanovo, Russia
e-mail: olmet07@yandex.ru

S.V. Fedosov
Department of Technosphere Safety,
Ivanovo State Polytechnical University, Ivanovo, Russia
e-mail: fedosov-academic53@mail.ru

© The Author(s) 2018   491
K.V. Anisimov et al. (eds.), *Proceedings of the Scientific-Practical Conference*
"Research and Development - 2016", https://doi.org/10.1007/978-3-319-62870-7_52

Keywords Cement composites · Mechanoacoustic processing · Polymer-fibrous composite materials · Comb copolymers · Fusible interlining · Elastic-deformation properties · Power reflecting material · Metallized coering · Pressurizing self-glued film material · Acrylic polymer

The present state of the domestic branches of light and the construction industry has a number of similar system problems requiring prompt solution. These include the high material and energy intensity of production; low levels of innovation and investment in the industry; high share of import materials and parts [1, p. 157]. Overcoming of technological backwardness in these production areas is impossible without the use of new composite materials that can improve the quality and reliability, increase service lives and reduce products material consumption. This raises the problem of creating scientific and technical basis for the development of technologies implemented in the garment industry and the construction industry for obtaining structural and functional materials with new improved performance characteristics based on the use of structurally modified polymer-inorganic composite materials, plasma surface modification of roll materials and polymer-cement composites, the proceeding of 3D-polymerization processes of the interphase layer composite materials formation with fibrous carrier nanoporous structure and nanodisperse mineral additives.

The range of tasks is to identify the specifics and general laws of changes in physical, chemical and physical properties of polymer-inorganic systems and effective methods of composite materials nanoengineering, characteristic features of modification of polymer-inorganic composite materials for light industry and construction industry.

Modern approaches to improvement of physical and mechanical properties of concrete are based on the formation of cement stone optimal structure, which can be achieved in particular by means of chemical modifiers that affect the physical and chemical processes of binders hardening [2, p. 55]. For this purpose the most commonly used in construction are plasticizers and hardening accelerators, of which calcium chloride is the strongest.

With the development of nanotechnology in recent years, much attention is paid to the perspective development of solid-state building materials by introducing additives in nanodispersed state and the creation of systems that incorporate nanosized elements as part of the structure. At the same time explanation of the known processes at the nanolevel can be referred to as an independent development of nanotechnology. An important place in similar processes is given to cement hydration, the initial stage of which is mixing the cement paste. According to [2, p. 55], the structural change of water by means of additives has a great influence on the process of hydration and crystalline neoplasms morphology.

Along with the chemical modification, the area of interest are the methods of mixing water activation providing its transfer to a metastable state and having a positive impact on the processes of hydration and structure formation in the obtained materials that result in improvement of their operational characteristics.

Influence of electroprocessing, electromagnetic and ultrasonic [2, p. 55] activations of water mixing on properties of cement composites is widely presented in literature. At the same time data on the condition of texturing agents in the mechanically activated solutions, on the extent of transformation of the structure of the formed materials to interrelations with changes of their strength characteristics are extremely limited, making it difficult to develop scientifically proved approaches to the achievement of desired effects of improving the physical and mechanical properties of concrete structures.

The aim of the first stage of work was to carry out by high-precision methods the quantitative assessment of the influence of mechanoacoustic processing calcium chloride hydrosol, that is widely used as the modifier of cement mortars, on a dimensional change of particles of the dispersed phase, porosity parameters of the formed cement stone and its strength characteristics.

Determination of the particles size in calcium chloride hydrosol by the method of dynamic light scattering [3, p. 138; 4, p. 136] on the Zetasizer Nano ZS analyzer by Malvern Instruments Ltd has shown that mechanoacoustic processing results in discretization of the dispersed phase to the size less than 1 nm, which is maintained during technologically acceptable period of time of not less than 24 h. The reinforcing effect of mineral additives introduced into structural compositions is associated, as a rule, with the appearance in cement slurries of multiple crystallization centers that increase formation rate and uniformity of the spatial structure of curable material [5, p. 215]. Considering the nature of proportionality growth of the particles number filling the volume unit at decrease of their size, it is logical to expect that the reduction in the dimensional parameter is followed by the particles number increase in the system by 9 orders of magnitude.

The assessment of porosity and specific surface of the cement stone samples by low-temperature (77 K) adsorption and desorption of nitrogen vapor on the NOVA Series 1200e gas sorption analyzer has shown (Fig. 1) the contribution of different diameter pores in the amount of porosity. So in cement stone the dominant contribution is provided by pores with diameter of 4 nm as well as 7.0 ± 1.0 and 11.5 ± 1.5 nm. Mechanoacoustic processing of calcium chloride solution for mixing cement stone results in reduction the number of the above mentioned sized pores in the obtained cement composite and appearance of the additional band of 5.0 ± 0.5 nm dominant pore size. This proves that the use for obtaining cement mass of the mechanically activated solutions of calcium chloride provides decrease in integral indexes of porosity of the formed cement stone reducing the size of the maximal diameter of pores by 1.8 times and leveling the distribution of indexes of specific surface area and the volume of pore spaces according to pores size. Decrease of the deficiency of the cement stone structure and especially the number of mosopores having large cross sectional dimensions (up to 160 nm) reduces probability of stress concentration in the weakened structural places under the influence of external loading promoting redistribution of efforts and increase of the material mechanical stability.

Developed in the project approaches to nanostructural modification of polymer-fiber composites are of practical importance for regulating mechanical

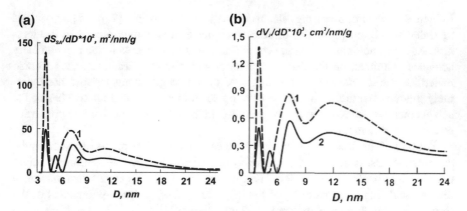

Fig. 1 Differential size curve distributions by the specific surface area (dS$_{SA}$/dD, m^2/nm/g) (**a**) and the pore volume (dV$_P$/dD, cm^3/nm/g) (**b**) in samples of cement stone 1 and 2

strength of various products of technical purpose (layer-frame structural materials, geotextiles, etc.) and domestic use. In production of light industry products, including textile, knitwear, leather and fur products, the most challenging technological task is to create a complex spatial form of products and ensure its safety in operation. For this purpose the sealing materials with thermoplastic adhesive are traditionally used (in terms of technology composites—prepregs) that duplicate the base material product, thereby providing the required level of shape stability of the package.

The prospects for improvement of polymer composites are primarily associated with using the methods of nanoscale molecular design of interphase layer. Broad opportunities for its specific organization open the development of synthetic approaches to obtaining the polymers of complex spatial architecture [6, p. 304; 1, p. 1328]. These include, in particular, molecular brush and comblike polymers. As shown in Fig. 2, the structure of macromolecules involves covalent attachment

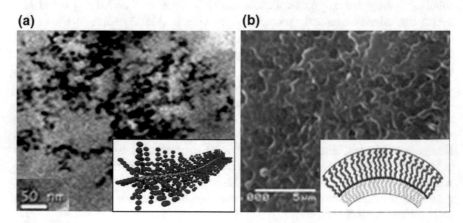

Fig. 2 Diagram of macromolecules structure and molecular brushes micrographs (**a**) and comblike polymers (**b**)

to the main chain of narrowly dispersed lateral radicals [7, p. 79; 8, p. 378]. Along with this the rigidity of the main valences chain is determined by the interaction of the lateral branches. Thus, flexibility of poly-α-olefins chain, polyalkylmethacrylates and polyalkylacrylates is reduced in 3 ... 4 times with increasing number of monomer units in the lateral branches of 1–10. For grafted block copolymers of polyimide and polymethylmethacrylate the equilibrium hardness is increased by tens of times in comparison with the index level for the initial aromatic polyimides [9, p. 1].

The realized in the project idea provides self-assembly of the composite materials interphase layer with the penetration of lateral branches of block copolymer binder into the pore structure of fibrous component. For this purpose the project implements for the first time a set of methods of nanoconstructing of composite materials and produced on their basis packages of shape forming product details, which involves the rational combination of the following complementary techniques:

- ultradispersing of the used polymer dispersions for their penetration into the nanoporous structure of fibrous carrier;
- nanolayer chemical modification of the fibrous carrier structure for regulating the penetration depth of polymerizing compositions and the area of hybrid formations development;
- polymerization process management and formation of polyblock, nanoporous and spatially oriented structures;
- the introduction of a rational number of nanodisperse reinforcing modifiers.

Both textile and leather materials have microfibrillar structure with submicroscopic pore system, as well as mesoporous spaces between the elements of supramolecular structure (fibrils). Their dimensions in the dry state are respectively 1 ... 2 and 10 ... 15 nm, and increase when swollen to 3 ... 7 and 25 ... 35 nm. Substances having a larger size cannot penetrate into the fiber structure and are located only on the surface of material and in its interfiber spaces. Figure 3 shows that the reinforcement effect is only achieved providing the penetration of reinforcement component of polymeric binder into the fibrous material internal pore spaces (curves 3, 4) and advantageously into the submicroscopic pore system (curve 5). To determine the dimensional parameters of nanoparticles in hydrosol of reinforcing dispersions the method of dynamic light scattering has been used.

For the development of internal volume in synthetic materials, effective method of localized surface saponification in the presence of interphase catalyst-carrier, ensuring the formation of nano-sized etch pits has been suggested [10, p. 105]. Assessment of changes in surface microrelief materials by scanning electron microscopy shows the presence of multiple point "defects" with transverse dimensions of about 50 nm and to 90 nm deep. By using the method of gas adsorption it has been determined that the main growth of free volume is due to the voids with a diameter of 35 nm, which makes nanomodification conditions favorable for introduction of dispersed forms of reinforcing polymer component into the fiber structure.

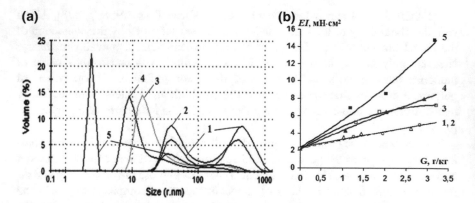

Fig. 3 Dependences of distribution by particle size of the dispersed phase volume in the preparations of reinforcing polymer dispersions (**a**) and the effect of their application on the stiffness of the garment materials duplicated package (**b**): *1, 2* preparations of surface and macroporous dislocation; *3, 4* preparations of mesoporous dislocation; *5* preparation penetrating into the fibrous carrier submicroscopic pores

The choice of reinforcing dispersions is based on the study results of copolymerization reactions with macromolecules thermoplastic adhesive (IR-spectroscopy method) and temperature intervals of their course in accordance with technological constraints (the method of differential scanning calorimetry) [11, p. 112]. According to the research results the matching pairs of adhesive and reinforcing components of polymeric binder as well as technological modes of producing the nanomodified prepregs have been determined. Reinforcement of gasket materials allows up to 20 times increase the stiffness of duplicate formative details used in the design of clothing, leather goods and footwear. The additional use of nanodisperse reinforcing additives allows up to 10-times increase the extent of discreteness in regulating the composite elastic-deformation properties and ensure the conformity of the gasket material to the designed volume and silhouette solution of products.

The substantiation of multilayer composite materials structures for special garments de-signing [12, pp. 157–161; 13, pp. 89–94] promotes creation of: (1) power reflecting materials with metal particles and development of multilayer composite heat-insulation materials with visually-optical and thermal masking properties; (2) sealing homogeneous multilayer film materials with high adhesive activity for impenetrability of details connections.

The updating metal technology of a textile basis is defined by its fibrous structure: ion-plazma metals dispersion is intended for chemical materials and proceed in low-temperature plasma soft conditions and in deep vacuum; alluvial drawing of a chemical composition with metal particles on textile cotton or mixed basis. Constructive decisions variants of heat-insulation materials are presented in Fig. 4a, b. The microporous membranous layer is executed from the thermoplastic polyurethane pitches with the sizes of holes $1.3–1.6 \times 10^{-6}$ m (variant *a*) and water solution of the film-forming polymer with pigmentary dyes addition (variant *b*). The metallized layer is executed from nitride of the titan and put on a thermoplastic polyurethane pitches material by thickness till 100 nm in number of 1–2 g/m^2 (variant *a*). The aluminium powder and acrylic aethers and pigmentary dye mixes is put on the basis instead of nitride of the titan (variant *b*).

The metallized covering of a material inside promotes a part of a thermal stream reflexion, which is transferred from a warm-blooded object to environment by radiation and takes away heat at the expense of evaporation [14, pp. 90–96; 15, pp. 14–17]. Textile materials with camouflaging and water-repellent finishings provide for the biological objects protection against a rain and snow. The camouflaging drawing is put on a textile basis. It can be executed as colour backgrounds, as environment backgrounds, where there are objects. It provides a visually-optical camouflage for materials in the afternoon.

The special sealing material is developed for realization of glutinous blocking technology. The thinnest layers macromolecular structurization of a film provides distinctions of their properties. The incomplete internal reflexion of IR-spectroscopy method is used as the basic studying method of a chemical compositions compound for the analysis of functional groups presence in the investigated system «an adhesive—a substratum» . It is used for an estimation of interactions character in the analyzed systems. Studying of interaction character and presence of chemical bonds of glutinous connection «a composite material—the self-glued film material» has allowed to develop and fill up system of scientific knowledge of self-glued materials application in development of glutinous technology. The formation diffusion molecular monolayer is confirmed by the microscopic researches (Fig. 5). They testify to dense thick contact as a result of phases rap-proachement up to distance of the order of 40 nm. The designed multilayer hermetic in the form of the set width rolled film possesses adhesive capacity to big substrata spectrum. It possesses constant residual stickiness property, at the most meets technological and economic requirements of products manufacture from materials with a defensive covering. The effective technology of glutinous hermetic sealing provides an impenetrability of special clothes connections for liquid environments.

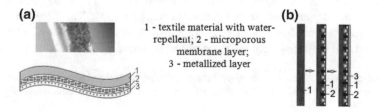

(a)

1 - textile material with water-repellent; 2 - microporous membrane layer; 3 - metallized layer

(b)

Fig. 4 The structure of materials containing metal particles

Fig. 5 Picture of the phase
boundary in the glue
joint «Adhesive film
material + fabric Action
Mistral»

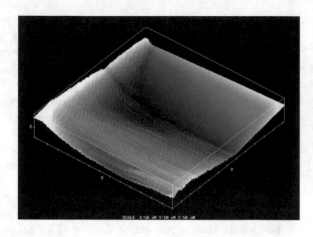

Conclusions

The specifics and general patterns of change in physical and physico-chemical
properties of polymer-inorganic systems and effective methods of composite
materials nanoengineering have been revealed. A quantitative assessment is given
of the impact of mechanoacoustic processing of calcium chloride hydrosol, that is
widely used as a modifier of cement slurries, on the change in nanoparticles size at
the dispersed phase, in parameters of nanoporous structure of cement stone formed
and its strength characteristics.

Scientific and technological groundwork for obtaining demanded in sewing
production prepregs and shape forming materials with high precision regulation of
elasticity-and-deformation properties has been created. The used methods of
nanostructural modification provide 3D-polymerization processes of the interphase
layer composite materials formation with nanoporous structure of fibrous carrier
and additional introduction of nanodisperse reinforcing additives.

The ways of updating of textile and film materials are proved. Additional
functional layers formation of composite materials is realized at the expense of
physical and chemical influences using on the macromolecular structure of these
polymers. Special materials with the improved power reflecting and hermetic
properties are created for special garments manufactures. The creation and the
introduction in manufactures of new composite materials provide a reception of
household and a special purpose garments with new functional characteristics.

Acknowledgments Researches are carried out (conducted) with the financial support of the state
represented by the Ministry of Education and Science of the Russian Federation. Agreement
(contract) 11.1798.2014/K.

References

1. Subbotin, A.V., Semenov, A.N.: Spatial self-organization of comb macromolecules. Polym. Sci. Ser. A. **49**(12), 1328–1357 (2007)
2. Koksharov, S.A., Bazanov, A.V., Fedosov, S.V., Akulova, M.V., Slizneva, T.E.: Analysis of the influence of the calcium chloride dispersity in mechanoactivated solution on structure and characteristics of cement stone. Build. Mater. (1–2), 55–61 (2016)
3. Berne, B.J., Pecora, R.: Dynamic Light Scattering. 376 p. Wiley, New York (1976)
4. Koksharov, S.A., Kornilova, N.L., Meteleva O.V.: Metod of solvent preperation for evalutiation of nano-dispersive objects by dynamic light scattering method. Izv. Vyssh. Ucebn. Zaved. Tekhnol. Text. Promyshlennosti. (1), 167–170 (2014)
5. Majorov, P.M.: Concrete mixes: a formulary for builders and producers of structural materials. Rostov-on-Don: Phoenix, 461 p (2009)
6. Plate, N.A., Shibayev, V.P.: Comb polymers and liquid crystals. M. Chimija. 304 c (1980)
7. Kolbina, G.F., Shtennikova, I.N., Kononov, A.I., et al.: Effect of the molecumar architecture of side radicals on the optical properties of comb-shaped polymers. Polymer Sci. Ser. C. **49**(1), 79–82 (2007)
8. Tsvetkov, V.N.: Rigid polymeric molecules L. 378 p. Nauka, Russia (1986)
9. Filippov, A.P., Belyaeva, E.V., Krasova, A.S. et al.: Synthesis and investigation of the solution behavior of graft block copolymers of polyimide and poly(methyl methacrylate). Polym Sci. Ser. A. **56**(1), 1–9 (2014)
10. Koksharov, S.A., Kornilov, N.L., Fedosov, S.V.: Modification of polyester fibers to create composite materials with adjustable rigidity. Izv. Vyssh. Ucebn. Zaved. Khimia i Khim. Tekhnolgija. **59**(6), 105–111 (2016)
11. Koksharov, S.A., Kornilov, N.L., Fedosov, S.V.: Developmend of rainforced composite materials with nanoporous textile carrier and the brush structure polymer interfacial layer. Rossijskij khivicheskij gurnal. **59**(3), 112–123 (2015)
12. Fedosov, S.V., Akulova, M.V., Koksharov, S.A., Meteleva, O.V.: Theoretical Foundations of heat and mass production of advanced technology materials textile and construction industries. Proceedings of higher education institution. Text. Ind. Technol. Sci. Tech. J. **6** (360), 157–161 (2015)
13. Chizhik, M.A., Rasskazova, M.N., Starikov, V.I.: Structural approach to modelling multi-component systems of fabric for products of clothing industry. Proceedings of higher education institution. Text. Indus. Technol. Sci. Tech. J. **6**(354), 89–94 (2014)
14. Belova, I.U., Veselov, V.V., Korolova, C.V.: Materials with metallon-dusting for protection from hazardous production factors. Proceedings of higher education institution. Text. Ind. Technol. Sci. Tech. J. **4**(346), 14–17 (2013)
15. Belova, I.Y., Veselov, V.V.: Technological aspects of products processing from composite materials containing special metal nanolayers. Proceedings of higher education institution. Text. Ind. Technol. Sci. Tech. J. **5**(345), 90–96 (2013)
16. Meteleva, O.V., Dyakonova, E.V., Bondarenko, L.I.: Self-glued material as the basis of formation of impenetrable connection in clothes Proceedings of higher education institution. Text. Ind. Technol. Sci. Tech. J. **5**(353), 105–108 (2014)

High-Speed Laser Direct Deposition Technology: Theoretical Aspects, Experimental Researches, Analysis of Structure, and Properties of Metallic Products

K.D. Babkin, V.V. Cheverikin, O.G. Klimova-Korsmik, M.O. Sklyar,
S.L. Stankevich, G.A. Turichin, A.Ya. Travyanov, E.A. Valdaytseva
and E.V. Zemlyakov

Abstract Additive technologies as an alternative to traditional methods arouse great interest in many industrial sectors. In recent years, many theoretical and experimental articles devoted to additive technologies and their applications have been published. In spite of it, now the area of outstanding issues is still remaining. The paper presents results of theoretical and experimental researches devoted to the stability of products formation from different metallic alloys with complex

K.D. Babkin · O.G. Klimova-Korsmik · M.O. Sklyar (✉) · S.L. Stankevich · G.A. Turichin ·
E.A. Valdaytseva · E.V. Zemlyakov
Institute of Laser and Welding Technologies, Peter the Great St. Petersburg Polytechnic
University, Saint Petersburg, Russia
e-mail: skmar.spb@gmail.com

K.D. Babkin
e-mail: babkin.kd@mail.ru

O.G. Klimova-Korsmik
e-mail: o.klimova@ltc.ru

S.L. Stankevich
e-mail: ssl.07@mail.ru

G.A. Turichin
e-mail: gleb@ltc.ru

E.A. Valdaytseva
e-mail: Ekaterina@ltc.ru

E.V. Zemlyakov
e-mail: e.zemlyakov@ltc.ru

V.V. Cheverikin · A.Ya. Travyanov
National University of Science & Technology (MISiS), Moscow, Russia
e-mail: cheverikin80@rambler.ru

A.Ya. Travyanov
e-mail: travyanov@mtr.misis.ru

© The Author(s) 2018
K.V. Anisimov et al. (eds.), *Proceedings of the Scientific-Practical Conference
"Research and Development - 2016"*, https://doi.org/10.1007/978-3-319-62870-7_53

geometry form, which were manufactured using high-speed direct laser deposition technology. The prospect of application of the method for the manufacturing of various materials details and products for various areas of engineering has been shown.

Keywords Additive manufacturing · High-speed direct laser deposition Surface stable formation · Mechanical properties · Metal powder Nickel-based alloys · Ultrafine structure

New manufacturing technologies are actively incorporated into most areas of engineering: aircraft industry, shipbuilding, medicine, engine construction, etc. The additive technologies have most popularity and become the foundation of a new industry, bringing together digital production, design and through manufacturing cycle. Methods of selective laser melting (SLM) are already integrated in the production both in Russia and in other countries [9, p. 290; 2, p. 36; 3, p. 42; 5, p. 98].

The most promising technology for industry of large-sized products from metal powder is high-speed direct metal (or laser) deposition (HSDMD, HSDLD, DMD, DLD). This technology is based on forming product geometry by 3D cladding of metal powder with laser beam as a heating source [12, p. 177; 10, p. 173]. DMD technology has high productivity and allows building parts for the few minutes, the common growing period for SLM technology is few hours [8, p. 410]. The additive technology of high-speed direct laser deposition application is perspective for the industry because production operations coordinate in uniform parallel process [1, p. 177].

The authors carried out comprehensive theoretical studies on high-speed direct laser deposition process, developed the technology of this process, and produced a working prototype of the equipment. With using these equipment is produced parts from different metal powders, which were investigation to microstructure and mechanical properties.

Theoretical Research

One of the main features of HSDLD technology is high productivity, which can attain 15 kg of metallic powder per hour: During growing process motion of the head, relatively product occurs with a high speed. Increase of cladding head motion speed with respect to the product leads to the development of active zone surface instabilities, causing the appearance of defects in the formation of products—quasi-periodic relief on the surface and interrupt the growth process. In the paper [11, p. 676], the stability of the process technology is described in details. The condition of process stability is as follows:

$$2\frac{\sigma H^2}{v_0 b^3 L} < \frac{\partial j}{\partial z},\tag{1}$$

where σ—the surface tension; H—depth of molten pool; v_0—linear velocity of the laser beam over the surface grown item; b—half the width molten pool; $\partial j/\partial z$—the gradient of mass flux density in the gas powder jet at the normal to the surface of the grown product. Figure 1 shows the stabilization process to the surface.

It should be noted that the depth of molten bath in cases 1–3 is related as follows:

$$H_1 > H_2 > H_3\tag{2}$$

Position 2 is level of stable formation of a surface, 1 and 3 of the acceptable process variations, respectively. When there is a change of the depth of molten pool (H) to H_1 level length L and width b increase. The same situation occurs, when there is a decrease of the depth the molten pool (H) to the H_3 level—the length L and width b decrease. For stabilization of surface position, there are necessary flux density decreases at the displacement to position 1, as a result, the thickness of the deposited layer decreases. When the depth of molten pool change to H_3 level thickness of the deposited layer increase, at the constant value of flux density for every 1–3 cases. Thus, the particle flux density determines the stability of the process, and for the fulfillment of this condition is to have the focus of gas powder jet below the working area of deposition.

High-speed growth process also leads to the presence of partially molten particles in the microstructure of deposited samples (Fig. 2a). In the work [4, p. 462; 7, p. 718], the heterophase process is shown on nickel-based alloy (Inconel 625). Figure 2b shows how it looks during the deposition process.

Heating gas powder jet laser affects the appearance of heterophase structures in the metal deposited products. Unmelted nuclei of particles is dropping in melt pool and save initial structure in structure of deposited part. In melt pool act hydrodynamic processes because of them unmelted powder particle move to the edge melt

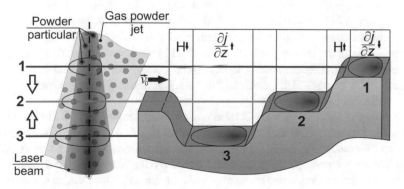

Fig. 1 Scheme of stabilization process

(a)   **(b)**

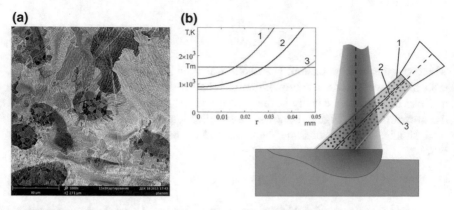

Fig. 2 Microstructure of deposited sample (**a**) result of simulation of heterophase process (**b**)

pool where crystallization occurs. Based on the simulation results, it was found that temperature field (Fig. 2b) of powder and particle radius of unmelted particles depends directly on the residence time of the particles in the laser radiation zone. Exposure time depends on the speed of the gas powder jet and powder particles trajectories of movement. The presence of heterophase process by HSDLD process allows to get the fine structure of metal products.

Materials and Methods

Theoretical and experimental studies of deposition processes were carried out at the Institute of Laser and Welding Technologies SPbPU (ILWT) using experimental setup (Fig. 3). The equipment consists of 5 kW Yb fiber laser LS-5 with external focusing optics, made on the base of IPG welding head, a powder disc feeder Medicoat AG with a COAX-11 jet, and a two-dimensional moving stage equipped with a revolving motor. Gas powder jets were formed by lateral and coaxial nozzles, designed with consideration of deposition stability condition. For deposition tool motion, a high precision 6 axis FANUC robot has been used.

Fractional composition of the powders used was varied in the range of 50–150 microns, the shape of the particles—spherical. The deviation from sphericity did not exceed 5%. Experimental studies were conducted on nickel-based alloys (Inconel 625, Inconel 718, GS6U, EI698P etc.), titanium (Grade 2, Grade 5, VT20, etc.), and steel (316L).

Metallographic studies were carried out on microscope DMI 5000 (Leica) with software Tixomet. Researches of chemical compound and the distribution of chemical elements are made on scanning electron microscope Phenom ProX and Mira Tescan microscope using console Oxford INCA Wave 500. To determine the mechanical properties, samples were tested on uniaxial tension, using universal testing machine Zwick/Roell Z250 Allround.

(a)　　　　　　　　　　　　　　　　　　　**(b)**

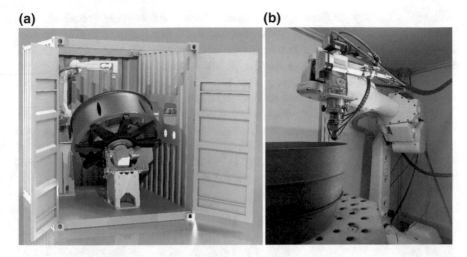

Fig. 3 Experimental setup: **a** scheme, **b** photo

Experimental Results

As a result of the work using the lateral and coaxial nozzles products with different geometry and from the different powder alloys were manufactured. Using different gas powder nozzles, high process productivity was achieved. Lateral feed with scanning has productivity of manufacturing part blanks above 18 kg/h (thin wall 3–20 mm); lateral feed by focused gas powder jet has productivity above 5 kg/h (wall 0.8–3 mm); coaxial feed by focused jet—productivity above 1 kg/h (wall 0.6–2 mm) (Fig. 4).

Process parameters were chosen for each group of alloys (iron-based, titanium-based, nickel-based, including a high-bar gamma phase). Samples of

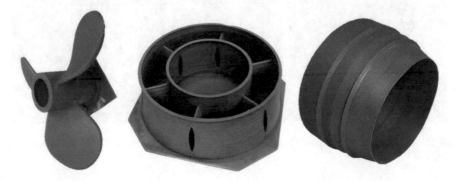

Fig. 4 Deposited samples

different geometry without pores, cracks, poor fusion were obtained. The microstructure of the obtained products contains the fine grains with size less than 30 microns for nickel and iron alloys and 70 micron for titanium alloys (Fig. 5). Features of formation structure for the different alloys using this technology are given in the article the authors. [6, p. 980; 7, p. 718]. Fracture of HSDLD samples is characterized by ductile destruction in comparison with cast samples.

In comparing with cast samples, the products are manufactured by high-speed direct laser deposition technology and have ultrafine structure. It provides high level of mechanical characteristics (Fig. 6). For example, tensile strength of samples, which were produced from GS6U alloy by HSDLD, is higher 20% in comparison with cast samples, elongation of the material is increased to 3 times.

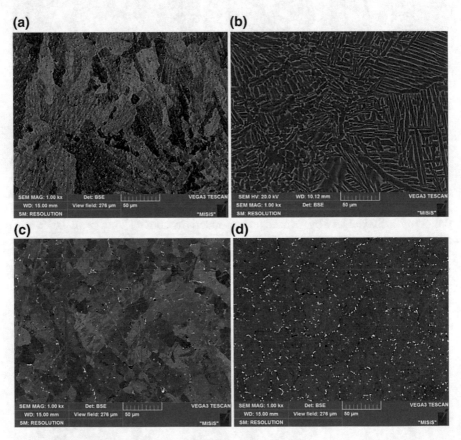

Fig. 5 Microstructure of deposited samples: **a** 316L steel, **b** titanium alloy VT20, **c** Inconel 625, **d** GS6U

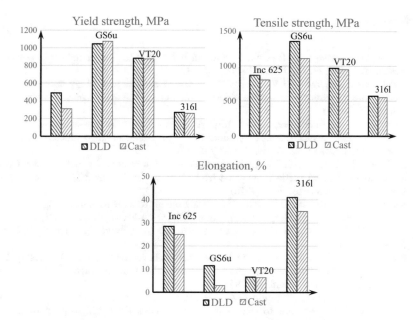

Fig. 6 Mechanical properties samples in deposited condition and cast condition

Conclusions

Direct laser deposition technology is a complex process with a large number of important parameters. The main requirements are stability of process formation of grown details and heterophase nature of the process. As a result, high-quality products with ultrafine structure could be achieved.

According the results of the project, first Russian equipment, which realize direct laser deposition technology, were manufactured. High-speed direct laser deposition is

- technology for manufacturing details with complex form from powder materials using 3D model;
- potential use different materials in a single part and obtaining details with gradient properties;
- dimensions of details are almost unlimited.

The process productivity is 10 times higher in comparison with layered synthesis technologies. Mechanical properties of the deposited parts at a level of hot rolled metal, there are no porosites, cracks.

Acknowledgements Project was done with the financial support of the Russian Ministry of education and science, FOP "Research and development on priority directions of scientific-technological complex of Russia for 2014-2020" Agreement (contract) no. 14.581.21.0010 December 01,2014 Unique project Identifier: RFMEFI5814X0010

References

1. Adova, I.B., Azimov, Yu.I. et al.: Theoretical bases of formation of industrial policy. In: Babkin, A.V. (ed) FGAOU VO SPbPU. 462 p (2015)
2. Ageev, R.V., Kondratov, D.V., Maslov, Y.V.: Use of additive technology in the design and manufacture of parts for aerospace objects. Polet. Obscherossijsliy nauchno-technicheskiy J. **6**, 35–39 (2013)
3. Dorochov, A.F., Abacharev, M.M., Additive technology in the production of shipboard power. Vestnik Astrachanskogo gosudarstvennogo technicheskogo universiteta, Seria: Morskaya technika I technologia. **2**, 42–47 (2015)
4. Glukhov, V., Turichin, G., Klimova-Korsmik, O., Zemlyakov, E., Babkin, K.: Quality management of metal products prepared by high-speed direct laser deposition technology. Key Eng. Mater. **684**, 461–467 (2016)
5. Kianiana, B., Tavassolib, S., Larsson, T.C.: The Role of Additive Manufacturing Technology in job creation: An exploratory case study of suppliers of Additive Manufacturing in Sweden. Procedia CIRP, 12th Global Conference on Sustainable Manufacturing. Emerging Potentials. **26**, 93–98 (2015)
6. Klimova-Korsmik, O.G., Turichin, G.A., Zemlyakov, E.V., Babkin, K.D., Petrovskiy, P.V., Travyanov, A.Ya.: Structure formation in Ni superalloys during high-speed direct laser deposition. Mater. Sci. Forum. **879**, 978–983
7. Klimova-Korsmik, O.G., Turichin, G.A., Zemlyakov, E.V., Babkin, K.D., Travyanov, A.Ya., Petrovskiy, P.V.: Structure formation in Ni superalloys during high-speed direct laser deposition. Phys. Procedia. **83**, 716–722 (2016)
8. Olakanmi, E.O., Cochrane, R.F., Dalgarno, K.W.: A review on selective laser sintering/melting (SLS/SLM) of aluminum alloy powders: Processing, microstructure, and properties. Prog. Mater. Sci. **74**, 401–477 (2015)
9. Seabra, M., Azevedo, J., Araújo, A., Reis, L., Pinto, E., Alves, N., Santos, R., Mortágua J.P.: Selective laser melting (SLM) and topology optimization for lighter aerospace components. Procedia Struct. Integrity. **1**, 289–296 (2016)
10. Turichin, G.A., Klimova, O.G., Zemlyakov, E.V., Babkin, K.D., Kolodyazhnyy, D.Yu., Shamray, F.A., Travyanov, A.Ya., Petrovskiy, P.V.: Technological aspects of high speed direct laser deposition based on heterophase powder metallurgy. Phys. Procedia. **78**, 397–406 (2015)
11. Turichin, G., Zemlyakov, E., Klimova, O., Babkin, K.: Hydrodynamic instability in high-speed direct laser deposition for additive manufacturing. Phys. Procedia. **83**, 674–683 (2016)
12. Wilson, Michael J., Piya, C., Shin, Y.C., Zhao, F., Ramani, K.: Remanufacturing of turbine blades by laser direct deposition with its energy and environmental impact analysis. J. Clean. Prod. **80**, 170–178 (2014)

Synthesis and Properties of Energetics Metal Borides for Hybrid Solid-Propellant Rocket Engines

S.S. Bondarchuk, A.E. Matveev, V.V. Promakhov, A.B. Vorozhtsov, A.S. Zhukov, I.A. Zhukov and M.H. Ziatdinov

Abstract In this paper, the problems of production and characterization of microsized metal borides (including aluminum, titanium, magnesium) are discussed. The preferences of application for high-energy materials are presented. The problems of chemical stability and chemical compatibility are discussed. A technique for production of metal borides is also described which is known as self-propagating high-temperature synthesis (SHS) and the subsequent mechanical treatment. The result is microsized borides which have an average size of around 5 microns with a sharp curve for distribution sizes. The purity is enough for use as fuel of high-energy materials hybrid solid-propellant rocket engines. The results of SEM, X-Ray, DSC, and TG analyses are also presented and discussed.

S.S. Bondarchuk (✉) · A.B. Vorozhtsov · A.S. Zhukov
Doctor of Physico-Mathematical Sciences, Tomsk State University, Tomsk, Russia
e-mail: isbi@mail.ru

A.B. Vorozhtsov
e-mail: abv1953@mail.ru

A.S. Zhukov
e-mail: Zhuk_77@mail.ru

A.E. Matveev
Tomsk State University, Tomsk, Russia
e-mail: cool.mr.c@mail.ru

V.V. Promakhov · I.A. Zhukov · M.H. Ziatdinov
Candidate of Engineering Sciences, Tomsk State University, Tomsk, Russia
e-mail: vvpromakhov@mail.ru

I.A. Zhukov
e-mail: gofra930@mail.ru

M.H. Ziatdinov
e-mail: ziatdinovm@mail.ru

S.S. Bondarchuk
Tomsk State Pedagogical University, Tomsk, Russia

© The Author(s) 2018   511
K.V. Anisimov et al. (eds.), *Proceedings of the Scientific-Practical Conference*
"Research and Development - 2016", https://doi.org/10.1007/978-3-319-62870-7_54

Keywords Metal borides · High-energy materials · SHS process
X-ray analysis · Properties · Oxidation

Safety requirements and the cost of placement into the required orbit determine the trends of improvement of propulsion systems of boosters. A possible compromise is associated with the use of hybrid engines including solid-propellant rocket engines (HSRE) which can also use liquid and gaseous oxidizers [1, 4]. Borides are advanced materials for use as fuel additives for energetic systems. The potential of use of metal powders (Al, Mg and others) is almost exhausted today. New additives with higher energy characteristics are required. Boron is the best alternative so far. Heat of combustion for boron is almost two times higher than for aluminum. Production technologies for boron and its compounds are well known and tested. Moreover, boron is nontoxic, occurs in nature in large quantities, and is produced on an industrial scale.

From the very beginning of the research in the field of combustion synthesis, a lot of promising systems from the practical point of view were determined, however, a self-sustaining process was not possible due to their insufficient exothermicity. The first approach was pumping of additional heat by means of preheating of initial mix in a resistance furnace. It was the production of intermetallides when the operation of increasing of initial temperature of reaction mix was used for the first time for SHS reactions. The increase of initial temperature of the mix up to 50–500 °C made it possible to synthesize aluminides of Ni, Co, Ti, Cr, Mo, and other metals in combustion mode. When such furnace SHS technology is used in practice, its characteristic advantages such as zero energy consumption, simplicity of equipment and low time consumption are reduced to zero.

The use of potentially high boron exothermicity becomes possible if it is used in the form of metal borides which also have high values of the heat of combustion. It was found that borides of Al and Mg as well as Ti and Zr are the most promising ones.

This paper describes a laboratory technology of production of borides of Al, Ti, Mg, etc., including double and mixed compounds. Elemental boron, boron carbide, and iron boride were used as boron source. The experimental results of XRD analysis, TGA, and particle size analysis obtained for synthesized powders are given.

Self-propagating high-temperature synthesis is the most suitable method for the development of new energy materials which makes it possible to produce ultrapure product with target chemical and phase composition by means of adjustment of synthesis parameters. Preliminary studies have indicated that it is possible to produce borides with high content of target phase. According to DTA data, the degree of oxidation of obtained powders exceeds 95%.

If we compare the heat release of different metals (Table 1), it appears clear that boron and lithium have the best characteristics for this purpose.

Lithium cannot be used due to its toxicity and low processability (low melting point and high chemical activity). Moreover, metallic lithium is rare and very expensive. Therefore, there are no alternatives are to boron regarding exothermicity and availability. The heat of combustion for boron is almost two times higher than that for aluminum and boron has a high availability in nature. However, due to specific properties of its oxide B_2O_3 (high boiling point and high viscosity of oxide

Table 1 Heat of combustion of energetic metals [2]

Element	Heat of combustion, kJ/g
Li	43.5
B	57.2
Mg	25.1
Al	31.4
Si	32.2
Ti	15.7
Zr	12.0

melt), the rate of oxidation of elemental boron in fuel composition is unacceptably low. Also, boron oxide causes an appreciable agglomeration which is an extremely undesirable phenomenon. Well-known fundamental works by A. Gany and other scientists demonstrate the possibilities of using boron as fuel, however, there are no widespread technologies making it possible to use this element as a component of high-energy materials [3].

Currently, the optimal solution for the task of improving the efficiency of metallic additives to fuel compositions would be a full or partial transition to the use of metal borides.

This paper provides an overview of the results following a comprehensive analysis of the properties of different micro- and nanoparticles which can be used as additives for HEM.

The use of potentially high boron exothermicity becomes possible if it is used in the form of metal borides which also have high values of the heat of combustion. The oxidation of boron particles unfortunately results in the formation of a solid layer of oxide B_2O_3 and the diffusion of oxygen through this layer is hindered. The oxidation of metal borides does not result in the formation of an oxide layer impenetrable for oxygen. Diffusion of oxygen through the faulted layer of complex oxides is facilitated and the degree and rate of oxidation is increased.

It was found that borides of aluminum and magnesium as well as titanium and zirconium are the most promising (Table 2). Values of the heat of combustion for these borides are given below. Heat of combustion of borides is significantly higher than the values for corresponding metals.

Vacuum heat technology is used for boride synthesis. At the same time, the formation of borides is associated with significant heat liberation. This energy is

Table 2 Heat of combustion of energetic borides

Borides	Heat of combustion, cal/g
AlB_2	9.430
AlB_{12}	12.160
MgB_2	9.050
TiB_2	5.700
$Mg_{0.5}Al_{0.5}B_2$	9.240
ZrB_2	4.230

often enough for the process in combustion mode according to self-propagating high-temperature synthesis (SHS). Large-scale furnace equipment is not necessary in this case, and most importantly, the synthesized materials have better performance characteristics.

At the same time, there are a lot of systems including borides which do not have enough heat release for SHS processes. In such cases, there are two possible variants of the SHS process: Energy can be pumped in from external sources or energy can be recovered. External energy can be introduced in the form of physical or chemical heat. In case of physical heating, the initial SHS mix is placed into an electric furnace and heated up to the required temperature and then an SHS reaction is initiated. In this case, both layer-by-layer and overall combustion modes can be provided. Preheating of exothermic mix or its part is widely used in metallothermic production of metals and master alloys in particular.

Another possible option of improving the exothermicity of the mix is by introducing additional chemical heat. This method is widely used in aluminothermy in the process of production of complex ferroalloys.

A.G. Merzhanov [6] stated the five most typical situations in the process of classification of chemical routes of SHS reactions. One of these types of routes was called "chemically independent routes in thermally coupled systems ("chemical furnace")". In these cases, chemical reactions proceed independently, however, the heat from the more exothermic reaction provides energy for the less exothermic one.

From the very beginning of the research in the field of combustion synthesis, a lot of promising systems from the practical point of view were determined, however, a self-sustaining process was not possible due to their insufficient exothermicity. The first approach was pumping of additional heat by means of preheating the initial mix in a resistance furnace. It was the production of intermetallides when the operation of increasing of initial temperature of reaction mix was used for the first time for SHS reactions. The increase of the initial temperature of the mix up to 50–500 °C made it possible to synthesize aluminides of Ni, Co, Ti, Cr, Mo, and other metals in combustion mode. When such furnace SHS technology is used in practice its characteristic advantages such as no energy consumption, simplicity of equipment, and low time consumption are reduced to zero.

The next step which extended the potential of SH synthesis for low-exothermicity systems was the invention of a so-called "chemical furnace". This term was coined by V.M. Maslov [5] for the process of synthesis of intermetallides in Nb–Al and Nb–Ge systems. Equiatomic mixture of Ni and Al powders with combustion temperature of 1640 °C was used as a material of "chemical furnace".

The principle of the "chemical furnace" could also be applied to the synthesis of compounds such as WC, NbC, SiC, B_4C, Al_4C_3, VC, Mo_2C, WB, WB_2, and others which do not have enough heat liberation for a general SHS process [7, 8]. In this case, mixtures with higher burning temperatures such as Ti or Zr powders with C or B (Table 3) can be used as materials for a "chemical furnace".

Table 3 Heat of formation of energetic borides

Borides	Heat of formation, cal/g
TiB_2	1006.6
ZrB_2	778.3
TiB_{12}	–
ZrB_{12}	542.8
AlB_2	–
AlB_{12}	–
MgB_2	289.4
MgB_{12}	223.1

In the case of insufficient heat liberation, the combustion synthesis can be combined with elements of the furnace synthesis. Energy is pre-pumped into the system providing a further combined SHS process. An initial mix of metal and boron is used for the implementation of boride combustion synthesis:

$$Ti + 2B \rightarrow TiB_2,$$
$$Al + 2B \rightarrow AlB_2,$$
$$Zr + 2B \rightarrow ZrB_2.$$

A laboratory-scale production technology for borides of Al, Ti, Mg, and other metals including double and mixed compounds has been tested. This has resulted the synthesized powders having particle sizes of $\delta_{50} \approx 10$ μm.

Production of borides using the SHS method involves three stages: preparation of the exothermic mix, combustion synthesis, and product processing. The first stage includes the operations of batching and mixing of initial powders in stoichiometric proportions and production of cylindrical samples with different diameters.

After this, the compressed tablets are placed inside a research SHS reactor with a sealed working chamber which is vacuumized and filled with an inert gas (Ar, He) to provide pressure level of 0.2 MPa. Ignition is then initiated by application of electrical pulse to a tungsten spiral with subsequent formation of a plane combustion front.

At the end of propagation of a synthesis wave and sample cool down, the SHS reactor is depressurized and the sample is processed and analyzed. Particle size distribution and phase composition of the powder produced by milling in a ball mill is determined along with the free boron content. Energy characteristics of the boride powder are then analyzed using a DTA (Differential Thermal Analysis) method.

The results of X-ray phase analysis of some synthesized borides are shown in Fig. 1 as an example. Particle size distribution as well as DTA (Fig. 2) data for AlB_2 were obtained.

According to X-ray phase analysis data, the content of the target phases in the powders studied comprised: 88.04% for $Al_{0.5}Mg_{0.5}B_2$ phase, 93.18% for AlB_2 and 98.43% for TiB_2. The average size for target phases did not exceed 30–40 nm.

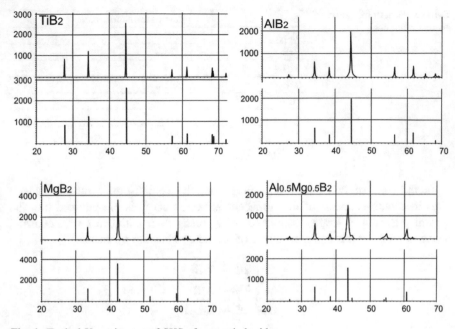

Fig. 1 Typical X-ray images of SHS of energetic borides

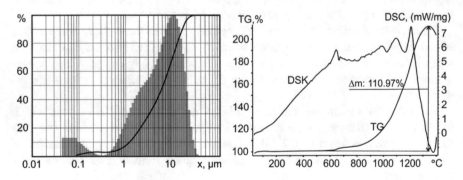

Fig. 2 Particle size distribution and DTA analysis data for AlB_2 powder produced by means of SH synthesis

The analysis of the particle size distribution in AlB_2 powders have indicated that the average particle size comprises 6.2 μm. The maximum size of the fraction was δ_{99}—24.9 μm. The structure of some powders (AlB_2, TiB_2, AlB_{12}) is shown in Fig. 3. According to DTA data, the sample mass increased ~2.17 times. The increase in the mass was thought to be due to the following reaction $2AlB_2 + 3O_2 = Al_2O_3 + B_2O_3$. Comprises ~2.26 UNCLEAR. It was also demonstrated that the degree of oxidation was 96%. General information on all received oxidation borides compared with aluminum and boron are shown in Fig. 4.

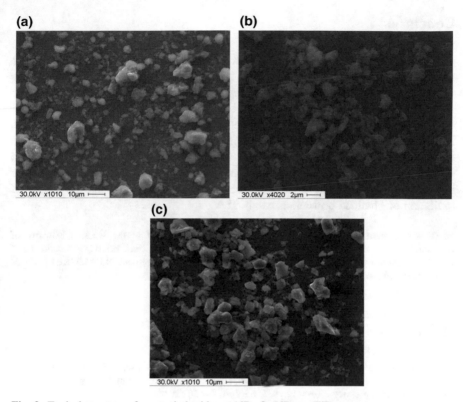

Fig. 3 Typical structure of energetic borides **a** AlB_2, **b** AlB_{12}, **c** TiB_2

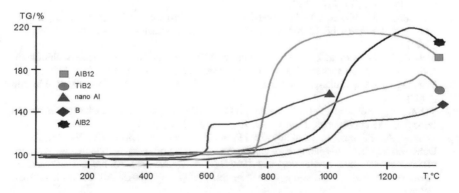

Fig. 4 Data TG analysis of boron, borides and nanoaluminum

Conclusion

This paper has presented the techniques for the production of borides which can be used as fuels for high energetic materials hybrid solid-propellant rocket motor. The main characteristics of nano- and microparticles have been shown and discussed.

Synthesis of aluminum borides is one of the most promising areas in the field of development of new energy materials. Self-propagating high-temperature synthesis is the most suitable method for these purposes, which makes it possible to produce ultrapure products with target chemical and phase composition by adjusting the synthesis parameters. Previous studies have indicated that it is possible to produce borides with high content of target phase. According to DTA data, the degree of oxidation of obtained powders exceeds 95%.

Acknowledgement Project was done with the financial support of the Russian Ministry of education and science, FOP "Research and development on priority directions of scientific-technological complex of Russia for 2014–2020". Agreement No 14.578.21.0034 5. June 2014. Unique project identifier: RFMEFI57814X0034.

References

1. Bondarchuk, S.S., Borisov, B.V., Zhukov, A.S.: Equations for calculation of parameters of reaching the steady-state mode of a solid propellant rocket engine for multi-component mixture of combustion products. Russ. Phys. J. **55**(9/3), 24–26 (2012)
2. David, R., Lide (ed.): CRC Handbook of chemistry and physics. CRC Press, Boca Raton (2005)
3. Gany, A.: Combustion of boron-containing fuels in solid fuel ramjets. Int. J. Energ. Mater. Chem. Propul. **2**, 91–112 (1993)
4. Gubertov, A.M., Mironov, V.V., Gollender, R.G. et al.: Processes in hybrid rocket engines. In: Koroteev, A.S. (ed.) FSUE Keldysh research center. Roscosmos, p. 405. Nauka, Moscow (2008)
5. Maslov, V.M., Borovinskaya, I.P., Ziatdinov, M.Kh.: Combustion of niobium-aluminium, niobium-germanium systems. Fiz. Goreniya Vzryva. **15**(1), 49–57 (1979)
6. Merzhanov, A.G., Mukasyan, A.S.: Solid flame combustion, p. 336. Torus Press, Moscow (2007)
7. Varma, A.S., Rogachev, A.S., Mukasyan, S.H.: Combustion synthesis of advanced materials: Principles and applications. Adv. Chem. Eng. **24**, 79–226 (1998)
8. Zhukov, A., Zhukov, I., Ziatdinov, M., Promakhov, V., Vorozhtsov, A., Vorozhtsov, S., Dubkova, Y.: Self-propagating high-temperature synthesis of energetic borides. In: Prospects of fundamental sciences development (PFSD-2016) proceedings of the xiii international conference of students and young scientists, vol. 1772(1), p. 020015. AIP Publishing, (2016, October)

Mechanical Treatment of ZrB_2–SiC Powders and Sintered Ceramic Composites Properties

S.P. Buyakova, A.G. Knyazeva, A.G. Burlachenko, Yu. Mirovoi
and S.N. Kulkov

Abstract The effect of mechanical treatment by planetary ball milling on the properties of hot pressed ZrB_2–SiC ceramics was studied. It has been shown that material densification after mechanical treatment is finished on initial stages of sintering process. Addition of SiC leads to essentially increasing of sample density up to 99% of a theoretical one for powder with 20% SiC, as compared with ZrB_2 not higher when 76%. It has been shown that all defects which were accumulated during mechanical treatment are annealed during hot pressure process and there are no any changes of coherently diffracting domain (CDD) values in sintered ceramics. The model was suggested to describe of three-layered porous composite synthesis at the conditions of hot isostatic pressing and investigate the porosity evolution during synthesis. Model takes into account the conjugate heat exchange between sintered materials and walls of the reactor.

Keywords High temperature · Hot pressed process · Milling
Modeling

Introduction

The refractory compounds are the basic components of materials used in high-temperature engineering, as thermal protection of space vehicles, in electronics, etc. Among materials with high melting temperature, the special attention is attracted

S.P. Buyakova (✉) · S.N. Kulkov
Institute of Strength Physics and Materials Science SB RAS, Tomsk, Russia
e-mail: sbuyakova@ispms.tsc.ru

S.N. Kulkov
e-mail: kulkov@ms.tsc.ru

A.G. Knyazeva · A.G. Burlachenko · Yu. Mirovoi
Tomsk Polytechnic University, Tomsk, Russia

S.N. Kulkov
Tomsk State University, Tomsk, Russia

© The Author(s) 2018   521
K.V. Anisimov et al. (eds.), *Proceedings of the Scientific-Practical Conference
"Research and Development - 2016"*, https://doi.org/10.1007/978-3-319-62870-7_55

with ceramic composites based on ZrB_2, however strong covalent bonds are the reason of low atom diffusion mobility that essentially stipulate sintered samples.

It is known [1] that addition of SiC leads to decreasing melting temperature of ZrB_2–SiC system and therefore one can obtain a higher density of sintered material. Besides this, ZrB_2–SiC ceramics are characterized a high thermochemical stability, including high stability to oxidation in conditions of extra-high temperatures.

The composites ZrB_2–SiC are usually obtained by sintering of powders under pressure at temperatures higher when 2000 °C [2] and for lowering of a sintering temperature, the powders undergo a mechanical treatment using high-energy (planetary) mills. In this case, subsequent sintering will be activated due to increasing of numbers defects, acceleration of diffusion processes and simplification of plastic flows during sintering [3], and process of sintering one can carry out under pressure, i.e., to realize a mode of hot pressing or SPS process [4]. Unfortunately, data on influence of mechanical treatment on properties of powders and the process of subsequent hot pressure are investigated not enough.

Today, new composite material synthesis attracts numerous investigators, and it is characterized by complex ideas and combined technologies. It connects with the necessity to obtain the materials with specific properties, for example, materials resistant to high thermal and mechanical loading simultaneous. Porous composites can be considered as a suitable example [5, 6]. Because the technologies are very complex and multiple-factor, mathematical modeling is used for the investigation. Theory and technologies of sintering are described for example in [7–10]. Numerous authors describe the temperature change in complex areas, residual stresses, density, porosity viscosity and mechanical properties evolution, the junction formation between individual particles, etc. The series of publications is devoted to the modeling of new material synthesis for the conditions of the heating combined with mechanical loading or to the study of possible stationary conversion regimes [11–14]. However, the coupling models taking into account the interrelation between the heat and mechanical processes and the conditions of conjugate heat exchange occurred not very often. Even so, when commercial software's are available, there are a lot of problems in the modeling [15]. Applications of coupling models [16, 17] to particular technology situations were described in [18, 19]. Generalized model of viscous elastic body of Maxwell type was used in [18] to describe the synthesis process combined with the extrusion. In [19], the material consolidation for the conditions of stress-assisted spark plasma sintering was studied taking into account conjugate heat exchange.

So, the aim of this paper is the study of influence of mechanical treatment of ZrB_2–SiC powders on their properties and properties of sintered by hot pressed ceramic composites and to develop simple model for the formation of porous ceramic composite consisting of three layers.

Materials and Experimental Procedure

The researches were carried out on powder mixtures ZrB_2 (d_{50} = 2.5 microns) SiC (d_{50} = 4.2 microns) with the SiC contents 10, 15 and 20 vol.%. The mechanical treatments of powders were made in a planetary mill at centrifugal acceleration approximately 30 g; the duration of processing was up to 20 min. The hot pressure of ceramic composites was carried out at temperature 1800 °C and pressure 50 MPa with isothermal sintering 30 min. A raw density of mixtures, phases and its structure and coherently diffracting domains (CDD) using x-ray with CuKα irradiation have been measured accordingly [5]. Scanning electron microscope observations operated at 20–30 kV were used to determine the structure and average grain size using Tescan VEGA-3SBH.

Results and Discussion

Mechanical Treatment of ZrB₂–SiC Powders in Planetary Milling

In Fig. 1, the dependence of relative density ρ_r/ρ_t (ρ_r raw density, ρ_t theoretical density) versus treatment time is presented. As one can see, for all powders, the increasing of treatment time was accompanied by increasing its raw density, and particles morphology are essentially changed from separate particles in the beginning state up to agglomerate formation at the end of treatment, Fig. 2. Before treatment, the smallest raw density had ZrB_2 powder, however after processing its density had increased essentially more, than raw densities of powder mixtures with SiC, these mixtures in an initial state had a very close values ρ_r/ρ_t, but after

Fig. 1 The dependencies of raw density of ZrB₂–SiC powders after mechanical treatment in planetary milling

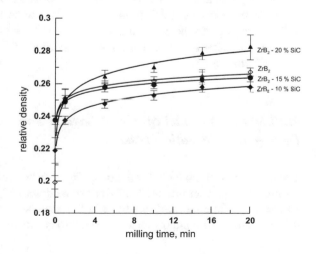

Initial   10 min   20 min

Fig. 2 The changing of powder morphology during milling time

Fig. 3 CDD versus milling times of powder and sintered ceramics

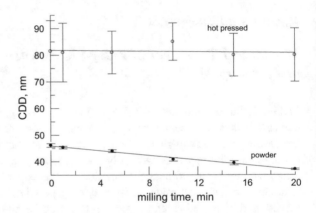

treatment, the higher raw density had powder with the maximal contents of SiC. The x-ray phase analysis of mixtures had shown that during treatment, there are no any changes, addition of SiC to mixtures leads to appearing of peaks belongs to it. With increasing of time treatment, we have found a broadening of peaks due to an increasing number of lattice defects and decreasing of CDD, or grain sizes from 46 up to 37 nm, Fig. 3.

Mathematical Model of Three-Layer Composite Synthesis During Hot Isostatic Pressing

The mathematical model corresponds to Fig. 4, where the reactor is presented. It is assumed that powder mixture behavior can be described similarly to viscous liquid. At the initial time moment, we have in the chamber three layers of powders with different chemical compositions and properties. Mechanical and thermal-physical

Fig. 4 Illustration to the
problem formulation

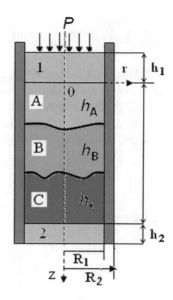

properties of green powders are known. The contact between layers is assumed as an ideal, in a first approximation. Macroscopic thermal stresses are small. Friction with walls is negligibly small too. The dependencies of the properties on porosity are taken into account. In turn, the porosity changes during the heating and loading. Heat exchange of powder mixture with the walls leads to the appearance of not uniform temperature field and can be the cause for not uniform composition change in the layers during synthesis. As a result, mathematical problem formulation will include two parts: mechanical and thermal-kinetical. Correctness of this approximation is confirmed by the results [18].

Mechanical part of the problem consists of one-dimensional motion equation and rheology relation

$$\rho \rho_s \left(\frac{\partial V}{\partial t} + V \frac{\partial V}{\partial z} \right) = \frac{\partial \sigma}{\partial z} \tag{1}$$

$$\sigma = \left(\zeta + \frac{4}{3} \mu_V \right) \frac{dV}{dz}, \tag{2}$$

where V is the component of velocity vector and σ is the component of stress tensor in the loading direction, ζ and μ_V are the volume and shear viscosity; ρ is relative density connecting with the porosity η by the relation $\rho = 1 - \eta$. Porosity changes correspondingly to some law depending on many physical parameters. Physical properties are different for each layer. Load is given. In the contact plane between plunger and flayer A, the loading condition takes a place

$$z = \xi(t) : -p(t) = \sigma = \left(\zeta + \frac{4}{3}\mu_V\right)\frac{dV}{dz}$$

This model is similar to [20], however, the thermal-kinetical part of the model takes into consideration the heat exchange between different materials including the losses to the plunger, the heating from the wall, etc. That is, we have two-dimensional problem containing the thermal conductivity equations for the plunger, bottom, and walls.

$$c_k \rho_k \frac{\partial T_k}{\partial t} = \frac{\partial}{\partial z}\left(\lambda_k \frac{\partial T_k}{\partial z}\right) + \frac{1}{r}\frac{\partial}{\partial r}\left(\lambda_k r \frac{\partial T_k}{\partial r}\right), \ k = 1, 2, 3 \qquad (3)$$

and for powder layers

$$c_k \rho \rho_{ks}\left(\frac{\partial T_k}{\partial t} + V\frac{\partial T_k}{\partial z}\right) = \frac{\partial}{\partial z}\left(\lambda_k \frac{\partial T_k}{\partial z}\right) + \frac{1}{r}\frac{\partial}{\partial r}\left(\lambda_k r \frac{\partial T_k}{\partial r}\right) + \sigma\frac{\partial V}{\partial z}, \ k = A, B, C$$

$$\qquad (4)$$

Here, c_k, ρ_k, λ_k—are the heat capacities, densities, and effective thermal conductivities of materials.

The conditions of ideal heat contact are assumed in the interfaces between different materials («1–2»; «1–3» and «2–3», and «A–B», «B–C» also etc.) The symmetry condition is correct for the axis $r = 0$. The radiation heat exchange with environment takes place on the surface of plunger. Lateral surfaces of press-tool are heated by radiation. The heater temperature T_W changes by given law that determines the powders mixtures rate. The moving boundary $\xi(t)$ is calculated during problem solution.

At the initial time moment, all properties and initial porosity are given, the velocity and stresses are zero. Then, the porosity θ evolution leads to the properties change and in a fist approximation, these relations are

$$c_k = c_{k0}(1 - \theta); \ \rho_k = \rho_{k0}(1 - \theta); \ \lambda_k = \lambda_{k0}\exp\left(-\frac{1,5\theta}{1-\theta}\right);$$

$$\mu_V = \mu_{V0}\frac{3(1-\theta)^2}{3 - \theta(1-\theta)}; \ \zeta_V = \mu_V \frac{4}{3}\frac{1-\theta}{\theta}$$

and we can study numerically the problem of three-layered composite sintering under pressure.

To realize numerically the proposed model, we develop a special algorithm based on implicit difference schemes and taking into account the difference mesh deformation in the calculation area with moving boundaries. The calculations showed that the result (porosity, properties, layer size, etc.) depends on the applied pressure, composition of green substances (powders), and heating regime. For example, different layer thickness evolutions are presented in Fig. 5 for identical

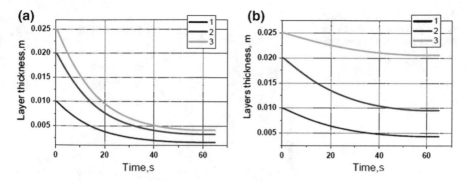

Fig. 5 Layer thickness versus time for **a** identical properties of layers (P = 40 MPa) and **b** for layers with different properties (P = 20 MPa). Layer A (1), B (2), C (3)

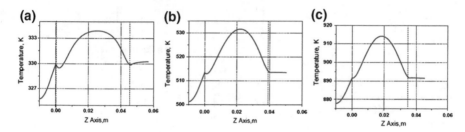

Fig. 6 Temperature along the axis Oz. Time t = 12 (**a**), 25 (**b**) and 45 s (**c**). The regime corresponds to Fig. 2b

(a) and different materials (b). It was found that the temperature field is inhomogeneous. It happens because materials have different properties, porosity evolution for various materials (layers) occurs with different kinetical parameters, there are viscous dissipation in the volume and heat losses from the sintered materials to the walls and bottom of the reactor. This corresponds to known experimental data [5, 6]. The temperature field is presented in Fig. 6 for three time moments (12, 25 and 45 s). One can see that the specimen size diminishes. It occupies the area between vertical dotted lines. The figures are presented in the coordinate system associated with interface 1-A.

Hot Pressed ZrB₂–SiC Ceramics After Mechanical Treatment of Powders

After hot pressure of ceramics, the phase content does not change, and increasing of treatment time in a planetary mill before sintering does not change a CDD on sintered materials, Fig. 3. This means that all defects are annealed during hot pressure process. In Fig. 7, the densities of hot pressed samples are shown.

Fig. 7 Relative density ρ/ρ_t of hot pressed ceramics versus milling time of powder

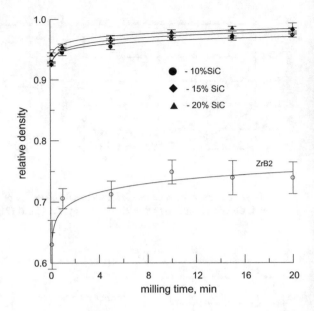

Fig. 8 The dependence of parabolic index n of sintering process versus SiC content in powder

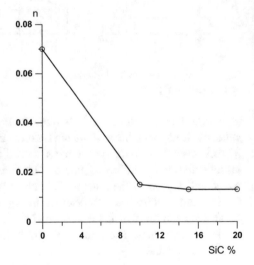

As one can see, addition of SiC leads to essentially increasing of sample density—its values up to 99% of a theoretical one for powder with 20% SiC, as compared with ZrB_2 not higher when 76%. This is agreeing well with data [21]. So, these raw densities of powders seem very well for sintering process of ceramics.

The experimental data of relative density changes of sintered materials are well described by simple function like $Y = aX^n$ where parameter n characterizes a speed of density change. Its may be easy to determine by re-plotting these dependencies in log–log coordinates, then the inclination of a straight line will be equal to this parameter. Thus, defined parameter "**n**" is shown in Fig. 8, and as one can see the addition of SiC to mixture leads to decreasing up to four time its value. This means besides the sample density after hot pressure, that material densification is finished on initial stages of sintering process.

Conclusions

- Mathematical problem of three-layered composite sintering under pressure was proposed in this work. Numerical realization of the problem was carried out. It was detected that the temperature field is nonhomogeneous due to difference in the properties of sintered materials and walls of the reactor. It was found that the porosity changes in a different way in the layers, and its changing are nonhomogeneous along the loading direction.
- It was shown that material densification after mechanical treatment is finished on initial stages of sintering process.
- Addition of SiC leads to essentially increasing of sample density up to 99% of a theoretical one for powder with 20% SiC, as compared with ZrB$_2$ not higher when 76%.
- It was shown that all defects which were accumulated during mechanical treatment are annealed during hot pressure process and there are no any changes of CDD values in sintered ceramics.

Acknowledgments Researches are carried out (conducted) with the financial support of the state represented by the Ministry of Education and Science of the Russian Federation. Agreement (contract) no. 14.607.21.0056 23.Sept 2014. Unique project Identifier: RFMEFI60714X0056.

References

1. Patel, M., Singh, V., Reddy, J.J., Prasad, V.B., Jayaram, V.: Densification mechanisms during hot pressing of ZrB 2–20vol.% SiC composite. Scripta Mater. **69**(5), 370–373 (2013)
2. Mashhadi, M., Khaksari, H., Safib, S.: Pressureless sintering behavior and mechanical properties of ZrB 2–SiC composites: Effect of SiC content and particle size. J. Mater. Res. Technol. **4**(4), 416–422 (2015)
3. Sreckovic, T.: Sintering of mechanically activated powders. Adv. Sci. Technol. **45**, 619–628 (2006)
4. Akin, I., Hotta, M., Sahin, F., Yucel, O., Goller, G., Goto, T.: Microstructure and densification of ZrB 2–SiC composites prepared by spark plasma sintering. J. Eur. Ceram. Soc. **29**(11), 2379–2385 (2009)
5. Buyakova, S., Burlachenko, A., Mirovoi, Y., Sevostiyanova, I., Kulkov, S.: The influence of ZrB2-SiC powders mechanical treatment on the structure of sintered ceramic composites. IOP Con. Ser. Mater. Sci. Eng. **140**(1), 012006 (2016)
6. Kulkov, S., Buyakova, S., Chartnikolaidou, M., Kocserha, I.: Rheology and porosity effect on mechanical properties of zirconia ceramics. J. Silicate Based Compos. Mater. **67**(4), 155–158 (2015)

7. Olevsky, E.A., Skorohod, V.: Some questions of sintering kinetics under external forces influence, pp. 97–103. Technological and Design Plasticity of Porous Materials. IPMS NAS, Ukraine, 97–103 (1988)
8. Olevsky, E.A.: Theory of sintering: From discrete to continuum. Mater. Sci. Eng., R: Rep. **23**, 41–100 (1998)
9. Skorokhod, V.V.: Science of sintering: Evolution of ideas, advances, current challenges, and new trends. I. From natural philosophy to physics of sintering. Powder Metall. Met. Ceramics **53**(9–10), 529–536 (2015)
10. Skorokhod, V.V.: Rapid rate sintering of dispersed systems: Theory, processing, and problems. Powder Metall. Met. Ceramics **38**(7–8), 350–357 (1999)
11. Amosov, A.P., Radchenko, V.P., Fedotov, A.F.: Effect of shell dimensions on compaction and shape change during SHS-pressing. Powder Metall. Met. Ceramics **43**(5), 229–235 (2004)
12. Smolyakov, V.K.: Structural mechanics of a substance in the wave of self-propagating high-temperature synthesis. Phys. Mesomech. **2**(3), 55–68 (1999)
13. Stolin, A.M., Stel'makh, L.S.: Mathematical modeling of SHS compaction/extrusion: An autoreview. Int. J. SHS **17**(2), 93–100 (2008)
14. Belyaeva, N.A., Stolin, A.M., Pugachev, D.V., Stel'makh Dokl, L.S.: Unstable modes of deformation in solid-phase extrusion of viscoelastic structured systems. Phys. Chem. **420**(2), 144–147 (2008)
15. Chung, S.H., Kwon, Y.-S., Park, S.J., German, R.M.: Modeling and simulation of press and sinter powder metallurgy. ASM Handbook **22**, 323–324 (2010)
16. Knyazeva, A.G.: The stationary modes of the reaction front and their stability for solid media with regard to chemically induced internal stresses and strains. Int. J. Energ. Propul. Mater. Chem. **5**(1–6) (2002)
17. Knyazeva, A.G.: Velocity of the simplest solid-phase chemical reaction front and internal mechanical stresses. Combust. Explosion Shock Waves. **30**(1), 43–53 (1994)
18. Knyazeva, A.G., Evstigneev, N.K.: The choice of rheological model describing intermetallide synthesis during extrusion through conical compression mould (in Russian). Vestn. PGTU, 59–71 (2010)
19. Sorokova, S.N., Knyazeva, A.G., Pobol, A.I., Goranskyi, G.G.: Mathematical modeling of pulsed electro contact sintering of carbide powder composition. Adv. Mater. Res. **1040**, 495–499 (2014)
20. Koval'chenko, M.S.: Theoretical bases of hot treatment of porous materials under pressure. Naukova Dumka, Kiev (1980)
21. Torayda, H., Yoshimura, M., Somiya, S.: Calibration curve for quantitative analysis of the Monoclinic-Tetragonal ZrO_2 system by X-ray diffraction. J. Am. Ceram. Soc. **67**(6) (1984)

Part III
Health and Ecology and Environment Sciences

The Influence of DCs Loaded with Tumor Antigens on the Cytotoxic Response of MNC Culture Patients with Oncology

A.P. Cherkasov, J.N. Khantakova, S.A. Falaleeva, A.A. Khristin,
N.A. Kiryishina, V.V. Kozlov, E.V. Kulikova, V.V. Kurilin,
J.A. Lopatnikova, I.A. Obleukhova, S.V. Sennikov, J.A. Shevchenko,
S.V. Sidorov, A.V. Sokolov and A.E. Vitsin

Abstract Currently, one of the most promising approaches for the treatment of oncological patients is the selective activation of T-cell antitumor immunity using immune cells. The generation of functionally active DCs in vitro constitutes a promising approach in the development of DC-based anticancer vaccines to mobilize patient defense systems, because their activation by tumor-specific antigens to induce cytotoxic responses, and their increased efficiency of antigen presentation to induce cytotoxic T lymphocytes (CTLs) via costimulatory molecules and cytokines can be controlled. The purpose of this study was to investigate the

A.P. Cherkasov (✉) · A.A. Khristin · A.V. Sokolov · A.E. Vitsin
City Clinical Hospital no. 1, Pirogov, Russia
e-mail: cherkassov333@mail.ru

A.A. Khristin
e-mail: AAlex.Khristin@gmail.com

A.V. Sokolov
e-mail: sokolowav2003@mail.ru

A.E. Vitsin
e-mail: gella2009@yandex.ru

J.N. Khantakova · S.A. Falaleeva · E.V. Kulikova · V.V. Kurilin · J.A. Lopatnikova ·
I.A. Obleukhova · S.V. Sennikov · J.A. Shevchenko
Federal State Budgetary Scientific Institution "Research Institute of Fundamental
and Clinical Immunology" Laboratory of Molecular Immunology, Novosibirsk, Russia
e-mail: KhantJN@gmail.com

S.A. Falaleeva
e-mail: kolenteonok@mail.com

E.V. Kulikova
e-mail: homka-88@inbox.ru

V.V. Kurilin
e-mail: 2221910@ngs.ru

J.A. Lopatnikova
e-mail: lopatnikova_j_a@ngs.ru

© The Author(s) 2018
K.V. Anisimov et al. (eds.), *Proceedings of the Scientific-Practical Conference
"Research and Development - 2016"*, https://doi.org/10.1007/978-3-319-62870-7_56

functional characteristics of peripheral blood DC subsets in colorectal cancer (CRR), breast cancer (BC), and non-small cell lung cancer (NSCLC) patients and the development of an antitumor cytotoxic response by mononuclear cells (MNCs) from patients using in vitro generated antigen-primed DCs.

Keywords Antigen-primed dendritic cells · Antitumor cytotoxic response Colorectal cancer · Breast cancer · Non-small cell lung cancer

Introduction

The generation of functionally active DCs in vitro constitutes a promising approach in the development of DC-based anticancer vaccines to mobilize patient defense systems, because their activation by tumor-specific antigens to induce cytotoxic responses, and their increased efficiency of antigen presentation to induce cytotoxic T lymphocytes (CTLs) via costimulatory molecules and cytokines can be controlled [5, p. 459; 1, p. 1]. DC-based anticancer vaccines are delivered using various methods, including the use of free peptides [10, p. 1868], tumor lysates [3, p. 2827], DNA [12, p. 1339; 8, p. 122] or RNA vaccines [9, p. 1], as well as DCs primed with various tumor antigens [6, p. 57]. Because tumors have a heterogeneous structure, the surface of cells comprising the tumor bears an individual set of TAAs. The use of tumor lysate as a source of tumor immunogens has the potential advantage of stimulating a response against a variety of known and unknown TAAs typical of both a particular tumor type and a particular patient. This method enables the induction of a polyclonal immune response, stimulating both helper CD4+ and cytotoxic CD8+ immune responses, thereby reducing the risk of the tumor escaping immune surveillance [7, p. 139]. The use of a tumor lysate reduces the time and effort spent identifying and synthesizing immunodominant peptide epitopes, enabling DCs to process tumor antigens naturally.

I.A. Obleukhova
e-mail: obleukhova.irina@yandex.ru

S.V. Sennikov
e-mail: sennikovsv@gmail.com

J.A. Shevchenko
e-mail: shevja80@gmail.com

N.A. Kiryishina · V.V. Kozlov
Novosibirsk Regional Clinical Oncology Center, Novosibirsk, Russia
e-mail: natahamed@mail.ru

V.V. Kozlov
e-mail: vadimkozlov80@mail.ru

S.V. Sidorov
Department of Surgery, Novosibirsk State University, Novosibirsk Oblast, Russia
e-mail: naf202@mail.ru

Materials and Methods

Study subjects. Peripheral blood and tumor samples were obtained from:

- 44 patients aged 42–83 years (mean = 66.8 years) with colorectal cancer at stages I–IV [I–II–30 patients (68.2%), III–IV–14 patients (31.8%)], including 21 men (47.7%), and 23 women (52.3%). Colorectal adenocarcinoma was confirmed in all patients.
- 16 patients with the histologically verified diagnosis of non-small cell lung cancer (NSCLC) stage IIA, IIB, and IIIA as well as tumor biopsy material obtained during surgery (14 males, 2 females; mean age 60.7±1.5 years).
- 20 patients aged 35–77 years (mean age, 57.9 years) with the histologically and immunohistochemically verified diagnosis breast cancer at stages I–II.

All patients provided their informed consent to participate in this study. Patient diagnostic and histology data were kindly provided by the medical institution (City Clinical Hospital No. 1 and Novosibirsk Regional Clinical Oncology Center, Novosibirsk, Russia), where patient examination and clinical surveillance (treatment) was performed.

DC preparation. Peripheral blood MNCs were isolated using standard Ficoll-Urografin density-gradient [2, p. 97]. Cells with enhanced adherence were isolated from the resulting MNC population using a short incubation period (2 h) in a 75-cm^2 culture flask, in 5% CO_2 at 37 °C. Cells were grown in RPMI-1640 complete medium supplemented with 10% fetal bovine serum (FBS), 40 µg/mL gentamicin, 200 U/mL penicillin, 2 mM L-glutamine, 5×10^{-5} M 2-mercaptoethanol, and 10 mM HEPES. The adherent MNC fraction was cultured in 48-well plates at 1×10^6 cells/mL in 0.5 mL complete medium. The adherent MNC fraction was supplemented with 50 ng/mL rhGM-CSF and 100 ng/mL rhIL-4 to produce immature DCs through a 4-day incubation. To obtain tumor antigen-loaded DCs, immature DCs were supplemented with autologous tumor cell lysate at a concentration of 100 µg/mL after 48 h of incubation. To generate DCs transfected with tumor cell RNA, we performed magnetic transfection using Promokine reagents (Germany), according to the manufacturer protocol. The culture was then supplemented with rhTNF-α (25 ng/mL) in fresh culture medium and an equivalent volume to generate mature cells over the following 24 h. The fraction of non-adherent cells was maintained in a 75-cm^2 culture flask at 2×10^6 cells/mL in RPMI 1640 complete medium until decantation.

Preparation of autologous tumor cells. A tumor sample was washed in RPMI-1640 with a doubled concentration of antibiotics. To obtain autologous tumor cells, a tumor fragment from each patient was crushed and left in 0.25% trypsin solution at +4 °C overnight. Warm complete RPMI-1640 was used to inactivate the enzyme. The cell suspension was filtered to remove large aggregates, washed twice, and frozen in FBS with 10% DMSO (Panreacsintesis, Spain). To obtain tumor cell lysates, a tumor fragment was mechanically homogenized and the resulting suspension was successively frozen at −70 °C and thawed at +37 °C through three cycles. Cells were pelleted by centrifugation and sterilized by passing through a 0.45-μm filter. Total protein in the lysates was determined by calculating the 260/280 nm absorbance ratio using a NanoDrop device (Thermo Scientific, USA).

Co-culture of DC and MNC. Co-culture of DC and MNC was carried out in several parallel cultures for subsequent functional tests under uniform conditions. The concentration of non-adherent cultured MNC was 1×10^6 cells/ml, the DC: MNC ratio was 1:10, and rhIL-12 (10 ng/ml; PeproTech, USA) and rhIL-18 (100 ng/ml; MBL, USA) were applied to stimulate Th1-polarization. Mononuclear cells used to assess cytotoxicity against autologous tumor cells were cultured for 4 days in complete RPMI-1640 in the presence or absence of recombinant cyto-kines. Mononuclear cells used for assessing perforin levels were cultured in com-plete RPMI-1640 in the presence or absence of recombinant cytokines for 2 days; the cultures were subsequently washed of growth factors, and cultured for a further 48 h.

Determination of perforin-positive cell count. The cells to be analyzed were washed once with PBS and fixed with 1% paraformaldehyde in cold PBS for 20 min. They were centrifuged and the pellet was resuspended in 0.2 ml PBS containing 0.2% Tween 20 (Panreacsintesis, Spain), and incubated for 20 min to permeabilize the cell membranes, after which the cells were centrifuged and incubated with fluorochrome-labeled monoclonal antibodies against perforin (perforin-FITC, BD) for 30 min. The cells were washed to remove excess antibody and the number of positive cells was determined by flow cytometry in the lym-phocyte region.

Determination of antitumor cytotoxic effect. Cytotoxicity was assessed by analyzing the lactate dehydrogenase (LDH) content in the conditioned medium obtained by co-culture of MNCs stimulated by transfected or control DCs (effector cells) and tumor cells (target cells), using a CytoTox 96® Non-Radioactive Cytotoxicity Assay (G1780, Promega, USA). The ratio effector cell: tumor cell was 10:1. Released LDH in culture supernatants is measured with a 30-min coupled enzymatic assay, which results in the conversion of a tetrazolium salt (INT) into a red formazan product. Visible wavelength absorbance data are collected using a standard 96-well plate reader. The amount of color formed is proportional to the number of lysed cells. To convert the concentration of LDH in the supernatant into percentage cytotoxicity, we applied the formula:

$$\% \text{ Cytotoxicity} = \frac{OD\,(\text{effector} + \text{target}) - OD\,(\text{effector}) - OD\,(\text{target})}{OD\,(\text{maximum target lysis}) - OD\,(\text{target})}$$

To determine the value OD (maximum target lysis), we lysed the tumor cells by Lysis Solution.

Statistical analysis. Statistical data were processed using the Statistica 6.0 program. The Friedman test and Newman–Keuls multiple comparison test were used to detect statistically significant differences. The Shapiro–Wilk test was applied to determine sample normality. The data are presented as the mean and standard error for normal distribution, and the median and interquartile range were used for the abnormal distribution.

Results

Cytotoxic T cells play a central role in the antitumor immune response because they directly lyse tumor cells and produce immunomodulatory cytokines, such as IL-2, TNF-α, GM-CSF, and IFN-γ, which indirectly affect malignant cells. The protective antitumor response involves the killing of tumor cells. Thus, we evaluated MNC cytotoxicity, which is activated via exposure of transfected DCs to autologous tumor cells, by quantifying the cytoplasmic LDH levels of lysed tumor cells. One of the mechanisms through which tumor cell death is triggered is the granule-dependent pathway, which is primarily mediated by perforin, granzyme, and granulysin. Perforin causes pore formation in the target cell membrane and results in cell death [13, p. 35; 14 p. 56]. To examine the role of the perforin-dependent mechanism in the lysis of tumor cells, we evaluated the effect of transfected DCs on perforin expression in co-cultures of MNCs from cancer patients.

We demonstrated in patients with **breast cancer** that the use of lysate-activated dendritic cells or dendritic cells transfected total RNA stimulates the cytotoxic response of the MNC culture that is expressed in an increased cytotoxic activity of MNCs against autologous cells (Figs. 1a and 2). The number of perforin-positive cells increased in the MNC culture stimulated lysate-activated dendritic cells (Fig. 1b).

In patients with **NSCLC,** DCs primed with an antigen's lysate and DCs transfected with RNA were shown to increase the cytotoxic activity of MNCs against autologous NSCLC cells, compared with the control group (MNC and MNC c-culture with non-primed DCs) (Fig. 3). In the total lymphocyte population after co-culturing lysate-primed DCs and RNA-transfected DCs increase the relative level of perforin-bearing cells in the lymphocyte population compared with control groups (Fig. 4).

In **colorectal cancer patients,** results indicate that cytotoxicity increased following use of immunogenic lysate-primed DCs, compared with all control groups. The cytotoxicity for the original MNC culture was 11.6%, whereas treatment with lysate-primed DCs increased this to 23.2% (Fig. 5).

Because tumor cell lysates contain a variety of relevant antigens, a wide range of TAAs is presented to T cells, inducing a pronounced immune response. By using complete tumor lysates in experimental models and clinical trials, high efficacy and low toxicity have been observed in various cancers, including colorectal [11, p. 6445; 4, p. 475]. However, the lysate may contain proteins suppressed the immune response [3, p. 2827] and residual tumor cells in patient organism can change their antigenic composition under the course of treatment that may also complicate the use of the lysate. We demonstrate that the use of dendritic cells to

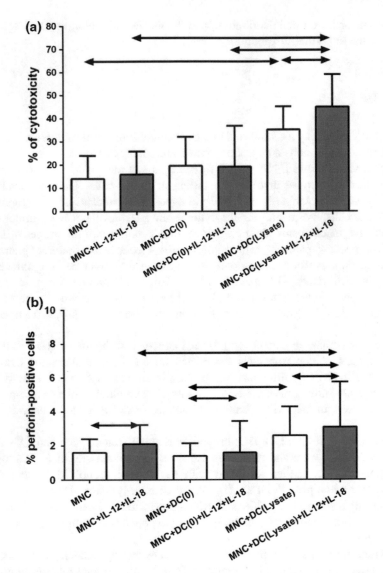

Fig. 1 Cytotoxicity against autologous tumor cells (**a**) and percentage of perforin-positive cells (**b**) co-cultured of MNCs and tumor lysate-loaded DCs ($n = 20$) in patients with breast cancer. Data are presented as median (Me) and interquartile range. *Arrows* indicate statistically significant differences. ($P < 0.05$). MNC—the MNC control culture; MNC + DC (0)—co-culture of MNCs and DCs unloaded with tumor lysate antigens; MNC + DC (Lysate)—co-culture of MNCs and DCs loaded with tumor lysate antigens

prime the lysate's antigens enables efficient activation of the effector functions of immune cells at various cancers. Our findings indicated an elevated cytotoxicity via the perforin pathway, which was corroborated by an increase in the ratio of

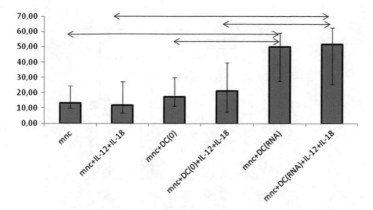

Fig. 2 Cytotoxicity against autologous tumor cells co-cultured of MNCs and tumor DCs, transfected total RNA ($n = 16$) in patients with breast cancer. Data are presented as median (Me) and interquartile range. *Arrows* indicate statistically significant differences. ($P < 0.05$). MNC—the MNC control culture; MNC + DC (0)—co-culture of MNCs and DCs unloaded with tumor antigens; MNC + DC (RNA)—co-culture of MNCs and DCs transfected with total tumor RNA

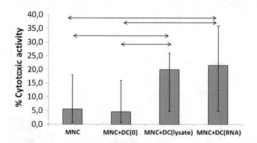

Fig. 3 The effect of antigen-primed DCs on the cytotoxic activity of MNCs against autologous NSCLC cells in vitro, $n = 14$. Data are presented as median (Me) and interquartile range. *Arrows* indicate statistically significant differences. ($P < 0.05$). MNC—the MNC control culture; MNC + DC (0)—co-culture of MNCs and DCs unloaded with tumor antigens; MNC + DC (Lysate)—co-culture of MNCs and DCs loaded with tumor lysate antigens; MNC + DC (RNA)—co-culture of MNCs and DCs transfected with total tumor RNA

perforin-positive cells within the total T-cell populations in the group that received RNA-transfected and lysate-primed DCs. Thus, we can speak about the activation of the cytotoxic potential of effector cells, determined by the number of perforin-positive cells.

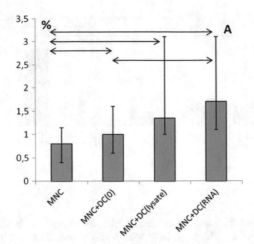

Fig. 4 The effect of antigen-primed DCs on the relative amount of perforin + lymphocytes in the co-culture of MNCs from NSCLC patients, $n = 16$. Data are presented as median (Me) and interquartile range. *Arrows* indicate statistically significant differences. ($P < 0.05$). MNC—the MNC control culture; MNC + DC (0)—co-culture of MNCs and DCs unloaded with tumor antigens; MNC + DC (Lysate)—co-culture of MNCs and DCs loaded with tumor lysate antigens; MNC + DC (RNA)—co-culture of MNCs and DCs transfected with total tumor RNA

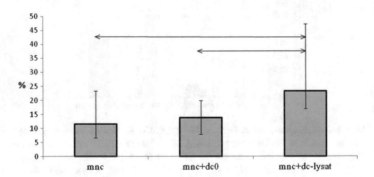

Fig. 5 Cytotoxic activity of colorectal cancer patients MNCs co-cultured with autologous DCs loaded with tumor lysates against autologous tumor cells ($n = 17$). Data are presented as median (Me) and interquartile range. *Arrows* indicate statistically significant differences. ($P < 0.05$). MNC—the MNC control culture; MNC + DC (0)—co-culture of MNCs and DCs unloaded with tumor antigens; MNC + DC (Lysate)—co-culture of MNCs and DCs loaded with tumor lysate antigens

Conclusions

Currently, there is an intensive development of approaches to induce antitumor immunity using DC-based vaccination technologies. In our opinion, investigates should be primarily concentrated on improving to deliver antigens to DCs to stimulate antitumor immune responses. In addition, the in vitro stimulation of an antitumor immune response, which involves the selection and use of various effective stimulators and inhibitors of immunosuppressive molecules to shift the response towards a Th1 response, is also important. In our research, we demonstrated that lysate-primed DC and DC transfected total RNA enhanced MNC antitumor activity, increasing tumor cell death and the percentage of perforin-positive lymphocytes at various cancers. In this regard, we assume that the use of antigens-primed DC in vitro can activate antigen-specific T cells for modulation antitumor response.

Acknowledgements Research is carried out with the financial support of the state represented by the Ministry of Education and Science of the Russian Federation. Agreement no. 14.607.21.0043. Unique identifier: RFMEFI60714X0043.

References

1. Benencia, F., Sprague, L., McGinty, J., Pate, M., Muccioli, M.: Dendritic cells the tumor microenvironment and the challenges for an effective antitumor vaccination. J. Biomed. Biotechnol. **2012**, 1–15 (2012)
2. Boyum, A.: Separation of leukocytes from blood and bone marrow. Scand. J. Clin. Lab. Invest. **21**, 97 (1968)
3. Dong, B., Dai, G., Xu, L., Zhang, Y., et al.: Tumor cell lysate induces the immunosuppression and apoptosis of mouse immunocytes. Mol. Med. Rep. **10**(6), 2827–2834 (2014)
4. Figdor, C.G., de Vries, I.J., Lesterhuis, W.J., Melief, C.J.: Dendritic cell immunotherapy: Mapping the way. Nature Med. **10**(5), 475–480 (2004)
5. Fricke, I., Gabrilovich, D.I.: Dendritic cells and tumor microenvironment: A dangerous liaison. Immunol. Invest. **35**, 459–483 (2006)
6. Janikashvili, N., Larmonier, N., Katsanis, E.: Personalized dendritic cell-based tumor immunotherapy. Immunotherapy **2**(1), 57–61 (2010)
7. Liu, L.N., Shivakumar, R., Allen, C., Fratantoni, J.C.: Delivery of whole tumor lysate into dendritic cells for cancer vaccination. In: Li, S. (eds.) Electroporation protocols, Methods Mol. Biol. vol. 423, pp. 139–153. (2008)
8. Marchini, C., Kalogris, C., Garulli, C.: Tailoring DNA vaccines: designing strategies against HER2-positive cancers. Front Oncol. **3**, 122 (2013)
9. McNamara, M.A., Nair, S.K., Holl, E.K.: RNA-based vaccines in cancer immunotherapy. J. Immunol. Res. **2015**, (2015)
10. Peres, Lde P., da Luz, F.A., Pultz Bdos, A., Brígido, P.C., de Araújo, R.A., Goulart, L.R., Silva, M.J.: Peptide vaccines in breast cancer: the immunological basis for clinical response. Biotechnol. Adv. **33**(8), pp. 1868–77. (2015)
11. Schnurr, M., Galambos, P., Scholz, C., et al.: Tumor cell lysate-pulsed human dendritic cells induce a T-cell response against pancreatic carcinoma cells: An in vitro model for the assessment of tumor vaccines. Cancer Res. **61**, pp. 6445–50. (2001)
12. Ugel, S., Facciponte, J.G., De, Sanctis F., Facciabene, A.: Targeting tumor vasculature: Expanding the potential of DNA cancer vaccines. Cancer Immunol. Immunother. **64**(10), 1339–1348 (2015)
13. Voskoboinik, I., Dunstone, M.A., Baran, K., Whisstock, J.C., Trapani, J.A.: Perforin: Structure, function, and role in human immunopathology. Immunol Rev. **235**(1), pp. 35–54. (2010)
14. Zhou, F.: Perforin: More than just a pore-forming protein. Int. Rev. Immunol. **29**(1), pp. 56–76 (2010)

Establishment of a Technological Platform for Pre-Clinical Evaluation of Biomedical Cellular Products in Russia

P.I. Makarevich, Yu P. Rubtsov, D.V. Stambolsky, N.I. Kalinina, Zh A. Akopyan, Y.V. Parfyonova and V.A. Tkachuk

Abstract After joint legislative efforts of scientific community and Government resulted in the adoption of Federal Law #180 "On biomedical cellular products (BMCP)" onset of first pre-clinical trials of cell therapies in Russia got into scope. Testing of BMCPs to assess their safety and obtain primary efficacy results is a cornerstone of development and translation to clinical trials. Thus, a task force consisting of Lomonosov Moscow State University, leading research groups and experts from both—regulatory and industrial entities have been established under a project funded by Russian Ministry of Science and Education. As far as Federal Law #180 is enforced starting January 1, 2017, completion of the project in December 2016 is a timely step to ensure the development of cell therapies and regenerative medicine in Russia. The present article gives an overview of the project in 2014–2016 and summarizes main results of the collaborative effort.

Keywords Federal law #180 · Biomedical cellular products · Tissue regeneration Cell transplantation · Biosafety · Manufacture · Evaluation · Pre-clinical testing Cell manipulation · Bioproduct storage and transportation · Cryoconservation Quality control · Mesenchymal stem cells · Progenitors of cardiac stem cells Platelet lysate · Genetically modified cells

Introduction

Modern biomedicine has reached a point when new revolutionary approaches based on the concept of regenerative medicine allow treatment of human disease not alleviated by existing drugs or surgical technique. Regenerative medicine employs the concept of the natural process of tissue/organ repair and renewal driven by specific cellular elements—namely stem and progenitor cells. Cultural approaches

P.I. Makarevich (✉) · Y.P. Rubtsov · D.V. Stambolsky · N.I. Kalinina · Z.A. Akopyan
Y.V. Parfyonova · V.A. Tkachuk
Lomonosov Moscow State University, Moscow, Russia
e-mail: pmakarevich@mc.msc.ru

© The Author(s) 2018
K.V. Anisimov et al. (eds.), *Proceedings of the Scientific-Practical Conference "Research and Development - 2016"*, https://doi.org/10.1007/978-3-319-62870-7_57

advanced rapidly and now allow to grow human cells in vitro, generate specific cell types using induced pluripotency protocol, create artificial organs/tissues, and to correct mutations in cells by genome editing (mediated by CRISPR/Cas9 system). Furthermore, in addition to systemic transplantation, there are multiple options to introduce therapeutic cells to patient's organism; these include combining cells with extracellular matrix, synthetic scaffolds or cell-derived secreted products to ensure engraftment and/or provide structural support where required (e.g., nerve repair, muscle defect filling or vascular repair).

However, unique properties of BCMP come along with major safety concerns including the risk of carcino-and tumorogenesis, ectopic tissue outgrowth, immunotoxic effects, and adverse reactions related to transplant rejection. Besides serious adverse reactions, limited knowledge of BMCP kinetics (survival and distribution) and pharmacology (including secondary and off-target effects) is another point that requires thorough analysis prior to first-in-human use. European and USA regulatory institutes have established a series of legal acts to control the field of cell therapy and increase its positive impact with reduction of potential harm [1, p. 2; 4, p. 2; 5, p. 2].

Initiation of clinical trials is crucial to advance in the field and develop efficient and safe BMCP to cure previously moribund and lethal conditions. However, abovementioned reasons clearly show that due to natural and scientific reasons evaluation of their safety and efficiency are challenging and requires efforts from both scientific community and regulatory entities. Emerging cell-and cell-based product medical technologies are novel for Russian Federation, therefore, prior to the introduction of Federal Law #180, the Ministry of Science and Education had initiated a complex project to establish a "technological platform" and develop reasonable regulation for procedures of pre-clinical testing of BMCP [2, p. 2].

Legal Field, Workgroup, and Project Goal

Until recent no specific "stem cell law" has existed in Russian Federation to regulate development, testing, manufacture, registration, and marketing of cell-based therapeutic products. Indeed, use of human cells (with exception of blood transfusion and organ transplantation) in the clinic has remained a "gray area" where successes of highly professional scientific groups were formally adjacent to procedures conducted by commercial clinics which offered cell-based therapy for various conditions. Lack of regulation led to use of poorly characterized cells without proper monitoring and control of results.

Federal Law #180 has been signed by the President of Russian Federation and will be enforced since January 1, 2017 to regulate all aspects of cell therapy in Russian Federation. This law regulates design, manufacture, quality and safety evaluation, transportation, clinical trial initiation and marketing of BMCP. These products consist or contain human cells and may contain modified human cells (not organs or human cells used for blood transfusion) obtained from human donors, which can be used for treatment by itself or in combination with other drugs or materials. Generally, BMCP is a new class of medicinal products covering the niche between drugs and transplanted organs/tissues. From one side BMCP has defined active components responsible for therapeutic efficiency, and, from the other side, they consist of cells which produce active components inside the body in a more physiological manner and interact with surrounding cells at the site of injection/introduction. The law defines major terms and limitations of BMCP production, handling, testing, and use.

Major issues covered by the law are presented below:

(1) Prohibition of embryonic material (including cloned embryos) use to generate BMCP and/or cell lines for BMCP production; the point arises from two main reasons: ethical concerns associated with use of abortive material and safety issues related to teratogenic potential of embryonic cells;

(2) Definition of BMCP classes includes autologous, allogeneic, genetically modified and combined—the latter are BMCPs containing both—cells and acellular component(s) such as drugs or medical devices;

(3) Requirements for BMCP production such as criteria for donor material, testing, handling, transportation, and evaluation of final product; organizations and companies manufacturing BMCP must comply with GMP standards for production of BMCP for clinical use;

(4) Procedures for BCMP registration and authorization for clinical trials of BMCP is stipulated; expertise of pre-clinical data is stated as a crucial point for first-in-human use and aggregated data of data on ethical, quality control and efficacy data are required for marketing of BMCP.

Project's main goal lies within the scope of the newly accepted Federal Law and arising demand for legal action to ensure effective pre-clinical testing of BMCP. It can be sub-divided into three major tasks:

(1) Establish a technological platform for pre-clinical evaluation of BMCPs;
(2) Propose and draft guidelines for conduct of pre-clinical evaluation of BMCP;
(3) Prepare programs and modules required for professional education of specialists in development, manufacture, and pre-clinical evaluation of BMCP

A consortium, consisting of Lomonosov Moscow State University, Pirogov Russian National Research Medical University, Gertsen Moscow Research Oncology Institute, Almazov Federal Medical Research Center, Russian Cardiology Research and Production Complex, Shumakov Federal Centre of Transplantation and Artificial Organs, Central Institution of Tuberculosis and Scientific Centre for Expertise of Medical Products was formed to develop the project. As a starting point, besides in-house developments of each organization we used the international experience as a reference, and based on that tried to develop practical recommendations for design and pre-clinical evaluation of certain types of BMCP.

Thus, the new law defines BMCPs, and how they are to be generated, tested, and used. It sets ethical, commercial, and technical limitations of the BMCP testing, production, and practical use. But these are mostly common words related to legislative part of the problem. In reality, there are some fundamental problems concerning testing of BMCP's safety and mechanisms of action. First of all, by definition BMCPs are human cell lines, which have to be biologically tested prior to clinical trials. This is not so simple since in most animal models transferred human cells will be immunologically rejected. Another evident problem is that it is hard to use the pre-clinical data obtained, for example, in small rodents to calculate numbers of cells that have to be used in humans due to physiological reasons. And there are many more examples of challenges of pre-clinical testing of BMCPs. Therefore, a need in the development of experimental protocols and recommendations regarding pre-clinical testing of BMCP exists.

Project Industrial Partner

CellThera Pharm is a pioneer national project realized by the Russian pharmaceutical company "Pharmstandard" incorporated on September 2013 is a full-cycle regenerative medicine company. At the disposal of the company has an R&D department, animal facility and a modern manufacturing facility for development and production of advanced medicinal products based on cell technology.

CellThera Pharm develops the products for unmet medical needs with innovative approaches based on scientific excellence to provide the Russian Federation public healthcare system with fully personalized modern advanced medicinal products manufactured locally for the treatment of severe and socially significant diseases:

- Personalized immunotherapy for cancer treatment
- Aesthetic surgery/reconstructive surgery

To date in the disposition of CellThera Pharm are following facilities with a total area of > 900 m^2 including:

- manufacturing site, based on the isolator technology
- R&D department
- animal facility

The core competencies of Cellthera Pharm are: Cell biology, immunology, Cells and tissue technologies, Molecular biology, Tech-transfer, and technology optimization, pre-clinical study and strong expertise in animal model development

The Cellthera's development strategy provides intensive development and consists both in the development of innovative products and the licensing of the best technologies that are currently present in the world.

Key Results and Product Description

The project's main goal was to develop a technological platform for pre-clinical evaluation of biomedical cell products. A consortium of 8 medical and scientific centers headed by Lomonosov Moscow State University accomplished this mission successfully in 2.5 years with regular reports to the Ministry. Overall, the project wraps up with a total of 8 patents pending, 5 papers in Scopus/WoS indexed journals and 6 conference proceedings to popularize and spread the results and current progress.

Member-organizations of the consortium used their scientific expertise and development experience conducting experiments required to elaborate parameters for BMCP evaluation of safety, efficacy, and distribution. Furthermore, a significant part of the project was devoted to the establishment of appropriate standards for BMCP manufacture and required quality control—these results were formulated as in-house manufacture regulations and standard operation procedures (SOPs).

BMCP from different cell types were evaluated: mesenchymal stem cells from adipose tissue and bone marrow (AD MSC and BM MSC, respectively), progenitors of cardiomyocytes (PC), and bio product based on platelet lysate (BPPL). Analysis of literature allowed the definition of criteria crucial for design, manufacturing, evaluation, in vitro and in vivo pre-clinical testing of each biomedical product. According to these criteria members of consortium thoroughly selected and tried various experimental protocols to determine the clinical efficiency of BMCP in animal models of human pathological conditions. Another set of experiments was used to evaluate stability, distribution in a body, and interaction of BMCP with surrounding tissue. The third line of tests was conducted to prove the biological safety of BMCP.

Laboratory protocols suitable for manufacturing of different BMCPs were generated. Different BMCPs were evaluated for efficiency in animal models. For example, BMCP based on AD MSC was tested in the model of myocardial infarction, and BMCP based on BM MSC—in the model of acute graft versus host disease caused by organ transplantation. AD MSC-mediated immune suppression was studied in the course of tuberculosis infection in mice. BMCPs derived from human MSC were studied for the ability to facilitate regeneration of joint cartilage, liver, and pancreas in rodents. BMCP from human PC passed tests for angio-, arteriogenesis, the effect on fibrosis, cardiomyocyte proliferation, inflammatory infiltration, and myocardium remodeling in a model of rat myocardium injury. BPPL influence on cultures of MSC and fibroblasts, as well as biosafety, was studied. Another BMCP designed for use on damaged skin and possessing dermatotropic properties were tested in mouse skin repair model.

Multiple biosafety tests and protocols were designed for abovementioned BMCP. As a result, now we have a collection of standard protocols and practical recommendation which will simplify life for those who will design, characterize, manufacture, and evaluate BMCP in Russia in the future.

The overall goal of the project was achieved by joint effort and draft of practical recommendations (Guidelines) for design and conduct of pre-clinical evaluation of

BMCP in general and particular examples of BMCPs. These Guidelines contain scientifically reasoned non-binding recommendations for evaluation of BMCPs pharmacology, kinetics (terms used on basis of lack of a suitable alternative) and safety including toxicity and secondary pharmacology. Much attention was paid to evaluation using homologous animal cell cultures in appropriate animal models required for certain situations of BMCP evaluation, especially in immune-modulating therapies.

Moreover, Guidelines were drafted as a partially harmonized regulation with internationally adopted regulatory documents and experience of FDA-and EMA-approved cell therapeutic drugs was taken into account during development.

Another important objective achieved was the development of educational programs for physicians and specialists in pre-clinical evaluation of BMCPs drafted to ensure proper standards for professionals. These programs will be required after enforcement of Federal Law #180 to propagate and educate healthcare workers and researchers endeavoring stem cell therapies as their primary field of work.

Results of the project will be in demand by the Industrial partner and whole field of regenerative medicine ranging from developers and researchers to regulatory authorities and federal government. Developed guidelines after approval will be used by the researchers, government-based and contract organizations to plan pre-clinical testing of developed BMCPs and fully evaluate their safety and biological properties crucial for efficacy in clinical settings.

Conclusions

It should be noted that despite BMCPs are used for treatment and can be to a certain extent compared with commonly used chemical-based drugs it has a lot of differences from low-molecular weight or antibodies used as therapeutic agents. BMCPs carry living cells and thus cannot be sterilized and have minimal (hours long) shelf-life. Furthermore, immune response to cellular antigens may induce toxicity and transplant rejection, thus, additional requirements for donor compatibility or autologous approach are enforced.

Complex nature of BMCP implies that pre-clinical testing one hand has to be in general framework of safety assessment and on the other has to take into account specific indications, routes of delivery and specific features of cells use for BMCP manufacture [p. 7, 3]. Thus, additional testing is to be applied to control kinetics (distribution and fate of delivered cells) and pharmacological effects. Adequate animal models are also an issue as far as some human-based BMCPs are immunogenic in other mammals and, thus, immunosuppressed or immunodeficient transgenic animal strains are to be used for testing [p. 7, 6].

Thus, to conclude, our project was a very timely attempt to establish a reasonable regulation of pre-clinical assessment for newly developed biomedical products relying on the potency of human cells as a "drug." We managed to create a methodology for the harvesting of donor cells, obtaining of cell cultures, maintenance, and storage of biomaterials, cytological and biochemical characterization of final products, biosafety and efficacy assessment. On this basis we stipulated another crucial point and drafted the general Guidelines that can be applied for BMCPs yet may require a certain extent of interaction with regulatory authorities to ensure certain clarity in ambiguous situations. Further challenges in the field are numerous ranging from marketing to clinical trial organizations and the scientific community is ready for further cooperation with federal government and industry to use its expertise for the sake of the progress of the biomedical field.

Acknowledgment Research being carried out with the financial support of the state represented by the Ministry of Education and Science of the Russian Federation. Contract no. 14.610.21.0001 03 Oct. 2014. Unique project Identifier: RFMEFI61014X0001.

References

1. Directive 2001/20/EC of the European parliament and of the council of 4 April 2001. Online resource: http://ec.europa.eu/health/human-use/clinical-trials/directive/index_en.htm. Accessed 27 Nov 2016)
2. Federal Law of Russian Federation: On biomedical cellular products. 23.06.2016 N180-FZ (current edition 27.11.2016)
3. Guidance for Industry: Preclinical Assessment of Investigational Cellular and Gene Therapy Products. Online resource: http://www.fda.gov/BiologicsBloodVaccines/GuidanceCompliance RegulatoryInformation/Guidances/CellularandGeneTherapy/ucm376136.htm. Accessed 26 Nov 2016
4. Guidance for Industry: Regulation of Human Cells, Tissues, and Cellular and Tissue-Based Products (HCT/Ps). Online resource: http://www.fda.gov/downloads/BiologicsBloodVaccines/ GuidanceComplianceRegulatoryInformation/Guidances/Tissue/ucm062592.pdf. Accessed 28 Nov 2016
5. Regulation (EC).: No 1394/2007 of the European Parliament and of the Council of 13 November 2007 on advanced therapy medicinal products. Online resource: https://www. eumonitor.eu/9353000/1/j9vvik7m1c3gyxp/vi7jgtb6qkzs. Accessed 28 Nov 2016
6. USA CFR 21, part 1271 « Human cells, tissues, and cellular and tissue-based products » . Online resource: http://www.accessdata.fda.gov/scripts/cdrh/cfdocs/cfcfr/CFRSearch.cfm? CFRPart=1271. Accessed 27 Nov 2016

Combination of Functional Electrical Stimulation and Noninvasive Spinal Cord Electrical Stimulation for Movement Rehabilitation of the Children with Cerebral Palsy

A.G. Baindurashvili, G.A. Ikoeva, Y.P. Gerasimenko,
T.R. Moshonkina, I.E. Nikityuk, I.A. Solopova, I.A. Sukhotina,
S.V. Vissarionov and D.S. Zhvansky

Abstract Movement and posture disability are appropriate to cerebral palsy (CP). The aim of the current study was to examine the hypothesis that the functional muscle electrical stimulation (FES) and transcutaneous spinal cord electrical stimulation (TSCS) combined with locomotor treadmill training improve posture and motor function in children with severe CP. Thirty-one children with CP (spastic

A.G. Baindurashvili · G.A. Ikoeva · I.E. Nikityuk · S.V. Vissarionov
Turner Pediatric Orthopedic Research Institute, Ministry of Public Health, Moscow, Russia
e-mail: turner01@mail.ru

G.A. Ikoeva
e-mail: ikoeva@inbox.ru

I.E. Nikityuk
e-mail: femtotech@mail.ru

S.V. Vissarionov
e-mail: turner01@mail.ru

Y.P. Gerasimenko
Pavlov Institute of Physiology Russian Academy of Sciences, Moscow, Russia
e-mail: yuryg@ucla.edu

T.R. Moshonkina
Pavlov Institute of Physiology Russian Academy of Sciences, LLC Cosyma, Moscow, Russia
e-mail: tmoshonkina@gmail.com

I.A. Solopova (✉) · D.S. Zhvansky
Institute for Information Transmission Problems RAS, Moscow, Russia
e-mail: solopova@iitp.ru

D.S. Zhvansky
e-mail: d.zhvansky@gmail.com

I.A. Sukhotina
First Pavlov State Medical University, Moscow, Russia
e-mail: irina.sukhotina@gmail.com

© The Author(s) 2018
K.V. Anisimov et al. (eds.), *Proceedings of the Scientific-Practical Conference "Research and Development - 2016"*, https://doi.org/10.1007/978-3-319-62870-7_58

diplegia) (7–13 years old; mainly levels III of GMFCS) participated in the study. The experimental group received muscle FES and TSCS at T11 and L1 spinal levels, combined with locomotor treadmill training, whereas the participants of the control group received locomotor treadmill training only. After treatment, the GMFM-88 score increased in 81% children of the experimental group and in 33% children of the control group. In the experimental group, there were a significant decrease of the stabilogram area in the eye opened condition and the significant decrease of forward shift of center of pressure projection in the sagittal plane in both eye opened and eye closed conditions, whereas in the control group any significant changes of stabilogram parameters did not observed. Knee torque and range of knee motion significantly increased in the experimental group. After electrical stimulation, the decrease of muscle co-activation in proximal and distal muscles occurred, whereas in the control group muscle co-activation decreased in proximal muscles only. Thus, improvement of motor functions and balance control system in children with severe CP in response to the combination of TSCS, FES and locomotor training revealed. Combination of these techniques can be used for the effective neurorehabilitation.

Keywords Cerebral palsy · Functional electrical stimulation · Transcutaneous spinal cord electrical stimulation · Posture · Rehabilitation

Introduction

Cerebral palsy (CP) is characterized by muscle weakness, spasticity of extremities, co-activation of antagonistic muscles, resulting in disruption of motor development-pathological stereotype of locomotion, and the impairment of the ability to maintain upright standing [6, 9, 26]. The goal of physical therapy for children with CP is to promote motor learning through motor and functional training with multiple sensory stimuli.

The main reason of motor impairment—a violation of brain structures, spinal cord changes are secondary. It was suggested that maturation of the spinal locomotor output is impaired in children with CP [4]. Postmortem examination of children with CP revealed abnormalities in motor centers of the brain, brainstem and in the motor nuclei of the cranial nerves as well as in rostral segments of the spinal cord [17]. Magnetic resonance imaging of the spinal cord in the CP subjects with spastic diplegia has significantly less area of the white matter than in healthy individuals in transverse sections at C6/C7 and T10/T11 segments. In the same segments, gray matter area of CP and healthy subjects did not differ [20]. A direct correlation between the degree of imbalance of inhibitory-excitatory connections of motoneurons and the severity of the motor disorders has been reported in CP subjects [5]. The main problem in regulation of motor functions by spinal neuronal networks is thought to be related to the deficit of supraspinal connections.

It is known that electrical stimulation applied to spinal cord at L2 spinal segment can induce EMG stepping patterns in leg muscles in spinal cord injured (SCI) patients [7]. We have shown that locomotor training combined with epidural electrical spinal cord stimulation (SCS) facilitated the recovery of voluntary leg movements in motor complete SCI patients [1, 14]. Recently, we developed a method of electrically activating the spinal circuitry via electrodes placed on the skin overlying the lower thoracic vertebrae and demonstrated that in healthy subjects the involuntary locomotor-like movements can be induced by transcutaneous SCS [11, 12]. The positive effect of using transcutaneous SCS (TSCS) was observed in SCI patients as well. It was demonstrated that this neuromodulatory strategy can induce voluntary locomotor-like leg and arm movements after motor complete paralysis [11]. These findings demonstrated that this noninvasive technology can activate spinal neuronal locomotor related networks without brain control in adult chronically paralyzed individuals.

In all cases, SCS induced motor improvement when combined with locomotor training. In recent years, locomotor training on a treadmill widely uses in the treatment of children with CP [16]. Functional electrical stimulation (FES) has been used to enhance the effect of locomotor training [15]. It was shown that in CP children FES assisted cycling or walking was tolerated well and resulted in increased muscle strength, balance improvement, normalization of biomechanical and innervation structure of locomotor act, increased cadence, power output, and heart rates [15, 19].

The aim of the work was to investigate the effect of locomotor training combined with TSCS and FES of legs muscles on postural and locomotor neuronal networks in children with CP. We hypothesized that TSCS combined with FES and locomotor training could be used to recalibrate abnormally developed neuronal networks in children toward a more functional state.

Methods

Thirty-one children with CP (spastic diplegia) (9.9 ± 1.7 years, mean ± SD) participated in the study (Table 1). All participants previously had surgical interventions to decrease the spasticity and contractures but no later than 2–5 years before the study. The Academic Council of the Turner Scientific and Research Institute for Children's Orthopedics approved this study in accordance with the Declaration of Helsinki. Parents of all participants provided informed consent in writing.

CP participants were mainly level III according to the classification of Gross Motor Function Classification System (GMFCS) [21] and mean leg spasticity 1.5 ± 0.7 scores on the Ashworth scale [2]. Children had poor control of vertical posture and were able to keep the upright posture independently for about 2 min. All participants were socialized and able to accurately perform the required tasks.

Table 1 Subject characteristics for each treatment group

Experimental group					Control group				
			GMFM-88					GMFM-88	
Child	GMFCS	Age	Pre	Post	Child	GMFCS	Age	Pre	Post
S1	II	7	245	249	C1	III	10	234	238
S2	II	10	245	245	C2	II	10	210	212
S3	III	9	220	249	C3	IV	10	195	196
S4	IV	11	219	219	C4	III	9	191	194
S5	III	9	210	212	C5	III	13	169	174
S6	III	7	207	237	C6	III	9	157	157
S7	II	9	191	204	C7	III	12	153	153
S8	III	11	188	192	C8	III	11	146	146
S9	III	11	162	166	C9	III	7	141	141
S10	III	9	157	171	C10	II	7	137	139
S11	III	12	148	152	C11	III	10	122	122
S12	III	9	146	150	C12	III	10	110	110
S13	IV	11	122	142	C13	III	8	106	106
S14	III	12	115	116	C14	IV	11	90	92
S15	III	12	110	114	C15	III	10	83	83
S16	IV	12	35	35					
mean	3.0 ± 0.6	10.1 ± 1.7	170 ± 56	178 ± 59*		2.9 ± 0.5	9.8 ± 1.7	150 ± 44	151 ± 55*

GMFCS gross motor function classification system; *GMFM-88* gross motor function measure-88

* ($p<0.05$) differences between pre and post treatment GMFM-88 scores

Exclusion criteria were severe contracture of the lower limbs, fractures, osteo-porosis, thromboembolic disease, cardiovascular system instability, convulsive readiness, and mental retardation.

Children were randomly assigned to one of two groups (Table 1). The experimental group ($n = 16$) received 15 sessions of locomotor training combined with TSCS and FES over a period of 3 weeks. During each training session, the child was placed in Lokomat device (Hocoma, Switzerland) (Fig. 1a). Game-like augmented performance feedback exercises were used to increase motivation and effort of participants of both groups during treatment and tests. Initially, the TSCS was delivered at L1 level for 5 min in upright posture. The participants were instructed to keep a normal upright posture. The level of body weight support was selected individually so that the child could stand while maintaining equilibrium. During the first 10 min of locomotor training, TSCS was applied at T11 vertebral level, followed by the combination of T11 and L1 stimulation for the next 10 min. Afterward, the stepping performance was continued for 20 min with FES of leg muscles assistant. Locomotor training was performed at a treadmill speed of ~ 1 km/h. The children of the Control group ($n = 15$) received only locomotor training with Lokomat (40 min) without any electrical stimulation for 15 sessions.

TSCS was delivered using a 2.5-cm round electrodes (Syrtenty®, China) placed midline at the T11, and L1 spinous processes as cathodes and two 5.0×8 cm^2 rectangular plates (Syrtenty®) placed symmetrically on the skin over the iliac crests as anodes. Biphasic rectangular 1.0 ms pulses (30 Hz), modulated frequency of 10 kHz were used. The main intensity of stimulation ranged from 10 to 50 mA for most children. Intensity was chosen individually, ranging from 5 to 10% below threshold of muscle contraction, and was well tolerated by all the participants.

FES method was described in detail previously [19]. A 16-channel stimulator (BiostimES-16, LLC Cosyma, Russia) was used for stimulation (rectangular monophasic pulses; 65 Hz; maximum pulse amplitude 80 mA, pulse width 100 µs). Stimulated muscles were mm longissimus thoracis et illiocostalis, gluteus medius, gluteus maximus, tibialis anterior, quadriceps femoris, extensor digitorum longus et brevis, fibularis longus, extensor hallucis longus et brevis. To accelerate electrode attachment and to prevent electrodes detach while treadmill training as for as for invariance of the position of the electrodes on the body during all procedures, we used a specially designed suit (Fig. 1a) (LLC Cosyma, Russia).

All CP children were tested before and after treatment. Evaluation of the level of motor functions was provided based on GMFM-88 (Gross Motor Function Measure) scale [22]. Calculation of isometric force in flexors and extensors of hip as well as of knee muscles performed using the L-FORCE test of Lokomat software. Evaluation of active range of the movements for flexion/extension in hip and knee joints was performed using the L-ROM test, as directed by Lokomat software. EMG muscle activity (rectus femoris (RF), biceps femoris (BF), lateral gastrocnemius (LG) and tibialis anterior (TA)) was recorded with a telemetric system (MEGA, Finland). EMG activity was tested during stepping on Lokomat belt at the subjects' maximal speed without body weight support, but supported from a horizontal rail, as needed. For assessment of EMG amplitudes, raw EMG signals were

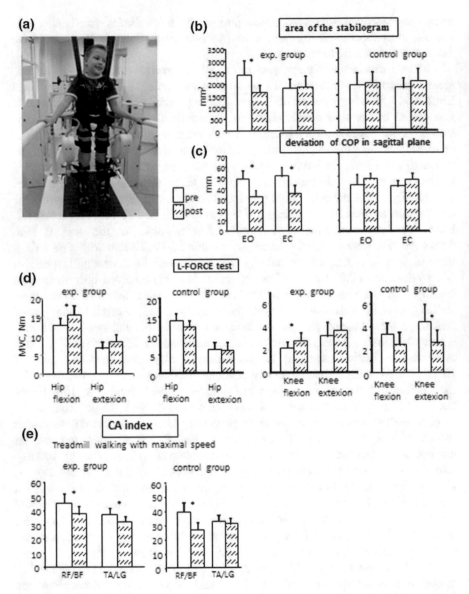

Fig. 1 The changes in stabilogram parameters, Lokomat tests and EMG in participants of both groups before and after treatment. **a** The participant of experimental group in suit produced by LLC Cosyma before procedure. **b** The stabilogram area. **c** Deviation of COP in sagittal plane. **d** Amplitude of maximal voluntary contraction (MVC) of hip and knee flexors and extensors. **e** CA values for hip (RF/BF) and ankle (TA/LG) muscle at treadmill walking with maximal speed. Descriptive statistics included means and standard error (SE). Asterisks denote significant differences in the motor functions before and after treatment ($p < 0.05$)

numerically rectified, low-pass-filtered with a zero-lag Butterworth filter at 20 Hz cutoff. Co-activation (CA) index of two agonist–antagonist muscle couples (RF/Bf, TA/LG) for both legs were calculated following the method described by Unnithan et al. [27]. Raw EMG data were normalized to the largest value of activation observed in each muscle across all trials for each subject. Integration of the over-lapping area between the two normalized linear envelopes defined the CA index. Because participants were diplegic the results for right and left legs were averaged. Lokomat and CA data obtained in control group are the same as have been pre-sented in manuscript "Effects of Spinal Cord Stimulation on motor functions in children with cerebral palsy" [25].

Stabiliometry performed on a "MBN-Biomechanic System" (Moscow, Russia). This system provides measurements of displacement of the center of pressure (COP) along x and y axes in frontal and in sagittal planes, respectively. The assessment was carried out at the standard functional test for 30 s under each condition of eyes open (EO) and eyes closed (EC). Glasses as needed corrected the vision. The length of COP motion path, area of the stabilogram, deviation of COP along x and y axes were analyzed.

Descriptive statistics included means and standard error (SE). The statistical analysis was performed by Statistica 10 software. The nonparametric Mann–Whitney U test was used to compare clinical scores and biomechanical values (EMG, L-FORCE, L-ROM, stabilometry parameters) before and after treatment. The Wilcoxon test was performed to compare the same parameters before and after treatment within each group. A level of $p < 0.05$ was accepted as statistically significant.

Results

No advice effects of TSCS or FES were detected. There were no significant group differences in the Ashworth score and GMFCS level (Table 1). Significant increase of the GMFM-88 score was revealed in both groups, but it was significantly higher in the experimental group than in control. The GMFM-88 score increased in 13 of 16 children of the experimental group and in 5 of 15 children of the control group (Table 1).

Initially, the children of experimental and control groups were similar in their ability to maintain upright position. Posture control violation accompanied by greater oscillations of the COP, a substantial increase of length of COP motion path and of the stabilogram area. In all patients, a typical COP forward shift in the sagittal plane was revealed . After treatment in the experimental group, there was a significantly ($p < 0.001$) decrease of area of the stabilogram in the EO condition, whereas in the control group it was not changed in any condition (Fig. 1b). In addition, the decrease of forward shift of COP projection in the sagittal plane in patients of the experimental group in both EC ($p < 0.01$) and EO ($p < 0.03$)

conditions was observed (Fig. 1c). Other analyzed stabilogramm parameters did not change significantly after treatment in both groups.

No differences between groups were found in the knee or hip torque (L-FORCE test) and range of joint movements (L-ROM test) before treatment. After treatment, neither maximal hip flexion nor extension torque changed in the control group (Fig. 1d), but in the experimental group, there was significantly increment in maximal hip flexion torque ($p < 0.02$). Maximal knee flexion and extensor torque increased after electrical stimulation, but this augmentation was significant only for flexion ($p < 0.01$). Unexpectedly in controls flexion and extension knee torques decreased, but significantly for extension ($p < 0.05$) (Fig. 1d). Range of hip movements (L-ROM test) increased in both groups, the change was significant ($p < 0.05$) in control group. In opposite knee range was unchanged in controls (from $17.8 \pm 2.6°$ pre to $16.7 \pm 2.7°$ post), but it increased in experimental group ($p < 0.01$) (from $13.2 \pm 2.4°$ pre to $18.3 \pm 2.9°$ post).

The mean treadmill stepping speed before treatment was similar in the experimental and the control groups: 1.4 ± 0.1 and 1.3 ± 0.2 km/h, respectively. We did not observe significant average speed changes in either group after treatment. Abnormal levels of muscle CA, i.e., showing simultaneous contraction of agonist and antagonist muscles [27], were observed in the experimental and the control groups (Fig. 1e). At the beginning of treatment, there were no differences in CA of either proximal or distal muscles between the experimental and the control group for treadmill stepping. After treatment, the decrease of CA index in proximal and distal muscles occurred in the experimental group, whereas in the control group CA index decreased in proximal muscles only (Fig. 1e).

Discussion

Previous studies of representative groups of children with CP, with mainly a level III deficit have shown maximal improvement of motor functions up to the age of 10 years with standard rehabilitation including mechanotherapy; afterwards motor skills did not increase and in some cases even decreased [13]. Using the GMFM-88 scale, we have shown that improvement occurred in less than 50% of the cases in the control group, while in the experimental group 81% of the subjects showed improved motor skills.

Improvement of balance control was not observed in the Lokomat trained group. Probably the reasons are as severity of disease as pure effectiveness of locomotor training for posture rehabilitation [3]. Opposite after combination of the FES and the TSCS significant positive changes revealed. Center of pressure stabilization and decrease of the pathological body forward inclination were observed. The effect may be connected with the activation of motor pools of muscles involved in maintaining the upright by TSCS [23]. Another reason —increase of strength of many muscles by FES and by this body weight support in upright posture was facilitated [8].

Degradation of motor activity in children with CP is linked to with pathology of supraspinal as well as of spinal networks connections during ontogenesis [17, 20]. One of the functional consequences of the abnormal CP development is the level of abnormal reciprocal inhibition of Ia afferents, presynaptic inhibition and nonreciprocal Ib inhibition a phenomenon largely attributable to spinal networks [10]. We have demonstrated that after TSCS combined with FES and step training the amplitude of movement in both the hip and knee joints increased (Fig. 1d). This observation is consistent with the reduction of co-activation of proximal and distal leg muscles (Fig. 1e). This suggests that TSCS, combined with FES and mechanotherapy, had positive influences on coordination not only for proximal, but for distal muscles as well. Previous studies have shown that improved locomotor performance is closely associated with improved coordination of motor pools [5]. TSCS at T11 and L1 vertebral levels of proximal and distal motor pools in paralyzed subjects are facilitated [24]. This effect is evident in spinal motor evoked potentials of proximal and distal leg muscles to single stimulation pulses [23]. We have found that epidural SCS combined with training can increase hand grip force in spinal patients with cervical injuries [18]. This suggests that similar interventions of training and cervical TSCS could also benefit for children with CP. Our findings of a stable improvement of motor functions and balance control system in children with severe CP in response to the combination of TSCS, FES and locomotor training suggest that some functional improvement of the spinal and supraspinal networks that control locomotor functions were induced. Further studies should receive a high priority.

Acknowledgements The research was conducted with the financial support of the state represented by the Ministry of Education and Science of the Russian Federation. Agreement No 14.576.21.0020 June 27, 2014. Unique project Identifier: RFMEFI57614X0020.

References

1. Angeli, C.A., Edgerton, V.R., Gerasimenko, Y.P., Harkema, S.J.: Altering spinal cord excitability enables voluntary movements after chronic complete paralysis in humans. Brain **137**(5), 1394–1409 (2014)
2. Bohannon, R.W., Smith, M.B.: Interrater reliability of a modified Ashworth scale of muscle spasticity. Phys. Ther. **67**(2), 206–207 (1987)
3. Borggraefe, I., Schaefer, J.S., Klaiber, M., et al.: Robotic assisted treadmill therapy improves walking and standing performance in children and adolescents with cerebral palsy. Eur. J. Paediatr. Neurol. **14**(6), 496 (2010)
4. Cappellini, G., Ivanenko, Y.P., Martino, G., MacLellan, M.J., Sacco, A., Morelli, D., Lacquaniti, F.: Immature spinal locomotor output in children with cerebral palsy. Front Physiol. **25**(7), 478 (2016)
5. Condliffe, E.G., Jeffery, D.T., Emery, D.J., Gorassini, M.A.: Spinal inhibition and motor function in adults with spastic cerebral palsy. J. Physiol. **594**(10), 2691–2705 (2016)
6. Damiano, D.L., Martellotta, T.L., Quinlivan, J.M., Abel, M.F.: Deficits in eccentric versus concentric torque in children with spastic cerebral palsy. Med. Sci. Sports Exerc. **33**(1), 117–122 (2001)

7. Dimitrijevic, M., Gerasimenko, Y., Pinter, M.: Evidence for a spinal central pattern generator in humans. Ann. NY Acad. Sci. **860**, 360–376 (1998)

8. Doucet, B.M., Lam, A., Griffin, L.: Neuromuscular electrical stimulation for skeletal muscle function. Yale J. Biol. Med. **85**(2), 201–215 (2012)

9. Engsberg, J.R., Ross, S.A., Olree, K.S., Park, T.S.: Ankle spasticity and strength in children with spastic diplegic cerebral palsy. Dev. Med. Child Neurol. **42**(1), 42–47 (2000)

10. Filloux, F.M.: Neuropathophysiology of movement disorders in cerebral palsy. J. Child Neurol. **11**(1 suppl), S5–S12 (1996)

11. Gerasimenko, Y.P., Lu, D.C., Modaber, M., Zdunowski, S., Gad, P., et al.: Noninvasive reactivation of motor descending control after paralysis. J. Neurotrauma **32**(24), 1968–1980 (2015)

12. Gorodnichev, R.M., Pivovarova, E.A., Puhov, A., Moiseev, S.A., Gerasimenko, Y.P., et al.: Transcutaneous electrical stimulation of the spinal cord: a noninvasive tool for the activation of stepping pattern generators in humans. Hum. Physiol. **38**(2), 158–167 (2012)

13. Hanna, S.E., Bartlett, D.J., Rivard, L.M., Russell, D.J.: Reference curves for the gross motor function measure: percentiles for clinical description and tracking over time among children with cerebral palsy. Phys. Ther. **88**(5), 596–607 (2008)

14. Harkema, S., Gerasimenko, Y., Hodes, J., Burdick, J., Angeli, C., et al.: Effect of epidural stimulation of the lumbosacral spinal cord on voluntary movement, standing, and assisted stepping after motor complete paraplegia: a case study. Lancet **377**, 1938–1947 (2011)

15. Harrington, A.T., McRae, C.G., Lee, S.C.: Evaluation of functional electrical stimulation to assist cycling in four adolescents with spastic cerebral palsy. Int. J. Pediatr. **2012**, 1 (2012)

16. Johnston, T.E., Watson, K.E., Ross, S.A., et al.: Effects of a supported speed treadmill training exercise program on impairment and function for children with cerebral palsy. Dev. Med. Child. Neurol. **53**(8), 742–750 (2011)

17. Levchenkova, V.D., Semenova, K.A.: Contemporary views of the morphological basis of infant cerebral palsy. Zhurnal nevrologii i psikhiatrii imeni SS Korsakova **112**(7), 4–8 (2012)

18. Lu, D.C., Edgerton, V.R., Modaber, M., AuYong, N., et al.: Engaging cervical spinal cord networks to reenable volitional control of hand function in tetraplegic patients. Neurorehabil Neural Repair. **30**(10), 951–962 (2016)

19. Nikityuk, I.E., Moshonkina, T.R., Shcherbakova, N.A., Vissarionov, S.V., et al.: Effect of locomotor training and functional electrical stimulation on postural function in children with severe cerebral palsy. Hum. Physiol. **42**(3), 262–270 (2016)

20. Noble J.J.: Musculoskeletal and Spinal Cord imaging in Bilateral Spastic Cerebral Palsy. Diss. King's College, London (2014)

21. Palisano, R., Rosenbaun, P., Walter, S., Russel, S., Wood, E., Galuppi, B.: Development and reliability of a system to classify gross motor function in children with cerebral palsy. Dev. Med. Child Neurol. **39**(4), 214–223 (1997)

22. Russell, D.J., Rosenbaum, P.L., Cadman, D.T., Gowland, C., Hardy, S., Jarvis, S.: The gross motor function measure: a means to evaluate the effects of physical therapy. Dev. Med. Child Neurol. **31**(3), 341–352 (1989)

23. Sayenko, D.G., Atkinson, D.A., Dy, C.J., Gurley, K.M., Smith, V.L., et al.: Spinal segment-specific transcutaneous stimulation differentially shapes activation pattern among motor pools in humans. J. Appl. Physiol. **118**(11), 1364–1374 (2015)

24. Sayenko, D.G., Atkinson, D.A., Floyd, T.C., Gorodnichev, R.M., Moshonkina, T.R., et al.: Effects of paired transcutaneous electrical stimulation delivered at single and dual sites over lumbosacral spinal cord. Neurosci. Lett. **609**, 229–234 (2015)

25. Solopova, I.A., Sukhotina, I.A., Zhvansky, D.S., Ikoeva, G.A., Vissarionov, S.V., Baindurashvili, A.G., Edgerton, V.R., Gerasimenko, Y.P., Moshonkina, T.R.: Effects of spinal cord stimulation on motor functions in children with cerebral palsy. Neurosci. Lett. **639**:192–198 (2017)

26. Stackhouse, S.K., Binder-Macleod, S.A., Lee, S.C.: Voluntary muscle activation, contractile properties, and fatigability in children with and without cerebral palsy. Muscle Nerve. **31**(5), 594–601 (2005)

27. Unnithan, V.B., Dowling, J.J., Frost, G., Volpe, Ayub B., Bar-Or, O.: Cocontraction and phasic activity during gait in children with cerebral palsy. Electromyogr. Clin. Neurophysiol. **36**, 487–494 (1996)

Bifunctional Recombinant Protein Agent Based on Pseudomonas Exotoxin A Fragment for Targeted Therapy of HER2-Positive Tumors

S.M. Deyev, O.M. Kutova, E.N. Lebedenko, G.M. Proshkina, A.A. Schulga and E.A. Sokolova

Abstract Four variants of bifunctional HER2-specific recombinant proteins (anti-HER2-toxins) were created composed of HER2-specific scFv antibody (4D5scFv) or HER2-specific DARPin (D29) as targeting module and *Pseudomonas* exotoxin A fragment (PE40) as toxic module, and distinguished by the way of bacterial expression (cytoplasmic or periplasmic). Physicochemical, immuno-chemical, and functional properties of the created recombinant proteins as well as their expression yields were analyzed and compared with the option of the most promising one, D29-PE40, for further implementation as an agent for targeted therapy of HER2-overexpressing tumors.

Keywords Recombinant protein · HER2 receptor · Targeted therapy *Pseudomonas* exotoxin A

S.M. Deyev · O.M. Kutova · E.A. Sokolova (✉)
Institute of Biology and Biomedicine, Lobachevsky University,
Nizhny Novgorod, Russia
e-mail: malehanova@mail.ru

S.M. Deyev
e-mail: deyev@mail.ibch.ru

O.M. Kutova
e-mail: papicat@rambler.ru

S.M. Deyev · E.N. Lebedenko · G.M. Proshkina · A.A. Schulga
Laboratory of Molecular Immunology, M.M. Shemyakin and Yu.A. Ovchinnikov Institute of Bioorganic Chemistry of the Russian Academy of Sciences (IBCh RAS), Moscow, Russia
e-mail: elebedenko@mail.ru

G.M. Proshkina
e-mail: galina.proshkina@gmail.com

A.A. Schulga
e-mail: schulga@gmail.com

Introduction

Development of new approaches and methods for cancer treatment presents one of the most important and rapidly developing areas of biology and medicine. The leading trend in this area is a personalized approach based on determining the molecular and genetic characteristics of the tumor in each particular case, and choosing the optimal method of treatment according to these characteristics. Due to progress in the study of the molecular basis of carcinogenesis, the targeted (or directed) therapy has been playing a greater role in recent decades, along with such main methods of cancer treatment as surgery and traditional chemotherapy. Targeted therapy implicates the use of bifunctional agents capable of selective binding to tumor cells expressing a specific molecular target (targeting function) and their elimination (therapeutic function). The tumor-selective action of such agents is intended to increase the therapeutic index values [9].

Membrane receptor HER2 (Human Epidermal growth factor Receptor 2) plays an important role in the process of epithelial cells malignant transformation, therefore being clinically relevant tumor marker. HER2-positive tumors are characterized by rapid growth and resistance to chemotherapy associating with a poor prognosis for the patient. HER2 overexpression typical for a number of carcinomas makes it also an advanced molecular target for targeted therapy that has been successfully proven [22].

Historically, the basis of targeted therapy is the use of monoclonal antibodies (MAbs). Their high affinity and specificity to target as well as antitumor action made it possible to create MAb-based drugs that have successfully passed clinical trials for the treatment of some tumors [9, 22]. There are two HER2-specific humanized MAbs used for targeted therapy of HER2-positive breast cancer, Trastuzumab (Herceptin) [7] and Pertuzumab (Omnitarg) [12], which are specific to different epitopes of the HER2 receptor. However, MAbs often fail to achieve optimum antitumor effect being used as monotherapy. Thus, prolonged use of Trastuzumab causes resistance to treatment in many patients [18], resulting in modifying the therapy strategy. Also, although a synergistic action of Trastuzumab and Pertuzumab, the latter is not effective against breast cancer cells resistant to Trastuzumab [19]. This problem has been addressed via conjugation of MAbs with other effector molecules: radioisotopes, low molecular and protein toxins, cytokines, etc., [20, 21]. In such immunoconjugates, antibody acts as a targeting component that delivers a cytotoxic one to target cells. Nevertheless, only one such HER2-specific immunoconjugate has been approved for the treatment of breast cancer, Trastuzumab emtansine (Kadsila), that is a conjugate of Trastuzumab and mitosis-inhibiting toxin of maytansinoid group [15]. The main difficulty of developing such immunoconjugates is chemical conjugation of targeting and toxic components (modules) that is characterized by a number of technical and scientific problems: lack of reproducibility and variability of the composition of the conjugates, possible decrease in the affinity of the targeting module or the effectiveness of the toxin, as well as admixture of nonconjugated molecules [11]. Solution to this problem is the usage of genetic engineering approach that enables to receive addressed bifunctional agents as recombinant proteins, i.e., fusion proteins comprising targeting and toxic modules [4, 23].

The targeting module of recombinant bifunctional agent may be presented by truncated antibody derivatives (scFv and dsFv antibodies, nanobodies, ect.), or non-immunoglobulin scaffold proteins, such as DARPins (Designed Ankyrin Repeat Proteins). As toxic module, bacterial or plant protein toxins are of interest, particularly *Pseudomonas* exotoxin A which is one of the most potent protein toxins known to date [14]. The mechanism of *Pseudomonas* exotoxin A action consists in catalytic ADP-ribosylation of eukaryotic elongation factor 2 with subsequent inhibition of protein translation. Multidomain structure of *Pseudomonas* exotoxin A and functional independence of its domains facilitate the genetic engineering when construction of recombinant bifunctional agent.

We have created four variants of bifunctional HER2-specific recombinant proteins (anti-HER2-toxins) composed of HER2-specific scFv antibody (4D5scFv) or HER2-specific DARPin (D29) as targeting module and *Pseudomonas* exotoxin A fragment (PE40) as toxic module, and distinguished by the way of bacterial expression (cytoplasmic or periplasmic). Physicochemical, immunochemical, and functional properties of the created recombinant proteins as well as their expression yields were analyzed and compared with the option of the most promising one for further implementation as an agent for targeted therapy of HER2-overexpressing tumors.

Research Description

Expression and Purification of Recombinant Proteins

The following plasmids were used for bacterial production of recombinant proteins:

pET22-scFv4D5-PE40 plasmid with coding sequence of anti-HER2-toxin composed of *Pseudomonas* exotoxin A fragment (PE40) and HER2-specific antibody 4D5scFv (4D5scFv-PE40, or anti-HER2-toxin №1), and providing its cytoplasmic expression in *E. coli* cells, under control of inducible *T7* promoter;

pET22-D29-PE40 plasmid with coding sequence of anti-HER2-toxin composed of *Pseudomonas* exotoxin A fragment (PE40) and HER2-specific non-immunoglobulin scaffold protein DARPin D29 (D29-PE40, or anti-HER2-toxin №2), and providing its cytoplasmic expression in *E. coli* cells, under control of inducible *T7* promoter;

pSD-scFv4D5-PE40 plasmid with coding sequence of anti-HER2-toxin composed of *Pseudomonas* exotoxin A fragment (PE40) and HER2-specific antibody 4D5scFv (4D5scFv-PE40/p, or anti-HER2-toxin №3), and providing its periplasmic expression in *E. coli* cells, under control of inducible *lac* promoter;

pSD-D29-PE40 plasmid with coding sequence of anti-HER2-toxin composed of *Pseudomonas* exotoxin A fragment (PE40) and HER2-specific non-immunoglobulin scaffold protein DARPin D29 (D29-PE40/p, or anti-HER2-toxin №4), and providing its periplasmic expression in *E. coli* cells, under control of inducible *lac* promoter.

Both vectors, pET22 and pSD, contain the ampicillin resistance gene, which allows to select bacterial cells expressing the target protein. All expression cassettes contain sequence encoding oligohistidine tag at the C-terminus of the protein to facilitate subsequent protein purification by immobilized metal affinity chromatography using Ni^{2+}-NTA-Sepharose. The pSD vector also contains ompA signal sequence, ensuring secretion of the target recombinant protein into the periplasmic space in order to both reduce its toxic effect on the cell and enhance the soluble fraction of the target protein.

Conditions used for expression of recombinant proteins №1–4 (strain, optical density of bacterial suspension at 550 nm before the induction (A_{550}), concentration of the inductor (isopropyl β–D-1-thiogalactopyranoside, IPTG), induction time and temperature) are presented in Table 1.

After protein expression, bacterial cells were lysed by ultrasound and centrifuged. Then the target proteins were purified from the cleared lysate in two steps by Ni^{2+}-chelate affinity chromatography using a 1 mL HisTrap FF column (GE Healthcare, USA) and ion exchange chromatography using a 1 mL Q Sepharose FF column (GE Healthcare, USA). Presence of the target proteins in elution fractions was controlled by denaturing electrophoresis in 12.5% polyacrylamide gel.

Examination of physicochemical and immunochemical properties of the recombinant proteins

The following procedures were carried out to test the structure of the proteins:

(1) verification of protein molecular weight by denaturing electrophoresis;
(2) checking the gene structure by sequencing;
(3) checking for presence of the address module 4D5scFv or D29 by immunoblotting using antibodies specific to 4D5scFv antibody or to DARPin D29, respectively;
(4) checking for presence of the toxic module by immunoblotting using antibodies specific to oligohistidine tag located at the C-terminus of the toxic module.

Table 1 Conditions of bacterial expression of four variants of anti-HER2-toxin

Anti-HER2 toxin	№1	№2	№3	№4
	4D5scFv-PE40	D29-PE40	4D5scFv-PE40/p	D29- PE40/p
E. coli strain	C41	BL21 (DE3)	SB536	BL21
A_{550}	0.6	0.4	0.5	0.6
IPTG (mM)	1	1	0.5	1
Time (h)	4	4	14–16	4
Temperature (°C)	28	37	28	37

The A_{280}/A_{260} ratio was determined spectrophotometrically. Checking the homogeneity of the proteins and their stability after 3 months storage was conducted by denaturing gel electrophoresis. Checking the content of bacterial endotoxins in the protein samples was conducted using limulous amoebocyte lysate test (LAL Test). Checking the specificity of interaction of the proteins with HER2 receptor was performed by ELISA using recombinant extracellular domain of the HER2 receptor as immobilized antigen.

Study of Functional Properties of the Recombinant Proteins

Equilibrium dissociation constant Kd of the recombinant proteins was determined on immobilized recombinant extracellular domain of the HER2 receptor by surface plasmon resonance using the BIAcore analyzer.

Cytotoxicity of the proteins was studied by the microculture tetrazolium test (MTT) [17] using adherent monolayer cell cultures expressing different amount of the HER2 receptor: SKBR-3 (HER2-overexpressing human breast adenocarcinoma, ATCC® HTB-30™) [13], SKOV-3 (HER2-overexpressing human ovarian adenocarcinoma, ATCC® HTB-77™) [8], CHO (Chinese hamster ovary with no HER2 expression, ATCC® CCL-61™) [16]. Data analysis and calculation of IC_{50} values were performed using the GraphPad Prism 6 software.

Antitumor effect of the recombinant protein D29-PE40 was estimated in vivo on intraperitoneal xenograft model of HER2-overexpressing human ovarian adenocarcinoma established in *nude* mice using ovarian cancer cells expressing far-red fluorescent protein TurboFP635. Monitoring of tumor growth was conducted by whole-body fluorescence imaging. To evaluate the efficacy of the anti-HER2-toxin, tumor growth inhibition coefficient (TGI) was calculated according to the following equation: $TGI\% = [(Fl_{control} - Fl_{experiment}) \times 100\%]/Fl_{control}$, where Fl is an integral fluorescence intensity in the peritoneal cavity at a selected time point.

Research Results

We have created four bifunctional recombinant proteins (anti-HER2 toxins) as potential agents for targeted therapy of HER2-overexpressing tumors. These recombinant proteins are composed of targeting module specific to HER2 receptor and toxic module based on *Pseudomonas* exotoxin A. Targeting modules used are: (i) single chain antibody 4D5scFv containing the variable domains of heavy and light chains of antibody Herceptin, which is widely used in clinical practice; (ii) innovative non-immunoglobulin scaffold polypeptide DARPin D29. DARPins present an alternative to antibodies as targeting molecules due to the following advantages: (i) smaller size of the protein, which facilitates the biochemical manipulation therewith; (ii) absence of cysteine residues in the structure and, as a result, aggregation stability and good expression; (iii) thermodynamic stability [5]. As a toxic module of the bifunctional recombinant proteins *Pseudomonas* exotoxin A fragment was used (PE40). This bacterial exotoxin is one of the most potent protein toxins known to date and has been successfully used for construction of a number of immunotoxins specific to different targets [2]. Two functional modules, targeting and toxic, were combined in a single recombinant protein by genetic engineering techniques. This approach to obtaining a bifunctional polypeptide has a number of advantages over commonly used chemical conjugation, primarily, precisely controlled composition, ease of production using bacterial producers and retention of functional properties of the modules. Bifunctional recombinant proteins were produced in bacteria that allows easy scaling of the technology for the laboratory and industrial production. In addition to the nature of the targeting module (4D5scFv or D29), the developed recombinant proteins vary in the way of expression in *E. coli*: cytoplasmic and periplasmic ones. The latter is aimed at increasing the level of recombinant protein expression in soluble form. Thus, four variants of anti-HER2-toxin were initially proposed (Table 2).

The recombinant proteins were expressed and purified. As was revealed by study of physicochemical and immunochemical properties of the produced proteins, each of them had expected molecular mass and structure, was highly pure, homogenous and storage stable as well as specific to HER2 receptor.

To quantify the affinity of the recombinants proteins for HER2 receptor, the equilibrium dissociation constant Kd was determined. All proteins were shown to have relatively equal Kd values lying in the low nanomolar range (Table 3) [24]. These particular Kd values are considered to be optimal when using targeted agents in vivo. Thus, very high-affinity scFv antibodies (with Kd values down to 10^{-11} *M*) demonstrate so-called «binding site barrier» effect that is stable binding of antibody to target receptors on tumor cells in the periphery of the tumor. As a result, such antibodies as well as targeted agents based on them poorly penetrate into the tumor tissue, so causing weak antitumor effect [1]. In the case of HER2-specific targeted

Table 2 Variants of anti-HER2 toxin

№	Protein	L (a.a.)	M (kDa)	Scheme of gene construct
1	4D5scFv-PE40	632	68	T7 → [4D5scFv][H][PE40][His₆][K]
2	D29-PE40	540	58	T7 → [D29][H][PE40][His₆][K]
3	4D5scFv-PE40/p	642	69.3	lac p/o → [ompA][His₆][4D5scFv][H][PE40][His₆][K]
4	D29-PE40/p	548	59	lac p/o → [ompA][His₆][D29][H][PE40][His₆][K]

L number of amino acid residues in the protein; *M* calculated molecular mass of the protein

bifunctional agents loaded with a potent toxic module, too high affinity also leads to a significant increase in systemic toxicity: first, due to more efficient binding to normal cells with low expression of the target receptor, and second, because of the intrinsic shedding of the HER2 extracellular domain (i.e., its cleavage by extracellular metalloproteinases, or sheddases). HER2-specific agents of higher affinity bind more efficiently to the free HER2 extracellular domain in the bloodstream then being eliminated by reticuloendothelial cells and leading to high hepatic toxicity [6]. So, nanomolar affinity of all produced recombinant proteins for HER2 predetermines their optimal distribution and pharmacokinetics if used for HER2-specific targeted therapy.

High affinity shown for all recombinant proteins to HER2 receptor determined to a large extent their high cytotoxicity against HER2-expressing eukaryotic cells. The recombinant proteins were shown to potently inhibit growth of HER2-overexpressing cells SKOV-3 and SKBR-3 with IC_{50} values lying in the low picomolar range. At the same time, HER2-negative CHO cells were slightly affected with IC_{50} values lying in the range of nanomolar concentrations (Table 3) [10, 24].

Such a great difference in IC_{50} values for target (HER2-positive) and normal cells indicates a strongly pronounced selectivity of toxic effect of the proteins on HER2-expressing cells. This provides a basis for high therapeutic index values of the created proteins when using them as HER2-specific targeted agents in vivo.

Thus, all the created recombinant proteins demonstrated high potential for selective elimination of HER2-overexpressing cancer cells. We suppose that the nature of the HER2-specific targeting module actually did not affect functional activity of the bifunctional agent. However, the D29-PE40 recombinant protein was produced in *E. coli* most efficiently with the yield of its cytoplasmic expression of 160 mg per liter of bacterial suspension that is several times higher than other protein variants (Table 3). Apparently, it is explained with DARP in structure that is free of cysteines. Cysteines residues are known to prevent a productive expression of the recombinant proteins in soluble and monomeric form due to (i) poor

Table 3 Comparative analysis of variants of anti-HER2 toxin

Anti-HER2 toxin	Property			
	1	2	3	
	Protein yield, mg per liter of bacterial culture	Kd (nM)	IC_{50} (nM)	
			HER2-overexpressing cells	HER2-negative cells
№1 (4D5scFv-PE40)	25	5.6	0.0010	18
№2 (D29-PE40)	160	11.3	0.0015	15
№3 (4D5scFv-PE40/p)	15.2	5.6	0.0010	23
№4 (D29-PE40/p)	37.5	11.3	0.0010	14

formation of intramolecular disulfide bonds under the reducing conditions of the *E. coli* bacterial cytoplasm that interrupts proper protein folding and (ii) possible formation of intermolecular disulfide bonds leading to protein aggregation. Indeed, DARPins mostly have high level of cytoplasmic expression in *E. coli* in soluble form reaching up to 30% of total protein [25].

In view of the aforesaid, we considered the D29-PE40 recombinant protein as the optimal variant of anti-HER2 toxin and evaluated its antitumor efficacy in vivo on the HER2-overexpressing ovarian tumors established as intraperitoneal xenografts in athymic *nude* mice. To create this xenograft model, we used intraperitoneal human ovarian cancer cells expressing far-red fluorescent protein TurboFP635. Creation of fluorescent cancer cell lines presents an advanced approach in experimental oncology ensuring tumor growth visualization in living animal by fluorescence whole-body imaging [3]. Dynamics of intraperitoneal tumor growth was assessed by integral fluorescence intensity in the peritoneal cavity. The mice treated with 50 µg D29-PE40 showed evident slowing down of tumor progression with the TGI value of 60% as compared to untreated (control) animals. The revealed antitumor effect against HER2-overexpressing tumor xenograft gives evidence of the therapeutic potential of the developed recombinant bifunctional protein D29-PE40 composed of HER2-specific DARPin and fragment of *Pseudomonas* exotoxin A.

Conclusions

The study resulted in creation of novel bifunctional HER2-specific recombinant protein (anti-HER2-toxin) based on HER2-specific DARPin and *Pseudomonas* exotoxin A fragment. The created anti-HER2 toxin was shown to selectively eliminate HER2-overexpressing cancer cells both in vitro and in vivo thus presenting a perspective agent for targeted therapy of HER2-positive tumors.

Acknowledgements Research was carried out with the financial support of the state represented by the Ministry of Education and Science of the Russian Federation. Agreement no. 14.578.21.0051, 23 Sent 2014. Unique project Identifier: RFMEFI57814X0051.

References

1. Adams, G.P., Schier, R., McCall, A.M., Simmons, H.H., Horak, E.M., Alpaugh, R.K., Marks, J.D., Weiner, L.M.: High affinity restricts the localization and tumor penetration of single-chain Fv antibody molecules. Cancer Res. **61**, 4750–4755 (2001)
2. Alewine, C., Hassan, R., Pastan, I.: Advances in anticancer immunotoxin therapy. Oncologist **20**(2), 176–185 (2015)
3. Balalaeva, I.V., Sokolova, E.A., Brilkina, A.A., Deyev, S.M., Petrov, R.V.: Far-red fluorescent cell line for preclinical study of HER2-targeted agents. Dokl. Biochem. Biophys. **465**, 410–412 (2015)

4. Becker, N., Benhar, I.: Antibody-based immunotoxins for the treatment of cancer. Antibodies **1**(1), 39–69 (2012)
5. Binz, H.K., Amstutz, P., Kohl, A., Stumpp, M.T., Briand, C., Forrer, P., Grutter, M.G., Pluckthun, A.: High-affinity binders selected from designed ankyrin repeat protein libraries. Nat. Biotechnol. **22**(5), 575–582 (2004)
6. Cao, Y., Marks, J.D., Huang, Q., Rudnick, S.I., Xiong, C., Hittelman, W.N., Wen, X., Marks, J.W., Cheung, L.H., Boland, K., Li, C., Adams, G.P., Rosenblum, M.G.: Single-chain antibody-based immunotoxins targeting Her2/neu: design optimization and impact of affinity on antitumor efficacy and off-target toxicity. Mol. Cancer Ther. **11**(1), 143–153 (2012)
7. Carter, P., Presta, L., Gorman, C.M., Ridgway, J.B., Henner, D., Wong, W.L., Rowland, A. M., Kotts, C., Carver, M.E., Shepard, H.M.: Humanization of an anti-p185HER2 antibody for human cancer therapy. Proc. Natl. Acad. Sci. U S A **89**(10), 4285–4289 (1992)
8. Dean, G.S., Pusztai, L., Xu, F.J., O'Briant, K., DeSombre, K., Conaway, M., Boyer, C.M., Mendelsohn, J., Bast Jr., R.C.: Cell surface density of p185(c-erbB-2) determines susceptibility to anti-p185(c-erbB-2)-ricin A chain (RTA) immunotoxin therapy alone and in combination with anti-p170(EGFR)-RTA in ovarian cancer cells. Clin. Cancer Res. **4**(10), 2545–2550 (1998)
9. Deyev, S.M., Lebedenko, E.N., Petrovskaya, L.E., Dolgikh, D.A., Gabibov, A.G., Kirpichnikov, M.P.: Man-made antibodies and immunoconjugates with desired properties: function optimization using structural engineering. Russ. Chem. Rev. **84**(1), 1–26 (2015)
10. Deyev, S.M., Lebedenko, E.N.: Supramolecular agents for theranostics. Bioorg Khim. **41**(5), 539–552 (2015)
11. Ducry, L., Stump, B.: Antibody-drug conjugates: linking cytotoxic payloads to monoclonal antibodies. Bioconjug. Chem. **21**, 5–13 (2010)
12. Franklin, M.C., Carey, K.D., Vajdos, F.F., Leahy, D.J., de Vos, A.M., Sliwkowski, M.X.: Insights into ErbB signaling from the structure of the ErbB2-pertuzumab complex. Cancer Cell **5**(4), 317–328 (2004)
13. Hynes, N.E., Gerber, H.A., Saurer, S., Groner, B.: Overexpression of the c-erbB-2 protein in human breast tumor cell lines. J. Cell. Biochem. **39**(2), 167–173 (1989)
14. Kreitman, R.J.: Immunotoxins for targeted cancer therapy. Aaps J. **8**(3), 532–551 (2006)
15. Lambert, J.M., Chari, R.V.: Ado-trastuzumab Emtansine (T-DM1): an antibody-drug conjugate (ADC) for HER2-positive breast cancer. J. Med. Chem. **57**(16), 6949–6964 (2014)
16. McCluskey, A.J., Olive, A.J., Starnbach, M.N., Collier, R.J.: Targeting HER2-positive cancer cells with receptor-redirected anthrax protective antigen. Mol. Oncol. **7**(3), 440–451 (2013)
17. Mosmann, T.: Rapid colorimetric assay for cellular growth and survival: application to proliferation and cytotoxicity assays. J. Immunol. Methods **65**, 55–63 (1983)
18. Nahta, R., Esteva, F.J.: Herceptin: mechanisms of action and resistance. Cancer Lett. **232**, 123–138 (2006)
19. Nahta, R., Hung, M.C., Esteva, F.J.: The HER-2-targeting antibodies trastuzumab and pertuzumab synergistically inhibit the survival of breast cancer cells. Cancer Res. **64**, 2343–2346 (2004)
20. Panowski, S., Bhakta, S., Raab, H., Polakis, P., Junutula, J.R.: Site-specific antibody drug conjugates for cancer therapy. MAbs **6**(1), 34–45 (2014)
21. Pastan, I., Hassan, R., FitzGerald, D.J., Kreitman, R.J.: Immunotoxin treatment of cancer. Annu. Rev. Med. **58**, 221–237 (2007)
22. Polanovski, O.L., Lebedenko, E.N., Deyev, S.M.: ERBB oncogene proteins as targets for monoclonal antibodies. Biochemistry **77**(3), 227–245 (2012)
23. Shapira, A., Benhar, I.: Toxin-based therapeutic approaches. Toxins **2**(11), 2519–2583 (2010)
24. Sokolova, E.A., Stremovskiy, O.A., Zdobnova, T.A., Balalaeva, I.V., Deyev, S.M.: Recombinant immunotoxin 4D5scFv-PE40 for targeted therapy of HER2-positive tumors. Acta Naturae **7**(4 (27)), 93–96 (2015)
25. Tamaskovic, R., Simon, M., Stefan, N., Schwill, M., Plückthun, A.: Designed ankyrin repeat proteins (DARPins): from research to therapy. Methods Enzymol. **503**, 101–134 (2012)

Development of Classification Rules for a Screening Diagnostics of Lung Cancer Patients Based on the Spectral Analysis of Metabolic Profiles in the Exhaled Air

A.V. Borisov, Yu. V. Kistenev, D.A. Kuzmin, V.V. Nikolaev,
A.V. Shapovalov and D.A. Vrazhnov

Abstract The pattern recognition technique was used for the development of classification rules for a screening diagnostics of lung cancer (LC) patients, based on the spectral analysis of metabolic profiles in the exhaled air, measured by the IR laser photoacoustic spectroscopy (LPAS). The study involved LC, chronic obstructive pulmonary disease, pneumonia patients, and healthy volunteers. The analysis of the measured spectra of exhaled air samples was based first on reduction of the dimension of the feature space using principal component analysis (PCA); thereafter the dichotomous classification was carried out using the support vector machine (SVM). The approaches to differential diagnostics based on the set of SVM classifiers usage are presented.

A.V. Borisov
Department of General and Experimental Physics, Tomsk State University,
Tomsk, Russia
e-mail: borisov@phys.tsu.ru

Yu.V. Kistenev (✉) · V.V. Nikolaev
Tomsk State University, Tomsk, Russia
e-mail: yuk@iao.ru

V.V. Nikolaev
e-mail: vik-nikol@bk.ru

D.A. Kuzmin
Siberian State Medical University, Siberian, Russia
e-mail: band107@mail.ru

A.V. Shapovalov
Department of Theoretical Physics, Tomsk State University, Tomsk, Russia
e-mail: shpv@phys.tsu.ru

D.A. Vrazhnov
Tomsklabs PTE LTD, Tomsk, Russia
e-mail: denis.vrazhnov@gmail.com

© The Author(s) 2018
K.V. Anisimov et al. (eds.), *Proceedings of the Scientific-Practical Conference*
"Research and Development - 2016", https://doi.org/10.1007/978-3-319-62870-7_60

Keywords Lung cancer · Noninvasive express diagnostics · Exhaled air
Volatile organic compounds · Laser photoacoustic spectroscopy
Support vector machine · Principal component analysis

Introduction

Lung cancer (LC) has been the most common cancer in the world for several decades. About 1.8 million of new cases were in 2012 (12.9% of the total), 58% of which occurred in the less developed regions. The disease remains the most worldwide common men cancer (1.2 million, 16.7% of the total) with the highest estimated age-standardized incidence rates in Central and Eastern Europe (53.5 per 100,000) and Eastern Asia (50.4 per 100,000). Notably, low incidence rates are observed in Middle and Western Africa (2.0 and 1.7 per 100,000 respectively). In case of women, the incidence rates are generally lower and the geographical pattern is a little different, mainly reflecting different historical exposure to tobacco smoking. Thus, the highest estimated rates are in Northern America (33.8) and Northern Europe (23.7) with a relatively high rate in Eastern Asia (19.2) and the lowest rates again in Western and Middle Africa (1.1 and 0.8, respectively) [4].

The growth of the mortality from LC is caused by late diagnostics of the disease. To solve this problem, the methods which provide registration of pathological changes in the molecular level (referred as metabolomics) before clinical manifestations should be designed. One of them—approach to diagnostics based on control of the volatile metabolites-markers in the exhaled air—is intensively developing. The additional advantages of such approach are non-invasiveness and suitability for mass screening studies.

It should be pointed out that mostly the molecular markers in the exhaled air are not highly specific [5, 7, 15]. In this case, the "profiling" approach, based on the set of markers control or profile of the absorption spectrum of breath sample as a "fingerprint" of the state, is more expedient to use [12].

Laser photoacoustic spectroscopy (LPAS) is one of the effective methods of exhaled air analysis [11]. In this report, we discuss the approaches of differential diagnostics of LC patients on a base of spectral analysis of exhaled air samples using IR LPAS and the methods of data mining.

The Experimental Base

The study involved the groups with lung cancer (LC) patients ($n = 18$); patients with chronic obstructive pulmonary disease (COPD) ($n = 22$), patients with pneumonia ($n = 21$); and a control group of healthy nonsmoking volunteers ($n = 39$). The interaction with the patients was limited by the sampling of a part of exhaled air into a disposable container. Protocol of the research was approved by the Ethic Committee of the Siberian State Medical University (Tomsk, Russia), Ref. Number 2882 at 24.11.2011.

The sampling procedure occurs before eating or 2 h thereafter. Prior to sampling, participants rinsed the mouth with running water without any special cleaning of the oral cavity. Then participant did some calm breaths through a sterile plastic tube into the sample container.

Registration of spectral characteristics of exhaled air probes (EAPs) was carried out using the LaserBreeze gas analyzer based on an LPAS method and OPO with a tuning range of 2.5–10.7 µm. The parameters of LaserBreeze gas analyzer are presented in [6].

The Data Analysis Methods

One of the key steps in the biomarkers analysis involves evaluation of latent dependencies in the variables data using reliable methods. To solve it, the principal component analysis (PCA) is frequently used which projects correlate variables into a lower number of uncorrelated variables termed the principal components. The mathematical background of PCA consists in decomposition of initial experimental data from a 2D matrix X $(I \times J)$ in the form of a matrix product [10]:

$$X = T \cdot P^t + E,$$

where T, P, E are the scores, loadings and residuals matrixes, respectively. The loadings matrix contains weight coefficients that characterize the contribution of features to a principal component. The scores matrix contains coordinates of the samples in the space of the principal components.

Most frequently used support vector machine (SVM) is for a two-stage (teaching and testing) binary classification. The application of SVM to the problem of data classification of object which should be assigned to one of two classes defines as follows:

$$(x_1, y_1), \ldots, (x_m, y_m) \in \mathbf{X} \times \{\pm 1\},$$

where X is a nonempty set; m is the number of objects in the training set; y_i is called a label or output data; and x_i are the objects under classification. Each classified object is a vector in n-dimensional space.

Thus, there is the task of some classifier rule building:

$$a(x) = \text{sign}\left(\sum_{j=1}^{n} w_j \cdot x^j - b\right) = \text{sign}(\langle \mathbf{w}, \mathbf{x} \rangle - b),$$

where operation $\langle \mathbf{w}, \mathbf{x} \rangle$ defines the scalar product of vectors, and vector $\mathbf{w} = (w_1, w_2, \ldots, w_n) \in \mathbb{R}^n$ and scalar threshold $b \in \mathbb{R}$ are the algorithm parameters.

The SVM method provides binary classification, i.e., it can separate objects only on two classes. For purposes of differential diagnostics, it is necessary to construct the classification rules on several classes. The statement of the problem can be formulated as follows.

Let there be N different classes, and each feature vector of the object under study belongs to one of them. A part of initial data can be used for construction of classification rules, the rest part will be for testing.

There are several approaches to solve this problem using binary classifiers [1]. The ideas were proposed by several researchers and are still used as the base.

According to the "One-or-None" (also known as "One-vs-All", "One-vs-Rest") method [16], we had to construct N independent binary classifiers, so that the i-th classifier will separate i-th class feature vectors from all other classes feature vectors. Evidently, this i-th classifier allows to determine whether the tested feature vector belongs to the i-th class. If the training set is fully separable, then after using of no more than N classifiers, we will get the answer to what class a feature vector from testing set belongs.

As mentioned above the strategy of "One-vs-All", includes training of N classifiers for the separation of each class. For every classifier the feature vectors belonging to the class under consideration correspond to the positive examples, all other feature vectors are considered as negative examples. At the stage of training, it should be drawing up the classification rule which will identify which class object under testing belongs. There are two main features to construct the classification rule.

The first method is based on enumeration of the labels of all classes. Under testing stage, we had to check the obtained labels for the object under study. It must be referred to only one class, if not, this object cannot be estimated using this classifier rule. This method can give ambiguous results, if several classifiers attributed the object to several classes.

The second method based on choosing the best from the full set. In this case, the labels of the class had to be a real value than in the stage of analysis the higher a specific class label value, the greater the likelihood that the object under study belongs to this.

According to the "One-vs-One" (also known as "All-vs-All") method [8], we had to construct $N(N-1)$ independent binary classifiers, each of which $f_{i,j}$ will separate i-th class feature vectors from j-th class feature vectors. Let, for definiteness, the classifier $f_{i,j}$ labels by "+1" the feature vectors of i-th class and by "−1" the feature vectors of j-class. Note that in this case $f_{i,j} = -f_{j,i}$. Then, the differential classification rule of feature vector x can be determined by the following formula:

$$f(x) = \arg \max_i \left(\sum_j f_{i,j}(x) \right).$$

Note, that each of these methods has its advantages and disadvantages. For example, methods "All-vs-All" demand less memory during the training phase, learn faster due to the smaller size of the training set, but their implementation is required to train $O(N^2)$ classifiers, when the method "One-vs-All" is required to train $O(N)$ classifiers.

There are also more complex methods for solving the problem of multiclass classification using SVM. However, Hsu and Lin [3] showed that among five investigated methods ("One-vs-One", "One-vs-All", Direct Acyclic Graphs (DAG) SVM [9], modification of "One-vs-One" by Vapnik [13] and Weston [14], the method of Crammer and Singer [2]) the most suitable from a practical point of view are "One-vs-One" and DAG SVM methods.

The "One-Vs-All" Classification Results

We used the spectral data of EAP from LC, COPD, pneumonia patients and healthy participants (10 feature vectors for every group in the teaching set). The volume of testing set was as follows: LC ($n = 8$), COPD (12), pneumonia ($n = 11$) patients, healthy participants (29).

Initially, we construct the classifies which had to separate the objects from one class from all other classes using SVM classifier with radial basis function (RBF) kernel. The optimal kernel parameters had been evaluated. The results of self-test classification accuracy of "One-vs-All" classifiers on test sets with the corresponding feature vectors are presented in the Table 1. The self-test approach was as follows. For example, classification accuracy of the classifier "Pneumonia vs All" was estimated using two groups from testing set: "Pneumonia" and "LC + COPD + Pneumonia + Healthy participants" and etc.

Below, we used experimental data after preprocessing by PCA and took into account the first five principal components. The results of classification by strategy of "One-vs-All" of feature vectors from testing set for the best parameters of RBF kernel of SVM classifier are presented in Table 2.

Table 1 Classification accuracy of the classifiers "One-vs-All"

	Pneumonia vs All	COPD vs All	Healthy vs All	LC vs All
Classification accuracy on test sets	98.84	45.89	52.72	93.98
RBF parameter	6.6367	0.1772	0.2487	0.0852

Table 2 Accuracy of classification by strategy of "One-vs-All"

Group	Quantity of the feature vectors in the testing set	Diagnosis		
		Set right	Set wrong	Did not set
Pneumonia	11	10	1	0
LC	8	8	0	0
COPD	12	7	4	3
Healthy volunteers	29	18	6	5

Thus, multiclass classification by strategy of "One-vs-All" is shown to provide not high accuracy, which in average is about 75%.

The "One-Vs-One" Classification Results

The "One-vs-One" method was realized on the same teaching and testing sets as above. We used preprocessing by PCA (up to 15 principal components were considered), then classification by SVM occurred. Table 3 shows the results of the pairwise classification in terms of the specificity and sensitivity. The random separation of initial data on teaching and testing sets in mentioned proportion was repeated 250 times. Then, results were averaged and presented in terms of mean value and dispersion.

These "One-vs-One" classifiers allow one to construct the rules for differential diagnostics. One of the possible approaches to this task is enumeration of classifiers for the feature vector of an object under study.

Below, the differential diagnostics rule was based on the result which was selected more times (see Table 4). Diagnosis did not set, if all possible results of classification (LC-COPD-Healthy-Pneumonia) for definite representative from the testing set met the same number of times.

Table 3 The pairwise SVM classification with RBF kernel of the testing set feature vectors in terms of the specificity and sensitivity

Pairwise classification	Kernel parameters	Sensitivity		Specificity	
		Mean	Dispersion	Mean	Dispersion
COPD-pneumonia	1.2041	0.95	0.0016	0.95	0.0012
Pneumonia-healthy volunteers	0.5641	0.96	0.0009	0.92	0.0019
COPD-healthy volunteers	1.2414	0.86	0.0022	0.83	0.0020
LC-pneumonia	0.7152	0.96	0.0014	0.93	0.0012
LC-COPD	1.2216	0.98	0.0003	0.94	0.0007
LC-healthy volunteers	0.2698	0.96	0.0011	0.90	0.0013

Table 4 Differential diagnostics based on the set of SVM classifiers usage

Group	Quantity of the feature vectors in the testing set	Diagnosis		
		Set right	Set wrong	Did not set
Pneumonia	11	11	0	0
LC	8	8	0	0
COPD	12	10	1	1
Healthy volunteers	29	26	1	3

Conclusions

The "profiling" approach, based on of the set of markers control or profile of the absorption spectrum of breath sample as a "fingerprint" of the state is presented. We used IR LPAS method to measure absorption spectra of exhaled air samples. The analysis of measured spectra was based first on reduction of the dimension of the feature space using PCA; thereafter the classification was carried out using SVM method. The latter provides binary classification, i.e., it can separate objects only on two classes. For purposes of differential diagnostics, it is necessary to construct the classification rules on several classes. To solve this problem, we used the "One-vs-All" and "One-vs-One" methods. The "One-vs-All" method was shown to provide not so high accuracy of classification in comparison with "One-vs-One" method on the same data set. The accuracy of classification by "One-vs-One" method based on spectral analysis of exhaled air of patients is high enough for using in routine practices especially for screening tests.

Acknowledgments Research is carried out with the financial support of the state represented by the Ministry of Education and Science of the Russian Federation. Agreement no. 14.578.21.0082 27.Nov. 2014. Unique project Identifier: RFMEFI57814X0082.

References

1. Aly, M.: Survey on Multiclass Classification Methods. Technical report, California Institute of Technology (2005)
2. Crammer, K., Singer, Y.: On the learnability and design of output codes for multiclass problems. Comput. Learn. Theory 35–46 (2000)
3. Hsu, C.W., Lin, C.J.: A comparison of methods for multiclass support vector machines. IEEE Trans. Neural. Netw 13(2), 415–425 (2002)
4. International agency for research on cancer All Cancers (excluding non-melanoma skin cancer) Estimated Incidence, Mortality and Prevalence Worldwide in 2012 [Electronic resource]. http://globocan.iarc.fr/Pages/fact_sheets_cancer.aspx [Site] (2012). Accessed 28 Nov 2016

5. Jatakanon, A., Lim, S., Kharitonov, S.A., Chung, K.F., Barnes, P.J.: Correlation between exhaled nitric oxide, sputum eosinophils, and methacholine responsiveness in patients with mild asthma. Thorax **53**(2), 91–95 (1998)
6. Karapuzikov, A.A., et al.: LaserBreeze gas analyzer for noninvasive diagnostics of air exhaled by patients. Phys. Wave Phenomena **22**(3), 189–196 (2014)
7. Kharitonov, S.A., Barnes, P.J.: Exhaled markers of pulmonary disease. Am. J. Respir. Crit. Care Med **163**(7), 1693–1722 (2001)
8. Milgram, J., Cheriet, M., Sabourin, R.: 'One against one' or 'one against all': which one is better for handwriting recognition with SVMs? In: 10th International Workshop on Frontiers in Handwriting Recognition (2006)
9. Platt, J.C., Cristianini, N., Shawe-Taylor, J.: Large margin DAGs for multiclass classification. Adv. Neural. Inf. Process. Syst **12**, 547–553 (2000). MIT Press
10. Pomerantsev, L., Ye, Rodionova O.: Concept and role of extreme objects in PCA/SIMCA. J. Chemom **28**(5), 429–438 (2014)
11. Stepanov, E.V.: Methods high-sensitivity gas analysis of biomarker molecules in studies of exhaled air. In: A.M. Prokhorov (ed.) Proceedings of General Physics Institute vol. 61, pp. 5–47 (2005)
12. Van der Schee, M.P., Paff, T., et al.: Breathomics in lung disease. Chest **147**(1), 224–231 (2015)
13. Vapnik, V.: Statistical Learning Theory. Wiley, New York, NY (1998)
14. Weston, J., Watkins, C.: Multi-class support vector machines. In: Verleysen M. (ed.) Proceedings of ESANN99, D. Facto Press, Brussels, pp. 219–224 (1999)
15. Zhang, J., Yao, X., Yu, R., Bai, J., Sun, Y., Huang, M., Adcock, I.M., Barnes, P.J.: Exhaled carbon monoxide in asthmatics: a meta-analysis. Respir. Res. l(11), pp. 50–60 (2010)
16. Zhao, X., Guan, S., Man, K.L.: An output grouping based approach to multiclass classification using support vector machines. Advanced multimedia and ubiquitous engineering. Vol. 393 of the series Lecture Notes in electrical engineering, pp. 389–395

Antitumor Effect of Vaccinia Virus Double Recombinant Strains Expressing Genes of Cytokine GM-CSF and Oncotoxic Peptide Lactaptin

G.V. Kochneva, O.A. Koval, E.V. Kuligina, A.V. Tkacheva
and V.A. Richter

Abstract In this study, the double recombinant vaccinia viruses were generated those express exogenous proteins: human granulocyte-macrophage colony-stimulating factor (GM-CSF) and the antitumor protein lactaptin in secreted and nonsecreted forms. We observed that recombinant VV-GMCSF-Lact with nonsecreted lactaptin exerted stronger cytotoxic activity than others in MDA-MB-231, BT-549 and BT-20 breast cancer cells with calculated CD_{50} of 0.005; 0.004 и 0.00083 PFU/cell correspondently. Strain VV-GMCSF-Lact also exhibited highest lytic activity in lung cancer cells H1299 and epidermoid carcinoma cells A-431. Normal MCF10A cells and diploid embryonic lung human cells LECH-240 were resistant to all recombinant vaccinia viruses. Strain VV-GMCSF-Lact showed the highest index of tumor selectivity in pairs normal/cancer cells: MCF10A/MDA-MB-231 (>2000) and LECH-240/H1299 (190). By flow cytometry, we demonstrated that all recombinants induced apoptosis in treated cancer cells but the rate of annexin V-positive cells was higher after treatment with VV-GMCSF-Lact than others. Thus nonsecreted lactaptin expression increased the toxicity of recombinant virus to cancer cells in the best way. It is likely that lactaptin expression inside the treated cells (without secretion outside)

G.V. Kochneva (✉)
State Research Center of Virology and Biotechnology "Vector", Novosibirsk, Russia
e-mail: kochneva@vector.nsc.ru

O.A. Koval · E.V. Kuligina · A.V. Tkacheva · V.A. Richter
Institute of Chemical Biology and Fundamental Medicine SB RAS, Novosibirsk, Russia
e-mail: o.koval@niboch.nsc.ru

E.V. Kuligina
e-mail: kuligina@niboch.nsc.ru

A.V. Tkacheva
e-mail: tkacheva_av@mail.ru

V.A. Richter
e-mail: richter@niboch.nsc.ru

K.V. Anisimov et al. (eds.), *Proceedings of the Scientific-Practical Conference "Research and Development - 2016"*, https://doi.org/10.1007/978-3-319-62870-7_61

intensifies apoptosis and as a consequence promotes the progression of apoptotic cells to secondary necrotic cells. These results demonstrate that recombinant VV-GMCSF-Lact has good oncolytic potential and stimulate further investigation of its anticancer activity in human tumor models in vivo and to use it in the development of anticancer therapeutic agents.

Keywords Recombinant vaccinia virus · Breast cancer cells · Apoptosis Lactaptin · GM-CSF · SCID mice · Tumor growth inhibition

Vaccinia virus (VACV) possesses many unique properties that place this virus at a leading position in molecular biology and genetic engineering. The ability to kill cancer cells is one of the fundamental biological properties of VACV, and was first reported by Levaditi C. and Nicolau S. in 1923 in the Annals of the Pasteur institute [16, p. 20]. Subsequent studies confirmed the oncolytic activity of VACV [14, p. 360]. However, over the next few decades' researchers tried to avoid complications caused by introducing the infectious virus into human organism. Progress in genetic engineering allowed changing the biological properties of VACV in a wide range, and this led to a surge of interest to oncolytic abilities of genetically modified VACV. Detailed analysis of whole bulk of the studies with genetically modified oncolytic VACV, some of which are under clinical trials, recently was published in several comprehensive reviews [4, p. 210; 5, p. 7; 1, p. 191].

The rational construction of a therapeutic VACV could be done using a virulent attenuated VACV strain with deletions of *tk* and *vgf* genes that would selectively target tumor cells without decreasing its oncolytic capacity [11, p. 2]. Two transgenes could be simultaneously inserted into the VACV genome to enhance the therapeutic efficacy of recombinant VACV—the *gm-csf* gene and the gene of cytotoxic protein lactaptin.

Lactaptin is a fragment of human milk kappa-casein (residues 57–134) that induces the death of cultured cancer cells. A recombinant analog of lactaptin, RL2, containing the complete amino acid sequence of lactaptin and corresponding to 23–157 of human kappa-casein, effectively induces apoptotic death in various mouse and human tumor cells (including breast tumor cells and primary endometrial cells) and has no effect on the viability of non-malignant mesenchymal stem cells (MSCs) [12, p. 178; 8, p. 2467; 2, p. 79; 10, p. 345]. RL2-induced apoptosis is accompanied by downregulation of BCL-2, activation of the executor caspase-3 and -7 and apoptotic fragmentation of DNA [9, p. 1]. The insertion of lactaptin sequence as a transgene into the deletion of *vgf* gene could attenuate the virulence of recombinant VACV against non-transformed cells as well as enhance its cytotoxic activity against cancer cells.

Here, we exploited VACV L-IVP strain that was used for anti-smallpox vaccination in Russia up to 1980 [13, p. 1; 16, p. 1]. Thus, L-IVP has a good medical history in Russia, which could provide advantages in clinical trials of new

L-IVP-based recombinant strains. We have previously demonstrated that genetically unmodified L-IVP possesses natural antitumor activity towards human and murine tumors [16, p. 1]. Recombinant VV-GMCSF-S1/3 in which the virus *tk* gene is inactivated by insertion of the human *gm-csf* gene was engineered earlier [3, p. 9]. This VV-GMCSF-S1/3 strain was used as a recipient for insertion of additional RL2 transgene into the deleted *vgf* gene region. We conducted a comparative study of both nonsecreted and secreted forms of lactaptin expressed as a part of a recombinant virus.

The objectives of this study were to generate new double recombinant VACV L-IVP strains expressing human GM-CSF and secreted or nonsecreted RL2 and to analyze its antitumor potential in vitro to choose the most promising construct for further research in vivo.

Recombinant VACVs were obtained via the transient dominant selection technique with the use of the puromycin resistance (*Pat*) gene as a selective marker [7, p. 4]. We constructed three recombinant strains containing a *gm-csf* gene insertion and different variants of lactaptin gene (RL2). Recombinant structures are shown in Fig. 1a. The VV-GMCSF-Lact strain encodes a nonsecreted form of RL2 that is produced only inside the infected cell [7, p. 4]. Strain VV-GMCSF-S(long)-Lact encodes chimeric RL2 protein with the signal peptide (MWLQSLLLLGTVACSIS) and the first 15 amino acids of GM-CSF (long signal sequence, S(long)) ligated to the N end. In the VV-GMCSF-S-Lact strain lactaptin presented as a chimera with a shorter leader fragment of GM-CSF, which does not contain the 15 N-terminal amino acids of the protein (signal sequence, S). Strains VV-GMCSF-S(long)-Lact and VV-GMCSF-S-Lact encode secreted forms of RL2 protein due to the presence of the GM-CSF leader peptide. We used two variants of GM-CSF leader peptide in order to clarify the need for the presence of downstream GM-CSF protein sequence to improve the oncotoxic properties of RL2. By the same method the control recombinant VV-GMCSF-dGF containing the GM-CSF transgene in the *tk* gene deletion and additional deletion of the *vgf* gene was constructed (Fig. 1a). This control recombinant VV-GMCSF-dGF provided our study with the correct estimation of the double recombinants oncolytic activity enhancement.

The structure of recombinant viruses was confirmed by both PCR assays and DNA sequencing of the *tk* and *vgf* loci. Specific primer positions are depicted in Fig. 1a. We observed that all double recombinants and control recombinant VV-GMCSF-dGF produced a 1760 b.p. fragment in the *tk* gene region that corresponded to the *gm-csf* gene sequence whereas DNA of the parental VACV L-IVP strain produced a 414 b.p. fragment (Fig. 1b). Using primers flanking the VGF region we amplified fragments of 710; 836; 791 b.p. using DNA of the recombinants VV-GMCSF-Lact; VV-GMCSF-S(long)-Lact; VV-GMCSF-S-Lact and a fragment of 423 b.p. using DNA of the VV-GMCSF-dGF, corresponding to different variants of lactaptin gene insertion and *vgf* gene deletion, respectively. A fragment of 584 b.p. was amplified from the DNA of the parental L-IVP strain using Up35 and Apa-L22 primers [7, p. 5].

Construction and verification of oncolytic VACVs

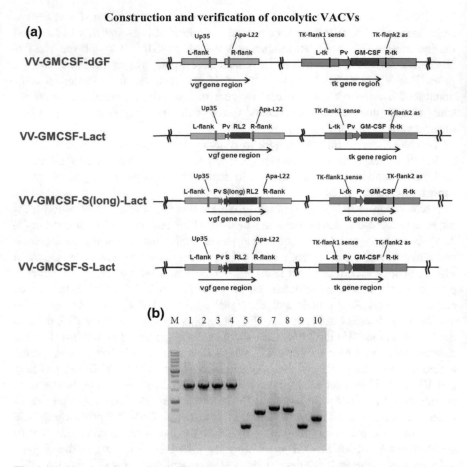

Fig. 1 Verification of recombinant VACVs structure. **a** schematic view of virus genomes with primer positions indicated. *Pv* VACV promoter; *S* GM-CSF signal peptide sequence; *S(long)* S + GM-CSF gene fragment (long signal); *L-flank* and *R-flank* sequences, flanking *vgf* gene [6]; *L-tk* and *R-tk* sequences, flanking *tk* gene [3]. **b** PCR identification of recombinant VACVs DNA with primers *TK-flank1 sense* and *TK-flank 2* as (Lanes *1–5*) and with primers Up35 x Apa-L22 (Lanes *6–10*). Lanes *5* and *10*—wild-type VACV (L-IVP); *1* and *6*—VV-GMCSF-Lact; *2* and *7*—VV-GMCSF-S(long)-Lact; *3* and *8*—VV-GMCSF-S-Lact; *4* and *9*—VV-GMCSF-dGF; *3* and *6*—VV-GMCSF-Lact. *M*—DNA molecular weight marker

The cytotoxic activity of recombinant VACVs in vitro was analyzed using XTT assay as described previously [6, p. 737] and measured as the 50% cytotoxic dose (CD_{50}), the virus concentration causing death of 50% of cells. Seven tumor cell lines of various origins were used to investigate the oncolytic activity of recombinant VACVs: breast cancer carcinomas BT-20, BT-549, MDA-MB-231, MCF-7, lung carcinoma A-549, non-small lung cell cancer H1299 and epidermoid carcinoma A-431. Summary results are presenting in Fig. 2. Breast carcinoma MCF-7 cells were more resistant to recombinant viruses than the other breast cancer cells. MCF-7 cells are the only estrogen-dependent breast cancer cells, in contrast to MDA-MB-231, BT-549 and BT-20 and possibly, lack of estrogen in cell culture medium reduces the malignancy potential of these cells, and together with it the replicative virus activity. Since our recombinant VACVs contain the inactivated *tk* and *vgf* genes, they are sensitive to the presence of precursors in DNA synthesis and proliferation rate of infected cells. In conditions of estrogen deficiency, MCF-7 cells are inferior in these parameters estrogen-independent cultures and their lysis requires 5–50 times more virus.

We observed that recombinant VV-GMCSF-Lact exerted stronger cytotoxic activity than others in MDA-MB-231, BT-549 and BT-20 breast cancer cells with calculated CD_{50} of 0.005; 0.004 и 0.00083 PFU/cell correspondently. Strain VV-GMCSF-Lact also exhibited the highest lytic activity in lung cancer cells H1299 and epidermoid carcinoma cells A-431 (Fig. 2). Thus, nonsecreted lactaptin expression increased the toxicity of recombinant virus to cancer cells in the best way.

Tumor selectivity of recombinant VACVs was investigated using two noncancer cell lines: MCF 10A normal epithelial breast cells and LECH-240 diploid embryonic cells of human lung. As given in Fig. 2, the CD_{50} of recombinant strains, including the control recombinant VV-GMCSF-dGF, was significantly lower than the same dose for diploid and normal cells ($P < 0.01$). In the case of normal

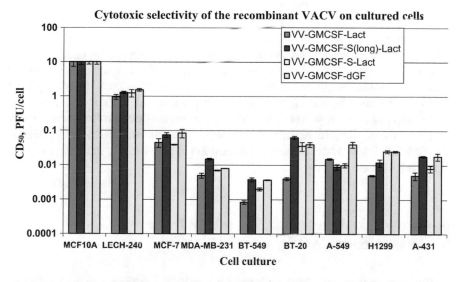

Fig. 2 Oncolytic and cytotoxic activities of recombinant VACV strains in vitro. Cells were grown in 96-well plates, infected with recombinant virus doses ranging from 0.001 to 10.0 PFU/cell (sequential tenfold dilutions) and incubated for 72 h. The 50% cytotoxic dose (CD_{50}) was determined for each cell line by an XTT assay. Data are expressed as mean ± SE

Table 1 Comparison of cytotoxic activities of recombinant VACVs in two pairs of cells: normal breast epithelium MCF10A/breast carcinoma MDA-MB-231 and diploid embryonic lung LECH-240/lung cancer H1299

Cell culture	CD_{50} (PFU/cell[a])			
	VV-GMCSF-Lact	VV-GMCSF-S-Lact	VV-GMCSF-S(long)-Lact	VV-GMCSF-dGF
MCF10A	>10	>10	>10	>10
MDA-MB-231	0.005	0.007	0.015	0.008
Selectivity index[b]	>2000	>1400	>650	>1250
LECH-240	0.95	1.24	1.28	1.53
H1299	0.005	0.012	0.025	0.025
Selectivity index[b]	190	103	51	61

[a]The virus titer was determined by plaque assay and CD_{50} expressed as number of plaque forming units (PFU) per cell
[b]Selectivity index was calculated for each virus as the ratio of CD_{50} values for normal and tumor cells

mammary epithelial MCF10A cells, we were unable to accurately determine the 50% cytotoxic dose for any of the virus strains as the maximum dose tested of 10 PFU/cell caused no toxic effect. The tumor selectivity index of recombinant viruses was calculated in pair MCF 10A/MDA-MB-231 or normal/cancer breast cells. The index value for all recombinant VACVs including control variant VV-GMCSF-dGF was more than 650 (Table 1) with the highest rate for VV-GMCSF-Lact strain (>2000).

Among all recombinant VACVs strain VV-GMCSF-Lact has a significantly higher cytotoxicity against H1299 lung cancer cells ($P < 0.05$). We calculated selectivity indexes paired cultures of lung embryo and human lung cancer cells— LECH-240/H1299 (Table 1). In this case, the value of selectivity indexes was lower than that of normal breast cells, specifically 190 in case of strain VV-GMCSF-Lact (Table 1). Apparently, this is due to the fact that embryonic cells have a relatively high proliferation activity, which, to some extent, makes them similar to cancer and increases the replicative activity of vaccinia virus recombinants. Recombinant VV-GMCSF-S(long)-Lact showed the lowest index of selectivity in relation to lung tumor cells (Table 1).

Features of Apoptosis and Necrosis Induced by Recombinant VACVs

The type of cancer cell death induced by newly constructed recombinant OVs coding the proteins with specific biological activity cannot be predicted precisely. We investigated apoptosis through the flow cytometry of plasma membrane phosphatidylserine exposure by using the BD Pharmigen Apoptosis Detection Kit (BD Biosciences). Staining with annexin V-FITC and propidium iodide (PI) allows separating apoptotic and necrotic cells. Phenotype annexin V+/PI− corresponds to

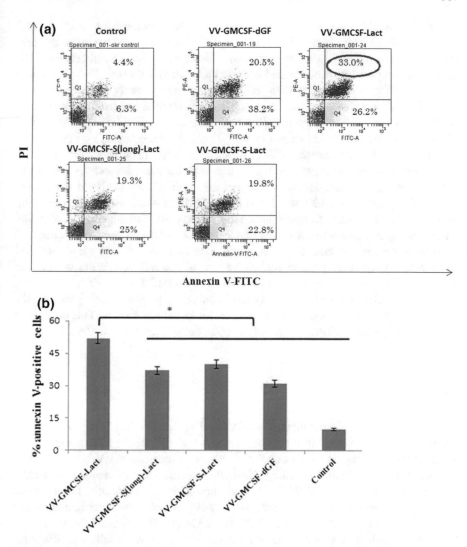

Fig. 3 Features of apoptosis of the MDA-MB-231 cells infected with recombinant VACVs. **a** MDA-MB-231 cells were treated with recombinant VACVs (1.5 PFU/cell) or with saline (control) for 48 h and then cells were harvested with trypsin and stained with annexin V-FITC and propidium iodide (PI) according to the manufacturer's protocol. Cell populations with the annexin V⁻/PI⁻ phenotype (Q3) were designated as living cells, annexin V⁺/PI⁻ (Q4) as apoptotic cells, and annexin V⁺/PI⁺ (Q2) as secondary necrotic cells. **b** *Bar graph* summarized the percentage of apoptotic cells from three independent experiments (*$p < 0.05$)

cells in a state of apoptosis, annexin V−/P+—cells in a state of necrosis and annexin V+/PI+—cells in the state of late apoptosis or secondary necrosis [15, p. 180].

MDA-MB-231 cancer cells were treated with recombinant VACVs (1.5 PFU/cell, 48 h) and then were analyzed for apoptosis and necrosis by flow cytometry. We observed that the apoptosis rate of virus-treated cells dramatically increased compared with nontreated cells and that strain VV-GMCSF-Lact induced more extensive cell death than other recombinants analyzed (Fig. 3a). In VV-GMCSF-Lact treated cells, the population of secondary necrotic cells annexin V^+/PI^+ was consistently higher (33%) than that in VV-GMCSF-S(long)-Lact or VV-GMCSF-S-Lact treated cells (25 and 22.8% correspondingly) whereas early apoptotic populations annexin V^+/PI^- differed slightly. Since apoptotic cells that have compromised plasma membrane integrity become subject to secondary necrosis (the phase that occurs after apoptosis in vitro), we analyzed the pooled annexin V^+ population which was larger in VV-GMCSF-Lact treated cells (Fig. 3b). It is likely that lactaptin expression inside the treated cells (without secretion outside) intensifies apoptosis and as a consequence promotes the progression of apoptotic cells to secondary necrotic cells. We found also that treatment of MDA-MB-231 cells with VV-GMCSF-Lact increased the size of the cell population with active caspase-3 and -7 in comparison with control VV-GMCSF-dGF [7, p. 9]. Thus, VACV-dependent expression of apoptosis-inducing proteins may promote the alteration of the route of death of infected cancer cells to apoptosis. This could be a helpful strategy to reinforce the oncolytic potential of recombinant VACVs.

Conclusions

We constructed three recombinant VACVs coding human GM-CSF and different forms of oncotoxic peptide lactaptin, secreted and nonsecreted. To estimate the contribution of lactaptin to the oncolytic activity of recombinant VACVs, we also constructed a recombinant coding GMCSF without the *vgf* gene. Our results showed that normal MCF10A cells and diploid embryonic lung human cells LECH-240 were resistant to recombinant VACVs. Among the investigated cancer cell lines, breast cancer cell lines MDA-MB-231 and BT-549 were the most sensitive to recombinant viruses. The VV-GMCSF-Lact strain exhibited significantly higher lytic activity in the majority of tumor cells tested. It also showed the highest index of tumor selectivity in pairs normal/cancer cells: MCF10A/MDA-MB-231 (>2000) and LECH-240/H1299 (190). We observed that the apoptosis rate of VV-GMCSF-Lact treated cancer cells significantly increased compared with other recombinants analyzed. Taking into account our results, it seems promising to further investigate the oncolytic action of VV-GMCSF-Lact in human tumor models in vivo and to use it in the development of anticancer therapeutic agents.

Acknowledgements Research is carried out with the financial support of the state represented by the Ministry of Education and Science of the Russian Federation. Agreement no. 14.604.21.0057 27. Jun 2014. Unique project identifier: RFMEFI60414X0057.

References

1. Chan, W.M., McFadden, G.: Oncolytic poxviruses. Annu. Rev. Virol. **1**, 191–214 (2014)
2. Fomin, A.S., Koval, O.A., Semenov, D.V., Potapenko, M.O., Kuligina, E.V., Kit, I.I., Richter, V.A.: The analysis of biochemical markers of MCF-7 cells apoptosis induced by recombinant analog of lactaptin. Russ. J Bioorg. Chem. **38**, 77–82 (2012)
3. Grazhdantseva, A.A., Sivolobova, G.F., Tkacheva, A.V., Gileva, I.P., Kuligina, E.V., Rikhter, V.A., Kochneva, G.V.: High-effective production of biologically active secreted human granulocyte macrophage colony-stimulating factor by recombinant vaccinia virus. Appl. Biochem. Microbiol. **52**(7), 7–13 (2016)
4. Kim, M.: Replicating poxviruses for human cancer therapy. J. Microbiol. **53**(4), 209–218 (2015)
5. Kochneva, G.V., Sivolobova, G.F., Yudina, K.V., Babkin, I.V., Chumakov, P.M., Netesov, S.V.: Oncolytic poxviruses. Mol. Genet. Microbiol. Virol. **27**(1), 7–15 (2012)
6. Kochneva, G.V., Babkina, I.N., Lupan, T.A., Grazhdantseva, A.A., Iudin, P.V., Sivolobova, G.F., Shvalov, A.N., Popov, E.G., Babkin, I.V., Netesov, S.V., Chumakov, P.M.: Apoptin enhances the oncolytic activity of vaccinia virus in vitro. Mol. Biol. **47**(5), 733–742 (2013)
7. Kochneva, G., Sivolobova, G., Tkacheva, A., Grazhdantseva, A., Troitskaya, O., Nushtaeva, A., Tkachenko, A., Kuligina, E., Richter, V., Koval, O.: Engineering of double recombinant vaccinia virus with enhanced oncolytic potential for solid tumor virotherapy. Oncotarget. (2016). doi:10.18632/oncotarget.12367. https://www.ncbi.nlm.nih.gov/pubmed/27708236
8. Koval, O.A., Fomin, A.S., Kaledin, V.I., Semenov, D.V., Potapenko, M.O., Kuligina, E.V., Nikolin, V.P., Nikitenko, E.V., Richter, V.A.: A novel pro-apoptotic effector lactaptin inhibits tumor growth in mice models. Biochimie. **94**, 2467–2474 (2012)
9. Koval, O.A., Tkachenko, A.V., Fomin, A.S., Semenov, D.V., Nushtaeva, A.A., Kuligina, E.V., Zavjalov, E.L., Richter, V.A.: Lactaptin induces p53-independent cell death associated with features of apoptosis and autophagy and delays growth of breast cancer cells in mouse xenografts. PLoS one **9**, e93921 (2014)
10. Koval, O.A., Sakaeva, G.R., Fomin, A.S., Nushtaeva, A.A., Semenov, D.V., Kuligina, E.V., Gulyaeva, L.F., Gerasimov, A.V., Richter, V.A.: Sensitivity of endometrial cancer cells from primary human tumor samples to new potential anticancer peptide lactaptin. J. Cancer Res. Ther. **11**, 345–351 (2015)
11. McCart, A., Bartlett, D., Moss, B.: Combined growth factor-deleted and thymidine kinase-deleted vaccinia virus vector. US Patent 7208313. (2007)
12. Semenov, D.V., Fomin, A.S., Kuligina, E.V., Koval, O.A., Matveeva, V.A., Babkina, I.N., Tikunova, N.V., Richter, V.A.: Recombinant analogs of a novel milk pro-apoptotic peptide, lactaptin, and their effect on cultured human cells. Protein J. **29**, 174–180 (2010)
13. Shvalov, A.N., Sivolobova, G.F., Kuligina, E.V., Kochneva, G.V.: Complete genome sequence of vaccinia virus strain L-IVP. Genome Announc. 4, e00372–16 (2016)
14. Thorne, S.H., Hwang, T.H., Kirn, D.H.: Vaccinia virus and oncolytic virotherapy of Cancer. Curr. Opin. Mol. Ther. **7**(4), 359–365 (2005)
15. Vermes, I., Haanen, C., Reutelingsperger, C.: Flow cytometry of apoptotic cell death. J Immunol. Methods **243**, 167–190 (2000)
16. Zonov, E., Kochneva, G., Yunusova, A., Grazhdantseva, A., Richter, V., Ryabchikova, E.: Features of the antitumor effect of vaccinia virus Lister Strain. Viruses **8**, 20 (2016)

Genome-Wide Association Studies for Milk Production Traits in Russian Population of Holstein and Black-and-White Cattle

A.A. Sermyagin, E.A. Gladyr, K.V. Plemyashov, A.A. Kudinov, A.V. Dotsev, T.E. Deniskova and N.A. Zinovieva

Abstract We performed the genome-wide association study of estimated breeding values for milk production traits in Russian Holstein and black-and-white cattle population. The join dairy cows' population of Moscow and Leningrad regions was used to create a common reference group of animals to obtain the genomic breeding values. We identified breeding and genetic parameters for milk yield for 305 days of lactation, milk fat and protein content, milk fat and protein yield. We found several high-significant conservative mutations associated with milk fat content (e.g., *DGAT1*, $P = 6.8 \times 10^{-22}$), as well as a 1.5 Mb locus on BTA14. Our results will be used to develop a genomic evaluation programs, aimed to improve economically important traits in dairy cattle in Russia.

Keywords Estimated breeding value · Genome-wide association studies (GWAS) Milk production · Single nucleotide polymorphism (SNP) · Quantitative trait loci Reference population

A.A. Sermyagin · E.A. Gladyr · K.V. Plemyashov · A.A. Kudinov · A.V. Dotsev · T.E. Deniskova · N.A. Zinovieva (✉)
L.K. Ernst Federal Science Center for Animal Husbandry, Moscow, Russia
e-mail: n_zinovieva@mail.ru

A.A. Sermyagin
e-mail: alex_sermyagin85@mail.ru

E.A. Gladyr
e-mail: elenagladyr@mail.ru

K.V. Plemyashov
e-mail: spbvniigen@mail.ru

A.V. Dotsev
e-mail: asnd@mail.ru

T.E. Deniskova
e-mail: horarka@yandex.ru

K.V. Plemyashov · A.A. Kudinov
RRIFAGB, St.Petersburg, Russia

Introduction

Improving the productive qualities of farm animals is the main goal of the most of breeding programs. The methods under development are targeted to collect and to accumulate not only phenotypic information from direct records of animal's productive qualities, but also the data of whole genome scanning. Presently, genomic evaluation assay was developed to increase the accuracy of estimated breeding values (EBVs) for productive and other economically important traits [5, p. 321]. In Russia, as well as in the world dairy farming, genomic prediction methods have been applied in cattle breeding, specifically, in the most numerous populations of Holstein and holsteinized black-and-white breed. The genome-wide associations studies (GWAS) performed on cows' population of Moscow region revealed several high-significant associations between 14 SNP localized on chromosomes 1, 5, 9, 13, 14, 17, 20, and 27 and an additive effect of more than 9% was observed [11, p. 188]. One of the validation tests for genes, responsible for a quantitative traits of dairy cattle, is the availability of accurate phenotypic data records, i.e., there are mutations that have been fixed under selection pressure and may explain about 40% of genetic variation [4, p. 387]. Respectively, sufficient number of animals, reliable information, considering the level of trait heritability should improve accuracy of the mapping of quantitative trait loci and thus the estimated genomic breeding values. Ongoing research on the analysis of genome-wide associations among different dairy cattle populations of the USA, Germany, the Netherlands, Australia and China identified a pool of common reference genes responsible for milk production traits and for milk fat content. Single nucleotide polymorphisms in *DGAT1*, *SCD1*, *GHR*, *EPS8*, *GPAT4* genes, and casein cluster genes (Hapmap24184-BTC-070077) were characterized by the highest significant values of the associations and an additive effects [10, p. 6; 16, p. 2; 6, p. 6519; 1, p. 869]. It is worth noting that pleiotropic effect of *DGAT1* gene on milk protein percentage and content, milk yield was observed in Holstein cattle population of Chinese origin [3, p. 6].

The GWAS for traits with low heritability (fertility, health) and strongly influenced by the genotype environment factors (exterior) are of special interest. Thus, the large number of SNP-effects for type traits was detected on chromosomes BTA11, BTAX, BTA10, BTA5, and BTA26 [2, p. 7]. Regarding fertility, the search is focused on the single semi-lethal mutations, which might be observed or be fixed, because only metabolic analysis can characterize the complex interaction of genotype and phenotype under the hormonal regulation [8, p. 6431]. The main restricting factor of wide implementation of genomic methods in animal breeding is often an insufficient population size or various objectives in breeding programs in different countries. For example, the join of Jersey cattle reference populations of Denmark and the USA are considered [12, p. 2] in one case, and a merger of the herds of different sizes, being bred under various environmental conditions, at the level of a country's regions is examined in the other case. In addition, issues of influence of single nucleotide mutations on the level of retirement and milk cow productivity are debated [14, p. 5804]. As part of a national program for dairy cattle improvement in New Zealand, it was indicated that the use of SNP data, which were more closely associated with milk production and reproduction traits, improved an accuracy of predictions for proven bulls by 1–2% [17, p. 663].

In this regard, studies, performed in Russian Holstein cattle population, are of some interest for an understanding the selection process commonality on a par with global peers, as well as to create own dairy cattle reference population.

The purpose of the study was to evaluate the genome-wide associations with estimated breeding values for milk yield and milk components in Holstein and black-and-white bulls from the different regions of Russia

Materials and Methods

Medium Density Bovine SNP50K v2 BeadChip (Illumina Inc., USA) was used for genotyping 477 individuals of Holstein and holsteinized Black-and-White breeds, which daughters had been lactating in 138 herds. The sample included 256 sires from the population of Moscow region, and 221 sires from the population of Leningrad region. The total number of daughters (primiparous) was 119,106 individuals. The following phenotyping traits were used for GWAS analysis: 305-day milk yield (MY), milk fat content (FC), and milk protein content (PC). Quality control for genotyping was carried out using Plink 1.9 software [9, p. 562]. After quality check, 466 bulls and 40279 polymorphic SNPs were selected for the analysis.

BLUP Sire Model was used for EBV calculation. The following equation was used:

$$Y_{ijk} = \mu + \mathrm{HYS}_i + \sum_k b_1 A_k + \sum_k b_2 \mathrm{DO}_k + \mathrm{Sire}_j + e_{ijk},$$

where Y_{ijk} is the k-th heifer trait index; μ is population constant; HYS_i is fixed effect of the i-th «herd-year-season» calving; ($i = 1, \dots, 3917$ factors); b_1 and b_2 are linear regression coefficients; A_k is first calving age of the k-th heifer; DO_k is days open of the k-th heifer; sire$_j$ is randomized effect of the j-th bull with normal distribution with a mean of 0, and a variance of $A\sigma_a^2$, where A is additive relationship matrix ($j = 1, \dots, 466$ individuals); e_{ijk} is unaccounted factor effect (0, σ_e^2).

EBV calculations were conducted using BLUPF90 software. Estimation of variance components was performed by the method of restricted maximum likelihood (REML), with the inclusion of additional features to the model: fat yield (FY), protein yield (PY), breedings per conception (BC), and days open (DO) [7, p. 21]. Estimation of genomic relationship matrix (G) was performed according to the algorithm developed by P.M. VanRaden [15, p. 4416] in the R programming language environment. The matrix consisted of elements presented by homozygous and heterozygous loci estimations: AA = 1, AB = 0, BB = −1. GEBVs were calculated as combination of SNP direct genomic value (DGV) and EBV (Parent Average) according to the GBLUP approach.

To identify associations of SNP-markers with milk production, traits regression analysis with pseudo-phenotypes or GEBV assessments implemented in Plink 1.90 were used (flags: –assoc –qt-means –adjust). To confirm the significant impact of SNPs and identify significant regions in the genome of cattle, several tests were used to check for null hypotheses by Bonferroni (threshold $P < 1.24 \times 10^{-6}$, $0.05/_{40279}$).

To search for the genes closely associated with economic traits, the National Center for Biotechnology Information (NCBI) database was used. Functional gene identification was performed using the Discover EggNOG 4.1 database (http://eggnogdb.embl.de/#/app/home). Data visualization was conducted using the qqman package and R programming language [13].

Results

The analysis of genetic differences between populations of bulls from Moscow and Leningrad regions shows their close relationship according to the fixation index (Fst = 0.00356). It was found that the heritability (h^2) of milk yield was = 0.180, indicating a relatively low proportion of additive genetic variation in the population of holsteinized Black-and-White breed (C_{Va} = 3.4%). The highest value of heritability was observed for milk fat yield—0.221, while for milk protein yield it was comparable to the total production of milk components (0.173). Heritability of the reproductive traits ranged from 0.015 to 0.039 and was largely due to paratypic (environmental or technological) factors (Table 1).

GWAS analysis for milk productive traits showed the following results of SNP associations that have significant impact on the additive value of a sire in terms of the daughter's milk yield (Fig. 1).

The highest number of highly significant SNPs for MY was observed on chromosomes 1, 2, 3, 11, 17, and 23, for FP—on chromosomes 9 and 14, and for PP—on chromosomes 9, 17, 20, and 23. Regression analysis revealed that 425 mononucleotide substitutions had significant effect on the assessment of the sires' EBV for MY, that actually corresponds to the lower threshold of reliability for genomic research ($p \leq 1.2 \times 10^{-6}$). In total, 77 significant mutations were

Table 1 Genetic (*below*) and paratypic (*above*) correlations between traits, heritability (on the diagonal)

Traits	MY	FP	MF	PP	MP	BC	DO
MY	**0.180**	−0.147	0.921	−0.163	0.915	0.190	0.226
FP	−0.158	**0.221**	0.235	0.279	−0.075	0.007	0.003
MF	0.920	0.236	**0.177**	−0.058	0.872	0.187	0.220
PP	−0.365	0.541	−0.161	**0.173**	0.077	0.015	0.008
MP	0.949	−0.071	0.910	−0.180	**0.142**	0.184	0.213
BC	0.334	−0.172	0.269	−0.208	0.308	**0.015**	0.562
DO	0.442	−0.137	0.375	−0.140	0.414	0.606	**0.039**

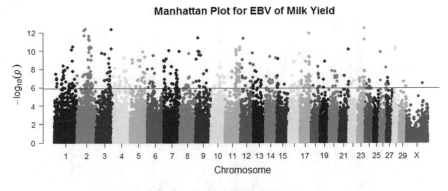

Manhattan Plot for EBV of Milk Yield

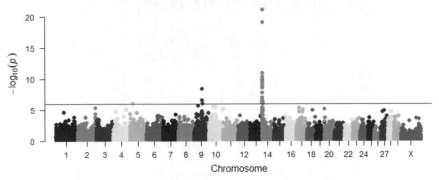

Manhattan Plot for EBV of fat percentage

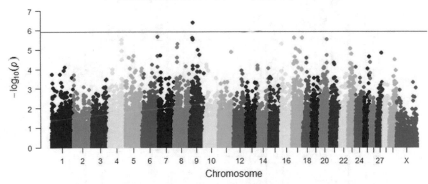

Manhattan Plot for EBV of protein percentage

Note: the upper line is threshold of Bonferroni criteria

Fig. 1 Distributions of significance regression coefficients for sires' EBV of daughter's milk traits

detected for FP, 34 of which were found on BTA14 that indicates a high probability of QTL detection in this region of genome with size of 1.42 Mb. The number of SNPs, which were significantly (P = 8.8 × 10⁻⁶) associated with PP was lower (14 SNPs). The impact of 17 highly significant mutations is shown in Table 2.

Table 2 Significant SNP and candidate gene for milk production traits

SNP	BTA	Position	Effect, X ± m	R^2	P-value	Closest gene	Distance from gene, b.p.
Milk yield							
Hapmap39230-BTA-56961	23	49900420	−95.8 ± 12.8	0.11	3.9×10^{-13}	*ECI2*	9153
ARS-BFGL-NGS-37839	2	59083405	+96.0 ± 12.9	0.11	4.3×10^{-13}	*SPOPL*	231520
						HNMT	292864
BTB-00154795	3	107386284	−112.5 ± 15.1	0.11	5.1×10^{-13}	*MACF1*	0
ARS-BFGL-NGS-32760	17	63394223	+105.0 ± 15.4	0.10	1.5×10^{-12}	*DTX1*	0
Fat percentage							
ARS-BFGL-NGS-4939	14	1801116	+0.028 ± 0.003	0.18	6.8×10^{-22}	*DGAT1*	0
ARS-BFGL-NGS-107379	14	2054457	+0.027 ± 0.003	0.17	6.9×10^{-20}	*PLEC*	266
Hapmap30086-BTC-002066	14	2524432	-0.019 ± 0.003	0.09	9.7×10^{-11}	*ZNF696*	0
ARS-BFGL-NGS-94706	14	1696470	+0.018 ± 0.003	0.09	1.0×10^{-10}	*VPS28*	0
ARS-BFGL-NGS-34135	14	1675278	+0.018 ± 0.003	0.09	1.9×10^{-10}	*CYHR1*	0
UA-IFASA-6878	14	2002873	+0.017 ± 0.003	0.08	4.2×10^{-10}	*GRINA*	15686
Protein percentage							
BTB-00383200	9	22919235	+0.009 ± 0.002	0.05	4.1×10^{-7}	*UBE3D*	142191
ARS-BFGL-NGS-32948	6	113822295	+0.006 ± 0.001	0.05	2.2×10^{-6}	*BOD1L1*	120881
Hapmap30759-BTA-123220	17	16415781	+0.008 ± 0.002	0.05	2.4×10^{-6}	*IL15*	0

The GWAS results for milk production traits showed the presence of 13 significantly associated polymorphisms, which is localized in functional genes. It was found that along with North American and European Holstein populations the SNPs, associated with quantitative traits were identified on chromosomes 2, 3, 6, 9, 14, 17, and 23. The inheritance complexity of such a comprehensive trait as the milk yield did not allow to identify unequivocal polymorphisms by its effect, but several of them were found to have molecular influence in the following substitutions in *ECI2, SPOPL, HNMT, MACF1*,and *DTX1* genes. It is known that percentage and content of milk fat are mostly influenced by *DGAT1* gene expression that was confirmed in our studies. In addition, the region, responsible for lipid synthesis and metabolic exchange, was detected in a quantitative trait locus on chromosome 14, at 253 kb between polymorphisms of *DGAT1* and *PLEC* (LOC786966) genes. Association analysis of milk protein content showed that this trait was determined by a small number of genes because of low variability as well as of complex nature of inheritance and synthesis of milk proteins. The surrounding polymorphisms in *UBE3D* and *BOD1L1* genes, responsible for protein metabolism and posttranslational modification of amino acid compounds on a par with cell control of development, are worth to be highlighted. Besides the mutation in *IL15* gene was associated with proliferation of T-lymphocytes and transduction mechanisms.

Conclusions

In general, we can state that significant influence of the reference mutations, responsible for metabolic processes of synthesis of lipids and proteins of milk, was confirmed on the basis of Holstein and holsteinized black-and-white cattle populations in the join reference groups of two regions of Russia. Regarding milk yield, we identified polymorphisms, influencing complex nature of metabolic processes: from histidine metabolism to posttranslational modification of cell structures. Our results show that further association studies of milk yield and milk protein content are required. The data will be used to improve the genetic evaluation of cattle in Russia.

Acknowledgements Studies were performed under the financial support of the Ministry of Education and Science of Russia, agreement No 14.604.21.0062. Unique identifier of the project: RFMEFI60414X0062.

References

1. Chamberlain, A.J., Hayes, B.J., Savin, K., Bolormaa, S., McPartlan, H.C., Bowman, P.J., Van Der Jagt, C., MacEachern, S., Goddard, M.E.: Validation of single nucleotide polymorphisms associated with milk production traits in dairy cattle. J. Dairy Sci. **95**. 864–875

2. Cole, J.B., Wiggans, G.R., Ma, L., Sonstegard, T.S., Lawlor, T.J.Jr., Crooker, B.A., Van Tassell, C.P., Yang, J., Wang, S., Matukumalli, L.K., Da, Y.: Genome-wide association analysis of thirty one production, health, reproduction and body conformation traits in contemporary U.S. Holstein cow. BMC Genomics **12**(408) (2011)

3. Fang, M., Fu, W., Jiang, D., Zhang, Q., Sun, D., Ding, X., Liu, J.: A multiple-SNP approach for genome-wide association study of milk production traits in Chinese Holstein cattle. PLoS ONE **9**(8) (2014) doi:10.1371/journal.pone.0099544

4. Goddard, M.E., Hayes, B.J.: Mapping genes for complex traits in domestic animals and their use in breeding programmes. Nat. Rev. Genet. **10**, 381–391 (2009). doi:10.1038/nrg2575

5. Gondro, C., Van der Werf, J., Hayes, B.: Genome-Wide Association Studies and Genomic Prediction, p. 566. Humana Press, Australia (2013)

6. Maurice-Van Eijndhoven, M.H.T., Bovenhuis, H., Veerkamp, R.F., Calus, M.P.L.: Overlap in genomic variation associated with milk fat composition in Holstein Friesian and Dutch native dual-purpose breeds. J. Dairy Sci. **98**, 6510–6521. http://dx.doi.org/10.3168/jds.2014-9196

7. Misztal, I., Tsuruta, S., Strabel, T., Auvray, B., Druet, T., Lee, D.H.: BLUPF90 and related programs (BGF90). In Proceedings of the 7th World Congress on Genetics Applied to Livestock Production, 28, 21–22 (2002)

8. Parker Gaddis, K.L., Null, D.J., Cole, J.B.: Explorations in genome-wide association studies and network analyses with dairy cattle fertility traits. J. Dairy Sci. **99**. 6420–6435. http://dx.doi.org/10.3168/jds.2015-10444

9. Purcell, S., Neale, B., Todd-Brown, K., Thomas, L., Ferreira, M., Bender, D., Maller, J., Sklar, P., de Bakker, P., Daly, M.J., Sham, P.C.: PLINK: A toolset for whole-genome association and population-based linkage analysis. Am. J. Hum. Genet. **81**(3), 559–575 (2007) (doi:10.1086/519795)

10. Qanbari, S., Pimentel, E.C.G., Tetens, J., Thaller, G., Lichtner, P., Sharifi, A.R., Simianer, H.: A genome-wide scan for signatures of recent selection in Holstein cattle. Anim. Genet. (2010). doi:10.1111/j.1365-2052.2009.02016.x

11. Sermyagin, A.A., Gladyr', E.A., Kharitonov, S.N., Ermilov, A.N., Strekozov, N.I., Brem, G., Zinovieva, N.A.: Genome-wide association study for milk production and reproduction traits in Russian Holstein cattle population. Agric. Biol. **51**(2), 182–193 (2016). (Sel'skokhozyaistvennaya Biologiya). doi:10.15389/agrobiology.2016.2.182eng

12. Su, G., Nielsen, U.S., Wiggans, G., Aamand, G.P., Guldbrandtsen, B., Lund, M.S.: Improving genomic prediction for Danish Jersey using a joint Danish-US reference population. In Proceedings, 10th World Congress of Genetics Applied to Livestock Production (2014)

13. The R Project for Statistical Computing: Free software environment for statistical computing and graphics. http://www.r-project.org

14. Tsuruta, S., Lourenco, D.A.L., Misztal, I., Lawlor, T.J.: Genotype by environment interactions on culling rates and 305-day milk yield of Holstein cows in 3 US regions. J. Dairy Sci. **98**, 5796–5805. http://dx.doi.org/10.3168/jds.2014-9242

15. Van Raden, P.M.: Efficient methods to compute genomic predictions. J. Dairy Sci. **91**, 4414–4423 (2008) (doi:10.3168/jds.2007-0980)

16. Wang, X., Wurmser, C., Pausch, H., Jung, S., Reinhardt, F., Tetens, J., Thaller, G., Fries, R.: Identification and dissection of four major QTL affecting milk fat content in the German Holstein-Friesian population. PLoS ONE **7**(7) (2012). doi:10.1371/journal.pone.0040711

17. Winkelman, A.M., Johnson, D.L., Harris, B.L.: Application of genomic evaluation to dairy cattle in New Zealand. J. Dairy Sci. **98**, 659–675. http://dx.doi.org/ 10.3168/jds.2014-8560

Overview of 17,856 Compound Screening for Translation Inhibition and DNA Damage in Bacteria

P.V. Sergiev, E.S. Komarova (Andreianova), I.A. Osterman,
Ph.I. Pletnev, A.Ya. Golovina, I.G. Laptev, S.A. Evfratov,
E.I. Marusich, M.S. Veselov, S.V. Leonov, Ya.A. Ivanenkov,
A.A. Bogdanov and O.A. Dontsova

Abstract Screening for new antibacterial compounds is an urgent need of medicinal chemistry. Understanding new antibiotics mechanism of action is needed for progression in the drug development pipeline. In the frame of the project supported by the Ministry of Science, we developed a reporter system which allows

P.V. Sergiev (✉) · I.A. Osterman · Ph.I. Pletnev · I.G. Laptev · S.A. Evfratov
Department of Chemistry, Lomonosov Moscow State University, Moscow, Russia
e-mail: petya@genebee.msu.ru

I.A. Osterman
e-mail: osterman@yandex.ru

Ph.I. Pletnev
e-mail: philippletnev@gmail.com

I.G. Laptev
e-mail: whiteswan92@gmail.com

S.A. Evfratov
e-mail: evfratov@gmail.com

E.S. Komarova (Andreianova)
Department of Bioengineering and Bioinformatics,
Lomonosov Moscow State University, Moscow, Russia
e-mail: ekaandreyanova@yandex.ru

A.Ya. Golovina
Belozersky Institute of Physico-Chemical Biology, Lomonosov Moscow
State University, Moscow, Russia
e-mail: malanka@yandex.ru

E.I. Marusich · M.S. Veselov · S.V. Leonov · Ya.A. Ivanenkov
Moscow Institute of Physics and Technology, Moscow, Russia
e-mail: mei@pharmcluster.ru

M.S. Veselov
e-mail: veselovmark@gmail.com

K.V. Anisimov et al. (eds.), *Proceedings of the Scientific-Practical Conference
"Research and Development - 2016"*, https://doi.org/10.1007/978-3 319-628 /0-7_63

an express, cost-effective and high-throughput screening for simultaneous detection of antibacterial activity, protein synthesis inhibition and induction of DNA damage SOS response. Automation of the screening process developed in the frame of this project allowed to screen up to 17,856 compound chemical library, supplied by the industrial partner of the project, Research Institute of Chemical Diversity. Among the tested compounds, DNA damaging agents appeared almost sixfold more frequently than those that inhibited protein synthesis. Several new families of antibacterial compounds were found among the tested set.

Keywords Antibiotic · Inhibitor · Ribosome · Protein synthesis
Bacteria · Translation · Reporter strain

After initial success of antibiotics discovery [4, 21], the problem of bacterial infections appeared to be solved. In a golden period of antibiotics, lasted from 1940s to 1960s many diverse families of natural antibacterial compounds were discovered and applied in clinical practice [8]. However, the first antibiotic resistant bacteria were revealed shortly after [25]. Since that time, the frequency of new antibacterial compound discovery is decreasing dramatically [8]. At the same time, the antibiotic resistance isolates of pathogenic bacteria are spreading and even microbes resistant to multiple antibiotics and more recently totally resistant bacteria are found [1].

The standard pipeline of antimicrobial drug discovery starts with screening for antibacterial activity, most often in a culture broth of soil microorganisms. Following initial demonstration of the activity, an active compound needs to be purified to homogeneity and its mechanism of action should by studied ab initio. Often, and in the recent years even predominantly, application of this pipeline results in rediscovery of known antibacterials [3, 8, 16]. The time spent on the purification of the active compound and ab initio studies of the mechanism of action make the whole procedure lengthy, costly and inefficient.

The problems in the application of the standard pipeline noticed by both academic scientists and large pharmaceutical companies abandoned the area of new antimicrobials discovery [8]. However, the problem of antibiotic resistance remains

---

S.V. Leonov
e-mail: sl@pharmcluster.ru

Ya.A. Ivanenkov
e-mail: yai@pharmcluster.ru

A.A. Bogdanov · O.A. Dontsova
Department of Chemistry and Belozersky Institute of Physico-Chemical Biology,
Lomonosov Moscow State University, Moscow, Russia
e-mail: bogdanov@genebee.msu.ru

O.A. Dontsova
e-mail: olga.a.dontsova@gmail.com

and growing worse year by year. Fresh ideas are needed to speed up antimicrobial compounds discovery. One of the possibilities is to combine initial screening for antibacterial activity with the built-in procedure to assess the mechanism of action [22]. High-throughput automation combined with the cost-efficient mechanism of action testing creates a possibility to revive antibacterial compounds discovery pipeline.

Usually, determination of the mechanism of action is time-consuming, expensive and not suitable for automation [22]. For example, the standard way is to test for the efficiency of incorporation of radioactive precursor molecules into DNA, RNA, and proteins. Experiments of this type require specialized protective environment to grow bacteria in radioactive culture medium and could not be done in a high-throughput form. Replacement of the radioactive precursor molecules with fluorescent ones [13] creates some advantage, but it is hardly suitable for initial screening due to the multistep lengthy implementation and the cost of reagents per test. A set of in vitro tests for partial reactions of protein biosynthesis is valuable for the screening of inhibitors [9], but very costly if applied to the high-throughput screening, due to the cost of the reagents and multistep procedure of the reaction setup.

An alternative to these methods is application of the reporter strains that respond to particular functional type of the inhibitor by upregulation of the gene, whose product is easy to detect [14, 22]. Among the set of possible reporter genes, those coding for fluorescent proteins are preferable due to the lack of any reagent requirements, making application of other reporters, such as luciferases [18], much more expensive. Several reporter systems designed to detect particular classes of antibiotics are available. For example, there are reporters aimed in detection of beta lactams [24], tetracyclines [7], and macrolides [2, 11]. Detection of specific classes of known antibacterials is of use, but it is not applicable for the discovery of new antibacterials [14, 22].

Previously in our laboratory, a reporter plasmid was created that was based on the application of modified tryptophan attenuator as a sensor for translation inhibitors [12]. The attenuator of transcription preceding the genes coding for tryptophan biosynthesis pathway was discovered by C. Yanofsky group in 1970s [26]. Ribosome slows down on the doublet of tryptophan codons in a course of leader peptide synthesis results in the formation of antiterminator secondary structure in between the ribosome and RNA polymerase. In contrast, steady and quick translation in the presence of sufficient concentration of tryptophanyl-tRNA results in the formation of terminator hairpin in the nascent transcript, leading to premature termination of transcription preventing expression of tryptophan biosynthesis genes. We substituted tryptophan codons with those coding for abundant aminoacid alanine and introduced compensatory changes to the RNA in order to preserve a formation of alternative regulatory secondary structures [12]. These mutations made attenuator insensitive to the concentration of tryptophan. Ribosome interacting antibiotics may stall translation while the ribosome traverse leader region of attenuator. This stalling would cause folding of antiterminator RNA structure resulting in upregulation of the following gene.

In the original reporter construct, we introduced CER fluorescent protein [20] downstream from the genetically modified tryptophan attenuator. Unregulated RFP fluorescent protein [10] was used as a control. This reporter construct was previously successfully applied for the screening of soil microorganisms culture broths provided by Gauze Institute of new antibiotics search [12]. This work resulted in identification of the mechanism of action for antibiotic amicoumacin A [17]. Antibacterial activity of this antibiotic was described previously [6], however, its mechanism of action was revealed only in our study [17].

In the frame of the current project supported by the Ministry of science, we improved the reporter construct. Due to the high background fluorescence of the rich culture media in the spectral area of CER protein fluorescence, we replaced CER protein gene with the gene of far red fluorescent protein Katushka2S [5] under a control of modified tryptophan attenuator (Fig. 1a). To test for two potential mechanisms of action simultaneously, we inserted *Escherichia coli sulA* promoter [23] in front of the RFP protein (Fig. 1a). This promoter is regulated by LexA transcriptional repressor, which is inactivated upon induction of the DNA damage SOS response [19]. The cells transformed by the resulting plasmid pDualrep2 [15] became a sensor for translation inhibitors which might be monitored by induction of Katushka2S expression and DNA damaging agents, such as topoisomerase II inhibitors, monitored by induction of RFP expression. These proteins possess readily distinguishable spectral properties that allow their separate detection via fluorescence scanner. A typical translation inhibitor erythromycin induces Katushka2S expression, while levofloxacin, an inhibitor of topoisomerase, induces expression of RFP (Fig. 1b). A number of antibiotics with known mechanism of action, such as erythromycin, roxithromycin, azithromycin, sulfanilamide, polymyxin, rifampicin, chloramphenicol, kanamycin, tetracycline, streptomycin,

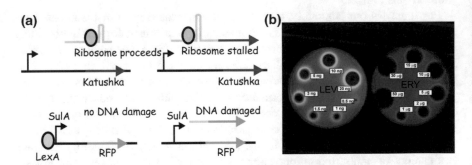

Fig. 1 Reporter system for detection of ribosome stalling and DNA damaging compounds. **a** Scheme of the reporter. *Upper panel* illustrates the organization of the sensor for translation inhibitors. On the *left* is the situation when ribosome is not inhibited. No Katushka2S expression is observed in this case. On the *right* is the situation when the ribosome is stalled by an antibiotic. In this case Katushka2S is upregulated. *Lower panel* corresponds to the sensor of DNA damage. On the *left* is situation when LexA represses transcription of the reporter in the absence of DNA damage. On the *right* is situation when LexA is inactivated when DNA is damaged. In this case RFP is upregulated

lincomycin, clindamycin, nalidixic acid, levofloxacin, ciprofloxacin, tobramycin, neomycin, etoposide, furagin, microcin B17, spectinomycin, etamycin A, hygromycin B, griseoviridin, tylosin, amicoumacin A, fusidic acid, puromycin, and gentamicin were tested against the created reporter strain [15]. It appeared that induction of RFP reporter is caused by all known topoisomerase inhibitors and in addition DNA precursor biosynthesis inhibitors. All substances that induced expression of Katushka2S belonged to translation inhibitors, although some ribosome-targeting antibiotics do not induce expression of the reporter gene. This result is expected, since some ribosome targeted antibiotics, such as aminoglycosides induce translation misreading, but not ribosome stalling.

The cells transformed by the pDualrep2 reporter were used for automated high throughput screening of the collection of 17,856 compounds provided by the industrial partner of the project, Research Institute of Chemical Diversity. The set was assembled from the core collection of Research Institute of Chemical Diversity aiming at maximization of chemical diversity. The screening was performed by automated liquid handling station Janus (Perkin Elmer). We spread the lawn of the *E. coli* reporter strain on top of 245 mm × 245 mm square agar plate. The tested compounds were dissolved in DMSO to the concentration 17 mg/ml. Total amount of 2 µl volume of each obtained solution were spotted on agar plates by a 96-channel pipetting head of Janus liquid handling station (Perkin Elmer). After overnight growth at the 37 °C, the plates are scanned by the fluorescence scanner (e.g., ChemiDoc, Bio-rad) at the wavelengths 553/574 nm (RFP) and 588/633 nm (Katushka2S). An example of a resulting plate scan is presented in Fig. 2a.

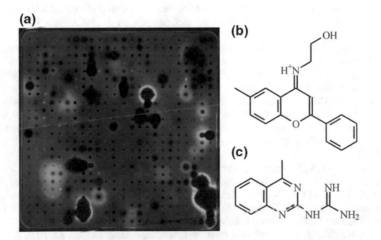

Fig. 2 Results of the screening with the developed reporter system. **a** A sample of the screening plate with the 576 compounds applied onto the *square* agar plate with the reporter strain. The image is the superposition of the same plate scanned at 553/574 nm (RFP) presented as *green* pseudocolor and at 588/633 nm (Katushka2S) presented as *red* pseudocolor. **b** An example of the lead compound identified in the screening, 2-hydroxyethyl-(6-methyl-2-phenylchromen-4- ilyden) azanium. **c** Another lead compound identified in the screening, 4-methyl, 2-guanidino quinazoline

In total, 184 compounds strongly inducing RFP and 32 compounds strongly inducing Katushka2S were found among the set of 17,856 compounds. Notably, DNA damaging agents are almost sixfold more abundant relative to translation inhibitors. Among the new lead compounds, we identified 2-hydroxyethyl-(6-methyl-2-phenylchromen-4- ilyden) azanium (Fig. 2b) and 2-guanidino quinazoline derivatives (Fig. 2c). These lead structures will be further optimized aiming at the design of new more potent and selective antimicrobial substance.

Conclusions

The system for high throughput screening of antimicrobial compounds was created. The system allows classification of antibacterial compounds into stalling protein biosynthesis, inducing DNA damage and the compounds with the mechanism of action unrelated to these two. The system was successfully tested on a large set of the compounds. Several new classes of translation inhibitors were found.

Acknowledgement Researches are carried out (conducted) with the financial support of the state represented by the Ministry of Education and Science of the Russian Federation. Agreement (contract) no. 14.607.21.0086 06. Nov 2014. Unique project Identifier: RFMEFI60714X0086.

References

1. Almeida, Da Silva, P.E., Palomino, J.C.: Molecular basis and mechanisms of drug resistance in Mycobacterium tuberculosis: classical and new drugs. J. Antimicrob. Chemother. **66**, 1417–1430 (2011)
2. Bailey, M., Chettiath, T., Mankin, A.S.: Induction of erm(C) expression by noninducing antibiotics. Antimicrob. Agents Chemother. **52**, 866–874 (2008)
3. Fabbretti, A., Gualerzi, C.O., Brandi, L.: How to cope with the quest for new antibiotics. FEBS Lett. **585**, 1673–1681 (2011)
4. Fleming, A.: On the antibacterial action of cultures of a penicillium, with special reference to their use in the isolation of B. influenzae. British J. Exp. Pathol. **10**, 226–236 (1929)
5. Gurskaia, N.G., Staroverov, D.B., Fradkov, A.F., Luk'ianov, K.A.: Coding region of far-red fluorescent protein katushka contains a strong donor splice site. Bioorg. Khim. **37**, 425–428 (2011)
6. Itoh, J., Omoto, S., Shomura, T., Nishizawa, N., Miyado, S., Yuda, Y., Shibata, U., Inouye, S.: Amicoumacin-A, a new antibiotic with strong antiinflammatory and antiulcer activity. J. Antibiot. **34**, 611–613 (1981). (Tokyo)
7. Kurittu, J., Karp, M., Korpela, M.: Detection of tetracyclines with luminescent bacterial strains Lumin. J. Biol. Chem. Lumin. **15**, 291–297 (2000)
8. Lewis, K.: Platforms for antibiotic discovery. Nat. Rev. Drug Discov. **12**, 371–387 (2013)
9. Lowell, A.N., Santoro, N., Swaney, S.M., McQuade, T.J., Schultz, P.J., Larsen, M.J., Sherman, D.H.: Microscale adaptation of in vitro transcription/translation for high-throughput screening of natural product extract libraries. Chem. Biol. Drug Des. **86**, 1331–1338 (2015)
10. Merzlyak, E.M., Goedhart, J., Shcherbo, D., Bulina, M.E., Shcheglov, A.S., Fradkov, A.F., Gaintzeva, A., Lukyanov, K.A., Lukyanov, S., Gadella, T.W.J., et al.: Bright monomeric red fluorescent protein with an extended fluorescence lifetime. Nat. Methods **4**, 555–557 (2007)

11. Möhrle, V., Stadler, M., Eberz, G.: Biosensor-guided screening for macrolides. Anal. Bioanal. Chem. **388**, 1117–1125 (2007)
12. Osterman, I.A., Prokhorova, I.V., Sysoev, V.O., Boykova, Y.V., Efremenkova, O.V., Svetlov, M.S., Kolb, V.A., Bogdanov, A.A., Sergiev, P.V., Dontsova, O.A.: Attenuation-based dual-fluorescent-protein reporter for screening translation inhibitors. Antimicrob. Agents Chemother. **56**, 1774–1783 (2012)
13. Osterman, I.A., Ustinov, A.V., Evdokimov, D.V., Korshun, V.A., Sergiev, P.V., Serebryakova, M.V., Demina, I.A., Galyamina, M.A., Govorun, V.M., Dontsova, O.A.: A nascent proteome study combining click chemistry with 2DE. Proteomics **13**, 17–21 (2013)
14. Osterman, I.A., Bogdanov, A.A., Dontsova, O.A., Sergiev, P.V.: Techniques for Screening Translation Inhibitors. Antibiot. Basel Switz. **5**, E22 (2016)
15. Osterman, I.A., Komarova, E.S., Shiryaev, D.I., Korniltsev, I.A., Khven, I.M., Lukyanov, D.A., Tashlitsky, V.N., Serebryakova, M.V., Efremenkova, O.V., Ivanenkov, Y.A., et al.: Sorting out antibiotics' mechanisms of action: a double fluorescent protein reporter for high throughput screening of ribosome and DNA biosynthesis inhibitors. Antimicrob. Agents Chemother. **60**, 7481–7489 (2016)
16. Payne, D.J., Gwynn, M.N., Holmes, D.J., Pompliano, D.L.: Drugs for bad bugs: confronting the challenges of antibacterial discovery. Nat. Rev. Drug Discov. **6**, 29–40 (2007)
17. Polikanov, Y.S., Osterman, I.A., Szal, T., Tashlitsky, V.N., Serebryakova, M.V., Kusochek, P., Bulkley, D., Malanicheva, I.A., Efimenko, T.A., Efremenkova, O.V., et al.: Amicoumacin a inhibits translation by stabilizing mRNA interaction with the ribosome. Mol. Cell. **56**, 531–540 (2014)
18. Prokhorova, I.V., Osterman, I.A., Burakovsky, D.E., Serebryakova, M.V., Galyamina, M.A., Pobeguts, O.V., Altukhov, I., Kovalchuk, S., Alexeev, D.G., Govorun, V.M., et al.: Modified nucleotides m^2G966/m^5C967 of *Escherichia coli* 16S rRNA are required for attenuation of tryptophan operon. Sci. Rep. **3**, 3236 (2013)
19. Reifferscheid, G., Buchinger, S.: Cell-based genotoxicity testing: genetically modified and genetically engineered bacteria in environmental genotoxicology. Adv. Biochem. Eng. Biotechnol. **118**, 85–111 (2010)
20. Rizzo, M.A., Springer, G.H., Granada, B., Piston, D.W.: An improved cyan fluorescent protein variant useful for FRET. Nat. Biotechnol. **22**, 445–449 (2004)
21. Schatz, A., Bugie, E., Waksman, S.A.: Streptomycin, a substance exhibiting antibiotic activity against gram-positive and gram-negative bacteria. Proc. Soc. Exp. Biol. Med. **55**, 66–69 (1944)
22. Sergiev, P.V., Osterman, I.A., Golovina, A.Y., Laptev, I.G., Pletnev, P.I., Evfratov, S.A., Marusich, E.I., Leonov, S.V., Ivanenkov, Y.A., Bogdanov, A.A., et al.: Application of reporter strains for new antibiotic screening. Biomeditsinskaia Khimiia **62**, 117–123 (2016)
23. Thomassen, G.O.S., Weel-Sneve, R., Rowe, A.D., Booth, J.A., Lindvall, J.M., Lagesen, K., Kristiansen, K.I., Bjørås, M., Rognes, T.: Tiling array analysis of UV treated *Escherichia coli* predicts novel differentially expressed small peptides. PloS one **5**, E. 15356 (2010)
24. Valtonen, S.J., Kurittu, J.S., Karp, M.T.: A luminescent *Escherichia coli* biosensor for the high throughput detection of beta-lactams. J. Biomol. Screen. **7**, 127–134 (2002)
25. Waksman, S.A., Reilly, H.C., Schatz, A.: Strain specificity and production of antibiotic substances: v. strain resistance of bacteria to antibiotic substances, especially to Streptomycin. Proc. Natl. Acad. Sci. USA **31**, 157–164 (1945)
26. Zurawski, G., Elseviers, D., Stauffer, G.V., Yanofsky, C.: Translational control of transcription termination at the attenuator of the *Escherichia coli* tryptophan operon. Proc. Natl. Acad. Sci. USA **75**, 5988–5992 (1978)

Shape of the Voltage–Frequency Curve Depending on the Type of the Object Detached from the QCM Surface

F.N. Dultsev

Abstract Analysis of the shapes of voltage–frequency curves depending on the type of object detached from the surface of the quartz crystal microbalance (QCM) is carried out. It is demonstrated that the shape of the curve depends not only on the size and shape of bio-object but also on the properties of the particle. For example, a detachment of hepatitis B virus is accompanied by the fragmentation of the bio-object, and signal shape is typical for this case. In addition to a voltage value which determines the bonding force, the signal shape is also characteristic for identification of bio-object.

Keywords QCM Sensor · Rupture event scanning · Bonding forces

Introduction

The method used in this work to measure bonding forces is based on the use of the quartz crystal resonator as a sensor. Quartz crystal resonators are usually used to measure mass, that is, as quartz crystal microbalance (QCM). The addition of some mass (m) on QCM surface (AT-cut quartz plate about 100 μm thick) causes a definite frequency shift Δf, which can be detected by the measuring equipment [1–3]. It turned out to be possible to broaden the measuring possibilities and to enhance the sensitivity of QCM-based sensors by applying the principle of the measurement of the acoustic signal generated during the rupture of bonds holding an object on the QCM surface, as proposed in [4]. The applications of this method to the determination of phages and bacteria were demonstrated in [5, 6].

In those works, QCM is not only a sensor but it also plays an active role with respect to a particle bound to its surface: the dynamic increase in the voltage at the AT-cut QCM electrodes (surface) causes an increase in the amplitude of shear

F.N. Dultsev (✉)
Institute of Semiconductor Physics, Siberian Branch of the Russian Academy of Sciences,
Novosibirsk State University, Novosibirsk, Russia
e-mail: fdultsev@isp.nsc.ru

© The Author(s) 2018
K.V. Anisimov et al. (eds.), *Proceedings of the Scientific-Practical Conference*
"Research and Development - 2016", https://doi.org/10.1007/978-3-319-62870-7_64

oscillations. Under smoothly increasing amplitude of oscillations (rupture event scanning), a particle bound to QCM surface will get detached due to inertial forces in a threshold manner, so that the force of bond rupture may be easily obtained from the threshold voltage. This procedure allows us to increase the sensitivity and to carry out reliable measurements of rupture forces of the order of 10 pN. Similarly to atomic force microscopy, rupture event scanning is a direct measurement method, it does not involve electromagnetic radiation, but rupture event scanning has much simpler instrumentation. The presence of an object on QCM surface causes a slight distortion of the shape of membrane oscillations. For the object mass 10^{-12} of the mass of the quartz plate, the frequency change is much smaller than that caused by the background temperature non-uniformities of quartz material. In spite of the fact that the presence or the absence of a small object on the surface cannot be recorded as mass change, the rupture moment is determined correctly because a transient process in the form of acoustic signal arises in the resonator plate. This signal points not only to the presence of analytes but also on their number, and their affinity to the receptor. In the present work, we will consider the determination of binding force on the basis of the analysis of the voltage–frequency dependence.

Experimental

AT-cut quartz plate 8.25 mm in diameter, plano-convex (curvature 0.2 dioptres) with the resonance frequency of 14.3 MHz (Morion, St. Petersburg, Russia) was used in the work.

The setup is shown schematically in Fig. 1. This is the simplest arrangement. Voltage is supplied from a signal generator (1) to the QCM (2). Frequency scanning around the resonance frequency is carried out. The voltage supplied to the QCM is increased after each scanning.

Fig. 1 Electrical circuit of the experimental setup. *1*—signal generator GSS-40, *2*—QCM, *3*—current transformer, *4*—amplifier units, *5*—analog-to-digital converter, *6*—personal computer

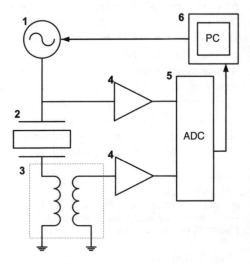

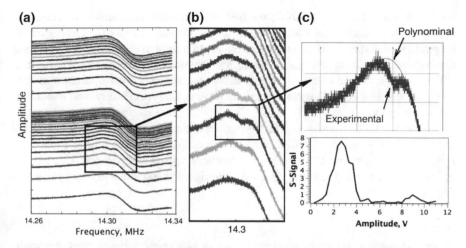

Fig. 2 The scheme explains the origin of S-signal

Voltage and the current passing through the QCM are supplied through the logarithmic amplifier (4) to the ADC (5). The software was made in MathLab. The rupture signal is recorded from the voltage–frequency dependence (VFD-method), as described in detail in [7]. Here we give a brief description of the essence of this method and illustrate it in Fig. 2a.

(a) description of the amplitude–voltage dependence for different voltage values applied to the QCM The voltage is increased step by step from bottom to top, from 0 to 10 V. Frequency change around the resonance frequency of the QCM is plotted along the X-axis. Distortions accompanying particle detachment (rupture) are exhibited at a definite voltage value (this region is marked in Fig. 2a and shown in more detail in Fig. 2b). S-signal is a sum of the absolute values of the integrals or the area between the experimental curve and the envelope polynomial (see upper Fig. 2c). A set of values is obtained. Figure 2c, lower part: dependence of S-signal on the amplitude of the voltage applied to the QCM. One can see that rupture occurs at the voltage of 2–4 V; the exact position of the maximum is determined by equating to zero the derivative of the analytical parabola $y = ax^2 + bx + c$, which is traced through the points near the maximum. The data for the unwinding of the double-stranded oligonucleotides are shown (the surface concentration is 1×10^{11} molecules/mm^2 [8].

Results and Discussion

Mathematical processing of the experimental curves depends on the shape of these curves, which in turn depends on the object bound to the QCM surface. Some typical examples for nano-objects on QCM surface are to be considered. In this work, we will not deal with surface modification because the modification is carried out in a special manner for each specific case. However, in any case, the surface is prepared so that the object of interest is bound to it selectively. Specifically, bound nano-object is held by a stronger bond, and rupture occurs during scanning at a higher voltage. As an example, Fig. 3 shows the separation of a mixture of two bacterial species, one of them bound to the surface specifically and the other non-specifically.

Below we will consider signal shapes depending on the shape, mass, and a number of nano-objects on QCM surface. The simplest shape of a nano-object is a sphere. At the moment of detachment, distortions appear near the resonance at the voltage–frequency dependence. As a rule, the signal shape looks like that shown in Fig. 4.

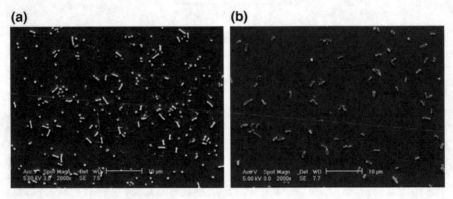

Fig. 3 (a) 20 μL of an equal mixture of *Staphylococcus aureus* (SA) and *Escherichia coli* (EC) at 5×10^5 bacteria mL^{-1} incubated for 1 h on the QCM then washed with cold PBS. (b) QCMs treated as in (a) then scanned from 0 to 4 V. No visible features could be discerned on a QCM coated with mAbs alone or a QCM left uncoated with protein [9]

Fig. 4 The shape of the signal accompanying the rupture of a spherical solid particle. Rupture signal is the area between the experimental and polynomial curves

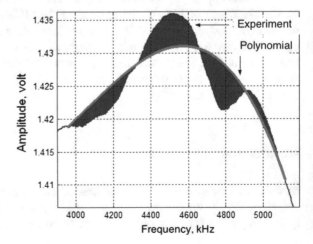

If particle mass is large, distortions are larger than those for a small particle. Testosterone linked with horseradish peroxidase was studied as an example. Testosterone provides specific bonding with QCM surface, while horseradish peroxidase has large size and mass. At the voltage about 2 V, the molecules get detached, and strong distortion is observed in the voltage–frequency dependencies (see Fig. 5). The shape of the voltage–frequency curve does not change during further scanning.

As a rule, particles are detached within the rather narrow voltage range. For the spherical particles of the same size, for example, aerosol particles for which the particle size distribution is shown in Fig. 6, distortions are observed at one voltage value, but signal shape deviates in the form of a smooth transition, see Fig. 7. This shape of the curve is typical for the high concentration of the spherical particles with narrow size distribution.

Fig. 5 The voltage–frequency curves for the detachment of molecules with large mass

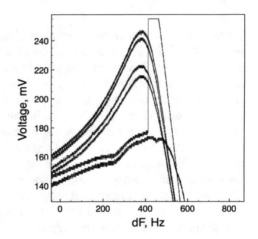

Fig. 6 Size distribution of aerosol particles obtained with the help of aerosol generator from the vapor of salicylic aldehyde

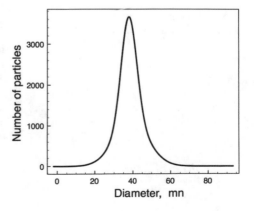

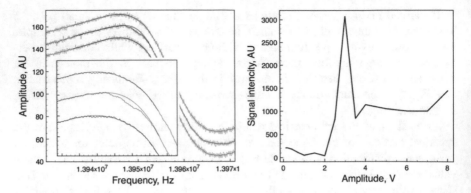

Fig. 7 Voltage–frequency dependencies (*left*) and the dependence of rupture signal on voltage (*right*) from which the bond strength may be determined. Here QCM surface with the attached amine groups NH_2 is used, specific bonding is observed, which suggests that there are carboxylic groups COOH on the surface of aerosol particles

Rupture signal for hepatitis B virus (HBsAg) is shown in Fig. 8. A specific feature of this signal in comparison, for example, with the rupture signal of herpes virus [5] is that HBsAg rupture signal has not a narrow peak but a broad peak. This shape suggests that rupture occurs within a broad voltage range. During repeated scanning, we observe the signal at a higher voltage, and the peak becomes narrower. We used different antibodies, but the signal behavior was the same. This observation may be explained by the structure of HBsAg. It was demonstrated in the studies of HBsAg particles by means of EPR examination of spin-labeled fatty acids [10] that fatty acids in the lipid bilayer are immobilized as a consequence of the binding role of polypeptides. On the basis of the data obtained, the mechanism of HBsAg particle formation through gemmation from the membrane of endoplasmic reticulum of the host cell was proposed.

Our studies also confirm that hepatitis virus is not a rigid nanoparticle; from the viewpoint of solid state physics, it is composed of nanoparticles weakly bound with

Fig. 8 Typical curve (rupture signal) for hepatitis B virus

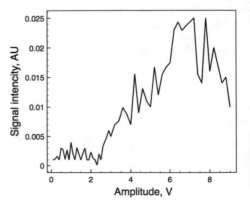

each other. So, it may be assumed that scanning causes the destruction of the virus, which we observe as a signal extended over the voltage axis.

The data for the unwinding of the double-stranded oligonucleotides (20 base pairs) are shown in Fig. 2. One can see that unwinding occurs not strictly at a definite voltage but within some interval (see Fig. 2c)—from 2 to 4 V.

Conclusions

It is demonstrated in the work how the properties of the particle detached from the QCM surface are connected with the shape of the rupture signal. A sharp single peak is observed as a result of the rupture of a rigid spherical particle. Distortions are observed on one voltage–frequency dependence, while peak broadening is observed in the case of the separation of long nano-objects (dsDNA) linked along their length. If a nanoparticle composed of separate smaller particles is detached from the surface, fragmentation of this particle occurs, and rupture signal as a set of separate peaks is observed.

QCM appears as an element functioning as an active resonator, absorbing surface, and signal recording unit. This allows us to use QCM as a tool to detect particles bound to the surface with different forces, in particular in medical diagnostics and in sensing. The time of sensor recovery is exactly measuring time because measurement involves purification of the sensor surface. An increase in sensitivity does not cause a decrease in the selectivity of the instrument.

Acknowledgements Research is carried out with the financial support of the state represented by the Ministry of Education and Science of the Russian Federation. Agreement No. 14.607.21.0125 from 27.10.2015. Unique project Identifier: RFMEFI60715X0125.

References

1. Sauerbray, G.Z.: Use of quartz vibrator for weighing thin films on a microbalance. Z. Phys. **155**, 206–212 (1959)
2. Mecea, V.M.: Quartz crystal microbalance to fundamental principles of mass measurements. Anal. Lett. **38**, 753–767 (2005)
3. Marx, K.A.: Quartz crystal microbalance: a useful tool for studying thin polymer films and complex biomolecular systems at the solution-surface interface. Biomacromolecules **4**(5), 1099–1119 (2003)
4. Dultsev, F.N., Ostanin, V.P., Klenerman, D.: "Hearing the bond breakage". Measurement of bond rupture forces using a quartz crystal microbalance. Langmuir **16**, 5036–5040 (2000)
5. Cooper, M.A., Dultsev F.N., Ostanin V.P., et al. Direct and sensitive detection of a human virus by rupture event scanning// Nat. Biotechnol. 2001. N 19. p. 833 − 837
6. Dultsev, F.N., Speight, R.E., Fiorini, M.T., et al.: Direct and quantitative detection of bacteriophage by "hearing" surface detachment using a quartz crystal microbalance. Anal. Chem. **73**, 3935–3939 (2001)

7. Dultsev, F.N., Kolosovsky, E.A., Mik, I.A.: New procedure to record the rupture of bonds between macromolecules and the surface of the quartz crystal microbalance (QCM). Langmuir **28**, 13793–137978 (2012)
8. Dultsev, F.N., Kolosovsky, E.A., Mik, I.A., et al.: QCM-based measurement of bond rupture forces in DNA double helices for complementarity sensing. Langmuir **30**(13), 3795–3801 (2014)
9. Cooper, M.A., Dultsev, F.N., Ostanin, V.P., Klenerman, D.: Separation and detection of bacteria using rupture event scanning. Analytica Chimica Acta. **702**, 233– 238 (2011)
10. Satoh, O., Imai, H., Yoneyama, T., et al.: Membrane structure of the hepatitis B virus surface antigen particle. J. Biochem. **127**, 543–550 (2000)

Complex Technology of Oil Sludge Processing

A.V. Anisimov, V.I. Frolov, E.V. Ivanov, E.A. Karakhanov, S.V. Lesin and V.A. Vinokurov

Abstract The technology of processing sludge includes electromagnetic activation of raw materials, catalytic cracking, hydrocracking, and oxidative desulfurization. Preliminary activation reduces electromagnetic temperature catalytic processes at 60–80 °C, increasing the yield of light products, reduce sulfur content in liquid products at 13–19% by weight, coke formation, and reduces gassing.

Keywords Oil sludge · Electromagnetic activation · Catalytic cracking
Hydrocracking · Oxidative desulfurization

At present, the refineries, the acute problem of recycling and disposal of sludge (PS) as well as available technological solutions aimed at processing of oily waste, cannot be the basis of a uniform method of processing sludge [1–3]. In the

A.V. Anisimov
Department of Petroleum Chemistry and Organic Catalysis,
Lomonosov Moscow State University, Moscow, Russia
e-mail: anis@petrol.chem.msu.ru

V.I. Frolov (✉) · V.A. Vinokurov
Gubkin Russian State University of Oil and Gas (National Research University),
Moscow, Russia
e-mail: fvi209@mail.ru

V.A. Vinokurov
e-mail: vinok_ac@mail.ru

E.V. Ivanov · S.V. Lesin
Department of Physical and Colloid Chemistry, Gubkin Russian State University
of Oil and Gas (National Research University), Moscow, Russia
e-mail: ivanov166@list.ru

S.V. Lesin
e-mail: lesinsv@gmail.com

E.A. Karakhanov
Lomonosov Moscow State University, Moscow, Russia
e-mail: kar@petrol.chem.msu.ru

© The Author(s) 2018
K.V. Anisimov et al. (eds.), *Proceedings of the Scientific-Practical Conference
"Research and Development - 2016"*, https://doi.org/10.1007/978-3-319-62870-7_65

development of technologies for processing and disposal of the NS the most important is the maximum possible extraction of hydrocarbons, so special attention is paid to methods aimed at obtaining a variety of commercial products, such as components of the engine and boiler fuel, secondary hydrocarbon feedstock of various technological processes of oil refining, secondary petrochemicals, road-building materials [1].

In recent years, proposed technologies for processing crude oil and oil waste using non-standard physical, chemical, and physical influences on them. In this case, the raw material is pretreated using acoustic energy, ultrasound, high-frequency (HF) and ultrahigh frequency (UHF) electromagnetic fields, ionizing radiation, exposure to plasma-chemical and electric current [1–17]. Efficient processing and recycling of the PS will solve the environmental problems in the storage of waste oil. Conducting such research confirms the global level of the project.

It is very promising for fuel products represented PS recycling technologies through their pre-wave treatment in combination with conventional catalytic methods. Using such technology would lead to the maximum recovery of liquid hydrocarbons with high destruction and subsequent conversion to fuel products. The most promising techniques using high-frequency electromagnetic emissions, which contribute to the effective separation of simultaneous PS phases, reduce the temperature of catalytic thermal processing, increase the depth of oil refining up to 85% and the production of high-quality fuel products. In carrying out this project developed a complex technology of processing raw materials nefteshlam-soderzhaschego including several processing steps as follows:

– Activation of the sludge by means of an electromagnetic field
– Catalytic cracking activated PS
– Hydrocracking PS
– Oxidative desulfurization activated PS.

Experimental Part

The raw material for the implementation of pilot studies on the impact of various external factors on the process of recycling the activated PS were selected sediments buffer ponds biological sewage treatment plants, having the following phase composition: 20.9% by weight of oil, water—69% by weight, fur impurity—10.1% by weight. After the separation of water and solids component of the oil extracted from the sediment buffer ponds, had an initial boiling temperature of—83.5 °C, final boiling point—680.4 °C. For the experiments, the oil component used was obtained after separation of water and solids extraction averaged sediment samples carbon tetrachloride with a water content of $1.5 \div 2.0$ mass %. Pre by simulated distillation was determined factional and group composition of the oil component of the bottom sediments, according to which they contain 40% by weight. paraffins; 32.5% by weight. mono-, bi-, tri-, and tetradenia naphthenes; 27.6% by weight aromatics. Electromagnetic activation watered sediment was carried out under optimal conditions found for the catalytic cracking process activated the oil component of dewatered sediment (frequency of electromagnetic radiation—50.5 MHz, power—0.55 kW, Activation time—6 h, activation temperature of—50 °C. Pilot studies activated the catalytic cracking process of petroleum PS conducted at 400–500 °C component, the mass hourly space velocity of 15 h^{-1} in order to reduce its viscosity PS preheated before entering the reactor to 70 °C. Liquid products were analyzed by GC and GC-spectrometry. The sulfur content in the raw compounds and the reaction products were determined by fluorescence spectrometer at a wave ARL PerformX2500. The catalyst used cracking catalyst NS brand "Grace" with an average particle size of 72 microns, a BET surface area of 350 m^2/g mass fraction of rare earth oxides 0.2% the masses, the mass fraction of Al_2O_3, 43% by mass, the mass fraction of Na_2O, 0.29% by weight. For identification of the main regularities of catalytic cracking, hydrocracking, and oxidative desulfurization activated PS oil component was the varied frequency of the electromagnetic radiation of electromagnetic power and time of activation to assess the effects on the composition and the total yield of the products of catalytic cracking, hydrocracking, and oxidative desulfurization.

The Discussion of the Results

Sludge processing technology development was carried out with pre-electromagnetic activation of raw materials on the basis of three processes—catalytic cracking, hydrocracking and oxidative desulfurization. For the selection of the optimal conditions of the electromagnetic activation sludge mathematical models were developed separately for the process of catalytic cracking, hydrocracking and oxidative desulfurization. For catalytic cracking in the presence of a catalyst 40% mass. LUX-2 (Russia)+60% mass. DA-250 (Grace, USA) under 500 °C temperature, the feed rate of 15.6 h^{-1} was obtained regression equation and calculated parameters OLS regression equation separately for a total yield of light oil for diesel and gasoline yield fractions. Optimization of electromagnetic activation of the catalytic cracking activated PS was carried out by a steep ascent on the response surface (for a total yield of light oil) and the optimum conditions of the electromagnetic activation PS process were found: the frequency of electromagnetic radiation—50.5 MHz, power—0.55 kW and activation time—6 h.

To carry out the catalytic cracking of heavy oil wastes was developed catalyst comprising a medium—mixture of the ordered mesoporous silica MCM-41 (17 wt %) Gamma alumina (49 wt%) and zeolite-Y (30 wt%) with the addition of lanthanum nonzero valence state in an amount of 3% by weight. ($La_2O_3/MCM-41 + \gamma-Al_2O_3$). In the catalytic cracking of a mixture of activated PS and non-activated vacuum gas oil (HS) HS conversion component in 520 °C more than 90%, the selectivity of the gasoline fraction in a MAT of 63.7%, a decrease in the sulfur content of broad fraction of light oil occurred at 37%.

Catalytic cracking was PS after preliminary treatment of the wave electromagnetic radiation with a frequency of 50.5 MHz (for the control of the characteristic frequencies of groups and individual frequencies of electromagnetic radiation signal generator parameters used data collector and signal analyzer VIBXPERT II and digital oscilloscope Agilent Technologies) and a capacity of 0.55 kW at 50 °C, pressure 0.4 MPa, treatment time clock 6. Catalytic cracking of feedstock was carried out in a continuous-flow reactor at a temperature of 500 °C and the mass feed rate of 15 h^{-1}. PS catalytic cracking results with and without activation of the activation of the pilot sample developed catalytic cracking catalyst are given in Table 1

As shown in Table 1 preactivation sludge followed by catalytic cracking catalyst on La/MCM-41/γ-Al$_2$O$_3$ allows cracking in comparison with the process without activating the lower sulfur concentrations in liquid products of 29.8%, cracking gas 4.4 wt% coke on the catalyst 1.0% wt; increase the yield of the gasoline fraction at 8.0% by weight of the diesel fraction to 4.0% by weight, the processing depth (the residue reduction) 7.6% by weight.

For hydrocracking of heavy oil and waste sludge was developed catalyst of general formula NiO–MoO$_3$/Al-HMS/γ-Al$_2$O$_3$, which is a support based on mesoporous aluminosilicate Al-HMS/γ-Al$_2$O$_3$ coated with a hydrogenation promoter NiO (5.1 % by weight) and MoO$_2$ (18.0%), the specific surface area of the catalyst—955 m^2/g. For the process of hydrocracking, PS activated in the presence of said catalyst at T = 400 °C, p = 5 MPa, t = 3 h were obtained regression equation and calculated parameters OLS regression equation separately for the total output of light oil, to exit gasoline and diesel fractions. Carried out optimization of the electromagnetic activation process hydrocracking PS activated by a steep ascent on the response surface (for a total yield of light oil) allowed for this process to find optimal conditions for electromagnetic activation NS: frequency electromagnetic radiation—47.5 MHz, power—0.35 kW activation time—3.5 h.

Hydrocracked oil sludge after it is subjected to a preliminary treatment by electromagnetic wave radiation at a temperature of 50 °C, 0.4 MPa pressure, 3.5 h treatment time. hydrocracking process was carried out in a steel autoclave of 45 ml, equipped with a magnetic stirrer. The autoclave was placed a sample hydrocracking catalyst NiO–MoO$_3$/Al-HMS/γ-Al$_2$O$_3$ amount of 0.2–1.5 g of 3 ml Hereinafter

Table 1 Results of sludge on the catalytic cracking test specimen developed catalytic cracking catalyst La/MCM-41/γ-Al$_2$O$_3$

Products	Un.	Output change		Change
		Before	After	
Light petrochemicals, including:	wt%	60.0	73.0	+13.0
Gasoline fraction	wt%	41.0	50.0	+9.0
The diesel fraction	wt%	19.0	23.0	+4.0
Coke	wt%	8.0	7.0	−1.0
Gases	wt%	17.1	12.7	−4.4
Balance	wt%	14.9	7.3	−7.6
The sulfur content	ppm	8790	6171	−29.8

feedstock after the electromagnetic activation, heated to 30–40 °C beforehand to obtain a liquid consistency. The autoclave was filled with hydrogen to a pressure of 9 MPa. And placed in an oven heated to 340 °C with vigorous stirring. For complete extraction fraction hydrocracking transmitted to the autoclave were added 4.0 ml of n-C_7H_{16} and stirred vigorously for 5 min. After centrifugation, the clear liquid in it an equal volume of 10% NaOH solution to remove the hydrogen sulfide dissolved therein. Thereafter, the aqueous phase was separated and the organic phase was washed with water, dried and analyzed by GLC. The sulfur content of the feedstock and the reaction products were determined on an analytical-based complex energy dispersive spectrometer JEOL JED-2300T. PS hydrocracking results with and without activation of the activation of the pilot pattern designed hydrocracking catalyst are given in Table 2.

As shown in Table 2 preactivation sludge followed by hydrocracking of the pilot sample developed catalytic cracking catalyst NiO–MoO_3/Al-HMS/γ-Al_2O_3 allows versus hydrocracking process without activating the lower sulfur concentrations in liquid products of 41.2%, lower cracked gas yield of 3.6% by weight, to increase the yield of light products to 19% by weight.

To remove the sulfur from the sludge along with hydrocracking can be used and the process of oxidative desulfurization for those cases where there are no available and cheap sources of hydrogen. For oxidative desulfurization of heavy oil and waste, sludge was developed catalyst comprising molybdenum peroxo, preparing directly in the same reactor, where the process is conducted by mixing desulfurizing sodium molybdate, sulfuric acid, and triethylbenzylammonium chloride.

Before desulfurization slime subjected to electromagnetic radiation with a frequency of 47.5 MHz, the power of 0.35 kW at 50 °C, under atmospheric pressure for 3.5 h. Validating oil sludge and activated oxidative desulfurization was then contacted with a catalyst system which is prepared mixing the dihydrate of sodium molybdate, 50% aqueous solution of hydrogen peroxide and a phase transfer agent in the form of quaternary ammonium salts–cetyltrimethylammonium bromide, combined in amounts to provide the following molar ratios: metal: sulfur contained in the oil sludge of 1:100, hydrogen peroxide: the sulfur contained in oil sludge 2:1, phase transfer: the sulfur contained in the oil sludge 1:20. The catalyst can withstand at least 6 cycles of continuous operation without substantial loss of activity and reduces the sulfur content in the resulting desulfurized sludge sample by 43%.

Table 2 Results of the pilot hydrocracking sludge sample developed hydrocracking catalyst NiO–MoO_3/Al-HMS/γ-Al_2O_3

Products	Un.	Output Change		Change
		Before	After	
Light petrochemicals, including:	wt%	22.0	39.0	+19.0
C_5-C_{11}	wt%	2.0	6.0	+4.0
C_{12}-C_{16}	wt%	20.0	33.0	+13.0
Gases	wt%	14.3	10.7	−3.6
Balance	wt%	63.7	50.3	−13.4
The sulfur content	ppm	85	50	−41.2

Table 3 Results of laboratory tests of experimental sample of oxidative desulfurization catalyst in the desulfurization sludge, initial sulfur content of 6500 ppm of, a catalyst—Sodium molybdate

№	The sulfur content in the experiment after the PS desulfurization ppm	
	Validating slime	Slime activated
1	4830	3700
2	4850	3800

In quantitative terms, using 25 g of sludge, 1.73 ml of a 50% aqueous solution of hydrogen peroxide, 0.264 g of cetyltrimethylammonium bromide and heated at 80 °C for 8 h with stirring the reaction mass at a speed of 35,000 rev/min. After the reaction, the water formed is distilled off, the remaining product was heated to 360 °C for 3 h, then cooled and analyzed for total sulfur content on the device "Atomic Emission Spectrometer microwave plasma AGILENT MP-AES 4100." The results of the process of liquid-phase oxidative desulfurization sludge are shown in Table 3.

As can be seen from Table 3, after the oxidative desulfurization of activated sludge in the presence of sodium molybdate content of sulfur in the final product is 22.5% lower compared to unactivated sludge. The desulfurized activated slime can then be used for further processing thermocatalytic.

Conclusions

The developed technology for processing oil sludge using electromagnetic treatment of raw materials, followed by catalytic cracking, hydrocracking, and oxidative obessrivaniem provides conversion of heavy oil waste into liquid hydrocarbon products. Preliminary activation reduces electromagnetic temperature catalytic processes, increasing the yield of light products, reduce sulfur content in liquid products, coke formation and reduce gassing.

The developed technology of catalytic oxidative desulfurization of petroleum residues and waste can reduce the sulfur content of two or more times and can be used for further catalytic thermal processing of oil residues and waste.

Acknowledgments Research is carried out (conducted) with the financial support of the state represented by the Ministry of Education and Science of the Russian Federation. Agreement (contract) no/ 14.577.21.0106, September 22, 2014. Unique project Identifier: RFMEF157714X0106.

References

1. Mazlova, E,A., Meshcheryakov, S.V.: Problems of waste sludge and methods of their processing. M.: Publishing House of the "noosphere", (2001). 56 p
2. Gridina, M.S.: The influence of the components of oily waste on the quality of products hydrotreating hydrocarbon fractions: dis… cand. tehn. Sciences. Samara, (2010). 135 p

3. Silva, L.J., Alves, F.C.: F.P.F. A review of the technological solutions for the treatment of oily sludges from petroleum refineries. Waste Manag Res. **30**(10), 1016–1030 (2012)
4. Chang, C., Shie, J.L., Lin, J.P.: Major products obtained from the pyrolysis of oil sludge. Energ. Fuels **14**(6), 1176–1183 (2000)
5. Khaydarov, P.R., Kudratova, S.K.: Development of engineering and technology of waste oil disposal. Publishing house "Young scientist», **11**, 125–127 (2014)
6. Hayrudinov, I.R.: Methods of processing of oil waste and sludge. M.: Publishing house "Chemistry", (1989). 425 c
7. Mazlova, E.A., Meshcheryakov, S.V.: Ecological characteristics of oil sludges. Chem. Technol. Fuels Oils **35**(1), 49–53 (1999)
8. Krestovnikov, M.P., Snegotsky, A.L.: The process of degradation of organic compounds and petrochemicals processing plant waste. RF patent **2246525**, (2005)
9. Aristarkhov, D.V., Krestovnikov, M.P., Snegotsky, A.L.: Method for processing of rubber products and high. RF patent **2273650**, (2005)
10. Kurochkin, A.K., Kurochkin, A.V.: Thermal non-catalytic cracking of hydrocarbon oils, in the absence of hydrogen. RF patent **2375409**, (2009)
11. Vinokurov, V.A., Frolov, V., Krestovnikov, M.P.: The study of the thermal cracking of used lubricating oils when wave action. Refining and Petrochemicals **12**, 8–16 (2011)
12. Vinokurov, V.A., Frolov, V., Krestovnikov, M.P.: Investigation of the effect of wave action on oil. Refining and Petrochemicals **3**, 3–8 (2012)
13. Vinokurov, V.A., Frolov, V., Krestovnikov, M.P.: Properties of the thermal cracking of heavy petroleum feedstock in the electromagnetic field. Ind. service **4**, 2–5 (2012)
14. Vinokurov, V.A., Kolesnikov, I.M., Frolov, V.I.: Influence of electromagnetic radiation on the thermal cracking of the activated sludge. Chem. Techno. Fuels Oils **1**(593), 34–39 (2016)
15. Lyubimenko, V.A., Frolov, V.I., Krestovnikov, M.P.: Mathematical modeling of the thermal cracking of oil sludge, activated by electromagnetic radiation. Chem. Technol. Fuels Oils **2** (594), 12–16 (2016)
16. Minnigalimov, R.Z., Nafikova, R.A.: Improving the technology for processing oil sludge. Oil Ind. **4**, 105–107 (2008)
17. Yagafarova, G.G., Nasyrova, L.A., Shakhov, F.A.: Environmental engineering in the oil and gas industry: a textbook for graduate students and researchers studying the ecology. Ufa. Publishing house UGNTU, (2007). 334 p

Comprehensive Ground-Space Monitoring of Anthropogenic Impact on Russian Black Sea Coastal Water Areas

V.G. Bondur and V.V. Zamshin

Abstract In this paper, we describe the developed methods and technologies, as well as the created research prototype of a ground-space regional monitoring system that was used for comprehensive experimental research of anthropogenic impact on Russian Black sea coastal water areas. Changes in significant water environment parameters (generation of additional spectral components of surface waves; changes in marine surface roughness affecting normalized radar cross-section; turbidity field anomalies affecting spectral brightness variations in various bands of electromagnetic spectrum, etc.) registered in satellite imagery of water areas under anthropogenic impact were revealed. It has been established that these effects were predominantly caused by deep wastewater discharges. Zones of anthropogenic pollution propagations, as well as pipe breakages, have been revealed. The validation of the obtained satellite imagery processing results has been conducted based on sea truth data carried out using buoys, as well as from boats and hydrophysical platform.

Keywords Remote sensing of the earth · Satellite monitoring
Satellite imagery · In situ measurements · Coastal water areas · Wave spectra
Anthropogenic impact

Introduction

One of the most urgent problems of sustainable nature management is the prevention of environmental pollution, including sea and ocean water area pollution. Anthropogenic impact on coastal water area ecosystems is essential among such problems [2, 4, 9, 21]. The abovementioned problems are crucial for coastal water

V.G. Bondur (✉) · V.V. Zamshin
State Scientific Institution "Institute for Scientific Research of Aerospace Monitoring
"AEROCOSMOS", Moscow, Russia
e-mail: vgbondur@aerocosmos.info

V.V. Zamshin
e-mail: office@aerocosmos.info

© The Author(s) 2018
K.V. Anisimov et al. (eds.), *Proceedings of the Scientific-Practical Conference*
"Research and Development - 2016", https://doi.org/10.1007/978-3-319-62870-7_66

areas of Russia, including the Black Sea shelf. This is due to intensive recreation activity, housing development, the start of hydrocarbon production on the shelf, as well as to planned construction of terminals, cross-country pipelines, and communication lines [22, 26]. One of the most efficient methods to solve these problems is the application of satellite monitoring methods and technologies combined with local sea truth measurements [1, 2, 4]. Therewith, it is necessary to perform comprehensive ground-space monitoring using various types of data, such as satellite optical and radar imagery processing results, results from in situ measurements of current velocity profiles, temperature, pollutant concentrations, hydrooptical, and hydrobiological parameters, etc., [1, 2, 4, 15, 16, 19, 24, 29].

In this effort, we present the results of experimental study of the Russian Black Sea coast using the developed methods and technologies of comprehensive ground-space monitoring of coastal water areas, as well as using the created research prototype of a regional system for collection and processing satellite and in situ data to monitor anthropogenic and natural impacts on the water environment for providing environmental safety and decreasing anthropogenic load on marine ecosystems.

The research is supported by the Ministry of Education and Science of the Russian Federation (unique project Identifier: RFMEFI57714X0110).

The Methods and Technologies Used, and the Monitoring System Research Prototype

In this research, the following developed methods have been applied:

- Linear and nonlinear methods for retrieving surface wave slope and elevation spectra using satellite optical imagery, based on the synthesis of retrieving operators which are parametrized spatial frequency filters taking into account linear and nonlinear components of the function of brightness field modulation by disturbed sea surface [5, 6, 14].
- An operator building method retrieving sea wave slope and elevation spectra using satellite imagery spectra, based on the parametrization of these operators and their synthesis by means of numerical modeling using the wavenumber power function having parameters, which depend on wave azimuth [5].
- A method of direct assessment of sea wave spatial spectra using wave buoy arrays. This method is based on the wavelet transform allowing for studying nonstationary phenomena [6, 23].
- A method of sea surface multi-polarized radar imagery processing, when normalized radar cross-section (NRCS) is given as a sum of Bragg scattering-related polarized scattering and non-polarized scattering related with wave breaking [20].
- A method of comparing sea surface structure statistical characteristics obtained using in situ methods and satellite data obtained using sea wave dispersion relationship.

The developed technologies for collection and comprehensive processing of various satellites and in situ data generated during coastal water area monitoring were used during the research. These technologies provide collection and systematization of satellite and in situ data in terms of time series; thematic processing providing analysis of such data time series for determination of significant water environment parameters and for detection of anomalies of anthropogenic origin; validation and verification of determined water environment parameters registered by satellite imagery using sea truth data.

In this effort, the sea truth and its interpretation methods played a significant part, including the methods to measure wave spectra using a string wave meter array, ADCP data processing methods, methods of laser location and sea surface Doppler radiolocation from the Stationary oceanographic platform, etc., which are described in [6, 22, 26].

To realize the methods and technologies developed, the research prototype of a regional system of coastal water area ground-space monitoring (Research Prototype) has been created. The flowchart of this Research Prototype is given in Fig. 1.

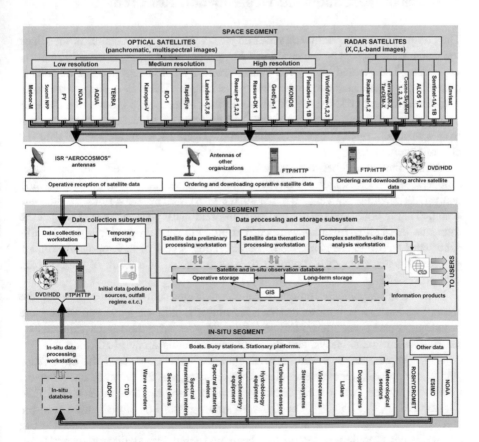

Fig. 1 The flowchart of the Ground-Space Monitoring Regional System Research Prototype developed

The system consists of:

- The Space segment, including various satellites, whose data are suitable for the monitoring system information support.
- The Ground segment, including data collection, processing, and storage subsystems serving for satellite data collection and analysis, data storage and handling, as well as for information product formation.
- The In Situ segment, including buoy stations, boats, stationary platforms, and additional ground information sources providing data on water environment conditions.

Individual system segments are interconnected by information communication, reception and transmission tools, which include antennas, FTP/HTTP channels, and DVD\HDD disks.

Features of the Conducted Experimental Research

The comprehensive experimental research was conducted at three sites of the Black sea water areas located near the cities of Sevastopol and Gelendzhik, and Katsively settlement. The studied sites were characterized by permanent sources of anthropogenic pollution, i.e., deep wastewater discharges. There is a deep wastewater outfall near the city of Sevastopol whose diffuser extends into the sea to a distance of ~ 3.3 km. This outfall discharges tens of millions of cubic meters of wastewater from a treatment plant per year. Similar anthropogenic sources are located near the city of Gelendzhik.

During the comprehensive experiments using the developed methods, technologies, and the Research Prototype, collection, systematization, processing, and analysis of satellite optical and radar imagery, as well as of various in situ data obtained using boats and buoys, and from the stationary platform (see Fig. 1) at the moments of satellite imaging were carried out. Information products with significant water environment parameters characterizing levels of anthropogenic impact on coastal water areas of the studied sites were created as a result of the obtained data processing.

Both anthropogenic pollutions escaping damaged outfall collectors near the shore, and random pollutions due to river run-offs, ship effluents, and other were investigated using satellite optical and radar imagery with the results validated by sea truth from boats, buoys, and the stationary oceanographic platform.

While implementing in situ experiments, we have considered the experience gained during comprehensive studies of water areas near the Oahu Island (Hawaii) influenced by deep wastewater discharges which were described in [2, 4, 7, 8, 11–13, 15, 17, 19, 24, 25, 29]. The most important feature of those studies was the revealed effect of generation and propagation of short-period internal waves causing surface wave modulation, which had been registered by satellite imagery [1, 9].

This was proved during the large-scale modeling in a hydrodynamic basin [10], as well as during various satellite and sea truth data processing [2, 4, 8]. In the framework of this effort, the abovementioned research has been continued and further developed, including due to the development of new methods and technologies, and application of the created research prototype of a ground-space monitoring system.

During the comprehensive monitoring of the Black sea coastal water areas near the cities of Sevastopol and Gelendzhik, and the Katsively settlement, the data were obtained using more than 20 types of equipment, including satellite optical panchromatic and multispectral systems aboard Resurs-P, WORLDVIEW; GEOEYE; Landsat 7,8, and other satellites, satellite synthesized aperture radars aboard TerraSAR-X, SENTINEL-1A, acoustic current velocity meters (ADCP), CTD, thermistor strings, turbidity gages, microstructure probes, etc.

To validate the developed remote sensing methods, the experimental research from the oceanographic platform near Katsively settlement was carried out.

Main Comprehensive Research Results

Performance assessment for the data obtained through satellite imagery processing has been carried out using sea truth data. For example, when assessing the adequacy of developed methods for wave spectra retrieval using satellite imagery spectra [5, 6, 14], sea surface slope and elevation spectra obtained through high-resolution satellite image processing were compared with wave spectra measured by a string wave meter array and through stereo photography from a hydrophysical platform [6].

Figure 2 shows the example results of comparing sea wave spectra retrieved from GEOEYE image fragment processing, obtained by a string wave meter array, and with Toba approximation [28].

As can be seen from Fig. 2, $\Psi(\omega)$ frequency spectra obtained using the developed remote methods [5, 6, 14] and those obtained in situ agree well. The degree of coincidence between these spectra obtained using $\Psi_{sat}(\omega)$ satellite and $\Psi_{cont}(\omega)$ in situ data is estimated by the measure of the deviation [6].

$$\Delta = \sqrt{\frac{1}{N}\sum_{n=1}^{N}\left(1 - \frac{\psi_{sat}(\omega_n)^2}{\psi_{cont}(\omega_n)^2}\right)^2}, \quad \text{which is } \Delta \approx 0.07$$

The Toba approximation is given in Fig. 2 [28]: $\Psi(\omega) = \alpha g u_* \omega^{-4}$, where $\omega = 2\pi/f$ is the cyclic frequency, u_* is the frictional velocity of the wind, α is the coefficient for two lines equal 0.06 and 0.11. The analysis of Fig. 2 shows good correlation between the measured frequency wave spectra and Toba approximation in the gravity wave equilibrium interval.

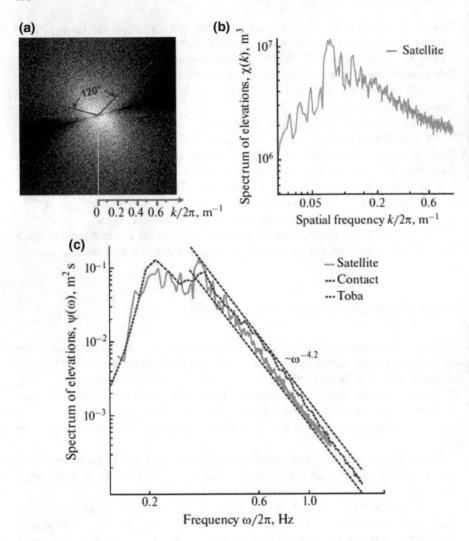

Fig. 2 Comparison of sea wave spectra retrieved from satellite and in situ data, and with the Toba approximation: **a** 2D spatial spectrum of sea surface slopes retrieved from a satellite image; **b** 1D spatial spectrum of elevations retrieved from satellite data; **c** superposition of the frequency spectra of the sea surface retrieved from a satellite image with a string wave meter array, and from the Toba approximation [28]

To detect surface manifestations of anthropogenic pollution caused by deep wastewater discharges, spatial spectral processing of high-resolution satellite optical and radar imagery was applied. The processing was based on the remote spatial frequency spectrometry method [1]. Such processing has allowed us to reveal the coastal water areas affected by short-period internal waves generated by deep wastewater discharges [1, 2, 4].

The example spatial spectral processing result for the fragments of GEOEYE optical high-resolution (0.5 m) image taken on September 10, 2015, for the coastal water area near Sevastopol where deep wastewater outfalls are placed, is given in Fig. 3a, b, c. The example spatial spectral processing result for the fragment of TerraSAR-X high-resolution (2.0 m) radar image taken on 6 October 2013 for the coastal water area near Gelendzhik where a deep wastewater outfall is placed, is given in Fig. 3d, e, f.

Figure 3a, b show the examples of 1 x 1 km^2 fragments of the processed optical satellite image and relevant 2D spatial spectra (left—the zones of deep outfall

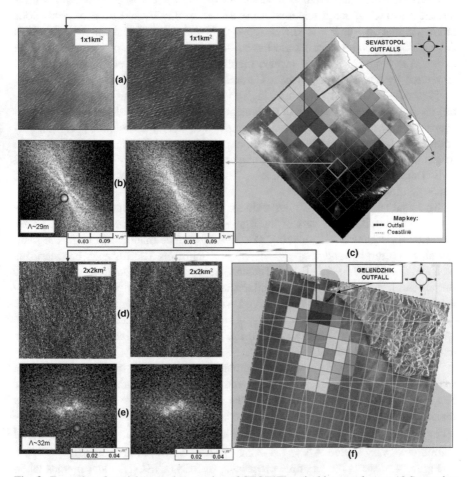

Fig. 3 Examples of spatial spectral processing of GEOEYE optical image taken on 10 September 2015 for the coastal water area near Sevastopol (a, b, c) and TerraSAR-X radar image taken on 6 October 2013 for the coastal water area near Gelendzhik (d, e, f): **a, b, d, e**—fragments of original satellite images and correspondent 2D spatial spectra (*left*—the zones of deep outfall surface manifestation, right—background); **c, f**—maps showing spatial distributions of deep outfall surface manifestations. "Quasimonochromatic" spectral maxima are seen in anomalous fragment spectra (relevant spatial period, $\Lambda \sim 30$ m)

surface manifestation near Sevastopol, right—background). As can be seen from the analysis of Fig. 3b, narrow symmetric spectral maxima (one is circled) is distinctly seen in the spectrum of the anomalous area. These spectral maxima correspond to certain periodic structures on the sea surface caused by propagating high-frequency internal waves generated by the deep outfall [1, 2, 4]. The spatial period of such structures is $\Lambda \sim 29$ m, and the width is $\Delta\Lambda = 2\text{--}3$ m. Thus, the condition of "quasimomochromaticity" $\Delta\Lambda \gg \Lambda$ is satisfied.

Figure 3d, e show the results of similar spatial spectral processing of TerraSAR-X radar image taken on 6 October 2013 for the coastal water area near Gelendzhik where an outfall is located. As for the optical satellite image, "quasi-monochromatic" spectral maxima are observed in the 2D spectrum of anomalous radar satellite image 2×2 km^2 fragment (in the area of a deep outfall). Such $\Delta\Lambda = 2\text{--}3$ m spectral maxima are the evidence of a periodic spatial structure with $\Lambda \sim 32$ m length. This is a surface manifestation of short-period internal waves generated by a wastewater jet near Gelendzhik.

Figure 3c, f show the maps of high-frequency internal wave surface manifestations in the vicinity of deep outfalls near Sevastopol and Gelendzhik. Processed fragments of optical and radar satellite images to which anomalous spectra correspond are marked with different colors depending on additional spectral harmonic intensities. Red color represents the highest level of anomaly manifestation, and light yellow represents the lowest one. The background is not painted. Red lines denote outfalls.

The analysis of Fig. 3c shows evident anthropogenic pollution caused by the main deep outfall (upper part of the figure) and by some lesser ones (lower) near Sevastopol (Herakleian Peninsula). The area of surface manifestations of the main outfall has almost round shape with ~ 5 km diameter. The distance between this area and the shore is $\sim 2.5\text{--}3$ km. The areas of surface manifestations of two lesser outfalls are merged. They have similar shapes with 1–2 km widths and are 0.5–4.5 km off shore. They can be observed within ~ 5 km along the shore.

The analysis of Fig. 3f shows that the surface anomaly detected through spatial spectral processing of satellite radar image processing near Gelendzhik outfall has an elongated shape with 14×18 km^2.

The processing of 152 image fragments with spectral maxima revealed in satellite imagery obtained in various moments during outfall monitoring near Sevastopol and Gelendzhik has shown that spatial periods corresponding to surface manifestations of internal waves generated by deep outfalls vary within $\Lambda \sim 20\text{--}80$ m ($\overline{\Lambda} \sim 35$ m is the average). The widths of "quasimonochromatic" spectral components vary within $\Delta\Lambda = 1.8\text{--}3.2$ m ($\overline{\Delta\Lambda} = 2.1$ m is the average).

Figure 4 presents the example processing of SENTINEL-1A radar images taken on 14 September and 2 October 2016 for the Gelendzhik coastal water area near the deep outfall. This figure shows the fragments of original radar satellite images (VV polarization) subjected to georeferencing and radio calibration (a,e), as well as the results of smoothing (median filter with 9 x 9-pixel scanning window) and brightness histogram equalization (b,f). Figure 4c, g present the results of thresholding and color-coding, and Fig. 4d, h present the maps of NRCS negative contrasts (2.5–4 dB) with pointed out outfall location near Gelendzhik.

Spectral informative features were applied during the processing multispectral satellite imagery [3, 13, 16]. To detect anomalies of water environment hydroop-tical characteristics attributed to outfall collector locations, spatial distributions of color index values were used [1, 3, 18]. Figure 5a shows the example fragments of time series for the processed satellite imagery obtained on 18 February, 10 and 25 May, 8 July, and 10 September 2015 by GEOEYE-1 and WorldView-2, 3 satellites for the coastal water area near Sevastopol.

Figure 5b presents an enlarged fragment of WorldView-3 (10 may 2015) processed image. There is a distinct optical anomaly adjacent from southeastward to the collector (in the area of its damage) in this image. Figure 5c shows a generalized sketch-map of all the anomalies which were registered in 2015-experiments. These anomalies were caused by wastewater escaping the damaged outfall collector. Phosphate concentration distributions measured by in situ sensors are also given in Fig. 5c. Increased phosphate concentration is observed in the near-surface water layer down to 5 m depth in the area of the collector's damage. Figure 5d show the results of sea truth boat measurements of light extinction index (LEI) at a wave-length of 370 nm (the 9-band sensor of directional light extinction index).

As we can see from the analysis of Fig. 5c, the detected outfall collector damage (denoted with an asterisk) is located at a distance of about 800 m offshore, whereas

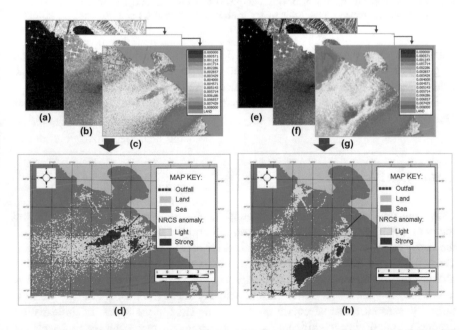

Fig. 4 Example processing (a, b, c, e, f, g) of SENTINEL-1A [27] radar images of Gelendzhik coastal water areas taken on 14 September 2016 (*left*) and 2 October 2016 (*right*), as well as distribution maps for anomalous NRCS values (d, h) due to deep outfall manifestations (d, h)

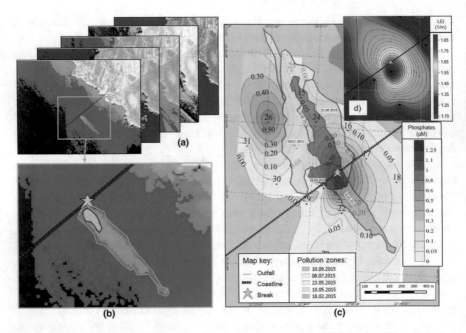

Fig. 5 Satellite optical multispectral imagery processing examples (18 February, 10 and 25 May, 8 July, and 10 September 2015, GEOEYE-1 and WorldView-2,3 satellites) for the coastal water area near Sevastopol (a, b). The map with superposed zones of anthropogenic impact due to outfall collector damage, created on the basis of sea truth hydrochemical (c) and hydrooptical (d) data

the outfall collector has the total length of 3.3 km. Phosphate concentration increase (Fig. 5c) and light extinction index increase from 1.15 to 1.9 1/m in the near-surface sea layer (Fig. 5d) are revealed in the vicinity of the damage.

Similar results were obtained by satellite and in situ observations near Gelendzhik.

Conclusion

Using the developed methods, technologies, and the Research Prototype of the regional ground-space monitoring system, the study of the anthropogenic impact on the ecosystems of the Black sea coastal water areas near the cities of Sevastopol, Gelendzhik, and Katsively settlement was carried out. During the comprehensive monitoring, more than 20 types of equipment were used, including satellite optical multispectral systems, acoustic current velocity meters, CTD, thermistor strings, turbidity sensors, microstructure probes, etc. Thirty-four types of information products quantifying significant water environment parameters and levels of anthropogenic impacts on the coastal water area ecosystems were generated based on the processing of large volume of data collected (more than 1 TB).

The comparison of collected remote and in situ data has verified the adequacy of developed remote sensing methods and methods for processing satellite imagery in various electromagnetic wave spectrum bands. For example, the measure of the difference between wave spectra retrieved using satellite imagery and those obtained by the string wave meter array is 0.07.

Spatial spectral processing of high-resolution (0.5–2.0 m) optical and radar images has allowed us to detect surface manifestations of anthropogenic pollution caused by deep outfalls near Sevastopol and Gelendzhik and to determine their dimensions (from 5 to 18 km). These manifestations can be detected by "quasi-monochromatic" spectral maxima (width, $\Delta\Lambda = 1.8–3.2$ m ($\overline{\Delta\Lambda} = 2.1$ m is the average) indicating the presence of quasiperiodic structures with spatial periods, $\Lambda \sim 20–80$ m ($\overline{\Lambda} \sim 35$ m is the average) on the sea surface caused by interaction of surface waves and short-period internal waves generated by deep outfalls.

Radar satellite VV polarization-imagery processing has enabled us to establish that anomalies of NRCS with prevalent negative contrasts (2.5–4 dB) appear periodically near the outfall collectors. These anomalies were caused by surface wave field transformation related with internal wave generation, as well as with direct impact of sleeks, i.e., smoothed stripes of floating-up wastewater jets.

On the basis of satellite multispectral imagery and sea truth data processing, we have detected the zones of intensive anthropogenic impacts related to deep wastewater discharges and damages of outfall collectors near Sevastopol and Gelendzhik. These deep wastewater discharges can be seen in satellite imagery in the form of anomalies having increased brightness and typical dimensions from 0.5 to 3 km, which are located at a distance of approximately 0.5 to 1.5 km from the shore. These results are confirmed by the data obtained during hydrochemical and hydrooptical measurements from aboard ships.

The analysis of the obtained results proves the efficiency of used satellite methods and technologies, as well as the efficiency of the created research prototype of a regional system of coastal water area ground-space monitoring.

Acknowledgement The Research is carried out with the financial support of the state represented by the Ministry of Education and Science of the Russian Federation. Agreement (contract) no. 14.577.21.0110 08. Sep 2014. Unique project Identifier: RFMEFI57714X0110.

References

1. Bondur, V.G.: Aerospace methods in modern oceanology New Ideas in Oceanology, vol. 1 Physics. Chemistry. Biology, pp. 55–117 + ill. Nauka, Moscow (2004)
2. Bondur, V.G.: Complex Satellite Monitoring of Coastal Water Areas 31st International Symposium on Remote Sensing of Environment. ISRSE, 7 p (2005)
3. Bondur, V.G.: Modern Approaches to Processing Large Hyperspectral and Multispectral Aerospace Data Flows. Izvestiya, Atmospheric and Oceanic Physics **50**(9), 840–852 (2014). doi:10.1134/S0001433814090060

4. Bondur, V.G.: Satellite monitoring and mathematical modelling of deep runoff turbulent jets in coastal water areas. In: Book Waste Water—Evaluation and Management, pp. 155–180. InTech, Croatia (2011). ISBN 978-953-307-233-3

5. Bondur, V.G., Dulov, V.A., Murynin, A.B., Ignatiev, V.Yu.: Retrieving sea-wave spectra using satellite-imagery spectra in a wide range of frequencies. Izvestia, Atmospheric and Oceanic Physics. **52**(6), 637–648 (2016)

6. Bondur, V.G., Dulov, V.A., Murynin, A.B., Yurovsky, Yu.Yu.: A study of sea-wave spectra in a wide wavelength range from satellite and in-situ data. Izvestia, Atmospheric and Oceanic Physics. **52**(9), 888–903 (2016). doi:10.1134/S0001433816090097

7. Bondur, V.G., Filatov, N.N.: Study of physical processes in coastal zone for detecting anthropogenic impact by means of remote sensing. Proceeding of the 7 Workshop on Physical processes in natural waters. 2–5 July 2003. Petrozavodsk, Russia, 98–103 2003

8. Bondur, V.G., Filatov, N.N., Grebenyuk, Yu.V., Dolotov, Yu.S., Zdorovennov, R.E., Petrov, M.P., Tsidilina, M.N.: Studies of hydrophysical processes during monitoring of the anthropogenic impact on coastal basins using the example of Mamala Bay of Oahu Island in Hawaii. Oceanology **47**(6), 769–787 (2007)

9. Bondur, V.G., Grebenuk, Y.V.: Remote indication of anthropogenic influence on marine environment caused by depth wastewater plume: Modeling, experiments. Issledovanie Zemli iz Kosmosa. **6**, 49–67 (2001)

10. Bondur, V.G., Grebenyuk, YuV, Ezhova, E.V., Kazakov, V.I., Sergeev, D.A., Soustova, I.A., Troitskaya, YuI: Surface manifestations of internal waves investigated by a subsurface buoyant jet: 1. The Mechanism of Internal-Wave Generation. Izvestiya, Atmospheric and Oceanic Physics **45**(6), 779–790 (2009)

11. Bondur, V.G., Grebenyuk, YuV, Sabinin, K.D.: The spectral characteristics and kinematics of short-period internal waves on the Hawaiian shelf. Izvestiya, Atmospheric and Oceanic Physics **45**(5), 598–607 (2009)

12. Bondur, V.G., Grebenyuk, Yu.V., Sabinin, K.D.: Variability of internal tides in the coastal water area of Oahu Island (Hawaii). Oceanology. **48**(5), 611–621 (2008)

13. Bondur, V.G., Keeler, R.N., Starchenkov, S.A., Rybakova, N.I.: Monitoring of the pollution of the ocean coastal water areas using space multispectral high resolution imagery. Issledovanie Zemli is Kosmosa **6**, 42–49 (2006)

14. Bondur, V.G., Murynin, A.B.: Methods for retrieval of sea wave spectra from aerospace image spectra. Izvestiya, Atmospheric and Oceanic Physics **52**(9), 877–887 (2016). doi:10.1134/S0001433816090085

15. Bondur ,V.G., Tsidilina, M.N. "Features of Formation of Remote Sensing and Sea truth Databases for The Monitoring of Anthropogenic Impact on Ecosystems of Coastal Water Areas." 31st International Symposium on Remote Sensing of Environment. ISRSE, 192–195 2005

16. Bondur, V.G., Vorobjev, V.E., Grebenjuk, Y.V., Sabinin, K.D., Serebryany, A.N.: Study of fields of currents and pollution of the coastal waters on the Gelendzhik Shelf of the Black Sea with space data. Izvestiya, Atmospheric and Oceanic Physics **49**(9), 886–896 (2013)

17. Bondur, V.G., Zhurbas, V.M., Grebenyuk, Yu.V.: Mathematical Modeling of Turbulent Jets of Deep-Water Sewage Discharge into Coastal Basins. Oceanology. **46**(6), 757–771 (2006)

18. Bondur, V.G., Zubkov, E.V. Showing up the small-scale ocean upper layer optical inhomogeneities by the multispectral space images with the high surface resolution. Part 1. The canals and channels drainage effects at the coastal zone. Issledovanie Zemli iz Kosmosa. **4**, 54–61 (2005)

19. Gibson, C.H., Keeler, R.N., Bondur, V.G., Leung, P.T., Prandke, H., Vithanage, D.: Submerged turbulence detection with optical satellites. In: Frouin, R.J., Lee, Z. (eds.) Proc. of SPIE, Coastal Remote Sensing, 1–8, Vol. 6680, 6680X, Aug. 26–27 (2007). doi:10.1117/12.732257

20. Hansen, M., Kudryavtsev, V., Chapron, B., Brekke, C., Johannessen, J.: Wave breaking in slicks: impacts on C-band quad-polarized SAR measurements. IEEE Geoscience and Remote Sensing Letters. (2015)

21. Israel, YuA, Tsyban, A.B.: Ocean anthropogenic ecology, p. 520. Flinta Nauka, Moscow (2009)
22. Ivanov, V.A., Katunina, E.V., Sovga, E.E.: Assessment of anthropogenic impacts on the ecosystem of the waters of the Herakleian peninsula in the vicinity of deep drains. Processes in GeoMedia 5(1), 62–68 (2016)
23. Leckler, F., Ardhuin, F., Benetazzo, A., Bergamasco, F., Peureux, C., Dulov, V.: Analysis and interpretation of frequency-wavenumber spectra of young wind waves. Journal of Physical Oceanography. American Meteorological Society (United States) 45(10), 2484–2496 (2015)
24. Keeler, R., Bondur, V., Gibson, C. Optical satellite imagery detection of internal wave effects from a submerged turbulent outfall in the stratified ocean. Geophysical Research Letters 32 (L12610), 1–5 (2005). doi:10.1029/2005GL022390
25. Keeler, R., Bondur, V., Vithanage, D.: Sea truth measurements for remote sensing of littoral water. Sea Technology,. 53–58 April, 2004
26. Kuklev, S.B., Zatsepin, A.G., Baranov, V.I., Ocherednik, V.V., Kukleva, O.N.: The results from monitoring of Gelendzhik marine dcep release outlet using Acoustic Doppler Current Profiler. Nauchny Almanakh. 2017 (in print)
27. Sentinels Scientific Data Hub [Electronic resource] URL: https://scihub.copernicus.eu/ (Assessed date: 2 Aug. 2016)./(Contains modified Copernicus Sentinel data 2016)
28. Toba, J.: Local balance in the air-sea boundary process. Oceanogr. Soc. Japan. 29, 209–225 (1973)
29. Vedernikov, V.I., Bondur, V.G., Vinogradov, M.E., Landry, M.R., Tsidilina, M.N.: Anthropogenic Influence on the Planktonic Community in the Basin of Mamala Bay (Oahu Island, Hawaii) Based on Field and Satellite Data. Oceanology 47(2), 221–237 (2007)

Determination of the Optimal Technological Conditions of Processing of the Alkali Alumosilicate

V.N. Brichkin, A.M. Gumenyuk, A.V. Panov and A.G. Suss

Abstract The future development of mineral raw material base in alumina's production is vary widely with resources development of low-grade alumina stock. It is actual for countries and regions with limit or stock out of traditional bauxite raw. Significant opportunities are presents the natural aluminosilicate as part of urtite, rischorrit, ijolite and the others alkali rocks and also the dump waste products of mine-mill consist of the alumosilicates incorporated overburden rocks and stocker's middlings. At the same time essential to the efficient use such materials is considerations of their chemical and mineral composition, mole ratio in calciferous-aluminosilicate dry mix, sintering temperature and other parameters that determine the recovery of valuable raw material components. The future development of raw material base for the production of alumina in the central part of Siberia is strongly associated with the development Goryachegorsk deposits of nepheline ores. Experimental results is allow to clarify chemical and mineralogical composition of sample of nepheline concentrate' obtained by Goryachegorsk field's ore beneficiation and to determine the optimal conditions for its processing with the extraction in an alkaline aluminate solution alumina and alkaline components. The obtained results establish the possibility of achieving valuable components extraction rates of more than 90%, which exceeds or matches the existing level for the same raw materials, and combined with the implementation of other resource-saving solutions can rely on high efficiency of the process.

V.N. Brichkin (✉) · A.M. Gumenyuk
Saint Petersburg Mining University, Saint Petersburg, Russia
e-mail: kafmetall@mail.ru

A.M. Gumenyuk
e-mail: kafmetall@mail.ru

A.V. Panov · A.G. Suss
United Company RUSAL's Engineering and Technology Centre, Moscow, Russia
e-mail: andrey.panov@rusal.com

A.G. Suss
e-mail: aleksandr.suss@rusal.com

© The Author(s) 2018
K.V. Anisimov et al. (eds.), *Proceedings of the Scientific-Practical Conference "Research and Development - 2016"*, https://doi.org/10.1007/978-3-319-62870-7_67

639

Keywords Alkali alumosilicate · Nepheline ores and concentrates
Alumina production · Lime-nepheline charge · Sintering · Modes
Performance · Quality · Experimental research

Introduction

Widely distributed in the earth's crust is the alkali aluminum silicates with a generalized stoichiometric formula (Na, K)$_2$O · Al$_2$O$_3$ · (2 ÷ 6) · SiO$_2$ characterized by the complexity of the material composition and varying content of the main components—aluminum oxide, silica and alkali liquor. A significant amount of alkali aluminum silicates found in plutonic rocks, that include nepheline syenite, rischorrites, ijolites, leucite, urtites, synnyrity, bolgarity and various feldspathic rocks. These rocks can be combined for the production of alumina ore and associated products [1, p. 121; 2, p. 95; 3, p. 33; 10, p. 167; 12, p. 653]. Virtually all mountain groups of these rocks are peculiar to large reserves and favorable mining conditions that allow for open cut mining, which causes a significant interest in these ores in countries with low arrearage of bauxite, such as Russia, USA, Canada, Venezuela, Mexico, Iran, Spain, Bulgaria and others.

Modern metallurgical complex of Russia produces about 40% of alumina from nepheline raw materials and a significant amount of associated products, including aluminum hydroxide non-metallurgical grades, Portland cement and building materials based on it, sand-lime brick, soda and potash products, mineral fertilizers, metal gallium and other materials. Basic characteristics of this production, including specific material flows are well known, and its high technical and economic indicators are a stimulus for the design and creation of new production facilities. The solution of the challenge of the national alumina short supply for primary aluminum invariably associated with the development of production on the basis of nepheline raw materials and its close substitute [2, p. 95; 3, p. 33; 10, p. 167]. The indispensable requirements for effective implementation of such plans are to solve the problems of the sum of long-term to ensure the production of raw materials, reduce energy costs and ensure an economically reasonable ratio of main and associated products, including products with high added value, and quality.

By now the basic tendencies and aspects development of raw material base for the production of alumina from high-silicon aluminum raw materials due to the identified resource urtits and rischorrits rocks of Khibiny's massif, involvement in the turnover of old rejects of apatite output and low-quality bauxite, the use of natural resources in Central and Eastern Siberia, the Table 1 [2, p. 95; 3, p. 33; 10, p. 167]. Major reserves in keeping the raw material base of aluminum production associated with the use of aluminum and ferroalloy dross foundry production, ashes from combined heat and power plants, waste water treatment and other raw materials of anthropogenic origin [1, p. 121].

There have been some progress in solution of the most painful for the processing technology of nepheline raw material issue associated to high consumption of fuel

Table 1 Comparative characteristics of the alkali aluminum silicates rocks, ores and concentrates

Material short text	Mass content of componentry, %						Molecular ratio of componentry		
	Al_2O_3	SiO_2	Fe_2O_3	CaO	Na_2O	K_2O	$\frac{Na_2O+K_2O}{Al_2O_3}$	$\frac{SiO_2}{Al_2O_3}$	$\frac{Fe_2O_3}{Al_2O_3}$
Kola nepheline concentrate	28.5	45.3	2.56	0.81	12.3	8.49	1.03	2.68	0.06
Kola urtite	21.06	42.0	5.4	6.1	10.4	5.3	1.09	3.39	0.163
Kola rischorrit	22.1	47.0	5.0	4.0	8.8	9.8	1.14	3.62	0.14
Ore Kiya-Shaltyrsk field	26.2	38.8	4.4	8	9.8	3	0.74	2.52	0.11
Ore Goryachegorsk deposits	23.6	42.7	7.7	9.5	8.2	1.5	0.64	3.08	0.208
Ore enrichment concentrate from Goryachegorsk field	27.4	43.5	2.2	8.4	10.1	1.7	0.674	2.70	0.05

the process and, consequently, the cost of it. Upon that economically feasible approaches, including technology of dry preparation of limestone-nepheline charge and the use of low-grade fuel technology on the basis of regional deposits of lignite and coal was estimated and implemented [9, p. 420; 11, p. 11].

Securing balance the production of primary and associated products which is associated with the utilization of nepheline (belite) sludge in the production of Portland cement not a simple question. This issue theoretical elaboration, made in the 80th and 90th years of the twentieth century, in recently has become a systematic study can count on its decision [7, p. 34].

Lastly the quality of the final product invariably associated with the possibility of obtaining an aluminum hydroxide and alumina not only meets the requirements of All Union State standard, but also superior in its characteristics the existing requirements. This creates a considerable reserve for the processing of low-grade and high-silicon raw material, the production of aluminum, high-purity and innovative products in-demand in high-tech industries. The real revolution in this area is associated with the development of the theory of synthesis of metastable solid solutions in the system $Na_2O–Al_2O_3–CaO–CO_2–H_2O$ and Usage for extensive purification aluminate solutions from inorganic impurities and production of special binding materials [4; 5; 6, p. 367].

The solution of all these problems is implemented as a reference to the use of existing ore base and prospective sources of raw materials, among which the most interesting is nepheline concentrate, obtained by enrichment of ore deposits Goryachegorsk currently central. Table 1. Represented data allow us to speak about sufficiently close chemical and modular compositions Kola nepheline concentrate and concentrate from ore beneficiation Goryachegorsk field that allows to count on obtaining satisfactory quality sintered. This is explained by arrangement of of data points sintered in the field of primary crystallization of dicalcium silicate in system $Na_2O \cdot Al_2O_3–Na_2O \cdot Fe_2O_3–2CaO \cdot SiO_2$ far beyond the field of low-melting compounds and compositions, which creates significant opportunity for the formation of an optimal phase composition and processing properties sintered [8, p. 102].

Experimental research indicators on extraction of Al_2O_3 and sum of alkali $(Na_2O + K_2O)$ in recalculation on Na_2O of flotation concentrate from Goryachegorsk deposit beneficiation of ore performed by the scheme that includes:

- fine crushing of raw materials and determining its chemical and mineralogical composition, including nepheline concentrate flotation, limestone Mozulsk mine, soda-potash mixture and circulating white slurry Achinsk alumina plant;
- calculation, preparation and chemical testing of homogeneous limestone-nepheline charge with a predetermined ratio, which is determined by the stoichiometry of the main processes proceeding during sintering treatment [8, p. 102];
- briquetting of charges in compliance with the constancy of the geometrical characteristics of briquettes, compacting pressure and the mass of the crude charge;

Table 2 Scheme of varying the chemical composition of the charge for the value of the alkaline and a limestone module

M_{lime}	M_{lime}/M_{al}		
	$M_{al} = 1.05$	$M_{al} = 1.1$	$M_{al} = 1.15$
1.9	1.9/1.05	1.9/1.1	1.9/1.15
2.0	2.0/1.05	2.0/1.1	2.0/1.15
2.1	2.1/1.05	2.1/1.1	2.1/1.15

- sintering of charges in compliance with the constant duration of temperature exposure, heat treatment and cooling of the sintered;
- leaching sintered in terms of modeling the production engineering process, along terms of preparation sintered, lixiviating, filtering and washing the slurry, and conducting chemical and phase analysis of nepheline sludge and aluminate solution.

Taking into account the ultimate value for the process evolution of sintered and further extraction of valuable components of the sintering temperature and the chemical composition of the charge [8, p. 102], it defined molar ratios, varying the scheme are given in Table 2:

$$M_{al} = Na_2O/(Al_2O_3 + Fe_2O_3 + SO_3) \text{ и } M_{lime} = CaO/ SiO_2,$$

where: M_{al}—alkaline unit charge; M_{lime}—limestone module; R_nO_m—the amount moles of the corresponding oxide in the charge.

This allowed to prepare the charge that fairly close in chemical composition to the compositions given in Table 2, and their actual component and modular composition is given in Table 3. The sintering batches was performed at a temperature isothermal exposure in 1260, 1280 and 1300 °C, the duration of which was 30 min.

In order to determine useful components extraction from the obtained sintered its lixiviating was performed under the following conditions:

- leaching temperature 75 ± 1 °C;
- time of lixiviating leaching 30 min;
- composition of soda-alkaline solution, g/dm^3: Na_2O_{tot}—53.4; Na_2O_k—40.3; Na_2O_{ang}—13.1; Al_2O_3—31.1; α_κ—2.1;
- partition size matches of the sintered 100% mesh size less 0,25 mm;
- correlation L: S = 2.7 or 18.5 grams of the sintered on 50 sm^3 solution;
- mixing was carried out with a propeller stirrer spinning speed of 230 rpm;
- after the leaching process slurry was filtered under vacuum and washed with hot water (about 12-fold amount relative to the weight of the slurry), after that samples to be taken for analysis and the remaining slurry was dried.

Sintered quantitative composition estimation demonstrates that the soluble phase as a solid solution and sodium aluminate ferrite (potassium) is contained in an

Table 3 Component and modular composition of charges on the basis of the flotation concentrate Goryachegorsk field

№№	Charge modules, settlement (actual)		Grade of concentration of components in the mixture, %				
	M_{al}	M_{lime}	Nepheline concentrate	Limestone	Soda-potash blend	White mud	Sum components
1	1.05 (1.02)	2.0 (2.03)	36.26	55.03	5.09	3.62	100
2	1.10 (1.13)	2.0 (2.04)	35.77	54.28	6.37	3.58	100
3	1.15 (1.16)	2.0 (2.04)	35.53	53.92	6.99	3.56	100
4	1.05 (1.01)	1.9 (1.94)	37.46	53.57	5.22	3.75	100
5	1.10 (1.12)	1.9 (1.93)	36.94	52.83	6.53	3.69	100
6	1.15 (1.16)	1.9 (1.94)	36.69	52.47	7.17	3.67	100
7	1.05 (1.03)	2.1 (2.14)	35.13	56.39	4.97	3.51	100
8	1.10 (1.10)	2.1 (2.13)	34.67	55.65	6.21	3.47	100
9	1.15 (1.17)	2.1 (2.06)	34.44	55.29	6.82	3.44	100

amount of $26.4 \pm 29.0\%$ as calcium orthosilicate submitted β-C_2S in amount 65.0 $\pm 60.0\%$. Contents of soda-lime silicate ($Na_2O \cdot CaO \cdot SiO_2$) is from 1.15 to 7.0%, and its presence has a negative influence on the extraction of alkali components in the solution, Fig. 1. All samples nepheline (belite) sludge was performed semi-quantitative X-ray analysis. Based on its results we can say that the basis of all the sludge is β-C_2S in an amount of about 90% with a small impurity phases α'-C_2S in the amount of $3 \pm 5\%$. This confirms the conclusion that loss of alkali lixiviating at a process is mainly linked to the presence of soda-lime silicate (NCS), and also partial decomposition α'-C_2S phases to form a number of X-ray amorphous products are not subject to quantify and lead to losses valuable components [2, p. 95; 8, p. 102; 10, p. 167].z

Whereupon the most favorable conditions for extracting alumina and alkalis in processing flotation concentrate from the ore beneficiation Goryachegorsk deposits are formed using a saturated batches, corresponding to the following actual values

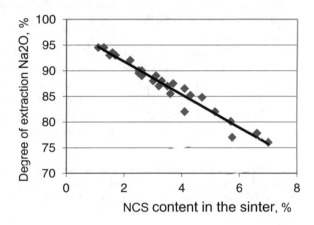

Fig. 1 The dependence of extraction of alkalis on the content of the soda-lime silicate (NCS) in the sinter

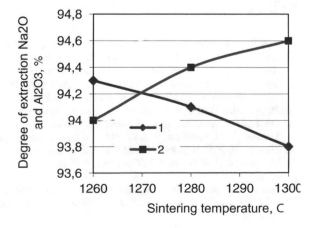

Fig. 2 Dependence of chemical extraction of Na_2O and Al_2O_3 in the aluminate solution on the sintering temperature of the limestone-nepheline charge: 1—Na_2O; 2—Al_2O_3

of charge units M_{al} = 1.01 and $M_{limestone}$ = 2.02, Fig. 2. Analysis of obtained results shows that for sintering concentrate of the Goryachegorsk ores most optimal are the following parameters of the process: sintering temperature—1280 ± 5 °C; $M_{limestone}$—1.9 ÷ 2.0; M_{al}—1.0 ± 0.05. With these parameters, the sintering process of recovery in the aluminate liquor (determined by solid phase) for in the lixiviating of Al_2O_3 = 94.4% and Na_2O = 94.1%, that is at the level of indicators of ore processing Chia-Shaltyrsk field.

Conclusions

1. It has been shown that the modular characteristics of limestone-based charge nepheline concentrate Goryachegorsk field and its sintering temperature within accepted ranges of variation are essential extracting alumina and alkali aluminate in solution.
2. It has been established experimentally that the indicators of the extraction of valuable components of nepheline concentrate, determined by the conditions of formation of phase composition sintered and its physical characteristics. It ensures the achievement of components extraction rates at the level of processing of ore base, which allows counting on the effective involvement in the production turnover of nepheline raw Goryachegorsk field.

Acknowledgements Research are carried out (conducted) with the financially support of the state represented by the Ministry of Education and Science of Russian Federation. Agreement (contract) no. 14.577.21.0208 02 nov.2015. Unique project Identifier: RFMEFI57715X0208.

References

1. Алюминиевое сырьё/ А.В., Акимова, О.С. Березнер, Н.В. Дудкин и др. // Государственный доклад «О состоянии и использовании минерально-сырьевых ресурсов Российской Федерации в 2011 году» . Центр «Минерал» ФГУНПП «Аэрогеология» , 2012. С. 121–129
2. Бричкин В.Н. Спекание известняково-нефелиновой шихты с добавкой рисчорритовых пород Хибинского массива / В.Н. Бричкин, М.В. Черкасова, А.М. Гуменюк // Вестник Иркутского Государственного технического университета, 2016. № 2. С. 95–99
3. Обогащение лежалых хвостов флотации апатит-нефелиновых руд / В.М. Сизяков, Ю.П. Назаров, В.Н. Бричкин, Е.В. Сизякова // Обогащение руд, 2016. №2. С. 33–40
4. Патент №.2560412, РФ. Способ обескремнивания алюминатных растворов / В.М. Сизяков, В.Н. Бричкин, Е.В. Сизякова, В.В. Васильев // Опубликовано: 20.08.2015 Бюл. № 23
5. Патент №.2560413, РФ. Способ глубокого обескремнивания алюминатных растворов / В.М. Сизяков, В.Н. Бричкин, Е.В. Сизякова, В.В. Васильев // Опубликовано: 20.08.2015 Бюл. № 23

6. Сизяков В.М. Модернизация технологии комплексной переработки Кольских нефелиновых концентратов на Пикалёвском глинозёмном комбинате // Цветные металлы—2010. Красноярск: ООО «Версо» , 2010. С. 367–378

7. Сизяков В.М. Повышение комплексности переработки нефелинового сырья на основе содовой конверсии белитового шлама / В.М. Сизяков, В.Н. Бричкин, Р.В. Куртенков // Обогащение руд, 2016. №1. С. 34–39

8. Сизяков В.М. Химико-технологические закономерности процессов спекания щелочных алюмосиликатов и гидрохимической переработки спеков // Записки Горного института, 2016. Т. 217. С. 102–112

9. Финин Д.В. Использование бурого угля на переделе спекания глиноземного комбината / Д.В. Финин, С.Н. Горбачев, М.А. Кравченя // Цветные металлы 2013. Сб. докладов 4го международного конгресса. Красноярск: «Версо» , 2013. С. 420–425

10. Черкасова М.В. Современные тенденции в переработке низкокачественного алюминиевого сырья и их влияние на развитие минерально-сырьевой базы производства глинозема / М.В. Черкасова, В.Н. Бричкин // Горный информационно-аналитический бюллетень (научно-технический журнал), 2015. №19. С. 167–172

11. Sine Bøgh Skaarup. Dry Sintering of Nepheline—A New More Energy Efficient Technology . Sine Bøgh Skaarup, Y.A. Gordeev, V.V. Volkov . Light Metals, 2014. pp. 111–116

12. Virginia T. McLemore. Nepheline Syenite. Industrial Minerals and Rocks, 2003. pp. 653–670

New Highly Efficient Dry Separation Technologies of Fine Materials

V.A. Arsentyev, A.M. Gerasimov, S.V. Dmitriev and A.O. Mezenin

Abstract During cleaning of high-ash coal mainly "wet" processes are used which require 5–10 tonnes water consumption per 1 tonne of coal. Arrangement of recycling water supply reduces demand in "fresh" water, but transportation of huge volumes of water slurry requires high-energy consumption. Dry cleaning of low-rank coal which has not been exposed to preliminary preparation is inefficient. It was suggested that to provide dry cleaning of high-ash coal it would be reasonable to expose it to chemical heat treatment first, and then to direct the treated coal mass for physical and mechanical cleaning to get the low-ash high-caloric product. It has been determined that in black coal exposed to medium temperature pyrolysis, as well as in brown coal, improvement of incombustible mineral fraction liberation is observed that facilitates further beneficiation with the use of a combination of high-intensity magnetic separation and triboelectrostatic separation. It has been determined that cleaned semicoke substantially exceeds both initial and cleaned coal by its qualities as a solid fuel, and tailings of semicoke dry cleaning can be utilised.

Keywords Low-rank coal · Dry cleaning · Dry processing · Semicoke
Magnetic separation · Triboelectrostatic separation

V.A. Arsentyev · A.M. Gerasimov · S.V. Dmitriev (✉) · A.O. Mezenin
Mekhanobr-Tekhnika Corp., Saint Petersburg, Russia
e-mail: dmitriev_sv@npk-mt.spb.ru

V.A. Arsentyev
e-mail: gornyi@mtspb.com

A.M. Gerasimov
e-mail: gerasimov_am@npk-mt.spb.ru

A.O. Mezenin
e-mail: mezenin_ao@npk-mt.spb.ru

© The Author(s) 2018
K.V. Anisimov et al. (eds.), *Proceedings of the Scientific-Practical Conference
"Research and Development - 2016"*, https://doi.org/10.1007/978-3-319-62870-7_68

Introduction

Mined coal is exposed to long processing flowsheet ending with its use as a fuel in the energy industry, metallurgy or chemical industry. There are several stages in this processing flowsheet at which the largest economic waste occurs and the environment is subjected to damage. For example, during coal cleaning mainly "wet" processes are used which require 5–10 tonnes water consumption per 1 tonne of coal. Arrangement of recycling water supply reduces demand in "fresh" water, but transportation of huge volumes of water slurry requires high-energy consumption.

Long-distance transportation of commercial coal is associated with expenses occurring due to movement of a relatively low-caloric product containing in addition from 15 to 25% of ballasting ash fraction.

It is a real disaster for the companies of the energy industry to store large volumes of coal combustion waste including ash and slag.

Considering the above processing flowsheet of coal usage after mining, one may pinpoint the following technical issues, a solution of which would significantly increase its efficiency:

- the transition from wet to dry coal cleaning at the places of coal mining;
- producing high caloric low-ash fuel during cleaning, which is suitable for use both in energy industry and metallurgy;
- separation of ash fraction of coal during its deep cleaning in the form which allows using it as raw material for commercial product manufacturing.

The accumulated experience shows that dry cleaning of high-ash low-rank coal which has not been exposed to preliminary preparation is inefficient [1, 2].

At the same time, there is a current steady trend in world practice to enhance coal beneficiation improving its qualities and coal product range expansion [3–5].

Analysis of previously performed studies shows [6–8] that on the basis of the task set, i.e. processing of coal without water use, the following processes are of the greatest interest:

- the Green Fields Coal Co. (USA) process, including deep drying and fine grinding of coal followed by its gravity separation in air cyclones collectors;
- the Convert Coal, Inc. (USA) process, including coal pyrolisis and separation of pyrrhotite by means of magnetic separation;
- the SynCoal (USA) process of the Rosehud SynCoal Partnership (USA) company, including coal pyrolysis and separation of ash by means of pneumatic separation;
- the Thermocoke process of the Sibtermo (RF) company including partial gasification of brown coal with the subsequent separation of ash by pneumatic separation.

It is worth mentioning that the main purpose of the above technological approaches was the production of water-free high caloric fuel on the basis of low-ash coal, mainly brown.

At the same time, testing of all the above technologies proved that during heat treatment of coal which corresponds to the *mode* of coal pyrolysis, physical and chemical transformations of coal mass take place which substantially influences its further processing, i.e. the following:

- the decrease of mechanical strength of coal due to moisture and volatile matters removal;
- exposure of particles of non-combustible mineral fraction along the boundaries of contact with the carbon part due to differences in physical and chemical properties resulting from heat impact;
- the increase of the calorific value of residual coal due to moisture and volatile matters removal;
- change of physical and chemical properties of ash forming minerals due to heat impact.

On the basis of the above, one may suggest that to provide for dry cleaning of high-ash coal, it is reasonable to expose it first to heat treatment, and then direct the treated coal mass for physical and mechanical cleaning to get the low-ash high-caloric product (coal char fuel).

This approach is shown in Fig. 1.

Implementation of the above chemical transformations provides new process potential in further processing of heat-treated coal, i.e. the following:

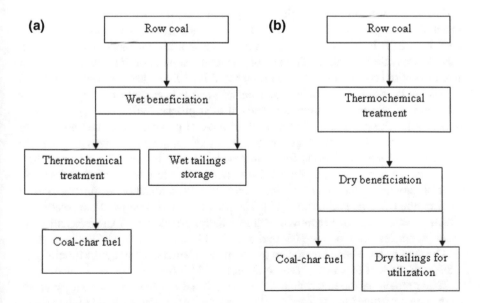

Fig. 1 High ash coal processing flowsheets: **a** Wet beneficiation; **b** Dry beneficiation

– the decrease of power consumption for crushing and grinding due to a decrease of mechanical strength;
– the possibility of deep removal of ash fraction from coal with the use of only dry cleaning processes;
– the possibility of agglomerated semicoke production;
– possibility to use burnt ash forming minerals fraction for the production of special binding agents, additives to concrete and construction materials.

Experimental

Analysis of the previous studies [6–8] shows that medium temperature pyrolysis at a temperature of 450–600 °C is the most reasonable method of heat modification of coal.
  Within this temperature range, the following process problems are solved:

– coal mass porosity increases and mechanical strength decreases, which provide good exposure of ash fraction at subsequent grinding;
– sulphide minerals being part of non-combustible fraction acquire higher magnetic susceptibility;
– clay minerals being part of non-combustible fraction lose crystal moisture, enlarge and lose the capacity to water regain;
– the generated carbon material—semicoke—has high calorific value;
– the volume of generated volatile fractions is sufficient for the provision of autothermal flow of medium temperature pyrolysis process.

  For the liberation of ash fractions in high-ash coal, it is usually required to grind material up to particle size less than 1 mm. To separate ash fractions with such particle size, it is possible to use pneumatic separation, magnetic separation and triboelectrostatic separation. The use of pneumatic separation for extraction of ash fraction out of coal has been studied rather well [9, 10] and has not been considered in this study. Studies on the use of magnetic and triboelectrostatic separation for extraction of ash fraction from coal proved their prospectivity [10–14]. The main difficulty for the processes of separation of mineral powders with the particle size less than 1 mm by their magnetic and electrical properties is provided by the availability of internal friction forces in powders which hinder effective separation of mineral particles. Studies proved that to overcome the internal friction forces in mineral powders, one may use the effect of vibrofuidization occurring during overlapping of certain vibrations [15]. The use of this effect allowed the creation of effective separators for separation of fine mineral powders by magnetic and electrical properties described in [16] and used in this study.
  The studies have been conducted on samples of hard coal having the humidity of 1.8%, ash content of 14.8%, devolatilisation of 34.7%.
  Experiments on the determination of the optimal temperature of coal pyrolysis have been performed in the standard vessel as per Fisher. On the basis of data given

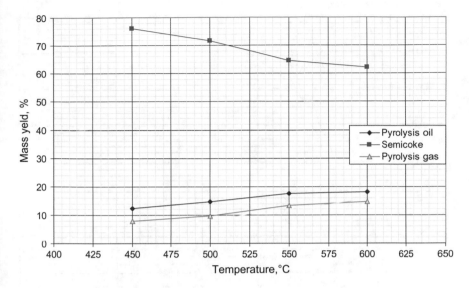

Fig. 2 The coal pyrolysis by Fisher method results

in Fig. 2, $t° = 550$ °C has been accepted as the optimal temperature of coal pyrolysis.

Semicoke sample for conducting process studies has been made at a laboratory pyrolysis unit having a 4 litres chamber, externally heated.

Pyroysis of the coal sample under test at a laboratory unit showed that at $t° = 550$ °C, semicoke yield makes 63.8%, the yield of pyrolysis oil is 11.4%, pyrolysis gas -20%, pyrogenetic water—4.7%, that is in good correspondence with the results received by Fisher's method.

Coal with particle size 10–20 mm has been treated. Samples of raw coal and semicoke were grinded in laboratory hammer grinder up to 0–2 mm particle size. The results of classification of grinded products by particle size and distribution of ash fraction by particle sizes are given in Fig. 3. The data in Fig. 3 shows that thermo chemical treatment of coal contributed to the substantial improvement of ash fractions liberation—recovery of ash fractions with grinding of semicoke to particle size 0–0.5 mm makes more than 90% if compared to 70% for raw coal.

Magnetic separation of raw coal and semicoke grinded up to particle size 0–0.5 mm was conducted in a laboratory magnetic separator in two steps—the first was at magnetic field induction of 0.35 T, the second—at 1.7 T.

Results of magnetic separation are given in Table 1 and show that during separation of coal exposed to heat treatment, recovery of ash fraction into low-intensity magnetic product increases almost threefold and recovery into high-intensity magnetic product increases twofold.

Triboelectrostatic separation of the non-magnetic product allows removal of about 30% ash fraction, that allows getting semicoke of the satisfactory quality (Tables 1 and 2).

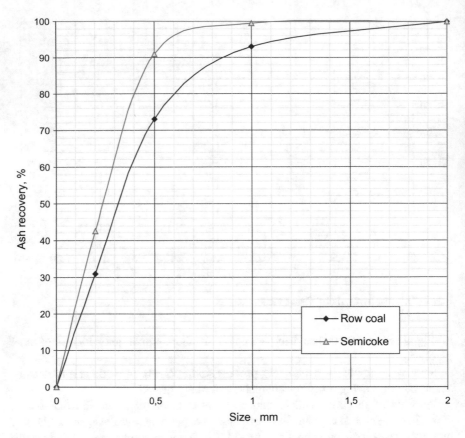

Fig. 3 Size and ash analysis of row coal and semicoke grinded to 0–2 mm size

Table 1 Results of raw coal and semicoke cleaning

Products	Raw coal			Semicoke		
	Mass yield (%)	Ash content (%)	Ash recovery (%)	Mass yield (%)	Ash content (%)	Ash recovery (%)
Low intensity	0.6	55.0	2.1	2.1	60.2	6.0
High intensity	3.1	67.2	13.3	9.2	63.4	27.8
Conductor	9.7	52.2	32.3	10.1	58.7	28.3
Non-conductor (concentrate)	86.6	9.4	52.1	78.6	10.1	37.9
Feed	100.0	15.6	100.0	150.0	21.0	100.0

Table 2 Comparative characteristic of studied products by ash specific yield

Parameters	Raw coal		Semicoke	
	Without cleaning	With cleaning	Without cleaning	With cleaning
Calorific capacity (MJ/kg)	12.55	15.06	23.72	25.94
Ash content (kg/kg)	0.158	0.100	0.195	0.100
Ash specific yield (kg/MJ·10^3)	12.6	6.6	8.2	3.9

Results Discussion

The results received to prove the practicability of using thermochemical modification of high-ash black coal to increase the efficiency of its dry cleaning with the use of physical and mechanical processes.

It has been determined that in black coal exposed to medium temperature pyrolysis, as well as in brown coal [8], improvement of incombustible mineral fraction liberation is observed that facilitates further beneficiation.

It has been proved that high-intensity magnetic separation and triboelectrostatic separation are effective methods of cleaning fine-grained high-ash coal. It has been determined that the efficiency of using these methods for non-combustible mineral fractionation increases after thermo chemical heat modification of high-ash hard coal. Combining high-intensity magnetic separation with triboelectrostatic separation increases the efficiency of high-ash coal and semicoke cleaning. To evaluate the efficiency of thermo chemical modification of coal and of cleaning both coal and products of its modification, it is reasonable to introduce such a parameter as ash specific yield per calorific capacity unit:

$$A_\mathrm{w} = \frac{A_\mathrm{c}}{W} \cdot 10^3, \tag{1}$$

where

A_w ash specific yield per calorific capacity unit, kg/MJ;
A_c ash content in product, kg/kg;
W product calorific capacity, MJ/kg.

Use of this parameter (Table 2) shows:

- combined dry cleaning of coal allows to decrease ash specific yield 1.5–1.7 fold;
- semicoke from high-ash coal without cleaning insignificantly exceeds initial coal by ash specific yield, i.e. 3.44 against 4.39;
- cleaned semicoke 2.5 times exceeds initial coal by ash specific yield.

Final tailing of semicoke dry cleaning containing more than 50% of combustible mineral fractions is raw material for the production of a binding agent for construction, i.e. cement analogue, which can be derived by means of their combustion without additional fuel and used for back filling of coal mines.

Conclusions

Cleaning high-ash coal without using water requires a new approach to the organisation of its conversion. To create conditions providing the possibility of dry cleaning of high-ash coal requiring its fine grinding for exposure non-combustible mineral fractions, it is reasonable to perform thermochemical modification of coal before cleaning. Medium temperature pyrolysis at $t° = 450–600$ °C is an effective method of modification of high-ash coal providing possibility of dry cleaning derived semicoke with the use of high-intensity magnetic separation and triboelectrostatic separation.

Acknowledgements Research being carried out with the financial support of the state represented by the Ministry of Education and Science of the Russian Federation. Agreement No. 14.579.21.0023. 05. Jun 14. Unique project Identifier: REMEF 157914X0023.

References

1. Maoming, F., et al.: Fine coal dry classification. Eur. J. Mineral Process. Environ. Prot. **3/2** (2003)
2. Soong, Y., Link, T.A., et al.: Dry beneficiation of Slovakian coal. Fuel Process. Technol. **72**, 185–198 (2001)
3. Dwari, R.K., Rao, H.K.: Dry beneficiation of coal—a review. Miner. Process. Extr. Metall. Rev. **28**(3), 177–234 (2007)
4. Katalambula, H., Gupta, R.: Low-grade coals: a review of some prospective upgrading technologies. Energy Fuels **23**(7), 3392–3405 (2009)
5. Domazetis, G., Barilla, P., James, B., Glaisher, R.: Treatments of low-rank coals for improved power generation and reduction in Greenhouse gas emissions. Fuel Proc. Technol. **89**(1), 68–76 (2008)
6. Skov, E.R., Neubauer, D.: Syncrude oil and upgraded syncoal production from mild temperature pyrolysis of LRC. International Freiberg Conference on IGCC and XTL Technologies, 9 May, Freiberg, Germany (2007)
7. Gong, X., Sh, Z.: Development and perspective of lignite modification technology. Proceedings on 17 International Coal Preparation Congress, Istanbul, Turkey. pp. 595–598 (2013)
8. Mikhalev, J., Islamov, S.: High efficiency and eco-friendly low-grade coal upgrade technology. Proceedings of 17 International Coal Preparation Congress, Istanbul, Turkey. pp. 707–709 (2003)
9. Sarunae, N., Ness, M., Bullinger, C., Mathews, J., Halleck, P.: A novel fluidized bed drying and density segregation process for upgrading low-rank coals. Int. J. Coal Prep. Util. **29**(6), 317–332 (2009)

10. Weinstein, R., Snoly, R., Oder, R.: Combining technology to make lignite into a premium fuel: using an integrated air and magnetic separation process. 18th Intetnational Low-Rank Fuels Symposium, June 24–26, Billings, Montana (2003)
11. Oder, R.: The Mag Mill: Innovation in dry coal cleaning technology enhancing environment compatibility and resource sustainability. Int. Symp. Clean Coal Technology, September 24–26, Taiwan, China (2012)
12. Turcaniova, L., et al.: The effect of microwave radiation on the triboelectrostatic separation of coal. Fuel **83**(14–15), 2075–2079 (2004)
13. Royaei, M., Joriani, E., Chehren, C.: Combination of microwave and ultrasonic irradiations as a pretreatment method to produce ultraclean coal. Int. J. Coal Prep. Util. **32**(3), 143–155 (2012)
14. Trigwell, S., Tennal, K.B., Mazumder, M.K., Lindquist, D.A.: Precombustion cleaning of coal by triboelectric separation of minerals. Part. Sci. Technol. **21**, 353–364 (2013)
15. Golovanevsky, V.A., et al.: Vibration-induced phenomena in bulk granular materials. Int. J. Miner. Process. **100**, 79–85 (2011)
16. Arsentyev, V.A., Dmitriev, S.V., Mezenin, A.O., Kotova, E.L.: Technology of recycling of fly ash at coal-fired power plants. Proceedings of 18th International Coal Preparation Congress, Saint-Petersburg, Russia (2016)

Hydrogenation Processing of Heavy Oil Wastes in the Presence of Highly Efficient Ultrafine Catalysts

A.E. Batov, Kh. M. Kadiev, M. Kh. Kadieva, A.L. Maximov and N.V. Oknina

Abstract The paper presents the results of studies aimed at obtaining experimental data for development of technological solutions to the production of marketable petrochemicals and petroleum products from heavy oil wastes using the process of hydroconversion and at designing engineering solutions in the area of preliminary treatment of heavy oil wastes for their further processing into marketable petrochemicals and petroleum products. The experiments have been conducted using a bench for heavy oil waste processing which combines pretreatment of oil wastes and subsequent hydroconversion processing in the presence of highly efficient ultrafine catalysts. Optimum conditions of oil sludge heavy residue hydroconversion (pressure, 7 MPa; temperature, 435 °C; feedstock space velocity, 1 h^{-1}; hydrogen: feedstock, 1000 nL/L; catalyst (Mo) content, 0.05 wt%; H_2O, 2 wt% (based on the feedstock) make it possible to achieve conversion of the 520 °C + fraction of feedstock of up to 67 wt% (per pass) or 90 wt% (based on the recycle stock).

Keywords Environmental control · Petroleum residues · Sludges
Heavy residue · Heavy oil feedstock · Hydrogen · Oil-Refining products
Hydroconversion · Ultrafine catalyst

A.E. Batov (✉) · Kh.M. Kadiev · M.Kh. Kadieva · A.L. Maximov · N.V. Oknina
"Elektrogorsk Institute of Oil Refining" Company, Moscow, Russia
e-mail: batov@ips.ac.ru

Kh.M. Kadiev
e-mail: kadiev@ips.ac.ru

M.Kh.Kadieva
e-mail: mkadieva@ips.ac.ru

A.L. Maximov
e-mail: max@ips.ac.ru

N.V. Oknina
e-mail: okninanv@rambler.ru

© The Author(s) 2018
K.V. Anisimov et al. (eds.), *Proceedings of the Scientific-Practical Conference*
"Research and Development - 2016", https://doi.org/10.1007/978-3-319-62870-7_69

Production, transportation, and refining of oil feedstock are inevitably associated with the appearance of diverse oil wastes and lead to considerable environmental pollution by marked amounts of heavy wastes such as sludge of various origin, vacuum residue, heavy residual fractions, still bottoms, residues in oil storage tanks, etc. [1–5]. These wastes are formed both in industrial controlled processes, such as oil refining from water, treatment of oil-containing effluents in treating facilities, during oil storage and transportation in various tanks and in emergencies with oil spillage.

Accumulation of oil wastes causes substantial pollution of the environment and involves the concentration of substantial environmental damage (it is believed that about 1 t of sludge is generated per 500 t of oil; this value is valid for developed countries). Disposal in sludge collectors, which are open earth capacities for sludge storage and which occupy vast surface areas, leads to alienation of agricultural lands and environmental pollution because of evaporation of petroleum products and their penetration into ground waters. Heavy aromatic hydrocarbons in sludges exhibit well-defined carcinogenic and mutagenic properties. The sludges and wastes are highly resistant to decomposition in the environment, and their components may be distributed over marked distances, being accumulated in animals, plants, soil, and water, thereby destroying the equilibrium of environmental systems and leading to the death of animals and plants and thus making environment unfit for life. Penetrating into the human body, these compounds accumulate in fat tissues and cause genetic mutations and keratosis of newborns. As a consequence, the neutralization and disposal of oil sludges is a pressing issue [5, 6].

Oil wastes are as rule heavy oil residues. The presence of water and solid impurities in them and the predominance of heavy asphaltic and resinous hydrocarbons and the products of physicochemical interaction of petroleum or petroleum products with oxygen, moisture, and mechanical impurities noticeably impede their qualified use [7–9].

Methods available for the disposal of oil sludges [10–18] do not often provide the necessary level of environmental protection against secondary pollutions and do not permit the efficient use of these resources. Modern approaches to the processing of oil wastes should envisage refusal not only of disposal (which is inapplicable from the point of view of environmental protection) but also of combustion. The processing of oil wastes should be directed at the recovery of the organic part of wastes and its subsequent processing. The most qualified method for processing of the organic part of heavy oil wastes is their chemical processing via hydroconversion [15]. This approach, on one hand, preserves the chemical potential of hydrocarbon components of heavy oil wastes and, on the other hand, makes it possible to transform carbonaceous components of oils wastes into motor fuels, thereby increasing the total depth of petroleum processing.

The authors of this paper propose a new approach to the processing of heavy oil wastes. The essence of this approach consists in the initial isolation of the organic part of oil wastes using a complex of pretreatment methods (extraction, filtration, distillation) and further hydroconversion processing of heavy hydrocarbon residue of oil wastes in the presence of highly efficient ultrafine catalysts that provide a high yield of distillate fractions.

For this purpose, a complex experimental bench for hydroconversion processing of heavy oil wastes was designed which allows pretreatment of oil wastes and their subsequent processing by the hydroconversion technology in the presence of highly efficient ultrafine catalysts.

The goal of this study is to obtain the experimental data for the development of technological solutions to the production of marketable petrochemicals and petroleum products from heavy oil wastes via the hydroconversion process and for designing engineering solutions in the area of pretreatment of heavy oil wastes for their further processing into marketable petrochemicals and petroleum products.

Experimental

Experiments on the processing of heavy oil wastes were performed on a complex experimental bench for the hydroconversion processing of heavy oil wastes (HPW) (Fig. 1). The bench is composed of an HPW pretreatment unit, an atmospheric–vacuum distillation unit, and a hydroconversion unit. The HPW pretreatment unit consists of an extractor, a settler, and a filter. The hydroconversion unit contains units of feedstock pretreatment and the introduction of additional components, a reactor unit, and a unit for separating target products.

Heavy residue of the organic part of oil tank sludge (HRS) and vacuum residue of West-Siberian oil refining was used as heavy oil wastes. HRS was obtained in the pretreatment unit of the complex experimental hydroconversion bench as described in [19].

Physicochemical properties of the feedstock and hydroconversion products were studied as described in [20].

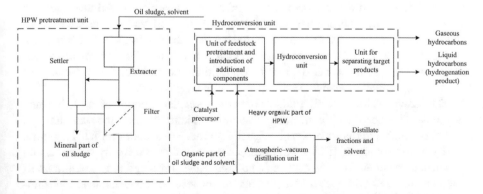

Fig. 1 Block flow diagram of complex experimental bench for hydroconversion processing of heavy oil residues

Table 1 Physicochemical properties of heavy oil wastes

Parameter	HRS	Vacuum residue
Density, kg/m³	968.9	1007.0
Coking behavior, %	12.25	15.1
Sulfur content, wt%	2.0	2.98
Fractional composition:	IBP 400 °C	IBP 400 °C
IBP-520 °C fraction.	27.8	9.25
520 °C + fraction	72.2	90.75
Group hydrocarbon composition:		
Paraffin-naphthene	47	14.7
Light aromatics	9.8	10.3
Medium aromatics	5.1	9.5
Heavy aromatics	18.8	34.3
Neutral resins	6.8	9.2
Acid resins	8.5	17.3
Asphaltenes	4.0	4.7

Ammonium paramolybdate $(NH_4)_6Mo_7O_{24} \cdot 4H_2O$ (APM) was used as a catalyst precursor. Before feeding in the reactor, the feed emulsion was preliminarily prepared, in which the dispersion medium and the dispersed phase were heavy oil waste and APM aqueous solution, respectively. The load of the APM aqueous solution with respect to feedstock was 0.05 wt% (based on Mo) and 2 wt% water. The efficiency of hydroconversion was estimated from the degree of conversion of feedstock fraction boiling above 520 °C (in what follows, 520 °C+) and from coke deposition on reactor walls (Table 1).

Discussion of Applied Results

Our experiments on the hydroconversion of vacuum residue showed that optimum conditions for the hydroconversion of West-Siberian oil vacuum residue are as follows: pressure in a reactor, 7 MPa; temperature in the reactor, 440 °C; hydrogen: feedstock, 1000 nL/L; feedstock space velocity, 1 h^{-1}; and the content of molybdenum in the reaction zone, 0.05 wt%. Under these conditions, the conversion of the 520 °C + fraction attains 55.5% at a coke yield of 0.05%.

The heavy residue of oil sludge contains a smaller amount of asphaltic and resinous hydrocarbons than the vacuum residue while the amount of paraffin-naphthenic hydrocarbons is higher than that in vacuum residue. Therefore, the hydroconversion of HRS should proceed more easily than the hydroconversion of vacuum residue. This fact made it possible to assume optimum conditions of vacuum residue hydroconversion (the maximum level of conversion at the minimum level of coking) as initial for the study of HRS hydroconversion.

Research into the effect of temperature on the hydroconversion of HRS (Table 2) revealed that the optimum conditions of HRS processing are as follows: a pressure of 7 MPa, a temperature of 435 °C, a feedstock space velocity of 1 h^{-1}, hydrogen: HRS, 1000 nL/L, and aqueous solution of ammonium paramolybdate as a catalyst precursor (Mo, 0.05 wt%; H_2O, 2 wt% based on the feedstock).

Under the optimum conditions of HRS hydroconversion, the conversion of the 520 °C + fraction is 67 wt%. In order to increase the depth of feedstock conversion, it is advisable to return heavy distillation residue (recycle stock) containing the 520 °C + fraction which was isolated from the hydrogenation product to the hydroconversion of vacuum residue. This trick makes it possible not only increase the final degree of conversion but also to reduce consumption of the fresh catalyst due to recycling. Conversions with the added recycle stock are calculated under the assumption that the 520 °C + fraction in the recycle stock constantly occurs in the system and enters into the 520 °C + fraction of the hydrogenation product.

The total conversion may be calculated via the following formula:

$$\eta = (C_0 - (C_1 - C_p))/C_0 * 100, \%$$

Table 2 Hydroconversion of heavy residue of the organic portion of oil sludge

Experiment		1	2	3	4
Variable parameter		Temperature, °C			
		430	435	440	445
Product yield, wt%					
Gas		1.62	1.81	1.90	2.21
Hydrogenation product		98.36	98.16	97.97	97.37
IBP-180 °C fraction		9.1	10.5	10.7	12.0
180–350 °C fraction		35.7	39.6	41.5	41.6
350–520 °C fraction		22.06	24.26	31.07	29.87
520 °C + fraction		31.5	23.8	14.7	13.8
Coke		0.02	0.03	0.13	0.42
Conversion of the 520 °C + fraction, %		56.4	67.0	79.6	80.9
Properties of products					
Hydrogenation product	ρ, kg/m^3	913	901	897	886
	S, %	1.60	1.57	1.50	1.43
IBP-180 °C fraction	S, %	0.40	0.38	0.36	0.35
	Iodine number g I_2/ 100 g	69.3	69.1	69.6	70.0
180–350 °C fraction	S, %	1.43	1.41	1.40	1.38
	Iodine number, g I_2/ 100 g	30.3	30.4	30.6	30.2
350–520 °C fraction	S,%	1.80	1.78	1.77	1.71

Feedstock: heavy residue of distillation of oil tank sludge organic part
Precursor: APM aqueous solution (Mo, 0.05%; H_2O, 2 wt% based on the feedstock)
Conditions: pressure, 7 MPa; volume velocity, 1 h^{-1}; hydrogen-containing gas: feedstock, 1000 nL/L

where C_0 is the amount of the 520 °C + fraction in the fresh feedstock (vacuum residue); C_p is the amount of the 520 °C + fraction; and C_1 is the amount of the 520 °C + fraction after hydroconversion.

The experimental studies, in this case, include the hydroconversion of HRS, atmospheric-vacuum distillation of the hydrogenation product, mixing of the obtained residue of hydrogenation product distillation, that is, the recycle stock containing the active catalyst, with vacuum residue, and introduction of the deficient amount of the precursor in the form of aqueous solution followed by mixture dispersing to obtain a molybdenum concentration in the reaction mixture of 0.05%. The properties of HRS obtained under the optimum conditions of hydroconversion are summarized in Table 3.

In the hydroconversion of HRS with the addition of the recycle stock, the conversion of the 520 °C + fraction was 90.2 wt% (Table 4) with a high yield of marketable distillate fractions.

Table 3 Composition and properties of recycle stock

Parameter	Value
Fractional composition	
IBP-180 °C fraction	0
180–350 °C	17.4
350–520 °C	37.6
520 °C+	45.0
Density, kg/m^3	0.985
Coking behavior, %	18.45
Insoluble in toluene, %	1.11
Sulfur content, wt%	2.38

Table 4 Hydroconversion of HRS with the recycle stock

Experiment		5	6	7	8	9
Time of experiment (regime time), h		24	10	10	10	10
Feedstock space velocity, h^{-1}		1	1	1	0,9	0,8
Taken: feedstock mixture						
including:						
HRS		98.0	68.6	58.8	49.0	49.0
Recycle stock	wt%	0	29.4	39.2	49.0	49.0
520 °C + fraction in the feedstock mixture	from HRS	70.8	49.5	42.5	35.4	35.4
	from recycle stock	0	13.2	17.6	22.1	22.1
	Total	70.8	62.8	60.1	57.4	57.4
Precursor		2.0	2.0	2.0	2.0	2.0

(continued)

Table 4 (continued)

Total:	100.0	100.0	100.0	100.0	100.0
Obtained:					
Hydrocarbon gas	1.79	1.69	1.55	1.91	1.97
Hydrogenation product	96.22	96.32	96.45	96.08	96.01
Density according to GOST 3900, kg/m^3	901	911	909	909	906
Coking behavior according to GOST 19932, %	13.06	14.10	14.11	14.73	14.80
Insoluble in toluene, %	0.68	1.11	1.15	1.18	1.20
Sulfur content, GOST 1437, wt%	1.57	1.60	1.52	1.41	1.39
including fractions:					
IBP-180 °C	10.6	10.1	10.3	10.8	10.8
180–350 °C	39.5	38.1	38.6	40.1	40.3
350–520 °C	21.12	21.72	21.75	18.98	19.61
520 °C+	25.0	26.4	25.8	26.2	25.3
including recycle stock	0	22.1	29.4	36.8	36.8
Densification products	0.03	0.03	0.04	0.05	0.06
Water	1.96	1.96	1.96	1.96	1.96
Total:	100.00	100.00	100.00	100.00	100.00
Conversion of the 520 °C + fraction per pass, %	64.7	57.9	57.1	54.4	55.9
Conversion of the 520 °C + fraction with account of recycle stock, %	64.7	73.4	80.8	88.3	90.8

Feedstock: Heavy residue of distillation of oil tank sludge organic part
Conditions: pressure, 7 MPa; T = 435 °C; hydrogen-containing gas: feedstock = 1000 nL/L.
Catalyst: MoS$_2$; Mo from recycle stock + Mo from catalyst precursor (APM aqueous solution; water, 2 wt%, [Mo$_{APM}$] = 0.05%−[Mo$_{precycle}$]% from recycle stock based on the taken feedstock

Conclusions

It has been shown that the optimum conditions of hydroconversion of oil sludge heavy residue (pressure, 7 MPa; temperature, 435 °C; feedstock space velocity, 1 h^{-1}; hydrogen: feedstock, 1000 nL/L; catalyst (Mo) content, 0.05 wt%; H$_2$O, 2 wt % (based on the taken feedstock)) make it possible to attain a conversion of the 520 °C+ fraction of the feedstock of up to 67 wt% (per pass) and up to 90 wt% (based on the recycle stock). The above experimental data may be used as a basis for the development of technological solutions to a new competitive method of heavy oil waste processing to marketable petrochemicals via hydroconversion conducted in the presence of highly efficient ultrafine catalysts.

Acknowledgments Research was supported by the Ministry of Education and Science of the Russian Federation. Agreement (contract) no. № 14.579.21.0052 22.09.2014. Unique Project Identifier: RFMEFI57914X0052.

References

1. Shperber, E.R.: Some kinds of wastes of oil refineries and their classification. Environ. Prot. Oil-And-Gas Indus. **2**, 27–32 (2011)
2. Hu, G., Li, J., Zeng, G.: Recent development in the treatment of oily sludge from petroleum industry: a review. J. Hazard. Mater. **261**, 470–490 (2013)
3. Elektorowicz, M., Habibi, S.: Sustainable waste management: recovery of fuels from petroleum sludge. Can. J. Civil. Eng. **32**, 164–169 (2005)
4. Shailubhai, K.: Treatment of petroleum industry oil sludge in soil. Trends Biotechnol. (1986)
5. Treatment of refinery wastes (oil sludge). Oil Gas J. **89**(l), 73–77 (1991)
6. Gron', V.A., Korostovenko, V.V., Shakhrai, S.G., Kaplichenko, N.M., Galaiko, A.V.: Problems of generation, processing, and disposal of oil sludges. Adv. Mod. Nat. Sci. **9**, 159–162 (2013)
7. Ulistkii, V.A.,Vasil'vistkii, A.E., Plushchevskii, M.B.: Industrial Wastes and Resource Saving. In: Sashko, M. (ed.) 368 pp (2006)
8. Mazlova, E.A., Meshcheryakov, S.V.: Oil sludge disposal and reclaiming problems. M.: Noosphere, 56 pp (2001)
9. Minigazimov, N.S., Pasvetalov, V.A.: Disposal and Treatment of Oil-Containing Wastes, p. 299. Ekologiya, Ufa (1999)
10. Shlepkina, Yu.S.: Analysis of methods for disposal of oil sludges. Advantages and drawback. Environ. prot. oil-and-gas ind. 32–34 (2009)
11. Velghe, I., Carleer, R., Yperman, J., Schreurs, S.: Study of the pyrolysis of sludge and sludge/disposal filter cake mix for the production of value added products. Bioresour. Technol. **134**, 1 (2013)
12. Shie, J., Lin, J., Chang, C., Wu, C., Lee, D., Chang, C., Chen, Y.: Oxidative thermal treatment of oil sludge at low heating rates. Energ. Fuels **18**, 1272 (2004)
13. Zheng, C., Wang, M., Wang, Y., Huang, Z.: Optimization of biosurfactant-mediated oil extraction from oil sludge. Bioresour. Technol. **110**, 338 (2012)
14. Je-Lueng, Shie, Jyh-Ping, Lin, Ching-Yuan, Chang, Shin-Min, Shih, Duu-Jong, Lee, Chao-Hsiung, Wu: Pyrolysis of oil sludge with additives of catalytic solid wastes. J. Anal. Appl. Pyrolysis **71**(2), 70–695 (2004)
15. Rocha, O., Dantas, R., Duarte, M., Silva, V.: Oil sludge treatment by photocatalysis applying black and white light. Chem. Eng. J. **157**, 80–85 (2010)
16. Xu, N., Wang, W., Han, P., Lu, X.: Effects of ultrasound on oily sludge deoiling. J. Hazard. Mater. **171**, 914–917 (2009)
17. Roldán-Carrillo, T., Castorena-Cortés, G., Zapata-Peñasco, I., Reyes-Avila, J., Olguín-Lora, P.: Aerobic biodegradation of sludge with high hydrocarbon content generated by a mexican natural gas processing facility. J. Environ. Manage. **95**, 93–98 (2012)
18. Kadiev, Kh., Dandaev, A.,Gyul'Maliev, A., Batov, A., Khadzhiev, S.: Hydroconversion of Polyethylene and tire rubber in a mixture with heavy oil residues. Solid Fuel Chem. **47**(2), 132–138 (2013)
19. Kadiev, Kh., Kadieva, M., Batov, A., Dandaev, A., Oknina, N., Maksimov, A.: Studies on preprocessing of reservoir oil sludges for further hydroconversion. Biosci. Biotechnol. Res. Asia **12** (Spl. Ed. 2), 473–483 (2015)
20. Oknina, N., Kadiev, Kh., Kadieva, M., Maksimov, A., Batov, A., Dandaev, A.: Physico-Chemical properties of oil sludges from reservoirs. Biosci. Biotechnol. Res. Asia **12** (Spl. Ed. 2), 497–505 (2015)

Development of Unified Import-Substituting Energy-Saving Technology for Purification of Roily Oils, Oil-Slimes, and Chemical and Petrochemical Effluents

V.V. Grigorov and G.V. Grigoriev

Abstract In the paper presented are the results of laboratory studies on roily oils purification from mechanical impurities and petroleum products using methods and devices designed at the JSC "SSC RF—IPPE." These studies were carried out within the framework of the Agreement on granting. These studies have demonstrated the effectiveness of nanostructured membranes for removal of mechanical impurities from water-oil emulsions and jet-film generator of air bubbles for floatation purification of emulsions from petroleum products. Nanostructured membranes are capable of completely removing mechanical impurities over 0.5 μm in size from model roily oils solutions resulting in neither impurities biofouling, nor pores plugging. Mechanical impurities are accumulated on the surface of the nanostructured membrane, and upon reaching their max permissible amount, regeneration of the membrane is carried out. As a result of regeneration, filtering element capacity is recovered up to 95–97% of its initial value. The possibility of production of finely divided air bubbles (less than 100 μm in size) by the jet-film generator was demonstrated and their high concentration in water was reached. Studies were carried out on the possibility of transport of finely divided air bubbles in the low speed (less than 2 m/min) water flow. The effectiveness of floatation purification method was tested under laboratory conditions by increasing amount of the air bubbles and decreasing their size. Taking into account the roily oils properties, the proposed methods of their purification seem quite promising.

Keywords Purification · Roily oils · Oil-Slimes · Chemical effluents Petrochemical effluents · Stratum water · Energy saving

V.V. Grigorov (✉) · G.V. Grigoriev
Joint-Stock Company "State Scientific Center of the Russian Federation—Institute for Physics and Power Engineering named after A.I. Leypunsky", Obninsk, Russia
e-mail: vgrigorov@ippe.ru

G.V. Grigoriev
e-mail: grig@ippe.ru

© The Author(s) 2018
K.V. Anisimov et al. (eds.), *Proceedings of the Scientific-Practical Conference "Research and Development - 2016"*, https://doi.org/10.1007/978-3-319-62870-7_70

Introduction

Oil-field development requires consumption of a considerable amount of natural or drain water for maintaining strata pressure in order to extend the period of the oil well blowing and significantly increase oil and gas recovery factors [1].

In particular, injection of water preliminarily treated with surface-active agents to the oil bed results in the decrease of the oil-water interfacial tension. This gives rise to fragmentation of the oil globules and formation of the low-viscous "oil in water" suspension, which needs lower pressure difference in the bore hole for its movement. Besides, oil-ground interfacial tension also decreases abruptly, thus facilitating oil displacement from pores and its sweep from the ground surface.

When the well products are withdrawn from the earth depths, the content of the impurities in the stratum water (which is emulsified) is quite low, namely: 10–20 mg/L. However, after the suspension is stratified into oil and water, the content of dispersed particles of oil and mechanical impurities in the separated water increases significantly: respectively, up to 4–5 g/L and 0.2 g/L.

The presence of oil drops and mechanical impurities in the effluent water results in the abrupt decrease of capacity of pay-out and intake beds. Therefore both pay-out and intake beds require additional treatment before the effluent water is supplied. It should be noted that supply of a large amount of foul water into the beds is accompanied by the clogging of pores, channels, and cracks, as well as a decrease of capacity of the injection wells, and, hence, the higher delivery pressure is required in formation pressure maintenance (FPM) system.

The basic impurities in water include oil, waterweeds, corrosion products, incrustation, sulfides, and bacteria with their products. If the water supplied to the bed is clean, then no clogging occurs (or it can be under active). Requirements to the stratum water used for FPM systems concern the following main three parameters: concentration of emulsified oil (petroleum products), size, and concentration of suspended solid particles (SSP), and its compatibility (in terms of both microbiology and chemistry) with reservoir rock. Permissible SSP size is a top-priority parameter.

The objective of studies is the development of the unified technological approaches based on jet-film separation of liquid–solid phases and new cermet membranes for setting up production of the globally competitive package for purification of roily oils, oil-slimes, and chemical and petrochemical effluents.

The Agreement on granting is implemented in stages. During 2015–2016 period, the following work has been carried out:

- analysis of scientific and technical publications, regulatory documents and other materials concerning the subject under study;
- justification of chosen areas of studies, and methods and tools of purification of roily oils, oil-slimes, and chemical and petrochemical effluents, including:

  - patent research;
  - comparative evaluation of the effectiveness of possible areas of studies;

- development of draft specification on liquid purification systems for petroleum refining, petrochemical, and chemical plants;
- creation of the prototype of purification system for roily oils, oil-slimes, and chemical and petrochemical effluents and preparation of the prototype for tests;
- development of test program and methodology;
- justification of the prototype design and model solution preparation procedures under laboratory conditions.

Overview of Existing Methods

In order to decrease fresh water consumption and stratum water utilization, effluent water is used in formation pressure maintenance (FPM) systems. Effluent water taken from the reservoir contains suspended material, and so it ranks among suspensions. Effluent water may contain oil drops and salts with up to 300 g/L or higher concentration. Particles of waterweeds, slime and iron compounds present in the supplied water would plug pore channels of pay-out bed, thus decreasing the capacity of the injection wells [2].

The principal method of suspension purification is based on gravity sedimentation of suspended matter. The effectiveness of sedimentation method depends on the rate of deposition of suspended particles, which determines required process time and capacity of depositing facilities. Suspended particles deposition rate increases with the decrease of suspension viscosity. This method is sufficiently effective, however, the procedure of fine purification of suspension from suspended particles is rather time-consuming.

Filtering effectiveness depends on the concentration of suspended particles in the flow, the size of particles [3], physical and chemical properties (density and viscosity) of water to be purified and chosen filtering method.

For the purpose of water fine purification from mechanical impurities the following equipment can be used [3]:

- vacuum filters, in which working pressure drop is provided by maintaining vacuum downstream filtering membrane;
- pressure filters, in which working pressure drop is provided by increasing pressure upstream filtering membrane.

Working pressure drop in vacuum filters is within 1 atm, while any pressure drop value can be provided in pressure filters and, therefore, such filters assure a high rate of mechanical impurities removal from the liquid.

Drum and disc vacuum filters designs are more preferable for trapping 150–300 μm particles by the membrane. As regards pressure filters, plate-and-frame filter-presses are used most often, which provide high filtering rate owing to high working pressure (up to 10 atm). The significant drawback of these filters is non-continuity of operation and considerable labor cost related to their dismantling and sediment unloading.

Activated macroporous carbon and carbon-mineral sorbents (DAK, BKZ, MIU coals) and other sorbing agents are used to decrease petroleum products content in the effluent water down to 0.5–0.05 mg/L values. однако, это дорогостоящий и дефицитный сорбент However this sorbent is expensive and scarce [4, 5]. Among other sorbents used for oil products removal from effluent water, silicon dioxides and aluminum silicates are most promising and available [6]. However, natural sorbents have low inherent oil absorption capacity (oil capacity), which can be increased by treating their surface with water-repelling compounds of various origin.

Results of analysis of engineering performance standard and trends of development of the research subject, as well as overview of patent and license context by the time of completion of patent information retrieval have proved the novelty of proposed unified import-substituting and energy-saving technology for purification of roily oils, oil-slimes, and chemical and petrochemical effluents, and therefore the patentability of this technology.

Outline of Studies Fulfilled

Unified technology based on jet-film separation of liquid and solid phases and new types of cermet membranes proposed within the framework of the Agreement for setting up production of the globally competitive package for purification of roily oils, oil-slimes, and chemical and petrochemical effluents will include units having the following functions:

– reduction of suspended solid particles content from 800 down to 10 mg/L;
– reduction of petroleum products content from 550 down to 10 mg/L.

Filtering elements with nanostructured membranes (see Fig. 1) will be used for water-oil emulsion purification from suspended solid particles. The advantages of these elements as compared to the other designs [7] include: (i) complete removal of suspended particles from water and achievement of ideal transparency of water downstream the membrane; (ii) high filtering fineness in terms of particles larger than 0.5 μm; (iii) peak specific permeability of membranes owing to extra well-developed nano-slot porosity; and (iv) possibility of multiple "sediment accumulation—discharge" cycling because of specific anti-adhesion properties of the membrane to sediment and assurance of long lifetime of the membrane.

The possibility of multiple accumulations of sediment and its further discharge from the surface of the membrane without filter dismantling is the key advantage of filtering element with nanostructured membrane [8]. Separated matter stays on the surface of the membrane and owing to its low adhesion to the membrane can be easily removed by the inverse hydro-impulse provided by the purified water.

In order to assure the possibility of using these filtering elements with nanostructured membranes for purification of water-oil emulsions, some modifications of filter design are required, including: (i) adoption of new structural material of nanostructured membrane assuring low adhesion to the sediment, preventing

Fig. 1 Design of filtering
element with nanostructured
membrane

biofouling and pores plugging; (ii) development of optimal technology of nanostructured membrane formation; and (iii) development of the new filtering element design taking into account specific features of oil-and-gas industry.

Laboratory tests of filtering elements with nanostructured membranes proved their effectiveness in the purification of water-oil emulsions from mechanical impurities. In order to determine hydraulic characteristic of water-oil emulsion membrane filtering unit, measurements of pressure drop as a function of liquid flow rate were made during tests: $\Delta P = f(q)$, where $\Delta P = P1 - P2$, $P1$—pressure at the inlet of membrane filter, $P2$—pressure at the filter outlet, q—flow rate. Characteristic $\Delta P = f(q)$ can be presented as a linear function $\Delta P = \alpha + \beta \cdot q$ with q as an argument. Results of measurements are given in Table 1 and plotted in Fig. 2. Based on this data, the appropriate surface area of filtering membrane was evaluated to meet the requirement on filter capacity according to the Agreement.

The effectiveness of filter in removing mechanical impurities (particles of over 5 mμ size) from water was determined using PAMAS device. Data on filtering effectiveness is presented in Table 2, and water samples are shown in Fig. 3.

The new jet-film air bubble generator design will be developed for water-oil emulsion purification from petroleum products. This generator is capable of producing air bubbles of about 100 μm in water thus providing larger contact surface

Table 1 Pressure drop ΔP on the mockup as a function of liquid flow rate q

No.	Flow rate, q, L/min	Inlet pressure, P1, atm	Outlet pressure, P2, atm	Pressure drop, $\Delta P = P1 - P2$, atm
1	6.4	0.5	0.2	0.3
2	7.31	1.0	0.5	0.5
3	9.06	1.5	0.7	0.8
4	10.88	2.0	1.0	1.0
5	11.12	2.5	1.3	1.2

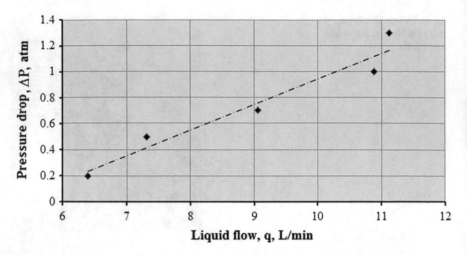

Fig. 2 Calibration curve showing pressure drop ΔP on the membrane mockup as a function of distilled water flow rate q

Table 2 Characteristics of water purification by membrane filtering element

	0.5–1 μm	5–10 μm	20–50 μm	100 μm	200 μm
Number of particles before purification	1,900,000–3,000,000	100,000–350,000	10,000–30,000	5000–10,000	1000–2000
Number of particles after purification	0–200	0	0	0	0

Fig. 3 Water samples taken before (1) and after (2) purification

and adsorption of petroleum products. This generator design would assure max water saturation with air and high effectiveness of petroleum products removal in the small size unit. Its operation is characterized by the following important technical advantages:

– increase of surface and time of air contact with water-oilemulsion and, hence, an increase of rate and depth of petroleum products removal; the possibility of automatic control of required air amount;
– possibility of automatic control of required air amount;
– reliability of operation, energy saving, and low metal consumption.

The results of laboratory tests of the jet-film air bubble generator are as follows:

– there is a sufficiently large number of the air bubbles of about 100 μm size in water;
– rate of the air bubbles drawing out from water (rate of shift of bubblessaturation zone boundary) is within the range of (0.3–0.9) m/h;
– in the case of the sectional arrangement of clarification tanks, over 90% petroleum products are removed in the first and the second sections, this corresponding to 25 min water-oil emulsion settling time as compared to 1 h settling time if no jet-film generator is used.

Photographs illustrating various stages of laboratory tests of bubbles generator are presented in Figs. 4, 5 and 6.

Fig. 4 Formation of petroleum products foam above jet-film bubble generator

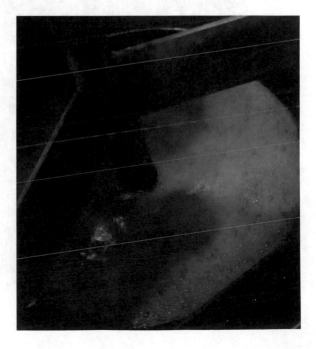

Fig. 5 Petroleum products emergence in the first and the second sections of clarification tank

(1)   **(2)**

Fig. 6 Water-oil emulsion before (**1**) and after (**2**) floatation

Conclusions

Based on the results of laboratory studies carried out by now within the framework of the Agreement on granting, the following conclusions can be drawn:

- basic drawbacks of the systems for purification of roily oils, oil-slimes, and chemical and petrochemical effluents include low performance and high weight-size parameters. In this view it is necessary to develop new technology of purification of roily oils, oil-slimes, and chemical and petrochemical effluents;
- application of nanostructured membranes is an advanced approach to removal of mechanical impurities (particles with size over 0.5 μm) from water–oil emulsion. Neither biofouling nor plugging of pores takes place during operation;
- use of jet-film bubbles generator results in significant reduction of the time required for petroleum products removal from water–oil emulsion.

In the course of further work to be carried out at the JSC "SSC RF—IPPE" it is planned to design and manufacture mock-ups of the devices for purification of roily oils from mechanical impurities and petroleum products, and to use them for complete justification of design and technological approaches and long-term performance of the equipment.

Acknowledgements Research work is carried out with the financial support from the State represented by the Ministry of Education and Science of the Russian Federation. Agreement No. 14.579.21.0120 05, Nov. 2015. Unique project Identifier: RFMEFI57915X0120.

References

1. Korshak, A.A., Shammazov, A.M.: Basic Principles of Oil and Gas Engineering. College textbook, LLC "Design Polygraph Service", Ufa (2001)
2. Vardanyan, M.A.: Afterpurification of oily effluent water by sorption method on expanded perlite and development of technology, Ph.D. thesis in Engineering Science, Erevan (2001)
3. Baker, Richard W.: Membrane Technology and Applications, 2nd edn. Membrane technology and research, Inc., Menlo Park, California (2000)
4. Martynov, P.N., Yagodkin, I.V., Askhadullin, R.Sh., Mel'nikov, V.P., Skvortsov, S.S., Posazhennikov, A.M., Grigoriev, G.V. Grigorov, V.V.: New category of nanostructured filtering materials in technologies of purification of liquid fluids in NPP with VVER. Basic Eng. Ind. **4**: 7–12 (2010)
5. Buzaeva, M.V.: Improvement of quality of effluent water purification from petroleum products. In: Proceedings of Samara Science Center of the Russian Academy of Sciences, vol. 2 (2005)
6. Sirotkina, E.E., Novoselova, L.Yu.: Materials for adsorption purification of water from oil and petroleum products. Chem. Benefit Sustainable Dev. **13** (2005)
7. Martynov, P.N., Papovyants, A.K., Mel'nikov, V.P., Grishin, A.G., Grigorov, V.V.: Preliminary purification of suspension-carrying liquids by cross-flow filtration. Res. J. Pharm. Biol. Chem. Sci. (2015)
8. Rachkov, V.I. Martynov, P.N., Grigorov, V.V., Denisova, N.A., Loginov, N.I., Melnikov, V. P., Miheev, A.S., Portnyanoy, A.G., Serdun' E.N., Sorokin, A.P., Storogenko, A.N., Ulyanov, V.V., Yagodkin, I.V.: Innovative technologies, developed in SSC RF—IPPE, News of higher educational institutions. Nucl. Power **1** (2014)

Development of Remote and Contact Techniques for Monitoring the Atmospheric Composition, Structure, and Dynamics

B.D. Belan, Yu. S. Balin, V.A. Banakh, V.V. Belov, V.S. Kozlov,
A.V. Nevzorov, S.L. Odintsov, M.V. Panchenko and
O.A. Romanovskii

Abstract Prototypes of a typical automated station for monitoring the atmospheric composition and state, as well as lidars and sodars, have been designed and manufactured within the project, which are to be the components of monitoring stations. Experimental tests of the instruments confirmed their designed-in specifications and functions. For contact measurements of air parameters, a complex air monitoring station and an aerosol multiwave diffusion spectrometer have been also designed and manufactured in accordance with technical assignment requirements. The measurement results have shown a continuous increase in the concentrations of

B.D. Belan (✉) · Yu.S. Balin · V.A. Banakh · V.V. Belov · V S. Kozlov · A.V. Nevzorov ·
S.L. Odintsov · M.V. Panchenko · O.A. Romanovskii (✉)
V.E. Zuev Institute of Atmospheric Optics SB RAS, Tomsk, Russia
e-mail: bbd@iao.ru

O.A. Romanovskii
e-mail: roa@iao.ru

Yu.S. Balin
e-mail: balin@iao.ru

V.A. Banakh
e-mail: banakh@iao.ru

V.V. Belov
e-mail: belov@iao.ru

V.S. Kozlov
e-mail: vkozlov@iao.ru

A.V. Nevzorov
e-mail: nevzorov@iao.ru

S.L. Odintsov
e-mail: balin@iao.ru

M.V. Panchenko
e-mail: pmv@iao.ru

K.V. Anisimov et al. (eds.), *Proceedings of the Scientific-Practical Conference*
"Research and Development - 2016", https://doi.org/10.1007/978-3-319-62870-7_71

carbon dioxide and nitrous oxide, as well as a renewed increase in the methane concentration in the layer from 0 to 7 km.

Keywords Atmosphere · Greenhouse gases · Aerosols · Laser sensing Lidar · Continuous in-situ measurements · Monitoring · Aircraft laboratory Satellite sounding · Sodar

Introduction

There is no proper network for complex air monitoring in Russia today, which covers the whole territory and answer modern requirements. The existing network of the Russian Hydrometeorological Service (Roshydromet) provides only hydrometeorological information and data on urban air pollution. However, these data alone are insufficient in the modern context, even despite the Roshydromet network density. Monitoring of the atmospheric composition, structure, and dynamics does not require such a dense network; however, the requirements for the equipment of the network stations are much higher. Thus, the new-generation Integrated Carbon Observation System (ICOS) [1] was recently created in Europe. This system was initially planned to be used for monitoring of only greenhouse gases on the basis of complex gradient measurements at tall masts (towers) and air sampling for the analysis with the usage of light airplanes. However, while developing the system, it became clear that this is insufficient. Since the atmosphere is a global chemical reactor and permanently interacts with the underlying surface, much more atmospheric admixtures are required to be measured, as well as their fluxes from the Earth's and ocean surfaces by both local and remote monitoring means.

All the works scheduled within the project can be divided into three directions.

1. Development and creation of new techniques and devices and improvement of processing algorithms.

  1.1. Design of a typical automated station for monitoring the atmospheric composition and state.
  1.2. Design and manufacturing of a diffusion spectrometer.
  1.3. Design of scanning polarization lidar "LOSA-M3".
  1.4. Design of the prototype of "Aerosol-3" lidar.
  1.5. Design of the prototype of "ST Ozon" lidar.
  1.6. Design of a multifrequency three-channel sodar.
  1.7. Development of algorithms for estimation of the wind speed and direction from lidar data.
  1.8. Development of algorithms for atmospheric correction of satellite images.

2. Metrological examination of the devices designed.
3. Continuation of the monitoring at existing stations and setups.

The aim of the project is the development of new techniques for monitoring the atmospheric composition, structure, and dynamics and design of new instruments, their metrological examination, and continuous measurements of currently operating systems for extension of long-term observation series.

Typical Up-to-Date Automated Station for Monitoring the Atmospheric Composition and State

The prototype of a monitoring station of atmospheric composition and state has been designed and manufactured. Its block diagram is shown in Fig. 1. The prototype consists of the following main parts: Gas-analysis unit; aerosol unit; actinometric unit; airlines of communications; meteorological unit; and interface, control, and data processing unit.

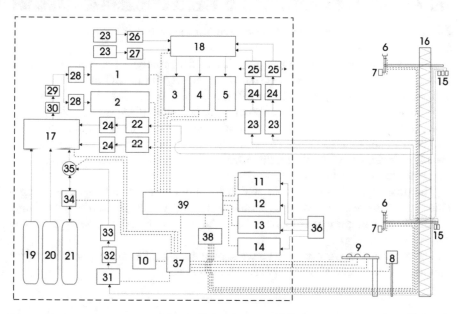

Fig. 1 Block diagram of a typical up-to-date atmospheric composition and state monitoring station: $CO_2/CH_4/H_2O$, CO, SO_2, NO_x, and O_3 gas analyzers (*1, 2, 3, 4,* and *5,* respectively); ultrasonic anemometers (*6*); temperature and humidity sensors inside a radiation protective case (*7*); automated precipitation gauge at a rod (*8*); pyranometers at actinometric holder (*9*); atmospheric pressure sensor (*10*); aerosol diffusion spectrometer (*11*); optical aerosol spectrometer (*12*); nephelometer (*13*); aethalometer (*14*); air samples (*15*); truss mast (*16*); airflow distributors (*17, 18*); tanks filled with calibration gas mixtures (CGM) (*19, 20*); tank filled with a reference gas mixture (RGM) (*21*); pumps (*22, 23*); water eliminators (*24*); receivers (*25*); thermostats with SO_2 and NO_2 microstream sources (*26, 27*); mass air flow controllers (*28*); controlling filter (*29*); cut-off valve (*30*); compressor (*31*); Nafion dryer (*32*); chemical dryer (*33*); high-pressure sensor (*34*); three-position electropneumatic distributor (*35*); aerosol isokinetic sampler (*36*); control and interface unit (*37*) UST to RS-232/422/485 converter (*38*); computer (*39*). Solid lines with arrow show airline communications; dashed lines, electronic communications

Fig. 2 General view of the experimental site, container and equipment inside it

The following components are provided in the prototype for monitoring trace gases: Greenhouse gas analyzer (CO_2, CH_4, and H_2O) (1); CO analyzer (2); gas analyzers for chemically active gases (SO_2, NO_x, and O_3) (3, 4, and 5, respectively); high-pressure tanks filled with calibration gas mixtures (CO_2, CH_4, and CO) (19 and 20); high-pressure tank filled with a reference gas mixture (21); calibration SO_2 and NO_2 microstream sources (26 and 27); airflow distributors (17 and 18); and sampling devices (15, 24, 25, 28, 29, 30, 32, 33, 34, and 35). Main specifications of the prototype are given in [2].

The prototype of an up-to-date station for monitoring the atmospheric composition and state was tested in experiments carried out in the territory of Large Experimental Complex of IAO SB RAS in August 15–30, 2015. The experimental site is shown in Fig. 2.

The experiments performed [2] have shown the complete correspondence of the prototype specifications to the technical assignment.

Aerosol Multiwave Diffusion Spectrometer

The aerosol (black carbon, BC) multiwave diffusion spectrometer has been designed for the study of absorbing properties of BC-containing atmospheric aerosols, which are generated during numerous natural and anthropogenic combustion processes. The BC diffusion spectrometer allows prompt local control and long-term monitoring of the mass concentration of absorbing matter and its size distribution inside submicron atmospheric aerosol on the basis of recorded signals of light scattered in the visible spectral region by a layer of particles deposited on an aerosol filter.

Fig. 3 General view of the BC diffusion spectrometer: Pump and airline communication (*1*), signal USB ports (*2*); IBM computer with software for data recording and control (*3*); eight-section diffusion battery (*4*); MDA-03 multiwave aethalometer (*5*); FAN-M multiwave nephelometer (*6*)

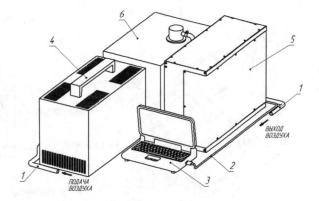

Figure 3 shows the BC diffusion spectrometer. Its specifications are given in [3].

The results of the experimental tests of the BC diffusion spectrometer in laboratory conditions and in the Large Aerosol Chamber of IAO SB RAS [3] witness that its specifications correspond to the requirements of the technical assignment.

Multiwave Scanning Polarization Lidar "LOSA-M3"

The LOSA-M3 lidar has been designed for measurements of optical and microphysical parameters of atmospheric aerosol on the basis of the analysis of multiwave lidar observation data [4]. The principle of operation of the lidar is the following: A directed laser pulse at wavelengths of 355, 532, and 1064 nm is sent into the atmosphere; aerosol-backscattered signal of elastic scattering at unshifted wavelengths and Raman scattering signal from molecular nitrogen at wavelengths of 387 and 607 nm and from water vapor at 407 nm are detected by a system of photodetectors, digitized by rapid receivers, and written on the computer. Then the aerosol optical parameters are retrieved on the basis of the signals recorded by the algorithms developed. The LOSA-M3 lidar is shown in Fig. 4.

The use of elastic and Raman scattering signals at several wavelengths allows retrieval of the aerosol optical parameters (attenuation and backscattering coefficients and mass concentration of aerosols in industrial emissions in the case of special calibration).

A transmit–receive unit of the lidar is assembled on a frame, where a laser with a collimator, near- and far-field receiving lenses, photodetectors, and a separate polarization receiver are mounted. The transmit–receive unit is mounted on a scanning rotating rod, which allows the lidar scanning in vertical and horizontal planes.

Figure 5 shows the lidar measurement results in the period of experimental tests of the prototype of the LOSA-M3 multiwave scanning polarization lidar at the IAO SB RAS test area in autumn 2015. The test of the LOSA-M3 lidar showed its correspondence to the requirements of the technical assignment.

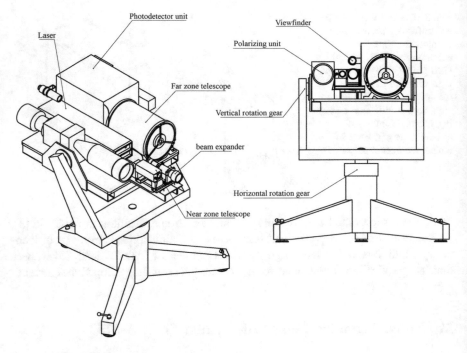

Fig. 4 LOSA-M3 lidar

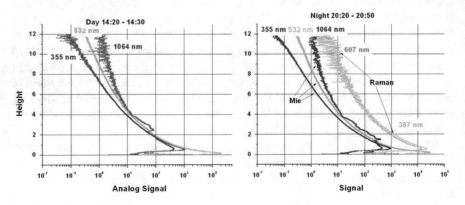

Fig. 5 Aerosol and Raman scattering signals in day- and nighttime

"Aerosol-3" Lidar

The Aerosol-3 lidar has been designed for measurements of optical and micro-physical parameters of atmospheric aerosol on the basis of the analysis of multi-wave lidar observation data. The lidar operates in the following way: A directed laser pulse at wavelengths of 355, 532, and 683 nm, which correspond to the third and second harmonic of an Nd:YAG laser and the first Stokes component of the 532 nm radiation conversion in hydrogen on the basis of stimulated Raman scattering (SRS) is sent into the atmosphere. The above wavelengths are implemented in one coaxial beam from one radiation source. This strongly simplifies the adjustment and operation of the three-frequency lidar and allows measurements in the routine mode.

The block diagram of the Aerosol-3 lidar designed is shown in Fig. 6.

During experiments, a sounding beam is directed to a flat roof of the Siberian Lidar Station (SLS) of IAO SB RAS parallel to the surface with the use of a rotating mirror. A roofed tunnel 1 m diameter and 6 m long is mounted at the roof edge. The tunnel is equipped with instruments that ensure creation of an aerosol cloud of a required density inside the tunnel and measure optical and microphysical parameters of the aerosol (scattering, extinction, and backscattering coefficients; particle radius, and volume concentration). Lidar measurements were carried out in

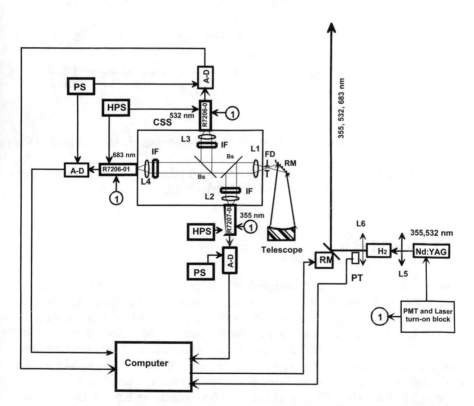

Fig. 6 Block diagram of the aerosol-3 three-wavelength lidar for stratospheric aerosol sounding: Solid-state laser (Nd:YAG), SRS conversion cell with H_2 (H_2), automated rotating mirror (RM), field stop (FS), spectral selection cell with a PMT (SSC), lenses (L), spectral line divider (SD), interferences filters (IF), amplifiers/discriminators (AD), power supplies (PS), high-voltage power supplies (HSU), optical transistor for photon counter actuation (OT)

Fig. 7 Elastic backscattering signals recorded at SLS in nighttime on June 13, 2016

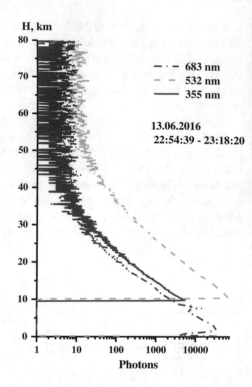

summer of 2016. Figure 7 shows the elastic backscattering signals recorded in the night time.

The experimental examination of the Arosol-3 lidar has shown the correspondence to the requirements of the technical assignment.

"ST Ozon" Lidar

The ST Ozon lidar has been designed for the study of ozone dynamics near the tropopause and stratospheric–tropospheric exchange in the upper troposphere–lower stratosphere. Figure 8 shows the block diagram of the lidar manufactured.

The fourth harmonics (266 nm) of the fundamental frequency of an Nd:YAG laser (LS-2134UT laser, LOTIS TII company, Minsk) is used as a laser radiation source, which is then SRS converted in hydrogen in the first (299 nm) and second (341 nm) Stokes components. During the experimental examination of the prototype of ST Ozon lidar at SLS of IAO SB RAS in summer of 2016, the lidar signals at 299/341 nm sounding wavelengths have been recorded in nighttime. They are shown in Fig. 9, as well as the ozone vertical profile retrieved from them. The experimental examination of the Arosol-3 lidar prototype has shown its correspondence to the requirements of the technical assignment.

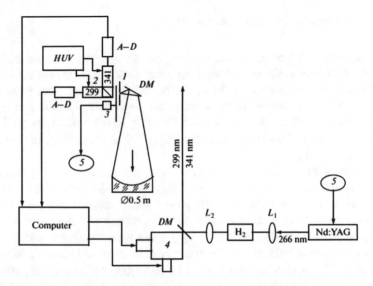

Fig. 8 Block diagram of ST Ozone lidar: Field stop (*1*), cell for spectral selection (*2*), mechanical shutter (*3*), adjustment unit of an output rotating mirror (*4*); system for synchronizing (*5*); rotating mirrors (RM); solid-state laser (Nd:YAG); SRS conversion cell with H_2 (H_2), amplifiers/discriminators (AD); high-voltage power supply units for the PMT (HSU); lenses (L_1 and L_2)

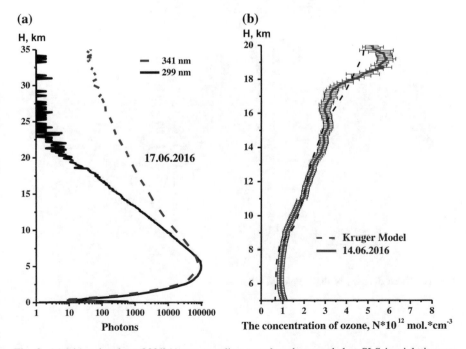

Fig. 9 (**a**) Lidar signals at 299/341 nm sounding wavelengths recorded at SLS in nighttime on June 14, 2016; (**b**) ozone vertical profile retrieved from signals (**a**)

Three-Channel Multifrequency Sodar

The "Volna-4 M-ST" sodar has been designed for the prompt remote acoustic diagnosis of altitude profiles of the wind vector and absolute values of the structural parameters of the temperature field in the boundary air layer. The prototype of Volna-4 M-ST three-channel Doppler sodar has been manufactured and used for the monitoring of wind speed $U(H)$ and direction $\phi(H)$ profiles, as well as the structural parameter of the temperature field $C_T^2(H)$. The block diagram of the sodar is shown in Fig. 10.

The Volna-4 M-ST sodar was tested at the experimental site in building A of IAO SB RAS. Along with the sodar tests, signals of sodar poraboloid accelerometers, which control the radiated signal power, were recorded independently, and the acoustic pressure amplitude at the sodar receiver exit was measured. Using the check measurement results, the $C_T^2(H, t)$ profiles were calculated and then averaged and compared with the sodar measurement results. Figure 11 shows the averaged $C_T^2(H)$ profiles measured by sodar in 08:00–08:10 (June 10, 2016) and check calculated. The coincidence of the profiles for this period is quite satisfactory and corresponds to the requirements for the sodar measurement accuracy for $C_T^2(H)$.

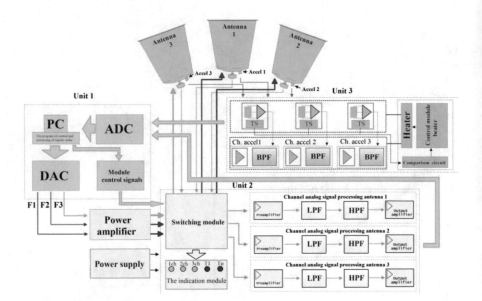

Fig. 10 Block diagram of the Volna-4M-ST sodar

Fig. 11 Comparison of the 08:00–08:10 averaged (June 10, 2016) profiles of the air temperature structural parameter $C_T^2(H)$ measured by the sodar and retrieved

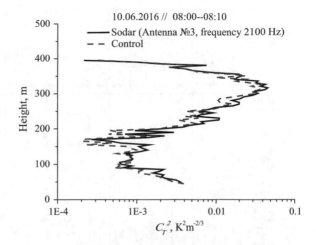

Continuation of the Monitoring at Existing Stations and Setups

During the project, atmosphere monitoring was carried out at all the available setups, including An-30 Optic-E and then Tu-134 Optic aircraft laboratories. During their flights, air was sampled in glass bulbs at altitudes of 0.5, 1, 1.5, 2.0, 3.0, 4.0, 5.5, and 7 km. The air samples were then analyzed at the laboratory of the National Institute for Environmental Studies (Japan) using the gas chromatography. The air sampling and measurements of the air gas composition have been carried out every month near 20s days under clear sky since July 1997. The sounding site is located to the southwest of Novosibirsk, to exclude the city effect. The flights are performed over the pine forest along the right bank of the Novosibirsk Reservoir, near Zyryanka and Ordynskoe settlements; they start at the point (54°35′ N, 82°40′ E). A series of continuous measurements over more than 18 years is accumulated by now. Though there were unsuccessful flights, the series analyzed includes 200 vertical profiles of CO_2, CH_4, and N_2O distributions. Figure 12 exemplifies long-term data on variations in the CO concentration over the region under study at altitudes of 0.5, 3.0, and 7.0 km.

Figure 12 shows that the CO concentrations increased from 1997 to 2015 at all the altitudes. However, the increase has a peculiarity at an altitude of 0.5 km in summer. The CO concentration varied insignificantly from 1997 to 2005, and it started increasing in 2005, even more rapid than at altitudes of 3.0 and 7.0 km.

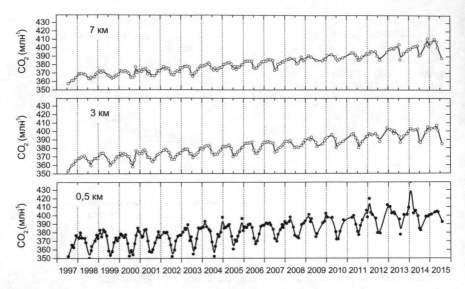

Fig. 12 CO concentrations over south of Western Siberia at altitudes of *0.5*, *3.0*, and *7.0* km

Conclusions

Instruments for the remote study of atmospheric composition and structure have been designed and tested within the project, including LOSA-M3, Aerosol-3, and ST Ozon lidars and Volna-4-ST sodar. The experimental examinations of the instruments confirmed the correspondence of their specifications and functional parameters to the technical assignment requirements. A typical automated station for monitoring the atmospheric composition and state and a multiwave diffusion spectrometer have also been designed and manufactured for contact measurements of air parameters. They also answer the technical assignment requirements. The project tasks include the continuation of air composition monitoring at currently operating setups. The measurement results showed the continuing increase in the carbon dioxide and nitrous oxide concentrations, as well as the renewed increase in the methane concentration in the layer from 0 to 7 km.

Acknowledgements Research is carried out with the financial support of the state represented by the Ministry of Education and Science of the Russian Federation. Agreement (contract) no. 14.604.21.0100, July 16, 2014. Unique project Identifier: RFMEFI60414X0100.

References

1. http://www.icos-infrastructure.eu
2. Matvienko, G.G., Belan, B.D., Panchenko, M.V. et. al.: Complex experiment on studying the microphysical, chemical, and optical properties of aerosol particles and estimating the contribution of atmospheric aerosol-to-earth radiation budget. Atmos. Meas. Tech. **8** (10), 4507–4520 (2015)
3. Kozlov, V.S., Shmargunov, V.P., Panchenko, M.V. et. al.: Seasonal variability of the black carbon size distribution in the atmospheric aerosol. Russ. Phys. J. **58** (12), 1804–1810 (2016)
4. http://www.iao.ru/en/resources/equip/lidars

Technology of Integrated Impact on the Low-Permeable Reservoirs of Bazhenov Formation

V.S. Verbitskiy, V.V. Grachev and A.D. Dmitrievskiy

Abstract Nowadays, the huge amount of the information concerning development and exploitation of hard-to-recover oil has been accumulated. The rational (sustainable) development of the Bazhenov formation is the main target for many years to come. The work is based on the research and has been performed within the framework of the Federal target program with the financial support of the Ministry of Education and Science of the Russian Federation (the unique identifier RFMEFI60714X0080). The aim of the research is to choose the best possible method of oil extraction. During this research, geological aspects of the Bazhenov formation, methods of influence, and field tests results were studied; methods impact on the productive part of the oil deposits; and research based on the extraction of shale oil. As the result of this literature review and patent research, the list of the main methods of influence on Bazhenov formation was formed. Among them, there are hydraulic formation fracturing, thermal methods, gas, and water-gas injection methods. The authors of the research work proposed a new way of development of oil Bazhenov formation, based on the results of mathematical modeling and computational and experimental studies, has created a unique laboratory facility for the integrated modeling of petrophysical and hydrodynamic characteristics of the reservoir model Bazhenov Formation. The project of the technical specification for the installation for vibration stimulation of the formation in the depression mode has been developed.

V.S. Verbitskiy (✉) · V.V. Grachev
Gubkin Russian State University of Oil and Gas
(National Research University), Moscow, Russia
e-mail: vsverbitsky@gmail.com

V.V. Grachev
e-mail: slavamgtu@yandex.ru

A.D. Dmitrievskiy
Oil and Gas Institute, Russian Academy of Sciences, Moscow, Russia
e-mail: direction@fcntp.ru

© The Author(s) 2018
K.V. Anisimov et al. (eds.), *Proceedings of the Scientific-Practical Conference
"Research and Development - 2016"*, https://doi.org/10.1007/978-3-319-62870-7_72

Keywords Scientific and technical solutions · Development of oil fields
The Bazhenov formation · Low permeability reservoirs · Fracturing
Research · Filtration · Oil reservoir collector · Model · Laboratory installations
Vibration exposure

Oil resources in Bazhenov formation, mainly in West Siberia, come to billion tons. Nowadays, there are no industrial methods of oil extraction from Bazhenov formation because of many complicating factors. The productive sediments of Bazhenov formation are source bed and are presented by two lithological reservoir types: Kerogen–clay–chert rocks—bazhenits, usually they present the main thickness of formation; and clayey limestones. Bazhenits can contain hydrocarbons in two substances. First in the form of light low-viscosity oil in voids and second, in the form of kerogen. The usual size of pore is 30–50 nm. Thus, there are collectors with pore sizes nanometers. The collector has an inhomogeneous structure. Therefore, in the Bazhenov formation there are super-permeability layers (super-k). It is well known that super-k leads to a low level of oil recovery due to water breakthrough during the waterflooding period. Thermal methods of development in Bazhenov formation have low oil recovery index (3–5%) [1, p. 4].

Carrying out hydraulic fracturing in the low-permeable matrix of bazhenits with clay rocks interlayering does not result in any positive effect. In the majority of cases short period of oil production from fractured matrix can be observed after conducted hydraulic fracturing followed by well oil rate decrease to minimum values (below commercial production) and as a result well might be shut in, Pressure maintenance is an option that can extend the period of production from bazhenits. Bazhenits are characterized with the high degree of heterogeneity and include tectonic faults, small shale layers, and that is why injection of gas is preferable for pressure maintenance [2, p. 3].

On the other hand, traditional techniques of pressure maintenance using gas injection cannot provide high ultimate recovery due to the following reasons:

(1) Poor sweep efficiency because of injected gas breakthrough through the super-k layers. It is well-known fact that there is even more negative impact of the heterogeneity on ultimate recovery when low-viscous gas is injected into the reservoir comparing to water injection.

(2) However, injected gas will not interreact with the bazhenits, which contain the main part of hydrocarbon reserves in form of light low-viscosity oil. The same way oil reserves are uncovered with waterflooding because of the breakthrough of water in high permeable fractures.

(3) Laboratory and field studies show that reservoir properties of bazhenits are significantly affected by the reservoir pressure changes. And this is the reason of a huge permeability reduction in the near well-bore area during the production that leads to rate decrease.

New technology that takes into account all the previous experience of production from bazhenits was created and, in addition unique laboratory facility that allows modeling of different combinations of impact on the bazhenits reservoirs was created.

Suggested technology includes the following: Drilling of horizontal production wells and vertical injection well in accordance with well pattern providing highest ultimate recovery, drilling of additional vertical production wells for creation of wave effects in the near well-bore areas; multistage hydraulic fracturing in the newly drilled wells, including simultaneous wave impact through the additional vertical production wells and injection of fluid into the reservoir (Fig. 1).

Technology includes the sequence of technological operations in alternate cycles, where each cycle consists of three stages: at the first stage fluid is injected into wells for T1 period, what results in reservoir pressure increase, dissolution of liquid hydrocarbons and their release from bound state in the kerogen containing matrix; meanwhile, horizontal producers and vertical injectors are shut in (Fig. 2a); at the second stage, injector and horizontal wells are shut in for the (T2-T1) period. Meanwhile, vertical wells are put into operation in the regime of vibrowave impact without production. Dissolution of liquid hydrocarbons continues and reservoir pressure flattens associated with intensive invasion of injected fluid into the kerogen containing matrix during (T2-T1) period (Fig. 2b). At the third stage, first, horizontal wells are consequently put into operation and after vertical production wells with simultaneous vibrowave impact on the reservoir; meanwhile, injector is shut in for (T3-T2) period (Fig. 2c). Hydrocarbons production continues to the minimum value of reservoir pressure which decreases during production; after that horizontal wells are shut in and vertical wells stop producing hydrocarbons and provide vibrowave impact and injection of fluid starts again (Fig. 2d). T1 period equals to the time necessary to increase reservoir pressure to its initial value, while (T2-T1) period duration is based on the field studies aimed at maximizing oil production from producers till T2 moment of time and T3 corresponds to the time when producing well oil rate reaches the planned minimum value. Produced water and

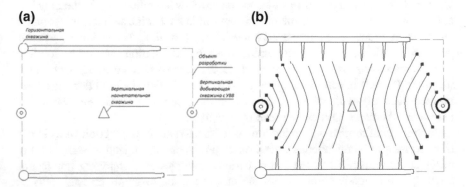

Fig. 1 Stages of preparation object oil field development of the Bazhenov Formation: (**a**) well drilling, (**b**) multistage hydraulic formation fracturing

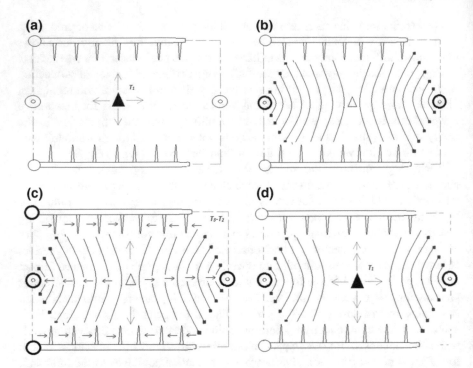

Fig. 2 Realization stages of integral action on reservoir and oil of Bazhenov formation:
(**a**) working fluid injection (producing wells are closed), (**b**) vibrowave impact with maximum
value of bottom hole pressure, (**c**) producing wells sequence starting, (**d**) working fluid injection
(horizontal wells are closed and vertical are working in circulation pattern)

gas are injected back into the reservoir what results in lower volumes necessary for
injection (Fig. 1).

Water injection starts in order to improve sweep efficiency when oil fraction
becomes smaller in produced volumes. Water may be considered as a working fluid
when polymers are injected in order to improve sweep efficiency. If the considerable amount of swelling clay minerals is present in bazhenits, saline water solutions
are used in order to minimize clay swelling, as well as saline water, gel, and
polymer solutions, like polyacrylamide water-based solution, and other solutions
with increased viscosity might be used. In the case of clay swelling, water alter-
nated gas (WAG) injection with various WAG ratios might be used that maximizes
ultimate recovery with minimum impact on the reservoir (WAG ratio selection is
based on laboratory studies and results of well testing).

In order to minimize oil production losses due to shut in production wells in case
of overpressure, simultaneous operation of producers and injection well as a pro-
ducer is recommended at the initial period of suggested technology application.
Such operation mode continues till the moment when reservoir pressure falls not
lower than hydrostatic pressure and production rates reach planned minimum value.

Conclusions

Laboratory and 3D simulation model studies are necessary to make the suggested technology widely applied while it is also important to keep static and dynamic models of bazhenits reservoirs updated.

Acknowledgments Research is carried out with the financial support of the state represented by the Ministry of Education and Science of the Russian Federation. Agreement no. 14.607.21.0080, October 20, 2014. Unique project Identifier: RFMEFI60714X0080.

References

1. Patent 2513963 Russian Federation, 21 B 43/16. The process of development of oil deposits in the sediments of the Bazhenov Formation IPNG RAN the applicant and the patentee: IPNG RAN—№ 2012142692/03; statement: 08.10.2012, published: 20.04.2014, bulletin № 11—9 p. : Illustrations
2. Dmitriyevsky A.N., Dmitriyevsky, A.N., Garifullin, R.I., Karpov, S.N., et al.: Results of complex technology simulation for low-permeable reservoir development. ARPN J. Eng. Appl. Sci. February 2016. **3**, 1918–1922 (2016)

Development of the First Russian Anammox-Based Technology for Nitrogen Removal from Wastewater

A.M. Agarev, A.G. Dorofeev, A.Yu. Kallistova, M.V. Kevbrina, M.N. Kozlov, Yu.A. Nikolaev and N.V. Pimenov

Abstract A pilot single-stage setup with the reactor volume of 20 m^3 was constructed for ammonium removal from the filtrate of thermophilically digested sludge. The setup was operated at temperatures of 20–37 °C, dissolved oxygen concentrations of 0.1–0.7 mg/L, pH of 5.7–8.5, hydraulic retention time of 12–36 h, and filtrate output of up to 30 m^3/day. The efficiency of nitrogen removal was 75–90%, nitrogen load was 0.9–1.1 kg N/(m^3·day), and the specific volumetric nitrogen removal capacity of the reactor reached 0.8–1.0 kg N/(m^3·day). The sludge retained activity at low pH (5.7) and enhanced nitrite concentration (up to 250 mg/L). A correlation was established between conductivity reduction of the treated liquid and nitrogen removal efficiency, and the formula for calculation of ammonium concentration using the conductivity was proposed.

Keywords Autotrophic anaerobic ammonium oxidation (Anammox) · Anammox bacteria · Filtrate of anaerobic digesters · Nitrogen removal · Wastewater · Wastewater treatment plants

A.M. Agarev · A.G. Dorofeev · M.V. Kevbrina · M.N. Kozlov · Yu.A. Nikolaev
Department of Advanced Equipment and Technologies, JSC Mosvodokanal, Moscow, Russia
e-mail: agarev_am@mosvodokanal.ru

A.G. Dorofeev
e-mail: dorofeev_ag@mosvodokanal.ru

M.V. Kevbrina
e-mail: kevbrina_mv@mosvodokanal.ru

M.N. Kozlov
e-mail: kozlov@mosvodokanal.ru

Yu.A. Nikolaev
e-mail: nikolaev_ya@mosvodokanal.ru

A.Yu. Kallistova · N.V. Pimenov (✉)
Winogradsky Institute of Microbiology, The Federal Research Centre "Fundamentals of Biotechnology" of Russian Academy of Sciences, Moscow, Russia
e-mail: npimenov@mail.ru

A.Yu. Kallistova
e-mail: kallistoanna@mail.ru

K.V. Anisimov et al. (eds.), *Proceedings of the Scientific-Practical Conference "Research and Development - 2016"*, https://doi.org/10.1007/978-3-319-62870-7_73

Introduction

At Moscow wastewater treatment plants the sludge is stabilized by anaerobic thermophilic digestion (52–55 °C) followed by thickening and dewatering of digested sludge. Significant degradation of volatile suspended solids (VSS) under thermophilic conditions results in high levels of ammonium (up to 700 mg N-NH$_4$/L) in liquid phase (filtrate) of digested sludge. This filtrate forms recycle water flow and increases the load of ammonium on the biological stage of wastewater treatment by up to 50%. Increasing ammonium concentrations in the liquid phase of digested sludge cause technological and economic problems.

The traditional nitrification/denitrification technology is inefficient for ammonium removal from the filtrate of industrial anaerobic digesters, since such filtrate has a low biochemical oxygen demand (BOD)/N ratios (0.7–1.7). The technologies based on microbial process of autotrophic anaerobic ammonium oxidation by nitrite are presently being introduced for treatment of wastewater with high concentrations of ammonium and low BOD values [7, P. 9]. These technologies combine two autotrophic processes: partial nitrification, which involves oxidation of half of ammonium to nitrite, and anaerobic ammonium oxidation by nitrite to molecular nitrogen (anammox) [1, 5, P. 8, 9]. The best-known anammox technologies applied at wastewater treatment plants worldwide are SHARON-ANAMMOX, DEMON, Canon, ANITA-Mox, DeAmmon, OLAND, etc. [8, P. 9]. High efficiency of nitrogen removal, the absence of organic carbon demand, low sludge yield, and 2–3 times lower nitrogen removal costs are the advantages of anammox technologies [6, 14, P. 9]. There are two principal schemes of anammox implementation: two-stage systems where processes of nitrification and anammox are spatially separated in different serially connected reactors, and single-stage systems consisting of one reactor where both processes go simultaneously. Single-stage systems are widely used due to their advantages over two-stage systems, particularly their higher potential rates of nitrogen removal [14, P. 9].

Joint Stock Company (JSC) Mosvodokanal is the largest company in Russia servicing Moscow wastewater treatment plants. The JSC Mosvodokanal in collaboration with Winogradsky Institute of Microbiology (The Federal Research Centre "Fundamentals of Biotechnology" of Russian Academy of Sciences) and within the framework of the agreement with Ministry of Education and Science of the Russian Federation has developed the first Russian anammox-based technology for ammonium removal from the filtrate of industrial anaerobic digesters and tested it in pilot-scale setup. The goal of the present work was the development, start-up, and optimisation of a pilot-scale single-stage setup for ammonium removal from the filtrate of digested sludge.

Methods

The schematic diagram of the pilot setup is shown in Fig. 1. The filtrate from dewatering centrifuges was passed through the grid (1) for removal of coarse particles and was then collected in the receiving tank (2). From this tank the filtrate was transferred to the 4-m^3 primary settler (3), then to the heat exchanger (4), and subsequently to the 20-m^3 main bioreactor (5). The setup was designed for carrying out the intended processes of water treatment within a broad range of operating temperatures (from 20 to 37 °C). The required operating temperature was maintained by the heat exchanger with an electric heater and biologically treated water as a coolant. The bioreactor was a completely stirring-type reactor. The sludge from the secondary settler (8) was also recycled to the reactor. Removal of ammonium and organic matter occurred in the bioreactor. The intended biochemical processes were carried out by suspended flocculated sludge and by the sludge immobilized on a stationary lamellar carrier. The bioreactor contained three blocks of the stationary lamellar carrier, the aeration system providing oxygen for the process, and the large-bubble system of regeneration for removal of excess biomass from the carrier. Air was supplied by a compressor (11). Oxygen concentration in the reactor was maintained at 0.1–0.7 mg/L and monitored using an oxygen sensor.

Hydraulic retention time (HRT) for filtrate was 12–36 h. From the bioreactor the sludge was transferred to a setup for intermediate aeration 2 m^3 in volume (6), then to 18-L sludge homogenizer (7), and to the secondary settler (4 m^3). Setups for intermediate aeration and homogenizer were required to prevent flotation of the activated sludge in the secondary settler under high nitrogen load. After settling, the treated filtrate was discharged, and return sludge was pumped back into the bioreactor. Some of the sludge (excessive) was removed from the setup.

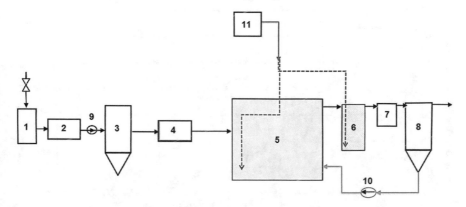

Fig. 1 Schematic diagram of the pilot setup. *1*—grid filter; *2*—receiving tank; *3*—primary settler; *4*—heat exchanger; *5*—bioreactor; *6*—setup for intermediate aeration; *7*—flow homogenizer; *8*—secondary settler; *9, 10*—pumps; *11*—compressor

Concentrations of suspended solids (SS), chemical oxygen demand (COD), BOD, NH_4^+, NO_2^-, NO_3^-, total suspended solids (TSS), and volatile suspended solids (VSS) were measured by common methods [12, P. 9].

Results

The above-described pilot setup with a 20-m^3 bioreactor was intended for ammonium removal from the filtrate of dewatering centrifuges of digested sludge from the Lyuberetskie wastewater treatment plant (LWWP), Moscow, Russia. The reactor was inoculated by anammox sludge from the reactor, described earlier [10, 11, P. 9]. During first days of reactor operation filtrate supply was 3–4 m^3/d, and then it was gradually increased. The planned filtrate supply rate of 20 m^3/day was reached after 70 days of setup operation. Dynamics of nitrogen compounds in the treated filtrate from the pilot setup (Fig. 2), showing a trend of decreasing ammonium and nitrate concentrations, reflects improvements of the technological process. Drastic fluctuations were caused by periodic changes in the operation mode due to adjustment of the systems for aeration, filtrate supply, temperature, etc. The concentrations of ammonium and nitrate were in opposite phases. The general picture indicated the start-up period of the setup.

The setup reached its full capacity, with nitrogen load of 0.9–1.1 kg N/(m^3·day) and specific volumetric nitrogen removal capacity of the reactor of 0.8–1.0 kg N/(m^3·day) after 12 months of operation. Nitrogen removal was the main target parameter. Nitrogen removal exceeding 70% was considered technologically sufficient (Table 1).

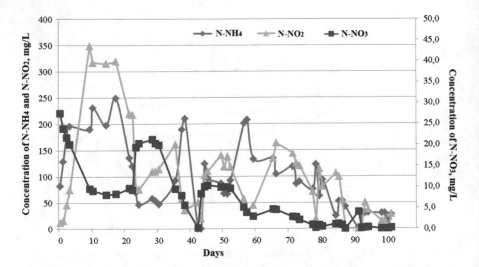

Fig. 2 Dynamics of nitrogen compounds in the treated filtrate of the pilot setup

Table 1 Composition of the incoming and treated filtrate of dewatering centrifuges of digested sludge from the LWWP used to feed the pilot setup, mg/L

Parameter	SS	COD	BOD	N-NH$_4$	N-NO$_2$	N-NO$_3$
Incoming filtrate	500–1500	700–2000	110–330	500–800	0	0
Outgoing filtrate	45	140	25–45	30–35	10–15	20–25
Nitrogen removal, %	92–97	80–95	80–90	All soluble forms of nitrogen —75–90%		

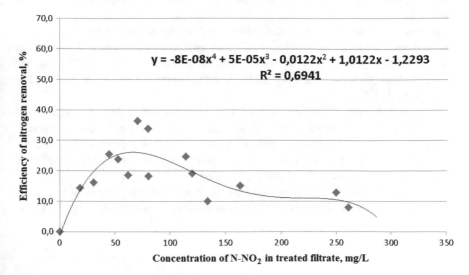

Fig. 3 Efficiency of nitrogen removal and amount mass of removed nitrogen depending on nitrite concentration

Implementation of the above-described technology resulted in a high efficiency of nitrogen removal and high reliability of the process. Thus, resistance of the activated sludge to elevated nitrite concentration (250 mg/L) was shown. In the single-stage setup, nitrites were formed in the course of the aerobic nitritation process, carried out by ammonium-oxidizing bacteria (AOB) [14, P. 9]. Nitritation is essential for efficient functioning of anammox reactors, since AOB provide anammox bacteria with nitrite, one of the major anammox substrates, which is absent in the digested sludge filtrate. While anammox bacteria have high affinity to both ammonium and nitrite, the anammox process is inhibited by nitrite concentrations over 100 mg/L (by nitrogen) [13, P. 9]. This threshold may be shifted to higher or lower concentrations [2, 3, 15, P. 8, 9] and depends on duration of the inhibitory action of nitrite [4, P. 8]. Recovery of anammox after nitrite removal from the medium indicates reversible inhibition [9, P. 9]. Calculations of the efficiency of nitrogen removal at different nitrite concentrations (Fig. 3) were carried out prior to the reactor reaching its designed capacity, i.e., when it did not reach its maximal output. Since at least 50% of the maximal activity was preserved (within the range from 50 to 90 mg N-NO$_2$/L) at nitrite concentrations up to 175 mg/L, and $\sim$30% of the maximal activity—at 250 mg/L, the technology was highly stable in this respect.

The presence of two extremes in Fig. 3 (at 50–90 and 225–260 mg N-NO$_2$/L) indicates heterogeneity of the anammox bacteria population. It is likely represented by two groups of bacteria with different nitrite optima. The optimum nitrite concentration was within the range of 50–100 mg N-NO$_2$/L for one of them and $\sim$250 mg/L for the other. Pyrosequencing of the 16S rRNA gene fragments revealed predominance of the genus *Candidatus* "Brocadia" in the activated sludge [10, P. 9].

Inverse correlation was revealed between nitrite concentration and pH, i.e., the medium was acidified at elevated nitrite concentrations. The optimal pH for the setup operation was 7.5–8.3 at nitrite concentrations of 50–100 mg/L. The pH minimum was pH 5.7 at 250 mg/L nitrite, and the maximum was pH 8.3–8.5 at less than 50 mg/L nitrite.

The conductivity of the reactor medium, an important factor in technological monitoring, was also investigated. Good correlation was found between relative decrease in conductivity (%) and efficiency of total nitrogen removal (the correlation coefficient was 0.63) (Fig. 4). This parameter may be used for online express analysis of the efficiency of the setup operation.

Conductivity was found to be associated with ammonium concentrations in the inflowing and treated filtrate (the correlation coefficient was 0.92) (Fig. 5). Using

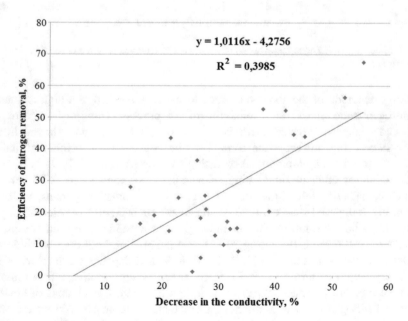

Fig. 4 Correlation of the decrease in the conductivity of the reactor medium and the efficiency of nitrogen removal

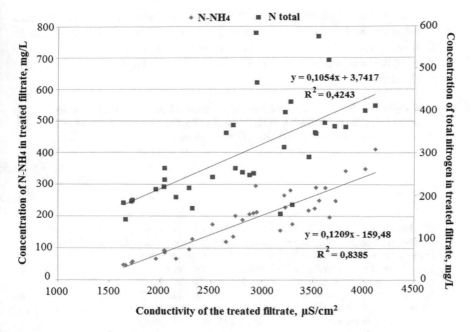

Fig. 5 Correlation of the treated filtrate and the concentrations of ammonium nitrogen and total mineral nitrogen at the outflow

the conductivity values, ammonium concentration may be calculated, which is important for controlling the technology:

NH_4 concentration (mg N/L) = C × 0.12−159.5,

where C is conductivity of the reactor medium ($\mu S/cm^2$).

Conclusions

Thus, the reported single-stage pilot setup carries out the target processes of ammonium removal from the filtrate of dewatering centrifuges of digested sludge. Flocculated free-floating sludge is involved in the oxidation of organic matter, sorption of suspended solids, and partial oxidation of ammonium to nitrite. In the inner layers of the biofilm immobilized on the carrier, autotrophic (anammox) and heterotrophic denitrification occurs. The setup carries out the target processes within the temperature range of 20–37 °C, oxygen concentrations from 0.1 to 0.7 mg/L, and pH 5.7–8.5. At hydraulic retention time of 12–36 h and filtrate output of 14–30 m^3/day, nitrogen load is as high as 0.9–1.1 kg N/(m^3·day), and specific volume capacity for nitrogen removal is 0.8–1.0 kg N/(m^3·day). The activated sludge was found to be stable at elevated nitrite concentration (250 mg/L) and decreased pH value (5.7). Using the conductivity of the medium for control of the efficiency of nitrogen removal and for calculation of ammonium concentrations in the incoming and outcoming filtrate was confirmed. The data obtained will be used for designing an industrial anammox bioreactor at the Lyuberetskie wastewater treatment plants, with a capacity of 18,000 m^3/day and nitrogen removal rate of up to 11 t/day.

Acknowledgements Research is carried out with the financial support of the state represented by the Ministry of Education and Science of the Russian Federation. Agreement no. 14.607.21.0018, June 5, 2014. Unique project Identifier: RFMEFI60714X0018.

References

1. Blackburne, R., Yuan, Z., Keller, J.: Partial nitrification to nitrite using low dissolved oxygen concentration as the main selection factor. Biodegradation **19**(2), 303–312 (2008)
2. Dapena-Mora, A., Fernandez, I., Campos, J.L., Mosquera-Corral, A., Mendez, R., Jetten, M. S.M.: Evaluation of activity and inhibition effects on anammox process by batch tests based on the nitrogen gas production. Enzyme Microb. Technol. **40**, 859–865 (2007)
3. Egli, K., Fanger, U., Alvarez, P., Siegrist, H., van der Meer, J., Zehnder, A.: Enrichment and characterization of an anammox bacterium from a rotating biological contactor treating ammonium-rich leachate. Arch. Microbiol. **175**, 198–207 (2001)
4. Fux, C., Huang, D., Monti, A., Siegrist, H.: Difficulties in maintaining long-term partial nitritation of ammonium-rich sludge digester liquids in a moving-bed biofilm reactor (MBBR). Water Sci. Technol. **49**, 53–60 (2004)
5. Hellinga, C., Schellen, A., Mulder, J., van Loosdrecht, M.C.M., Heijnen, J.J.: The SHARON process: an innovative method for nitrogen removal from ammonium-rich wastewater. Water Sci. Technol. **37**(9), 135–142 (1998)
6. Jetten, M.S.M., Strous, M., Van de Pas-Schoonen, K.T., Schalk, J., Van Dongen, U.G.J.M., Van de Graaf, A.A., Logemann, S., Muyzer, G., van Loosdrecht, M.C.M., Kuenen, J.G.: The anaerobic oxidation of ammonium. FEMS Microbiol. Rev. **22**(5), 421–437 (1999)
7. Kuenen, J.G.: Anammox bacteria: from discovery to application. Nat. Rev. Microbiol. **6**, 320–326 (2008)
8. Lackner, S., Gilbert, E.M., Vlaeminck, S.E., Joss, A., Horn, H., van Loosdrecht, M.C.M.: Full-scale partial nitritation/anammox experiences—an application survey. Water Res. **55**, 292–303 (2014)
9. Lotti, T., van der Star, W.R.L., Kleerebezem, R., Lubello, C., van Loosdrecht, M.C.M.: The effect of nitrite inhibition on the anammox process. Water Res. **46**, 2559–2569 (2012)
10. Mardanov, A.V., Beletskii, A.V., Kallistova, A.Yu., Kotlyarov, R.Yu., Nikolaev, Yu.A., Kevbrina, M.V., Agarev, A.M., Ravin, N.V., Pimenov, N.V.: Dynamics of the composition of a microbial consortium during start-up of a single-stage constant flow laboratory nitritation/anammox setup. Microbiology. **85**(6), 681–692 (2016)
11. Nikolaev, Yu.A., Kozlov, M.N., Kevbrina, M.V., Dorofeev, A.G., Pimenov, N.V., Kallistova, A.Yu., Grachev, V.A., Kazakova, E.A., Zharkov, A.V., Kuznetsov, B.B., Patutina, E.O., Bumazhkin, B.K.: *Candidatus* "Jettenia moscovienalis" sp. nov., a new species of bacteria carrying out anaerobic ammonium oxidation. Microbiology. **84**(2), 256–262 (2015)
12. Standard methods for the examination of water and wastewater.: American Public Health Association, American Water Works Association, Water Environment Association. In: Rice, E.W., Baird, R.B., Eaton, A.D., Clesceriand, L.S., Bridgewater, L. (eds.). Washington, D.C. 22nd ed. 1496 p (2012)

13. Strous, M., Kuenen, J.G., Jetten, M.S.M.: Key physiology of anaerobic ammonium oxidation. Appl. Environ. Microbiol. **65**, 3248–3250 (1999)
14. Van Hulle, S.W.H., Vandeweyer, H.J.P., Meesschaert, B.D., Vanrolleghem, P.A., Dejans, P., Dumoulin, A.: Engineering aspects and practical application of autotrophic nitrogen removal from nitrogen rich streams. Chem. Eng. J. **162**, 1–20 (2010)
15. Xiao, Y., Xiao, Q., Xiang, S.: Modeling of simultaneous partial nitrification, anammox and denitrification process in a single reactor. J. Environ. Anal. Toxicol. **4** (2014). doi:10.4172/2161-0525.1000204

Pulse-Detonation Hydrojet

**S.M. Frolov, K.A. Avdeev, V.S. Aksenov, F.S. Frolov, I.A. Sadykov,
I.O. Shamshin and R.R. Tukhvatullina**

Abstract Geometrical configuration and operational parameters of a valveless
pulse-detonation hydrojet have been determined based on extensive numerical
simulations using 2D two-phase flow equations. The theoretical propulsive performance of such a hydrojet in terms of the specific impulse was shown to be on the
level of modern liquid propellant rocket engines and amount 350–400 s. Based on
the results of numerical simulation a valveless pulse-detonation hydrojet operating
on liquid hydrocarbon fuel (regular gasoline) and gaseous oxygen has been
designed and fabricated. For firing the hydrojet, a special test rig with flowing water
was designed and assembled. Experiments showed that the measured values of the
specific impulse varied within the range from 255 to 370 s which overlaps the
theoretical range, thus demonstrating the predictive capabilities of the numerical
approach.

S.M. Frolov (✉) · I.O. Shamshin · R.R. Tukhvatullina
Semenov Institute of Chemical Physics, Moscow, Russia
e-mail: smfrol@chph.ras.ru

I.O. Shamshin
e-mail: igor_shamshin@mail.ru

R.R. Tukhvatullina
e-mail: tukhvatullinarr@gmail.com

K.A. Avdeev · F.S. Frolov · I.A. Sadykov
Semenov Institute of Chemical Physics, Center for Pulsed Detonation Combustion,
Moscow, Russia
e-mail: kaavdeev@mail.ru

F.S. Frolov
e-mail: f.frolov@chph.ru

I.A. Sadykov
e-mail: ilsadykov@mail.ru

V.S. Aksenov
National Research Nuclear University MEPhI, Moscow, Russia
e-mail: v.aksenov@mail.ru

© The Author(s) 2018
K.V. Anisimov et al. (eds.), *Proceedings of the Scientific-Practical Conference
"Research and Development - 2016"*, https://doi.org/10.1007/978-3-319-62870-7_74

Keywords Pulse-Detonation hydrojet · Experiment · Three-Dimensional calculation · Thrust · Specific impulse · Bubbly liquid · Valveless hydrojet Two-Phase flow

Introduction

In our patent [4] and articles [1–3] we suggested to replace conventional screw propellers used in boats and ships by a pulse-detonation hydrojet. Such a propulsion device is composed of a pulse-detonation tube inserted in a properly shaped water guide submerged in water. The detonation tube is cyclically filled with a fuel mixture which is cyclically ignited to generate a detonation wave. High-momentum detonation (shock) wave enters cyclically a water guide and involves bubbly water therein in accelerated motion toward the exit nozzle due to increased compressibility of water saturated with gas bubbles of a previous operation cycle. The objective of this communication is to investigate possible geometrical configurations and operational parameters of such a hydrojet and to assess its propulsive performance using both numerical simulation and experiments.

Numerical Simulation

Below we consider bubbly liquid that consists of two phases, namely, dispersed gas phase (subscript 1) whose volume fraction is α_1 and carrier liquid phase (subscript 2) with volume fraction α_2. The following simplifying assumptions are adopted: (1) gas compressed in a bubbly liquid is not dissolved in the liquid while the liquid is not evaporated inside bubbles; (2) the liquid phase density depends solely on the liquid temperature T_2; (3) flow of the bubbly liquid is laminar; (4) effects of gravity, lifting and friction forces at bounding surfaces on the relative phase motion in a bubbly liquid are ignored.

The mathematical model of the two-phase flow is based on partial differential equations of mass, momentum, and energy conservation derived within the framework of mutually penetrating continua [5, 6]:

$$\frac{\partial \alpha_i \rho_i}{\partial t} + \nabla \cdot \alpha_i \rho_i \mathbf{v}_i = 0$$

$$\frac{\partial \alpha_i \rho_i \mathbf{v}_i}{\partial t} + \nabla \cdot \alpha_i \rho_i \mathbf{v}_i \mathbf{v}_i = -\alpha_i \nabla p_i + \nabla \cdot \alpha_i \tau_i + \mathbf{M}_{ij} \tag{1}$$

$$\frac{\partial \alpha_i \rho_i h_i}{\partial t} + \nabla \cdot \alpha_i \rho_i \mathbf{v}_i h_i = \nabla \cdot \alpha_i \mathbf{q}_i + \nabla \cdot \alpha_i \tau_i \cdot \mathbf{v}_i + \alpha_i \frac{\partial p_i}{\partial t} + \mathrm{H}_{ij},$$

where t is the time, i and j are the indices of phases, ∇ is the differential operator with respect to radius vector $\mathbf{r}$, ρ is the phase density, $\mathbf{v}$ is the phase velocity vector, p is the phase pressure, τ is the phase tensor of viscous stress, $\mathbf{q}$ is the phase heat flux, terms $\mathbf{M}_{ij}$ and H_{ij} describe the interphase momentum and energy exchange, respectively, h is the phase total enthalpy given by

$$h_i = h_{i,0} + \int_{T_{i,0}}^{T_i} c_{p,i}\, dT + \frac{1}{2}\mathbf{v}_i\mathbf{v}_i. \tag{2}$$

here, $c_{p,i}$ is the phase specific heat at constant pressure, T is the phase temperature, and index 0 denotes the initial values of variables.

The set of Eqs. (1), (2) is supplemented with the relationships for fluxes τ_i, $\mathbf{q}_i$, $\mathbf{M}_{ij}$, and H_{ij}:

$$\tau_i = \mu_i \left[\left(\nabla\mathbf{v}_i + \nabla\mathbf{v}_i^T\right) - \frac{2}{3}\nabla\cdot\mathbf{v}_i \right]$$

$$\mathbf{q}_i = \frac{\kappa_i}{c_{p,i}}\nabla h_i$$

$$\mathbf{M}_{12} = C_D \frac{A\rho_2\,|\mathbf{v}_{12}|\mathbf{v}_{12}}{8}; \quad \mathbf{M}_{21} = -\mathbf{M}_{12}$$

$$H_{12} = Nu\frac{\kappa_2 A(T_1 - T_2)}{d_1}; \quad H_{21} = -H_{12}$$

where μ is the phase dynamic viscosity, κ is the phase thermal conductivity, $\mathbf{v}_{12} = \mathbf{v}_1 - \mathbf{v}_2$ is the relative velocity of phases, d_1 is the bubble diameter, $A = \frac{6\alpha_1}{d_1}$ is the total interphase surface area in a unit volume of bubbly liquid, C_D and Nu are the hydrodynamic drag coefficient and Nusselt number, which in general depend on Reynolds number of relative motion of phases $\mathrm{Re}_{12} = \frac{\rho_2\mathbf{v}_{12}d_1}{\mu_2}$ and liquid Prandtl number $\mathrm{Pr}_2 = \frac{c_{p,2}\mu_2}{\kappa_2}$[6]:

$$C_D = \min\left[\frac{24}{\mathrm{Re}_{12}}(1 + 0.15\mathrm{Re}_{12}^{0.687}), \frac{72}{\mathrm{Re}_{12}}\right]$$

$$Nu = 2 + 0.6\mathrm{Re}_{12}^{0.5}\,\mathrm{Pr}_2^{0.33}.$$

The set of Eqs. (1) and (2) and supplementary relationships for fluxes contain 12 dependent variables α_1, α_2, ρ_1, ρ_2, $\mathbf{v}_1$, $\mathbf{v}_2$, p_1, p_2, h_1, h_2, T_1, and T_2. To close the statement of the problem we add four more relationships:

$$\rho_2 = \rho_2(T_2); \ \alpha_2 = 1 - \alpha_1; \ p_1 = \rho_1 R T_1; \ p_2 = p_1. \tag{3}$$

The resultant equations are also supplemented by initial and boundary conditions for the listed variables and their derivatives.

In [3], we carried out an a priori analysis of stability of Eqs. (1) for isothermal case with nonzero fluxes. It has been proven that allowance for momentum fluxes (τ_1, τ_2) within phases at $p_2 = p_1$ makes the evolution problem well-posed.

The multi-phase balance Eqs. (1) can be expressed for every phase k = 1, 2 in the form of a general differential equation for the mean flow variable $\phi_k(\mathbf{x}, t)$ which reads

$$\underbrace{\frac{\partial \alpha_k \rho_k \phi_k}{\partial t}}_{R} + \underbrace{\nabla \cdot \alpha_k \rho_k \mathbf{v}_k \phi_k}_{C} - \underbrace{\nabla \cdot \alpha_k \Gamma_{\phi k} \nabla \phi_k}_{D} = \underbrace{\nabla \cdot \alpha_k \mathbf{S}^A_{\phi k} + S^V_{\phi k}}_{S} . \tag{4}$$

The term R on the left-hand side denotes the rate of change of the transported quantity ϕ_k. The term C stands for the convective transport rate, D represents the diffusive transport, where $\Gamma_{\phi k}$ stands for the diffusion coefficient, and S on the right-hand side denotes the specific sources or sinks of ϕ_k. The latter term consists of the volumetric term $S^V_{\phi k}$ and the surface term $\nabla \cdot \alpha_k \mathbf{S}^A_{\phi k}$ accounting for the diffusion flux, which is not included in the diffusion flux in D. The interphase exchange terms, $\mathbf{M}_{12}$ and $\mathbf{H}_{12}$ in Eq. (1), are part of the volumetric source $S^V_{\phi k}$.

After applying the Gauss theorem for a grid cell P surrounded by its neighbors P_j, and with the outward surface (cell–face) vectors $\mathbf{A}$, the discretized control volume equation can be written as

$$\underbrace{\frac{\mathrm{d}}{\mathrm{d}t}(\alpha_k \rho_k \phi_k V_{\mathrm{cel}})_P}_{R} + \sum_{f=1}^{n_f} \underbrace{(\alpha_k \rho_k \mathbf{v}_k \mathbf{A} \phi_k)_f}_{C_f} - \sum_{f=1}^{n_f} \underbrace{(\alpha_k \Gamma_{\phi k} \nabla \phi_k \mathbf{A})_f}_{D_f}$$
$$= \sum_{f=1}^{n_f} \underbrace{\left(\alpha_k \mathbf{S}^A_{\phi k} \mathbf{A}\right)_f}_{S^A_\phi} + \underbrace{\left(S^V_{\phi k} V_{\mathrm{cel}}\right)_P}_{S^V_\phi}, \tag{5}$$

where subscript P denotes the cell-center and f face-center values; C_f and D_f are the convective and diffusion transport through face f, respectively; n_f is the number of cell faces surrounding grid cell P, and V_{cel} is the cell volume. It is assumed that P is bounded by piecewise smooth surfaces. All dependent variables, such as volume fraction, density, velocity, pressure, and enthalpy, are evaluated at the cell center. The cell–face-based connectivity and interpolation practices for gradients and cell–face values are introduced to accommodate an arbitrary number of cell faces. A second-order midpoint rule is used for integral approximation and a second-order linear approximation for any value at the cell face. The cell gradients can be calculated by using either the Gauss theorem or a linear least-square approach.

The convection is solved by an upwind scheme. The time derivative R is discretized by the implicit first-order accurate Euler (two levels) scheme.

Figure 1 shows the schematic of the valveless pulse-detonation hydrojet comprising a detonation channel 1 of 20 mm high and 400 mm long and a shaped water guide 2 of length 600 mm and width 80 mm with intake 3 and nozzle 4. This hydrojet configuration was obtained as a result of extensive parametric study.

Figure 2 shows a two-dimensional (2D) computational domain (CD) of dimensions 430 x 1950 x 1 mm with adopted boundary conditions. Initially, the flow everywhere in the CD except for the interior of the detonation tube is the homogeneous flow of bubbly water with low gas volume fraction ($\alpha_1 = 10^{-2}$). The detonation tube with one closed (left) and one open (right) ends is initially filled with a quiescent fuel—air mixture at pressure 0.1 MPa and temperature 293 K. At the inlet (left boundary of CD), the mass flow rate of water equal to 2.33 kg/s is set to simulate the homogeneous approach stream velocity of 5 m/s. At the upper boundary of CD, a constant pressure of 0.1 MPa is fixed. At the outlet (right boundary of CD), the Dirichlet boundary conditions with constant pressure (0.1 MPa) are adopted. At the symmetry plane (lower boundary of CD), zero gradients normal to the boundary are set. At the rigid walls of the hydrojet the velocity nonslip conditions are adopted.

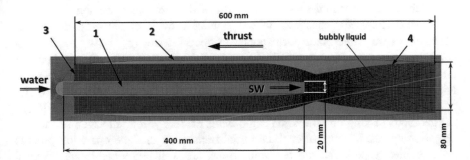

Fig. 1 Schematic of a pulse-detonation hydrojet: *1*—detonation tube, *2*—water guide, *3*—intake, *4*—nozzle

Fig. 2 Computational domain: *1*—water inlet, *2*—symmetry plane, *3, 4*—constant pressure boundaries

Once started, the calculation is continued until a steady-state flow is established in the entire CD including the inner flow in the water guide. Thereafter, the first detonation pulse is triggered in the detonation tube by temporarily applying a proper value of the mass flow rate (0.03 kg/s) of detonation products with the Chapman–Jouguet temperature of 2500 K at its left end to obtain a detonation wave propagating at a constant velocity of about 1700 m/s along the tube. After the detonation wave travels along the detonation tube for 0.3 ms approaching the right (open) end of the tube, the left inlet boundary is instantaneously replaced by a rigid wall, thus generating a rarefaction wave running toward the open end. When the detonation wave reaches gas–water interface it is partly reflected from it and partly transmitted through it into the compressible bubbly water as a shock wave. Further flow dynamics includes the propagation of the transmitted shock wave through the nozzle accompanied by shock-induced acceleration of water towards the nozzle exit, followed by the expansion of a cloud of hot gaseous detonation products in the water guide. After complex wave—water flow interactions inside the water guide—the cloud of gas detonation products is separated from the detonation tube exit and is convected downstream in the nozzle and outside. At this instant, the left boundary of the detonation tube is instantaneously replaced by the inlet boundary to fill the tube with fresh fuel mixture at a proper mass flow rate and to start the next operation cycle. Contrary to the first cycle, in all subsequent cycles a detonation wave, when reaching the gas–water interface, is transmitted into the bubbly water with bubbles of gaseous detonation products of the previous operation cycle.

Figure 3 shows the snapshots of one operation cycle of the pulse-detonation hydrojet, where white color corresponds to "pure" water ($\alpha_1 = 10^{-2}$) and black color corresponds to "pure" detonation products ($\alpha_2 = 10^{-6}$). Clearly, the first and the last snapshots are quite similar to each other thus representing the repetitive initial conditions for subsequent cycles at overall operating frequency of 9–10 Hz.

Figure 4 shows the calculated time history of the instantaneous force F (N/m in 2D geometry) acting on all internal rigid walls of the hydrojet in seven successive operation cycles. Clearly, after three initial transient cycles the operation process becomes nearly periodic: the last four cycles are well reproducible and show a positive mean force acting on the hydrojet (the force is directed against the approaching water stream). The peak positive force attains 20 kN/m. One can distinguish three main stages of force development in each cycle: the first stage with a peak positive force caused by pressure rise in the detonation tube followed by a peak negative force caused by reflection of the shock wave from the contracting portion of the water guide nozzle; the second relatively long stage of shock wave propagation in the diverging part of the nozzle followed by extension of the gas cloud therein with positive force; and the third stage with negative force during which a new portion of water fills the water guide. The theoretical specific impulse $I_{sp,calc}$ (dimension in s) defined as the integral-mean force $\overline{F}_{calc}$ (N/m) in Fig. 4 divided by the mass flow rate of fuel mixture $\dot{m}$ (kg/(m·s) in 2D geometry) and acceleration of gravity, g (m/s^2),

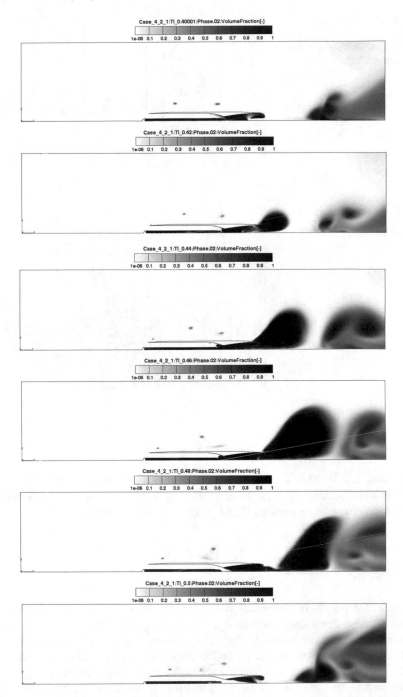

Fig. 3 Spatial and temporal evolution of gas volume fraction in one operation cycle of the pulse-detonation hydrojet

Fig. 4 Time history of
instantaneous force acting on
the pulse-detonation hydrojet
in seven successive cycles

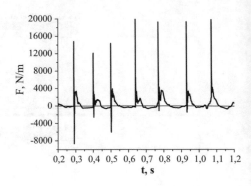

Fig. 5 Schematic of the test
rig

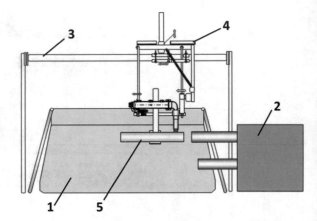

$$I_{\text{sp,calc}} = \overline{F}_{\text{calc}} / \dot{m}g$$

is equal to 350–400s. This means that the valveless pulse-detonation hydrojet of the considered design is capable of producing a positive thrust with a specific impulse on the level of modern liquid propellant rocket engines.

Experimental Studies

Based on the results of numerical simulation we have designed and fabricated a valveless pulse-detonation hydrojet operating on liquid hydrocarbon fuel (regular gasoline) and gaseous oxygen. For firing the hydrojet, a special test rig was designed and assembled. Figures 5 and 6 show the schematic and photographs of the test rig with flowing water. The test rig includes (see Fig. 5) water pool 1, pump 2 with suction and pressure tubing, support beam 3, thrust-measuring frame 4 with a load cell, and with suspended hydrojet 5.

Fig. 6 Photographs of the test rig (**a**) and thrust-measuring system with a suspended pulse-detonation hydrojet

Figure 7 shows the example of load cell record in one of the experimental runs. It shows the time history of instantaneous force acting on the hydrojet operating in the pulse-detonation mode at a frequency of 3 Hz in a water stream of 5 m/s entering the pool from the pump. The mass flow rate of near-stoichiometric gasoline–oxygen mixture is $\dot{m} \approx 8$ g/s in this run. The mixture was cyclically ignited in the detonation tube by a standard automotive spark plug and the arising deflagration was transitioned to a detonation at a short distance. Based on the record of Fig. 7 one can estimate the experimental specific impulse $I_{sp,exp} \approx 300$ s, which is close to the theoretical value of $I_{sp,calc}$ of 350–400 s. Note that the registered force of hydrodynamic drag acting on the hydrojet during a water purging stage (prior to triggering the pulse-detonation operation process) was negligible as compared to the forces arising during hydrojet firing.

Table 1 summarizes the results of seven experimental runs with different operation frequencies (cycle timing) in the pulse-detonation mode, ranging from 6 to 1 Hz. In all runs, the approach velocity of the water stream was 5 m/s, water in the water guide was artificially aerated with air bubbles to have $\alpha_1 \approx 0.04$ and the mass flow rate of fuel mixture was $\dot{m} \approx 8$ g/s. The last column in Table 1 shows the estimated values of the experimental specific impulse $I_{sp,exp}$. Clearly, the measured values of $I_{sp,exp}$ vary within the range from 255 to 370 s which overlaps the theoretical range of 350–400 s.

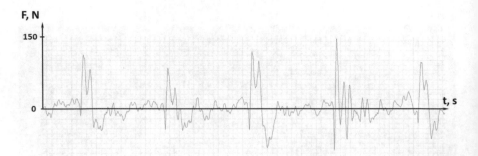

Fig. 7 Load cell record in one of the experimental runs with the pulse-detonation hydrojet operating at a frequency of 3 Hz in a water stream of 5 m/s

Table 1 Results of seven experimental runs

Run	Cycle timing, ms	Cycle frequency, Hz	Force amplitude[a], N	Specific impulse $I_{sp,exp}$ [a], s
1	170	6	131	370
2	200	5	141	345
3	250	4	134	340
4	330	3	110	305
5	400	2.5	101	260
6	500	2	98	255
7	1000	1	104	280

[a]*Remark* shown in the column is the mean value over 5–7 successive cycles

Conclusions

Geometrical configuration and operational parameters of a valveless pulse-detonation hydrojet have been determined based on extensive numerical simulations using 2D two-phase flow equations. The theoretical propulsive performance of such a hydrojet in terms of the specific impulse was shown to be on the level of modern liquid propellant rocket engines and amount 350–400 s. Based on the results of numerical simulation a valveless pulse-detonation hydrojet operating on liquid hydrocarbon fuel (regular gasoline) and gaseous oxygen has been designed and fabricated. For firing the hydrojet, a special test rig with flowing water was designed and assembled. Experiments showed that the measured values of the specific impulse varied within the range from 255 to 370 s which overlaps the theoretical range, thus demonstrating the predictive capabilities of the numerical approach. Further efforts will be focused on improving the propulsive performance of the valveless hydrojet and on computer-aided design of an efficient valved hydrojet.

This work was supported by the Ministry of Education and Science of Russian Federation (contract ID RFMEFI60914X0001).

Acknowledgments Researches are carried out with the financial support of the state represented by the Ministry of Education and Science of the Russian Federation. Agreement no. 14.609.21.0001, June 5, 2014 Unique project Identifier: RFMEFI60914X0001.

References

1. Avdeev, K.A., Aksenov, V.S., Borisov, A.A., Tukhvatullina, R.R., Frolov, S.M., Frolov, F.S.: Numerical simulation of the momentum transfer from the shock waves to the bubbly media. Combust. Explosion **8**(2), 57–67 (2015)
2. Avdeev, K.A., Aksenov, V.S., Borisov, A.A., Tukhvatullina, R.R., Frolov, S.M., Frolov, F.S.: Numerical simulation of momentum transfer from the shock wave to the bubbly environment. Khim. Fiz. **34**(5), 34–46 (2015)
3. Avdeev, K.A., Aksenov, V.S., Borisov, A.A., Frolov, S.M., Frolov, F.S., Shamshin, I.O.: Momentum transfer from a shock wave to a bubbly liquid. Khim. Fiz. **34**(11), 27–32 (2015)
4. Frolov, S.M., Aksenov, V S., Frolov, F.S., Avdeev, K.A.: Pulse detonation engine (variants) and the way to create hydrojet thrust. PCT/RU2013/001148 dated 23.12.2013. (2013) http://www.idgcenter.ru/patentPCT-RU2013-001148.htm
5. Kutateladze, S.S., Nakoryakov, V.E.: Heat and Mass Transfer and Waves in Gas–Liquid Systems. Nauka Publ, Novosibirsk (in Russian) (1984)
6. Nigmatulin, R.I.: Dynamics of Multiphase Media. Vol. 1. Hemisphere, N.Y (1990)

Development of Technological Process of Matrix Conversion of Natural and Associated Petroleum Gases into Syngas with Low Content of Nitrogen

V.S. Arutyunov, A.V. Nikitin, V.I. Savchenko, I.V. Sedov, O.V. Shapovalova and V.M. Shmelev

Abstract There is an acute need in more wide use of huge world resources of natural gas as a petrochemical raw material. But solving of this global task is hardly possible without developing of more effective methods for conversion of natural gas into the most important gas–chemical intermediate—syngas. This paper describes recent results on a principally new type of reformers based on the non-catalytic conversion of hydrocarbons into syngas in volumetric (3D) matrix burners. The use of enriched air and oxygen results in production of syngas with low content of nitrogen for petrochemical applications, including production of methanol, syncrude oil, and others. The effective recuperation of heat of produced syngas inside the matrix cavity permits to operate at optimal values of oxygen excess coefficient $\alpha = 0.32$–0.36, thus making it possible to obtain in such simple non-catalytic process very high yield of nitrogen-free syngas with concentration of H_2 more than 50% and that of CO more than 30%.

V.S. Arutyunov (✉) · A.V. Nikitin · O.V. Shapovalova · V.M. Shmelev
Semenov Institute of Chemical Physics, Russian Academy of Sciences, Moscow, Russia
e-mail: v_arutyunov@mail.ru

A.V. Nikitin
e-mail: ni_kit_in@rambler.ru

O.V. Shapovalova
e-mail: shapovalova.oksana@gmail.com

V.M. Shmelev
e-mail: shmelev.05@mail.ru

V.I. Savchenko · I.V. Sedov
Institute of Problems of Chemical Physics of Russian Academy of Sciences, Chernogolovka, Moscow Region, Russia
e-mail: savch1152@mail.ru

I.V. Sedov
e-mail: igor.v.sedov@gmail.com

© The Author(s) 2018   721
K.V. Anisimov et al. (eds.), *Proceedings of the Scientific-Practical Conference
"Research and Development - 2016"*, https://doi.org/10.1007/978-3-319-62870-7_75

Keywords Natural gas · Associated petroleum gas · Syngas · Partial oxidation Combustion limits · Volumetric matrix burners

Introduction

Huge world resources of natural gas, about two orders of magnitude above that of oil [1], is the main source of energy and hydrocarbons for the nearest future. Because modern civilization strongly depends on the wide use of liquid hydrocarbon fuels and different petrochemicals, it will be necessary to replace many of petrochemical processes by gas chemical technologies to use the abundant gas resources as petrochemical raw material. But modern gas chemistry technologies are too complex and costly and cannot compete today with petrochemistry in the production of liquid motor fuels and other important chemicals [2]. The main problem is that practically all multi-tonnage gas chemical processes need preliminary conversion of natural gas into syngas (the mixture of H_2 and CO), which can be more easily converted into liquid hydrocarbons and chemicals. This very complex, energy, and capital-intensive stage consumes up to 70% of all expenditures for obtaining final products [3]. Therefore, global gas chemistry urgently needs more simple and effective technologies for natural gas conversion into syngas. Nowadays, the absence of such technologies seriously restricts the more wide use of natural gas as a petrochemical raw material. In spite that a number of alternative technologies for natural gas conversion to syngas were announced in literature during last decades [3], up to now no one of them has shown its practical consistency. This paper describes recent results obtained by development of principally new approach to solve this very important global task, which is based on the use of volumetric (3D) matrix burners [4–8].

Basic Principles of Matrix Conversion

Matrix conversion of hydrocarbon gases into syngas is based on so-called "flameless" combustion, when flame front is stabilized on the surface of a solid porous or perforated planar matrix. Such type of combustion is widely used for many technological applications, first of all as a source of IR radiation [9, 10]. In this case, the flame front is stabilized at some distance above the surface, with its temperature being low enough, ~ 1000–$1200\ °C$, due to intense convective and radiation heat transfer from the flame front to the surface. The surface heated in this manner intensely radiates in the IR region and can be used as an effective source of IR radiation for many practical applications. However, intense convective and radiation loss from the flame front significantly narrow the combustion limits.

To widen the combustion limits and ensure the combustion of lean gas mixtures with low NOx emissions, it was suggested to use 3D permeable matrixes with a closed inner cavity [11–14]. It was demonstrated that, in such matrix burners, the stable combustion of very lean mixtures at an oxygen excess coefficient of $\alpha = [O_2]/2[CH_4] > 2$ can occur at a specific combustion power of up to 30–40 W/cm^2. These conditions provide very low concentrations of nitrogen oxides and carbon monoxide. Theoretically, estimated temperatures of the flame front and that at the entrance and working surfaces of such 3D matrixes were found to be in good agreement with the available experimental data [11, 12].

The same organization of the combustion process makes it possible to widen the combustion limits for rich mixtures as well, thus providing necessary conditions for an effective conversion of hydrocarbons into syngas [4–8]. The principal scheme of a syngas reformer based on a 3D permeable matrix is shown in Fig. 1. After passing the mixer, a homogeneous fuel–oxidizer mixture of specified composition is fed through permeable walls of the 3D matrix burner into its inner cavity, where it burns near the surface. Due to a very intense convective and radiant transfer of heat from flame front and combustion products to the matrix walls and then to the incoming fresh gas, the temperature of the gas mixture entering the flame front increases up to 400–600 °C, while that of flame front and combustion products decreases to 1000–1200 °C. Alongside with the absence of radiation losses in the closed cavity of the matrix, it lowers the limit of stable combustion of rich methane–oxidizer mixtures to a value of oxygen excess coefficient $\alpha = 0.32$–0.36 and thus enables to attain yields of H$_2$ and CO very close to the thermodynamically equilibrium values.

Fig. 1 Principal scheme of matrix burner with a deep volumetric (3D) matrix. *1*—external shell, *2*—volume with a fuel–air mixture, *3*—gas outlet, *4*—side walls and bottom made from perforated ceramic, *5*—cap made from perforated ceramic

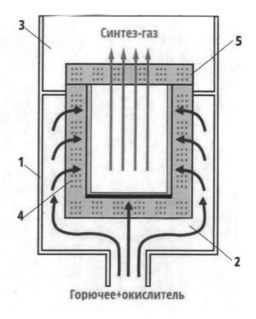

Pilot Installations for Matrix Conversion Testing

Several series of laboratory scale investigations that were thoroughly described in [4–8] have shown principal features and possibilities of matrix conversion. The purpose of this study was the pilot testing and demonstration of matrix conversion ability to produce nitrogen-free syngas for subsequent use in methanol and Fisher–Tropsch synthesis.

Two pilot installations were used in this study. The first one was designed for testing the process at atmospheric pressure with input gas flows up to 20 m³/h (Fig. 2). Two matrix cavities were formed by two pairs of flat square porous plates with dimensions 250 x 250 mm made from metallic foam or metallic wool.

The second installation was designed for testing the process at enhanced pressures with input gas flows up to 10 m³/h (Fig. 3). It has one matrix cavity formed by the pair of flat porous circular plates with diameter 200 mm. Plates were also made from metallic foam or metallic wool. After several series of testing of different matrixes in both installations, it was concluded that the best choice is metallic wool matrixes with the width of about 8 mm fabricated by pressing of coiled Nichrome wire with 0.1–0.2 mm diameter. In both cases, the round stainless steel apertures with a diameter of 100 or 150 mm were installed before matrix to decrease marginal effects, especially for rectangular matrixes. Behind matrixes thin stainless steel screens were situated to reflect IR radiation of flame front back to the matrix, thus

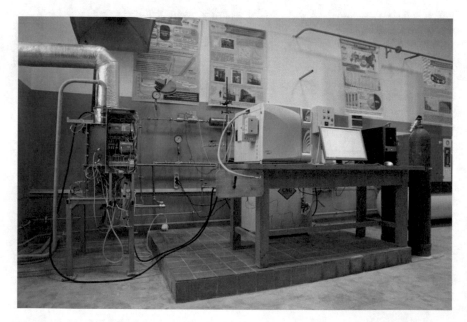

Fig. 2 Pilot installation for matrix conversion of natural gas at atmospheric pressure. Input natural gas capacity up to 20 m³/h

Fig. 3 Pilot installation for matrix conversion of natural gas at enhanced pressures. Input natural gas capacity up to 10 m³/h

enhancing the recuperation of heat. In all testing, cylinders with natural gas used for fueling city buses and technical grade oxygen were used.

Results and Discussion

The first series of testing was accomplished at atmospheric pressure with pilot installation shown in Fig. 2 with atmospheric air as an oxidant. It was shown the possibility of stable conversion of natural gas by atmospheric air at values of oxygen excess coefficient α as low as 0.30 and even lower. But the best results were obtained at values of $\alpha = 0.34$–0.36. At lower values of oxygen excess coefficient, the conversion of natural gas decreases thus decreasing the yield of syngas. The oxidation by air at optimal values of oxygen excess coefficient results in the production of syngas with concentration of H_2 up to 25%, of CO up to 15% and H_2/CO ratio about 1.6. At this, the conversion of both methane and oxygen was above 95%. These figures are very close to thermodynamically equilibrium figures for these conditions. Therefore, matrix conversion by atmospheric air results in relatively easy production of inexpensive syngas which can be used, for example, in simple small-scale technologies for natural or associated petroleum gas conversion into liquid hydrocarbons just in the field conditions [15, 16].

Series of testing under enhanced pressures up to 5 atm with pilot installation presented in Fig. 3 have shown the principal possibility of matrix converters operation at enhanced pressure. Nevertheless, significant difficulties induced by very intensive heat release on matrix surface at enhanced pressures that lead to matrix overheating make it desirable to look for some alternative solutions concerning the design and operation of matrix converter at enhanced pressure.

The results of several series of testing with both pilot installations presented in Figs. 2 and 3 with enriched air and technical oxygen are presented in Table 1 and in Figs. 4, 5, and 6.

As it can be seen from Table 1 and Fig. 4, enriching of oxidizer by oxygen leads to a steadily increasing of hydrogen and carbon monoxide concentrations in syngas. This increase is directly proportional to oxygen concentration and can be simply the consequence of decreasing of nitrogen concentration. The concentrations of principal products increase up to values about 54% for H_2 and about 31% for CO. The concentration of CO_2 at that changes insignificantly, from 3.6% at oxidation by air to slightly above 5% at oxidation by oxygen. Taking into account the decreasing of nitrogen concentration, it means that the carbon selectivity of syngas production changes insignificantly with the increase of oxygen content in oxidizer in spite of the increase of the specific heat release with the decrease of nitrogen content in products. The conversion of oxygen is practically always complete, and that of methane is up to 88% at oxidation by oxygen, while practically complete at oxidation by air. Some structural improvements of converter construction are now

Table 1 Experimental results of natural gas conversion to syngas with air, enriched air and technical oxygen in pilot installations presented in Figs. 2 and 3 at values of oxygen excess coefficient $\alpha = 0.34$–0.36

O_2, %	Concentration of products, %						H_2/CO	Conversion, %	
	H_2	CO_2	O_2	N_2	CH_4	CO		CH_4	O_2
21.0	22.2	3.6	0.6	47.1	0.5	13.8	1.61	97.0	95.2
21.0	23.0	3.6	0.6	48.0	0.6	14.4	1.60	96.6	95.3
21.0	23.7	3.7	0.6	45.1	1.0	14.2	1.67	94.3	95.0
24.4	25.0	3.6	0.8	45.9	0.6	15.6	1.61	97.1	94.9
25.4	19.9	3.9	1.6	49.3	1.9	12.8	1.55	91.4	90.5
25.4	19.6	4.2	1.3	49.8	1.9	12.6	1.56	91.9	92.3
47.3	27.8	4.7	0.6	29.2	9.9	19.2	1.45	74.6	97.8
47.3	33.0	3.8	0.2	31.1	6.2	20.9	1.58	83.9	99.2
60.5	32.1	3.7	0.4	20.4	8.4	23.1	1.39	81.5	98.8
100	51.8	5.5	0.2	0.5	9.2	30.3	1.71	84.1	99.6
100	53.4	5.3	0.0	0.4	7.9	30.6	1.74	86.2	100.0
100	54.0	5.2	0.0	0.4	7.2	31.2	1.73	87.0	100.0
100	51.9	5.5	0.0	0.4	9.3	30.8	1.68	84.1	99.9
100	54.5	5.1	0.0	0.3	6.4	31.4	1.74	88.4	100.0
100	53.8	5.1	0.0	0.3	7.2	31.7	1.70	86.7	100.0
100	53.0	5.2	0.0	0.4	8.0	30.9	1.72	85.9	100.0

Fig. 4 Dependence of the concentration of principal syngas components on oxygen content in oxidizer at oxygen excess coefficient $\alpha = 0.34$–0.36

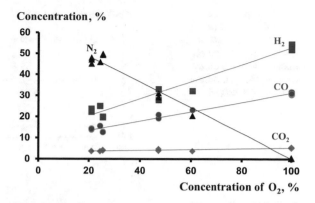

Fig. 5 Dependence of the concentration of principal syngas components on oxygen excess coefficient α at methane conversion by oxygen

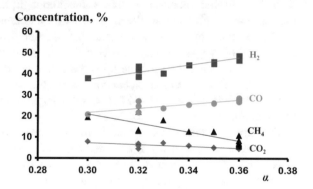

Fig. 6 Dependence of the H_2/CO ratio at methane conversion by oxygen on oxygen excess coefficient α

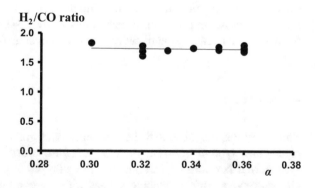

under consideration to reach the same figures as at the oxidation by air. The main purpose of incomplete methane conversion by oxygen is the overheating of matrixes.

Very important figure for stable operation of converter is the optimal choice of the oxygen excess coefficient α. Figure 5 represents the dependence of the concentration of principal syngas components on oxygen excess coefficient at methane

Fig. 7 Dependence of the concentrations of principal syngas components on the time of pilot testing at methane conversion by oxygen. Oxygen excess coefficient $\alpha = 0.36$

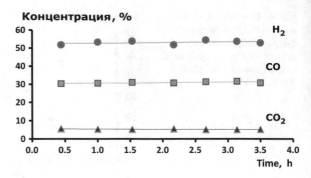

conversion by oxygen. As it can be seen, the optimal value of this ratio is about 0.36. The further increase of this value although increases the conversion of methane can involve the overheating of matrix and thus instability in its operation.

One of the important parameters that characterize the attractiveness of any technology of syngas production is H_2/CO ratio. As it was experimentally revealed, at matrix conversion of methane by oxygen this parameter is practically independent on the value of oxygen excess coefficient (Fig. 6). At optimal values of oxygen excess coefficient $\alpha = 0.34$–0.36 this ratio slightly increases with increasing of oxygen content in oxidizer (Table 1).

Stability of obtained parameters of matrix conversion of natural gas into syngas was tested in several series of time testing. Each series lasted during several hours of continuous operation at stable process conditions. As it can be seen in Fig. 7, all parameters were stable enough during testing period.

Therefore, the possibility of stable long-run matrix conversion of natural gas (methane) by oxygen with obtaining nitrogen-free syngas with parameters well suit for its further use in processes of methanol or Fischer–Tropsch synthesis, as well as for many other practical applications, was demonstrated.

Conclusions

This study has demonstrated the possibility and good practical prospects of principally new type of non-catalytic matrix reformers for natural gas conversion into syngas. Matrix reformers can operate with all types of oxidizers using air, enriched air, and oxygen, thus producing syngas with given nitrogen content depending on demand, including nitrogen-free syngas for petrochemical applications. The optimal values of oxygen excess coefficient $\alpha = 0.34$–0.36 lead to the gain high conversion of reagents and obtain in such simple non-catalytic process very high yield of nitrogen-free syngas with a concentration of H_2 more than 50%, that of CO more than 30%, and H_2/CO ratio about 1.7.

Acknowledgments Research is conducted with the financial support of the state represented by the Ministry of Education and Science of the Russian Federation. Agreement (contract) no. 14.607.21.0037, June 5, 2014 Unique project Identifier: RFMEF160714X0037.

References

1. Kvenvolden, K.A.: Chem. Geol. **71**, 41 (1988)
2. Hobbs, H.O. Jr., Adair, L.S.: Analysis shows GTL viable alternative for US gas producers. Oil Gas J. 68–75 (2012). Accessed 6 Aug 2012
3. Arutyunov, V.S.: Oxidative conversion of natural gas. KRASAND (URSS), 640 p. Moscow, (2011) (in Russian)
4. Arutyunov, V.S., Shmelev, V.M., Lobanov, I.N., Politenkova, G.G.: A generator of synthesis gas and hydrogen based on a radiation burner. Theoretical. Found. Chem. Eng. **44**(1), 20–29 (2010)
5. Arutyunov, V.S., Shmelev, V.M., Sinev, M.Yu., Shapovalova, O.V.: Syngas and hydrogen production in a volumetric radiation burners. Chem. Eng. J. **176–177**, 291–294 (2011)
6. Shapovalova, O.V., Young Nam Chun, Arutyunov, V.S., Shmelev, V.M.: Syngas and hydrogen production from biogas in 3D matrix reformers. Int. J. Hydr. Eng. **37**(19), 14040–14046 (2012)
7. Arutyunov, V.S., Shmelev, V.M., Rakhmetov, A.N., Shapovalova, O.V., SÄtrekova, L.N., Zakharov, A.A. Oxidative conversion of hydrocarbon gases in the surface combustion mode. Russ. Chem. Bull. (International Edition). **62**(7), 1504–1509 (2013)
8. Arutyunov, V.S., Shmelev, V.M., Rakhmetov, A.N., Shapovalova, O.V.: 3D matrix burners: a method for small-scale syngas production. Ind. Eng. Chem. Res. **53**(5), 1754–1759 (2014)
9. Bouma, P.H., Goey, L.P.H.: Premixed combustion on ceramic foam burners. Comb. Fl. **119**, 133 (1999)
10. Nemoda, S., Trimis, D., Zivkovic, G.: Numerical simulation of porous burners and hole plate surface burners. Therm. Sci. **8**, 3 (2004)
11. Shmelev, V.M.: Combustion of natural gas at the surface of a high-porosity metal matrix. Russ. J. Phys. Chem. B. **4**, 593 (2010)
12. Shmelev, V.M., Nikolaev, V.M., Arutyunov, V.S.: Energy-efficient combustors based on volumetric matrixes. Gazokhimiya **4**(8), 28 (2009) (in Russian)
13. Arutyunov, V.S., Shmelev, V.M., Rakhmetov, A.N., Shapovalova, O.V., Zakharov, A.A., Roschin, A.V.: New approaches to development of low-emission combustion chambers for gas turbine engines. Int. Sci. J. Altern. Energy Ecol. (ISJAEE). **6**(128), 105–120 (2013)
14. Rakhmetov, A.N., Shmelev, V.M, Arutyunov, V.S.: Low emission combustion chambers for gas turbine engines based on permeable volumetric matrixes. Combust. Plasmachemistry **2**, 83–91 (2013)
15. Jess, A., Popp, R., Hedden, K.: Fischer–Tropsch-synthesis with nitrogen-rich syngas. Fundamentals and reactor design aspects. Appl. Catal. A. **186**, 321–342 (1999)
16. Arutyunov, V.S., Savchenko, V.I., Sedov, I.V.: On the prospects of field gas-chemical technologies based on nitrogen containing syngas. Neftegazokhimia. № **4**, 12–21 (2016) (in Russian)

Printed in the United States
By Bookmasters